Ronald E. Wrolstad, M. Monica Giusti and
Wilhelmina Kalt (Eds.)

# Anthocyanins

**MDPI**

This book is a reprint of the Special Issue that appeared in the online, open access journal, *Molecules* (ISSN 1420-3049) from 2014–2015, available at:

http://www.mdpi.com/journal/molecules/special_issues/anthocyanins

*Guest Editors*
Ronald E. Wrolstad
University Department of Food Science and Technology
Oregon State University
USA

M. Monica Giusti
Food Science Department
Ohio State University
USA

Wilhelmina Kalt
Agriculture and Agri-Food Canada
Canada

| *Editorial Office* | *Publisher* | *Managing Editor* |
| --- | --- | --- |
| MDPI AG | Shu-Kun Lin | *Jing Gao* |
| St. Alban-Anlage 66 | | |
| Basel, Switzerland | | |

**1. Edition 2016**

MDPI • Basel • Beijing • Wuhan • Barcelona • Belgrade

ISBN 978-3-03842-228-0 (Hbk)
ISBN 978-3-03842-229-7 (PDF)

# Table of Contents

## Section 1: Anthocyanins in Fresh and Processed Foods

## Section 2: Grape Anthocyanins and Wine Quality

## Section 3: Anthocyanin Biosynthesis and Regulation

V

## Section 4: Anthocyanin Composition and their Biological Properties

# List of Contributors

**Sheiraz Al Bittar** University of Avignon, INRA, UMR408, Avignon 84000, France.

**Carlos Areche** Departamento de Química, Facultad de Ciencias, Universidad de Chile, Casilla 653, Santiago 7800024, Chile.

**Jesús Ayuso** Department of Physical Chemistry, University of Cadiz, Puerto Real 11510, Spain.

**Latifa Azaroual** Faculty of Sciences, Abdelmalek Essaâdi University, Tetouan 93000, Morocco.

**Gerardo F. Barbero** Andalusian Center for Wine Research; Department of Analytical Chemistry, University of Cadiz, Puerto Real 11510, Spain.

**Jesús Manuel Barrón-Hoyos** Programa de Doctorado en Ciencias de los Alimentos, Universidad de Sonora, Blvd. Luis Encinas y Rosales s/n, Hermosillo, Sonora 83000, Mexico.

**Carmelo G. Barroso** Department of Analytical Chemistry; Andalusian Center for Wine Research, University of Cadiz, Puerto Real 11510, Spain.

**Hana Bártíková** Department of Biochemical Sciences, Charles University in Prague, Faculty of Pharmacy, Heyrovského 1203, Hradec Králové 50005, Czech Republic.

**Iva Boušová** Department of Biochemical Sciences, Charles University in Prague, Faculty of Pharmacy, Heyrovského 1203, Hradec Králové 50005, Czech Republic.

**Anghel Brito** Laboratorio de Productos Naturales, Departamento de Química, Facultad de Ciencias Básicas, Universidad de Antofagasta, Av. Coloso S-N, Antofagasta 1240000, Chile.

**Noelia Briz-Cid** Nutrition and Bromatology Group, Analytical and Food Chemistry Department, Faculty of Food Science and Technology, University of Vigo, Ourense Campus, Ourense E-32004, Spain.

**Beatriz Cancho-Grande** Nutrition and Bromatology Group, Analytical and Food Chemistry Department, Faculty of Food Science and Technology, University of Vigo, Ourense Campus, Ourense E-32004, Spain.

**Kaylyn L. Carpenter** Department of Biology, High Point University, University Station 3591, High Point, NC 27262, USA.

**Ceferino Carrera** Andalusian Center for Wine Research, University of Cadiz, Puerto Real 11510, Spain.

**Guo Cheng** College of Enology, Northwest A&F University, Yangling 712100, Shaanxi, China.

**Olivier Dangles** University of Avignon, INRA, UMR408, Avignon 84000, France.

**Luminiţa David** "Babeş-Bolyai" University, Faculty of Chemistry and Chemical Engineering, 11, Arany Janos Str., 400028 Cluj-Napoca, Romania.

**Alessandra Del Caro** Department of Agriculture, University of Sassari, Viale Italia 39, Sassari 07100, Italy.

**Michael Dossett** BC Blueberry Council (in partnership with Agriculture and Agri-Food Canada-Pacific Agri-Food Research Centre), 6947 Hwy #7, P.O. Box 1000, Agassiz, BC V0M 1A0, Canada.

**Anayansi Escalante-Aburto** Programa de Doctorado en Ciencias de los Alimentos, Universidad de Sonora, Blvd. Luis Encinas y Rosales s/n, Hermosillo, Sonora 83000, Mexico.

**Ana Fernandes de Oliveira** Department of Agriculture, University of Sassari, Viale Italia 39, Sassari 07100, Italy.

**Marta Ferreiro-González** Department of Physical Chemistry; Department of Analytical Chemistry, University of Cadiz, Puerto Real 11510, Spain.

**María Figueiredo-González** Nutrition and Bromatology Group, Analytical and Food Chemistry Department, Faculty of Food Science and Technology, University of Vigo, Ourense Campus, Ourense E-32004, Spain.

**Juan de Dios Figueroa-Cárdenas** Centro de Investigación y Estudios Avanzados (CINVESTAV—Unidad Querétaro), Libramiento Norponiente#2000, Fraccionamiento Real de Juriquilla, Querétaro, Querétaro 76230, Mexico.

**Chad E. Finn** United States Department of Agriculture, Agricultural Research Service, Horticultural Crops Research Unit (HCRU), Corvallis, OR 97330, USA.

**Carolina Fredes** Departamento de Ciencia de los Alimentos y Tecnología Química, Facultad de Ciencias Químicas y Farmacéuticas, Universidad de Chile, Santos Dumont 964, Independencia, Santiago 8380494, Chile.

**Wan Gao** State Key Laboratory of Bioactive Substance and Function of Natural Medicines, Institute of Materia Medica, Chinese Academy of Medical Sciences and Peking Union Medical College, Beijing 100050, China.

**Elena I. Gordeeva** Institute of Cytology and Genetics, Siberian Branch, Russian Academy of Sciences, Lavrentjeva ave. 10, Novosibirsk 630090, Russia.

**Roberto Gutiérrez-Dorado** Programa Regional del Noroeste para el Doctorado en Biotecnología, Universidad Autónoma de Sinaloa, Av. de las Américas y Blvd. Universitarios s/n, Culiacán, Sinaloa 80010, Mexico.

**Md. Abul Hasnat** College of Biomedical & Health Science, Department of Applied Biochemistry, Konkuk University, Chungju 380-701, Korea.

**Fei He** Center for Viticulture and Enology, College of Food Science and Nutritional Engineering, China Agricultural University, Beijing 100083, China.

**Yan-Nan He** College of Enology, Northwest A&F University, Yangling 712100, Shaanxi, China.

**Wu-Yang Huang** Department of Functional Food and Bio-active compounds, Institute of Farm Product Processing, Jiangsu Academy of Agricultural Sciences, Nanjing 210014, China.

**Nicole M. Hughes** Department of Biology, High Point University, University Station 3591, High Point, NC 27262, USA.

**Pavla Jedličková** Department of Biochemical Sciences, Charles University in Prague, Faculty of Pharmacy, Heyrovského 1203, Hradec Králové 50005, Czech Republic.

**Teng-Fei Ji** State Key Laboratory of Bioactive Substance and Function of Natural Medicines, Institute of Materia Medica, Chinese Academy of Medical Sciences and Peking Union Medical College, Beijing 100050, China.

**Jeong Eun Jo** College of Biomedical & Health Science, Department of Applied Biochemistry, Konkuk University, Chungju 380-701, Korea.

**Timothy S. Keidel** Department of Biology, High Point University, University Station 3591, High Point, NC 27262, USA.

**Edward J. Kennelly** Department of Biological Sciences, Lehman College and The Graduate Center, The City University of New York, 250 Bedford Park Boulevard West, Bronx, NY 10468, USA.

**Elena K. Khlestkina** Institute of Cytology and Genetics, Siberian Branch, Russian Academy of Sciences, Lavrentjeva ave. 10, Novosibirsk 630090, Russia; Novosibirsk State University, Pirogova St. 2, Novosibirsk 630090, Russia.

**Da Hye Kim** College of Biomedical & Health Science, Department of Applied Biochemistry, Konkuk University, Chungju 380-701, Korea.

**Jungmin Lee** United States Department of Agriculture (USDA), Agricultural Research Service (ARS), Horticultural Crops Research Unit (HCRU) Worksite, Parma, ID 83660, USA.

**Yoon Mi Lee** College of Biomedical & Health Science, Department of Applied Biochemistry, Konkuk University, Chungju 380-701, Korea.

**Cheng Li** State Key Laboratory of Bioactive Substance and Function of Natural Medicines, Institute of Materia Medica, Chinese Academy of Medical Sciences and Peking Union Medical College, Beijing 100050, China.

**Xiao-Xi Li** Center for Viticulture and Enology, College of Food Science and Nutritional Engineering, China Agricultural University, Beijing 100083, China.

**Zheng Li** Food Science and Human Nutrition Department, Institute of Food and Agricultural Sciences, University of Florida, Gainesville, FL 32611, USA.

**Ali Liazid** Department of Analytical Chemistry, University of Cadiz, Puerto Real 11510, Spain.

**Beong Ou Lim** College of Biomedical & Health Science, Department of Applied Biochemistry, Konkuk University, Chungju 380-701, Korea.

**Ya-Mei Liu** National Technical Research Centre of Veterinary Biological Products, Jiangsu Academy of Agricultural Science, Nanjing 210014, China.

**Kateřina Lněničková** Department of Biochemical Sciences, Charles University in Prague, Faculty of Pharmacy, Heyrovského 1203, Hradec Králové 50005, Czech Republic.

**Michèle Loonis** INRA, University of Avignon, UMR408, Avignon 84000, France.

**Jaime López-Cervantes** Centro de Investigación e Innovación en Biotecnología Agropecuaria, Instituto Tecnológico de Sonora, 5 de Febrero 818 Sur, Col. Centro, Ciudad Obregón, Sonora 8500, Mexico.

**David C. Manns** Department of Food Science, New York State Agricultural Experiment Station, Cornell University, 630 W. North St., Geneva, NY 14456, USA.

**Luca Mercenaro** Department of Agriculture, University of Sassari, Viale Italia 39, Sassari 07100, Italy.

**Taira Miyahara** Department of Biotechnology and Life Science, Tokyo University of Agriculture and Technology, 2-24-16 Nakacho, Koganei, Tokyo 184-8588, Japan.

**Bianca Moldovan** "Babeş-Bolyai" University, Faculty of Chemistry and Chemical Engineering, 11, Arany Janos Str., 400028 Cluj-Napoca, Romania.

**Nathalie Mora** University of Avignon, INRA, UMR408, Avignon 84000, France.

**Ignacio Morales-Rosas** Programa de Doctorado en Ciencias de los Alimentos, Universidad de Sonora, Blvd. Luis Encinas y Rosales s/n, Hermosillo, Sonora 83000, Mexico.

**Giovanni Nieddu** Department of Agriculture, University of Sassari, Viale Italia 39, Sassari 07100, Italy.

**Yuzo Nishizaki** Department of Biotechnology and Life Science, Tokyo University of Agriculture and Technology, 2-24-16 Nakacho, Koganei, Tokyo 184-8588, Japan.

**Yoshihiro Ozeki** Department of Biotechnology and Life Science, Tokyo University of Agriculture and Technology, 2-24-16 Nakacho, Koganei, Tokyo 184-8588, Japan.

**Olga I. Padilla-Zakour** Department of Food Science, New York State Agricultural Experiment Station, Cornell University, 630 W. North St., Geneva, NY 14456, USA.

**Miguel Palma** Andalusian Center for Wine Research; Department of Analytical Chemistry, University of Cadiz, Puerto Real 11510, Spain.

**Qiu-Hong Pan** Center for Viticulture and Enology, College of Food Science and Nutritional Engineering, China Agricultural University, Beijing 100083, China.

**Mehnaz Pervin** College of Biomedical & Health Science, Department of Applied Biochemistry, Konkuk University, Chungju 380-701, Korea.

**Melissa C. Pihl** Department of Biology, High Point University, University Station 3591, High Point, NC 27262, USA.

**Néstor Ponce-García** Programa de Doctorado en Ciencias de los Alimentos, Universidad de Sonora, Blvd. Luis Encinas y Rosales s/n, Hermosillo, Sonora 83000, Mexico; UAEMex Campus Universitario "El Cerrillo". El Cerrillo Piedras Blancas s/n, Toluca, Estado de Mexico 50200, Mexico.

**Luca Pretti** Porto Conte Ricerche Srl, S.P. 55 Porto Conte/Capo Caccia, Tramariglio-Alghero (SS) 07041, Italy.

**Benjamín Ramírez-Wong** Programa de Doctorado en Ciencias de los Alimentos, Universidad de Sonora, Blvd. Luis Encinas y Rosales s/n, Hermosillo, Sonora 83000, Mexico.

**Raquel Rial-Otero** Nutrition and Bromatology Group, Analytical and Food Chemistry Department, Faculty of Food Science and Technology, University of Vigo, Ourense Campus, Ourense E-32004, Spain.

**Paz Robert** Departamento de Ciencia de los Alimentos y Tecnología Química, Facultad de Ciencias Químicas y Farmacéuticas, Universidad de Chile, Santos Dumont 964, Independencia, Santiago 8380494, Chile.

**Ana Ruiz-Rodríguez** Andalusian Center for Wine Research; Department of Analytical Chemistry, University of Cadiz, Puerto Real 11510, Spain.

**Gavin L. Sacks** Department of Food Science, New York State Agricultural Experiment Station, Cornell University, 630 W. North St., Geneva, NY 14456, USA.

**Nobuhiro Sasaki** Iwate Biotechnology Research Center, 22-174-4, Narita, Kitakami, Iwate 024-0003, Japan.

**Beatriz Sepúlveda** Departamento de Ciencias Químicas, Universidad Andrés Bello, Campus Viña del Mar,  Quillota 980, Viña del Mar 2520000, Chile.

**Olesya Y. Shoeva** Institute of Cytology and Genetics, Siberian Branch, Russian Academy of Sciences,  Lavrentjeva ave. 10, Novosibirsk 630090, Russia.

**Jesús Simal-Gándara** Nutrition and Bromatology Group, Analytical and Food Chemistry Department, Faculty of Food Science and Technology, University of Vigo, Ourense Campus, Ourense E-32004, Spain.

**Mario J. Simirgiotis** Laboratorio de Productos Naturales, Departamento de Química, Facultad de Ciencias Básicas, Universidad de Antofagasta, Av. Coloso S-N, Antofagasta 1240000, Chile.

**Lenka Skálová** Department of Biochemical Sciences, Charles University in Prague, Faculty of Pharmacy, Heyrovského 1203, Hradec Králové 50005, Czech Republic.

**Ya-Lun Su** State Key Laboratory of Bioactive Substance and Function of Natural Medicines, Institute of Materia Medica, Chinese Academy of Medical Sciences and Peking Union Medical College, Beijing 100050, China.

**Li-Li Sun** State Key Laboratory of Bioactive Substance and Function of Natural Medicines, Institute of Materia Medica, Chinese Academy of Medical Sciences and Peking Union Medical College, Beijing 100050, China.

**Barbora Szotáková** Department of Biochemical Sciences, Charles University in Prague, Faculty of Pharmacy, Heyrovského 1203, Hradec Králové 50005, Czech Republic.

**Patricia Isabel Torres-Chávez** Programa de Doctorado en Ciencias de los Alimentos, Universidad de Sonora, Blvd. Luis Encinas y Rosales s/n, Hermosillo, Sonora 83000, Mexico.

**Ai-Guo Wang** State Key Laboratory of Bioactive Substance and Function of Natural Medicines, Institute of Materia Medica, Chinese Academy of Medical Sciences and Peking Union Medical College, Beijing 100050, China.

**Jian Wang** Department of Functional Food and Bio-active compounds, Institute of Farm Product Processing, Jiangsu Academy of Agricultural Sciences, Nanjing 210014, China; College of Food Science and Technology, Nanjing Agricultural University, Nanjing 210095, China.

**Jun Wang** Center for Viticulture and Enology, College of Food Science & Nutritional Engineering,  China Agricultural University, Beijing 100083, China.

**Xing-Na Wang** Department of Functional Food and Bio-active compounds, Institute of Farm Product Processing, Jiangsu Academy of Agricultural Sciences, Nanjing 210014, China.

**Tai-Xin Yue** College of Enology, Northwest A&F University, Yangling 712100, Shaanxi, China.

**Meng-Meng Zhang** State Key Laboratory of Bioactive Substance and Function of Natural Medicines, Institute of Materia Medica, Chinese Academy of Medical Sciences and Peking Union Medical College, Beijing 100050, China.

**Zhen-Wen Zhang** Shaanxi Engineering Research Center for Viti-Viniculture; College of Enology, Northwest A&F University, Yangling 712100, Shaanxi, China.

# About the Guest Editors

**Ron Wrolstad** earned his B.S. in Food Technology from Oregon State University and his Ph.D. in Agricultural Chemistry at the University of California at Davis. He joined the faculty of Oregon State's Food Science and Technology Department in 1965 where he taught a course in food chemistry and did research on the composition and quality of fruits and vegetables. Anthocyanin pigments were a significant thread in his research, which addressed issues on color quality of fresh and processed fruits and vegetables, fruit juice authenticity, natural colorants, and the antioxidant properties of fruits and vegetables. He is an Editor for the journal *Food Chemistry*.

**M. Monica Giusti** is a Professor at the Food Science and Technology Department, at The Ohio State University, and a graduate faculty member of the Universidad Nacional Agraria La Molina (UNALM), Perú. Her research is focused on the chemistry and functionality of flavonoids, with emphasis on anthocyanins. Areas of research include anthocyanin incidence and concentration in plants, stability and interactions with food matrices, novel analytical procedures, and bioavailability, bio-transformations and potential bioactivity of these plant pigments. Dr. Giusti received a Food Engineer degree from UNALM and Master's and Doctorate degrees in Food Science from Oregon State University.

**Wilhelmina Kalt** conducts research on berry crops with a focus on the health benefits of their anthocyanins. She has conducted horticultural and food research to determine the factors affecting the content of bioactive components in plant materials. She has fractionated berry extracts to assess the bioactivity of their flavonoid fractions. She has also conducted animal and human studies to examine the digestive absorption of anthocyanins and their effects on health. Wilhelmina is located at the Agriculture and Agri-Food Canada research centre in Nova Scotia, Canada. She collaborates with scientists in the fields of neuroscience, aging, vision, gluco-regulation and cardiovascular research.

# Preface to "Anthocyanins"

The number of research articles on anthocyanin pigments have escalated dramatically in the last 20 years. While PubMed shows 230 anthocyanin publications in the decade from 1982 through 1991, there were 753 from 1992 through 2001, and 3043 from 2002 through 2011. Anthocyanin pigments have long intrigued scientists, and earlier investigations documented the dynamic nature of their chemistry and their role in the color quality of foods, particularly wine because of its high economic value. Historically botanists have investigated these pigments in chemotaxonomic and horticultural research to understand the role of anthocyanins in the color quality of flowers and in fruit ripening. More recently, the widely-publicized "French Paradox" made the public aware of the epidemiological evidence that the French, despite a diet high in saturated fats, had a lower than predicted rate of coronary heart disease compared to people in several Western countries with similar risk factors. It was suggested that the consumption of flavonoid-rich foods including anthocyanins that are abundant in red wine and other fruit-based foods might account, at least in part, for the phenomena. These findings have stimulated an explosion of investigations on various phytochemicals, their bioactivities and their possible role in human health. As part of an early working hypothesis, it was suggested that the antioxidant properties of plant food phytochemicals could be a positive predictor of possible health benefits. Numerous investigations revealed that there was a high correlation specifically between the anthocyanin content of some vegetables, fruits and especially berries and their antioxidant activity in vitro. However, determining the in vivo significance of anthocyanin antioxidant activity in human health has been more difficult since studies have shown that anthocyanins, in their native food forms, are rapidly lost after intake. Notwithstanding there remains abundant in vivo evidence from closely-controlled animal studies, and an increasing amount of human clinical evidence that anthocyanins do indeed provide beneficial health effects. Complementary mechanistic studies have shown that anthocyanins can affect a variety of physiological processes in a beneficial manner. Most encouraging perhaps is recent epidemiological evidence indicating that anthocyanins specifically are associated with a reduced risk of cardiac events, type 2 diabetes and cognitive decline in free-living human populations.

Ronald E. Wrolstad, M. Monica Giusti and Wilhelmina Kalt
*Guest Editors*

# Section 1:
# Anthocyanins in Fresh and Processed Foods

# Influence of Temperature and Preserving Agents on the Stability of Cornelian Cherries Anthocyanins

Bianca Moldovan and Luminiţa David

**Abstract:** Cornelian cherry (*Cornus mas L.*) fruits are known for their significant amounts of anthocyanins which can be used as natural food colorants. The storage stability of anthocyanins from these fruit extracts, at different temperatures (2 °C, 25 °C and 75 °C), pH 3.02, in the presence of two of the most widely employed food preserving agents (sodium benzoate and potassium sorbate) was investigated. The highest stability was exhibited by the anthocyanin extract stored at 2 °C without any added preservative, with half-life and constant rate values of 1443.8 h and $0.48 \times 10^{-3}$ h$^{-1}$, respectively. The highest value of the degradation rate constant ($82.76 \times 10^{-3}$/h) was obtained in the case of anthocyanin extract stored at 75 °C without any added preservative. Experimental results indicate that the storage degradation of anthocyanins followed first-order reaction kinetics under each of the investigated conditions. In aqueous solution, the food preservatives used were found to have a slight influence on the anthocyanins' stability.

Reprinted from *Molecules*. Cite as: Moldovan, B.; David, L. Influence of Temperature and Preserving Agents on the Stability of Cornelian Cherries Anthocyanins. *Molecules* **2014**, *19*, 8177–8188.

## 1. Introduction

Cornelian cherry (*Cornus mas L.*) is a species of dogwood native to Southern Europe and Southwest Asia. The fruit is an oblong, red drupe, 2–3 cm long, containing a single seed, edible, but astringent when unripe. Fresh cornelian cherry fruits contain twice as much ascorbic acid (vitamin C) as oranges, being also rich in sugar, organic acids and tannins [1]. Cornelian cherry fruits also contain significant amounts of anthocyanins which are known to possess antioxidant and anti-inflammatory effects. The most popular application of cornelian cherries is in different drinks, gels and jams, but they can also be eaten fresh, dried whole or pickled. The use of Cornelian cherries for the medical treatment of gastrointestinal disorders and diarrhea has been reported [2]. The anti-bacterial, anti-histamine, anti-allergic, anti-microbial and anti-malarial properties of the fruits are also known [3]. In Europe, Cornelian cherry fruits were reported to have food and cosmetic applications [4]. Because of their health benefits, there are several reports about Cornelian cherry fruits, especially regarding their physical and chemical properties, as well as their polyphenolic, ascorbic acid and anthocyanin contents [5].

Anthocyanins are a class of naturally occurring phenols, being the largest group of water-soluble pigments in plants. Many edible plants are sources of anthocyanins [6–8]. These compounds play a significant role in the color of many fruits, flowers, vegetables and products derived from them. In recent years, various important biological activities, such as antioxidant, antimutagenic, anticancer, anti-inflammatory and antiobesity properties of anthocyanins have been reported [9–12]. The bright color of anthocyanins (orange, red, purple, blue), ensures a high potential of being used as natural colorants, as a healthy alternative to synthetic dyes. Color directly affects the appearance and the consumer acceptability of the fruits and their derived products.

The color of anthocyanins depends essentially on the different structural forms in which they can be found, these structures being strongly influenced by the pH value. At pH values between 4 and 6 (typical for fresh and processed fruits) a mixture of equilibrium forms of anthocyanins: red flavylium cation (I), blue anhydrous quinoidal base (IV), colorless carbinol pseudobase (II) and yellow chalcone (III) coexists (Scheme 1).

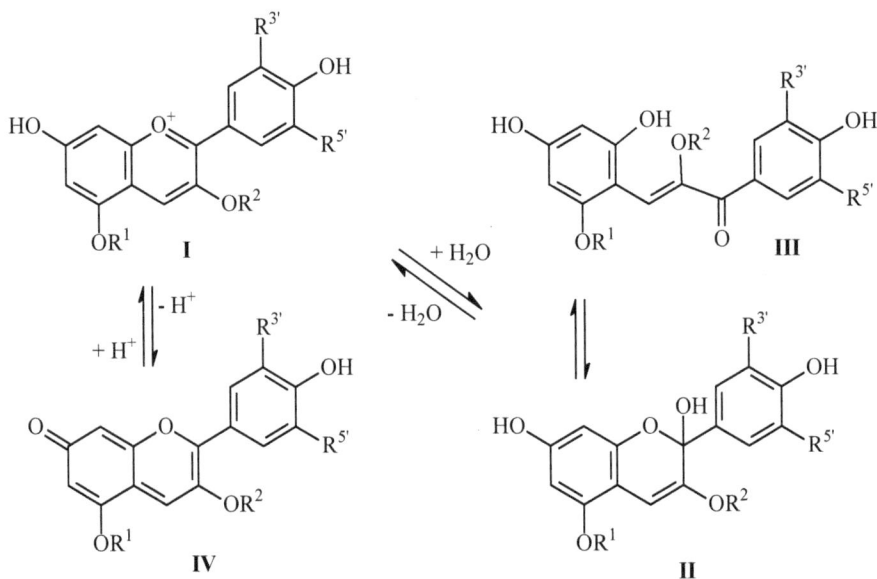

$R^1$ = H or glycosyl

$R^2$ = glycosyl

$R^{3'}$, $R^{5'}$ = H, OH or OCH$_3$

Scheme 1. Chemical structures of anthocyanins at different pH values.

Anthocyanins easily convert to undesirable colorless or brown compounds as a consequence of their high reactivity. The anthocyanin stability can be influenced by many factors, the most important being temperature. The light, pH value, presence of oxygen, ascorbic acid, sugars, hydrogen peroxide and enzymes also affect the stability of these natural pigments [13–18].

Thus, investigation of anthocyanins degradation and measurement of their content at various intervals of storage offers useful experimental information for the food industry. However, to date, no information is available in the literature on the degradation kinetics of Cornelian cherry anthocyanins. The accurate determination of the degradation kinetics for these compounds during storage or thermal processing is essential for predicting changes that may occur in food products containing these anthocyanins.

Sodium benzoate and potassium sorbate are used as food preservatives due to their antimicrobial properties. They are widely used in foods such as soft drinks, jams and fruit juices. Although the effect of these compounds on the inactivation of bacterial pathogens in fruit juices was reported [19], the influence of these synthetic food preserving agents on anthocyanin degradation has been little investigated. Thakur and Araya tested the effect of sodium benzoate and sorbate on the stability of blue grape anthocyanins during storage at 15–35 °C [20], indicating that the stability of the anthocyanins after 60 days was higher in samples preserved with sorbate than with benzoate.

The objective of the present study was to evaluate the influence of temperature and these two commonly used synthetic food preservatives on the stability of Cornelian cherry anthocyanins. The investigated conditions (temperature and nature of an added preservative) ensure a high compatibility with processing techniques often applied in the food industry.

## 2. Results and Discussion

The influence of temperature and food preserving agents on the stability of anthocyanins from the Cornelian cherry fruits extract during storage was investigated. The determined values for the kinetic parameters (kinetic rate constants and the half-life values) are summarized in Table 1.

The content of anthocyanins from Cornelian cherries aqueous extract during storage at different temperatures (2 °C, 22 °C and 75 °C) were plotted as a function of time. The initial total content of the extract was 68.68 ± 0.088 mg/L. The linear regression of the total anthocyanins content of Cornelian cherry fruits extracts during storage confirmed that the degradation process of these pigments followed first order reaction kinetics. These results are in agreement with previously reported literature data [15,21–24] that indicated first order reaction kinetics for the storage and thermal degradation of anthocyanins from various sources.

**Table 1.** Kinetic parameters of degradation of anthocyanins from Cornelian cherriesextracts in different conditions.

| Sample | Temp. (°C) | $k \cdot 10^{-3}$ (h$^{-1}$) [1] | $t_{1/2}$ (h) [2] |
|---|---|---|---|
| Crude extract | 2 | 0.48 (0.9632) | 1443.75 [a] |
| Extract+sodium benzoate | 2 | 0.57 (0.9733) | 1215.78 [b] |
| Extract+potassium sorbate | 2 | 0.65 (0.9903) | 1066.15 [c] |
| Crude extract | 22 | 0.87 (0.9188) | 796.55 [a] |
| Extract+sodium benzoate | 22 | 0.93 (0.9739) | 745.16 [b] |
| Extract+potassium sorbate | 22 | 1.13 (0.9703) | 613.27 [c] |
| Crude extract | 75 | 82.61 (0.9922) | 8.38 [a] |
| Extract+sodium benzoate | 75 | 78.84 (0.9908) | 8.78 [a] |
| Extract+potassium sorbate | 75 | 77.35 (0.9912) | 8.95 [a] |

[1] Numbers in parentheses, $R^2$, are the determination coefficients; [2] Values within a column with different superscript letters are significantly ($p < 0.05$) different in the same temperature group.

The thermal stability of the extracts was evaluated. As expected, the increase of storage temperature (22 °C) resulted in a 1.8 times faster degradation as compared to refrigerated storage (at 2 °C) while, at 75 °C, the degradation rate was 172.1 times higher.

The total content of anthocyanins from Cornelian cherries extracts stored at 2 °C was plotted as a function of time (Figure 1). By comparing the half-life values, one can conclude that, the presence of food preservatives displayed a slightly destabilizing effect on the anthocyanins from the investigated extracts.

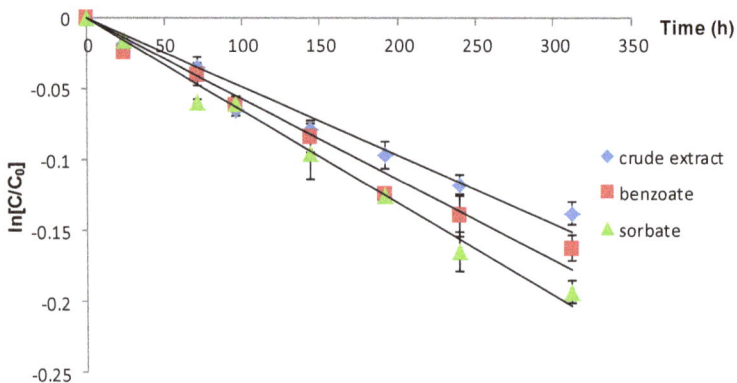

**Figure 1.** Influence of different food preservatives on the anthocyanin stability during storage at 2 °C (vertical lines represent SD, $n = 4$).

However, the difference between the two added food preservatives was not significant, the degradation process being 1.14 fold faster in the presence of potassium sorbate as compared to sodium benzoate.

As observed in the case of refrigerated storage, the Cornelian cherry anthocyanins stored at 22 °C showed the same degradation profile (Figure 2). In this case, storage of the extracts in the presence of potassium sorbate resulted in faster degradation compared to storage with added sodium benzoate, the value of the half-life ratio being 1.21.

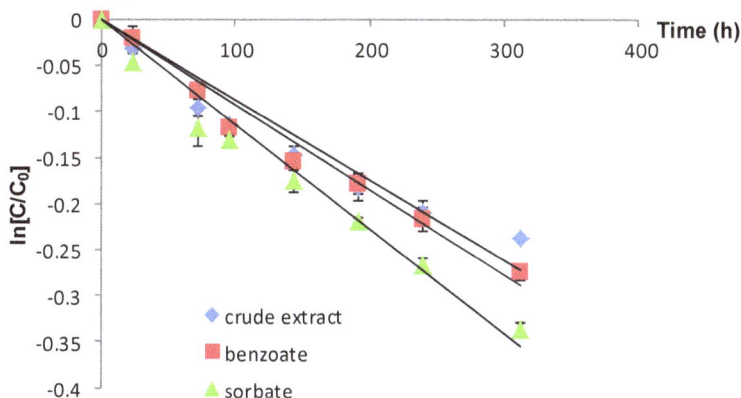

**Figure 2.** Influence of different food preservatives on the anthocyanin stability during storage at 22 °C (vertical lines represent SD, $n = 4$).

As expected, the increase of temperature at 75 °C resulted in an accelerated degradation of anthocyanins. During high temperature storage, as applied at 75 °C, the destabilizing effect of sodium benzoate and potassium sorbate on the anthocyanin pigments (observed at lower storage temperatures) was not evident. In contrast, the added food preserving agents had almost no influence on the stability of these pigments (Figure 3), the half-life values being practically the same for all the investigated extracts (8.4 ÷ 8.95 h).

The effect of temperature on the kinetics of the degradation process was determined by fitting the rate constants to an Arrhenius type equation.

The anthocyanin degradation rate constants obtained for each extract were plotted as a function of temperature (Figure 4).

The calculated activation energies ($E_a$), frequency factors ($K_o$) and the temperature coefficients ($Q_{10}$) are given in Table 2.

Since high activation energy reactions are more sensitive to temperature, the anthocyanins in the extract proved to be more susceptible to degradation by exposure to elevated temperatures. The calculated $E_a$ values ranged from 54.09 to 58.55 kJ/mol. The highest influence of the temperature on the stability of the investigated compounds (the highest value of $E_a$) was observed for the anthocyanins stored in the crude extract, while the pigments stored in the presence of potassium sorbate exhibited lower susceptibility to thermal degradation, presenting the lowest value of the $E_a$. The low differences between the activation energy values suggested that the added synthetic preservatives exhibited a slightly influence on the stability of anthocyanins.

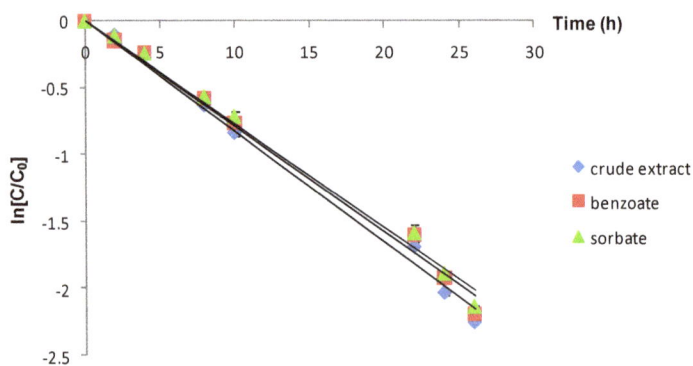

**Figure 3.** Influence of different food preservatives on the anthocyanin stability during storage at 75 °C (vertical lines represent SD, $n = 4$).

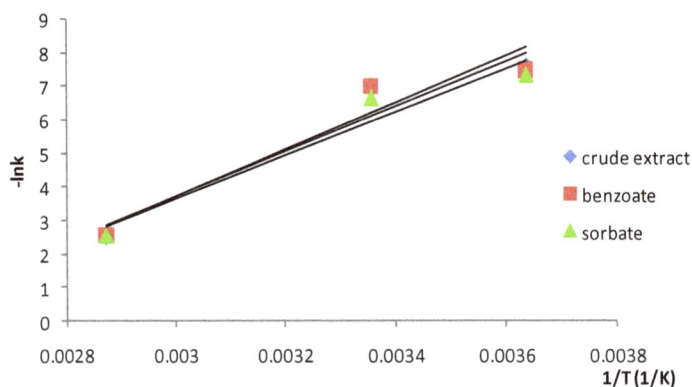

**Figure 4.** The Arrhenius plots for degradation of anthocyanins in Cornelian cherries extracts.

**Table 2.** Effect of temperature on the degradation of anthocyanins from Cornelian cherry fruits extracts.

| Solvent | $E_a$ (kJ/mol) [a] | $K_o$ (h$^{-1}$) | $Q_{10}$ | |
|---|---|---|---|---|
| | | | 2–22 °C | 22–75 °C |
| Crude extract | 58.55 (0.9307) | $3.72 \times 10^7$ | 1.346 | 2.361 |
| Extract+sodium benzoate | 56.21 (0.9226) | $1.57 \times 10^7$ | 1.277 | 2.311 |
| Extract+potassium sorbate | 54.09 (0.9446) | $7.9 \times 10^6$ | 1.318 | 2.219 |

[a] Numbers in parentheses, $R^2$, are the determination coefficients.

In order to evaluate the dependence of degradation rate on temperature, the temperature coefficient $Q_{10}$ (the change of degradation rate upon a temperature increase of 10 K) was calculated.

Higher $Q_{10}$ values for storage temperatures of 22–75 °C were obtained, indicating that anthocyanins are more sensitive to temperature elevations at high storage temperatures compared to low storage temperatures (2–22 °C) where the $Q_{10}$ values were ranged from 1.277 to 1.346, whereas the differences were insignificant. The lowest temperature coefficient value (1.277 at 2–22 °C) was obtained for the anthocyanins stored in the presence of sodium benzoate. Almost the same $Q_{10}$ values were obtained for the degradation of anthocyanins stored at 2–22 °C for all the investigated extracts (crude or with added preservative), proving that the influence of the added food preserving agents was not significant. Storage at 22–75 °C resulted in higher $Q_{10}$ values the calculated temperature coefficients for this storage interval presenting almost the same value for all the investigated extracts. Higher $Q_{10}$ values indicate that at high storage temperatures (22–75 °C) anthocyanins are more sensitive to temperature elevations than at low storage temperatures (2–22 °C).

All these results clearly indicate that low storage temperatures are required to inhibit the degradation process of these pigments from Cornelian cherry extracts.

Significantly different temperature coefficients $Q_{10}$ for the two investigated temperature intervals, may be due to a possible change in the reaction mechanism of the degradation of Cornelian cherries anthocyanins at elevated temperatures, such as 75 °C, compared to low temperature degradation process. The high ascorbic acid content of Cornelian cherry fruits [1] could accelerate the degradation of anthocyanins. The loss of anthocyanins caused by ascorbic acid (AA) occurs due to the free radical oxidative cleavage of the pyrilium ring in which AA acts as molecular oxygen activator. At high temperatures, AA itself undergoes a degradation process, generating degradation products which are also responsible for anthocyanins degradation [25]. Since no degradation studies were performed on the Cornelian cherry anthocyanins, the determined $t_{1/2}$ values are compared to the literature reported data for the degradation process of anthocyanins obtained from other fruits. Wang and Xu reported that the $t_{1/2}$ value for anthocyanin degradation in blackberry juice at pH 2.86 was 4.7 h at 80 °C [23]. In blood orange juice concentrate, the reported $t_{1/2}$ value at 4 °C was 55.7 days [26]. Compared to this value, our results are in the same range ($t_{1/2}$ = 60.2 days at 2 °C). The major anthocyanins in Cornelian cherries are cyanidin-3-O-galactoside, pelargonidin-3-O-galactoside and delphinidin-3-O-galactoside [3]. However, the major anthocyanins in blackberry are cyanidin-3-O-glucoside, cyanidin-3-O-rutinoside and cyanidin-3-O-malonyl-glucoside [27,28]. Therefore, the different stability of the anthocyanins might be due to the varying composition of the fruit extracts; the major constituents of these extracts being sugars, ascorbic acid, and flavonoids, which are known to be intrinsic factors that influence anthocyanins degradation [29]. It could be concluded that Cornelian cherry anthocyanins are more stable than anthocyanins from other sources (e.g., blackberry), proving a quite good stability during storage and heating, indicating a potential use of these pigments as natural colorants in food industry.

## 3. Experimental

### 3.1. Materials

#### 3.1.1. Plant Material

Samples of Cornelian cherry fruits were purchased in August 2012 from a local market in Cluj-Napoca, Romania. Fruits were packed in polyethylene bags and kept frozen at −18 °C before being subjected to extraction of anthocyanins.

#### 3.1.2. Chemicals and Reagents

Potassium chloride, sodium acetate, acetic acid and HCl conc., were purchased from Merck (Darmstadt, Germany). Sodium benzoate and potassium sorbate were purchased from Chimopar (Bucharest, Romania). All chemicals and reagents were of analytical grade and were used without further purification. The distilled water was obtained using a TYPDP1500 Water distiller (Techosklo LTD, Držkov, Czech Republic).

### 3.2. Methods

#### 3.2.1. Preparation of Anthocyanin Extract

Fifty grams of frozen Cornelian cherry fruits were crushed in a mortar. Thirty five g of fruit puree were transferred to an Erlenmeyer flask and distilled water (200 mL) and concentrated HCl (0.25 mL) were added. The mixture was stirred for 1 h at room temperature and then filtered. The residue was washed twice with extraction solvent (acidified water, 20 mL). The filtrate was quantitatively transferred to a 250 mL volumetric flask and made up to 250 mL with solvent. The pH of the extract was determined using a Hanna Instruments (HI) 99161 pH-meter and the measured value was 3.02.

#### 3.2.2. Determination of Anthocyanin Content

The total anthocyanin content was determined using optical spectroscopy, by the convenient method of Giusti and Wrolstad [30]. This method is based on the structural changes of the pigments as a function of pH. At pH = 1, the red to purple oxonium form predominates while at pH = 4.5 the major structural form is the colourless hemiketal. The difference in absorbance of the anthocyanin solutions at these two pH values permits an accurate and rapid determination of total monomeric anthocyanin content in the sample matrix. The two desired pH values were reached using two buffer systems: potassium chloride buffer (0.025 M, pH = 1.0) and sodium acetate buffer (0.04 M; pH = 4.5).

Aliquots of Cornelian cherry fruits extract (5 mL) were transferred to a 10 mL volumetric flask, made up to 10 mL with corresponding buffer (pH = 1 and pH = 4.5) and allowed to equilibrate for 15 min. The absorbance of each equilibrated solution was then measured at 506 (the wavelength where the maximum of absorbance occurs = $\lambda_{VIS\ max}$) and 700 nm (for haze correction), using an UV-VIS Perkin Elmer Lambda 25 double beam spectrophotometer.

Pigment content was calculated using a molar extinction coefficient of 26,900 L/mol·cm and a molecular weight of 449.2 g/mol (cyanidin-3-glucoside). Results were expressed as mg cyanidin-3-glucoside equivalents·$L^{-1}$ extract. Visible spectra of samples were recorded by scanning the absorbance between 400 and 700 nm. Quartz cuvettes of 1 cm path length were used. Absorbance readings were performed against distilled water as a blank. All the measurements were carried out at room temperature (~22 °C).

The total anthocyanin content (expressed as cyanidin-3-glucoside equivalents), was calculated from the experimental data, using the following equation [30]:

$$TA = \frac{A \cdot MW \cdot DF \cdot 1000}{\varepsilon \cdot l} \tag{1}$$

where: TA = total anthocyanin content (mg/L) ; A = absorbance, calculated as: [Equation (6)]

$$A = (A_{pH\ 1.0} - A_{pH\ 4.5})_{506\ nm} - (A_{pH\ 1.0} - A_{pH\ 4.5})_{700\ nm} \tag{2}$$

MW = molecular weight; DF = dilution factor; l = path length; $\varepsilon$ = molar extinction coefficient; 1000 = conversion factor from grams to milligrams.

Four determinations (n = 4) were performed for each analysis and the average values of total anthocyanin content were used for kinetic parameters determination.

## 3.2.3. Degradation Studies

The influence of temperature on the storage stability was studied at 2 °C, 22 °C and 75 °C. Each extract was divided into 50 mL portions and kept away from light (well capped to avoid evaporation) at 2 °C (in refrigerator), at room temperature (22 °C) and in a thermostatic water bath, preheated to 75 °C, respectively (±1 °C).

To test the influence of food preservatives on the thermal stability of anthocyanins from Cornelian cherry fruits extract, sodium benzoate or potassium sorbate was dissolved in the fruits extract at a final concentration of 1 g/L.

Changes in total anthocyanin content of the samples were measured in order to evaluate the stability of the pigments in the investigated extracts. Samples were analyzed at 0, 1, 3, 4, 6, 8, 10 and 13 days for all extracts except for those stored at 75 °C, which were sampled at 0, 2, 4, 8, 10, 22, 24 and 26 h. The storage intervals were different for the last thermal treatment due to the differences in anthocyanin degradation rates.

### 3.2.4. Degradation Kinetics

The kinetics for the degradation reaction of the investigated anthocyanins can be expressed by the equations:

$$\ln[TA] = \ln[TA_0] - kt \tag{3}$$

$$t_{1/2} = -\ln 0.5/k \tag{4}$$

where: $[TA]$ = total anthocyanin content (mg/L) at time $t$; $[TA_0]$ = initial total anthocyanin content (mg/L); $k$ = reaction rate constant $(h^{-1})$; $t$ = reaction time (h); $t_{1/2}$ = half−life (h).

The effect of temperature on the kinetics of the degradation process was determined by fitting the rate constants to an Arrhenius type equation [Equation (5)]:

$$k = K_0 e^{-Ea/RT} \tag{5}$$

where: $k$ = rate constant $(h^{-1})$; $K_0$ = frequency factor $(h^{-1})$; $E_a$ = activation energy (kJ/mol); $R$ = universal gas constant (8.314 J/mol·K); $T$ = absolute temperature (K).

The $Q_{10}$ temperature coefficient was calculated according to Equation (6):

$$Q_{10} = \left(\frac{k_2}{k_1}\right)^{10/(T_2 - T_1)} \tag{6}$$

where: $Q_{10}$ = the temperature coefficient $(K^{-1})$; $k_{1,2}$ = rate constant $(h^{-1})$ at temperature $T_{1,2}$ (K).

### 3.3. Statistical Analysis

Data are reported as mean values of at least four experiments. Results were analyzed using one-way variance analysis (ANOVA). Analysis of variance was performed using XLSTAT Release 10 (Addinsoft, Paris, France). Differences at $p < 0.05$ were considered statistically significant.

## 4. Conclusions

The total anthocyanin content of Cornelian cherries and the storage stability of these compounds indicated that these fruits can be used as an important source of natural red pigment for the food industry. The results of the present study have provided detailed information on the degradation kinetic parameters of anthocyanins during storage and heating. Increasing the temperature resulted in higher degradation rate constants: the degradation rate of anthocyanins from Cornelian cherries extract at 22 °C was 1.8 times faster than at 2 °C, while at 75 °C this process was 172 times faster than at 2 °C. Comparison of the rate constants and half-life values showed that the anthocyanin stability was slightly influenced by the kind of the added organic food preservative.

**Acknowledgments:** This research was supported by the Ministry of Education, Research, Youth and Sports, Romania (project No. 147/2011 PN-II-PT-PCCA-2011-3-1-0914).

**Author Contributions:** Luminița David, as leading author, has been involved in designing the research, interpreting the data, writing and revising the paper. Bianca Moldovan has been involved in performing the research, analyzing data as well as drafting the paper.

**Conflicts of Interest:** The authors declare no conflict of interest.

## References

1. Seeram, N.P.; Schutzki, R.; Chandra, A.; Nair, M.G. Characterization, quantification and bioactivities of anthocyanins in *Cornus* species. *J. Agric. Food Chem.* **2002**, *50*, 2519–2523.
2. Celik, S.; Bakirci, I.; Sat, I.G. Physico-chemical and organoleptic properties of yogurt with cornelian cherry paste. *Int. J. Food Prop.* **2006**, *9*, 401–408.
3. Vareed, S.K.; Reddy, M.K.; Schutzki, R.E.; Nair, M.G. Anthocyanins in *Cornus alternifolia, Cornus controversa, Cornus kousa* and *Cornus florida* fruits with health benefits. *Life Sci.* **2006**, *78*, 777–784.
4. Polinicencu, C.; Popescu, H.; Nistor, C. Vegetal extracts for cosmetic use: Extracts from fruits of *Cornus mas*. Preparation and characterization. *Clujul Med.* **1980**, *53*, 160–163.
5. Pantelidis, G.E.; Vasilakakis, M.; Manganaris, G.A.; Diamantidis, G. Antioxidant capacity, phenol, anthocyanin and ascorbic acid contents in raspberries, blackberries, red currants, gooseberries and Cornelian cherries. *Food Chem.* **2007**, *102*, 777–783.
6. Mazza, G. *Natural Food Colorants: Science and Technology*; Marcel Decker: New York, NY, USA, 2000; pp. 289–314.
7. Reyes, L.F.; Cisneros-Zevallos, L. Degradation kinetics and colour of anthocyanins in aqueous extracts of purple- and red-flesh potatoes (*Solanum tuberosum* L.). *Food Chem.* **2007**, *100*, 885–894.

8.  Prodanov, M.P.; Dominguez, J.A.; Blazquez, I.; Salinas, M.R.; Alonso, G.L. Some aspects of the quantitative/qualitative assessment of commercial anthocyanin-rich extracts. *Food Chem.* **2005**, *90*, 585–596.

9.  Yoshimoto, M.; Okuno, S.; Yamaguchi, M.; Yamakawa, O. Antimutagenicity of Deacylated Anthocyanins in Purple-fleshed Sweetpotato. *Biosci. Biotechnol. Biochem.* **2001**, *65*, 1652–1655.

10. Tsuda, T.; Horio, F.; Uchida, K.; Aoki, H.; Osawa, T. Dietary cyanidin 3-*O*-β-D-glucoside-rich purple corn color prevents obesity and ameliorates hyperglycemia in mice. *J. Nutr.* **2003**, *133*, 2125–2130.

11. Smith, M.A.L.; Marley, K.A.; Seigler, D.; Singletary, K.W.; Meline, B. Bioactive Properties of Wild Blueberry Fruits. *J. Food Sci.* **2000**, *65*, 352–356.

12. Crisan, M.; David, L.; Moldovan, B.; Vulcu, A.; Dreve, S.; Perde-Schrepler, M.; Tatomir, C.; Filip, A.G.; Bolfa, P.; Achim, M.; *et al.* New nanomaterials for the improvement of psoriatic lesions. *J. Mater. Chem. B* **2013**, *1*, 3152–3158.

13. Cevallos-Casalas, B.A.; Cisneros-Zevallos, L. Stability of anthocyanin-based aqueous extracts of Andean purple corn and red-fleshed sweet potato compared to synthetic and natural colorants. *Food Chem.* **2004**, *86*, 69–77.

14. Özkan, M.; Yemenicioğlu, A.; Asefi, N.; Cemeroğlu, B. Degradation kinetics of anthocyanins from sour cherry, pomegranate, and strawberry juices by hydrogen peroxide. *J. Food Sci.* **2002**, *67*, 525–529.

15. Hernández-Herrero, J.A.; Frutos, M.J. Degradation kinetics of pigment, colour and stability of the antioxidant capacity in juice model systems from six anthocyanin sources. *Int. J. Food Sci. Technol.* **2011**, *46*, 2550–2557.

16. Garzon, G.A.; Wrolstad, R.E. Comparison of the stability of pelargonidin-based anthocyanins in strawberry juice and concentrate. *J. Food Sci.* **2002**, *67*, 1288–1299.

17. Moldovan, B.; David, L.; Chisbora, C.; Cimpoiu, C. Degradation kinetics of anthocyanins from European Cranberrybush (*Viburnum opulus L.*) fruits extracts. Effects of temperature, pH and storage solvent. *Molecules* **2012**, *17*, 11655–11666.

18. Moldovan, B.; David, L.; Donca, R.; Chisbora, C. Degradation kinetics of anthocyanins from crude ethanolic extract from sour cherries. *Stud. Univ. Babes-Bolyai Chem.* **2011**, *56*, 189–194.

19. Gurtler, J.B.; Bailey, R.B.; Geveke, J.D.; Zhang, H.Q. Pulsed electric field inactivation of *E. coli* O157:H7 and non-pathogenic surrogate *E. coli* in strawberry juice as influenced by sodium benzoate, potassium sorbate, and citric acid. *Food Control.* **2011**, *22*, 1689–1694.

20. Thakur, B.R.; Arya, S.S. Studies on stability of blue grape anthocyanins. *Int. J. Food Sci. Technol.* **1989**, *24*, 321–326.

21. Chisté, R.C.; Lopes, A.S.; DeFaria, L.J.G. Thermal and light degradation kinetics of anthocyanin extracts from mangosteen peel (*Garcinia mangostana L.*). *Int. J. Food Sci. Technol.* **2010**, *45*, 1902–1908.

22. Kirca, A.; Özkan, M.; Cemeroğlu, B. Effects of temperature, solid content and pH on the stability of black carrot anthocyanins. *Food Chem.* **2007**, *101*, 212–218.

23. Wang, W.D.; Xu, S.Y. Degradation kinetics of anthocyanins in blackberry juice and concentrate. *J. Food Eng.* **2007**, *82*, 271–275.

24. Amaro, L.F.; Soares, M.T.; Pinho, C.; Almeida, I.F.; Pinho, O.; Ferreira, I.M. Processing and storage effects on anthocyanin composition and antioxidant activity of jams produced with Camarosa strawberry. *Int. J. Food Sci. Technol.* **2013**, *48*, 2071–2077.

25. Jackman, R.L.; Yada, R.Y.; Tung, M.A. A review: Separation and chemical properties of anthocyanins used for their qualitative and quantitative analysis. *J. Food Biochem.* **1987**, *11*, 279–308.

26. Kirca, A.; Cemeroğlu, B. Thermal degradation of blood orange anthocyanins. *Food Chem.* **2003**, *81*, 583–587.

27. Fang-Chiang, H.J.; Wrolstad, R.E. Anthocyanins pigment composition of blackberries. *J. Food Sci.* **2005**, *70*, 198–202.

28. Rommel, A.; Wrolstad, R.E. Blackberry juice and wine: Processing and storage effects on anthocyanin, color and appearance. *J. Food Sci.* **1992**, *57*, 385–391.

29. Cao, S.Q.; Liu, L.; Lu, Q.; Xu, Y.; Pan, S.Y.; Wang, K.X. Integrated effects of ascorbic acid, flavonoids and sugars on thermal degradation of anthocyanins in blood orange juice. *Eur. Food Res. Technol* **2009**, *228*, 975–983.

30. Giusti, M.M.; Wrolstad, R.E. *Current Protocols in Food Analytical Chemistry*; Wiley: New York, NY, USA, 2001; pp. 1–13.

**Sample Availability:** *Sample Availability*: Samples of extracts are available for the next 2 months (the anthocyanins from the extracts totally degraded after this time period) from the authors.

# Anthocyanin Characterization, Total Phenolic Quantification and Antioxidant Features of Some Chilean Edible Berry Extracts

Anghel Brito, Carlos Areche, Beatriz Sepúlveda, Edward J. Kennelly and Mario J. Simirgiotis

**Abstract:** The anthocyanin composition and HPLC fingerprints of six small berries endemic of the VIII region of Chile were investigated using high resolution mass analysis for the first time (HR-ToF-ESI-MS). The antioxidant features of the six endemic species were compared, including a variety of blueberries which is one of the most *commercially significant* berry crops in Chile. The anthocyanin fingerprints obtained for the fruits were compared and correlated with the antioxidant features measured by the bleaching of the DPPH radical, the ferric reducing antioxidant power (FRAP), the superoxide anion scavenging activity assay (SA), and total content of phenolics, flavonoids and anthocyanins measured by spectroscopic methods. Thirty one anthocyanins were identified, and the major ones were quantified by HPLC-DAD, mostly branched 3-*O*-glycosides of delphinidin, cyanidin, petunidin, peonidin and malvidin. Three phenolic acids (feruloylquinic acid, chlorogenic acid, and neochlorogenic acid) and five flavonols (hyperoside, isoquercitrin, quercetin, rutin, myricetin and isorhamnetin) were also identified. Calafate fruits showed the highest antioxidant activity (2.33 ± 0.21 μg/mL in the DPPH assay), followed by blueberry (3.32 ± 0.18 μg/mL), and arrayán (5.88 ± 0.21), respectively.

Reprinted from *Molecules*. Cite as: Brito, A.; Areche, C.; Sepúlveda, B.; Kennelly, E.J.; Simirgiotis, M.J. Anthocyanin Characterization, Total Phenolic Quantification and Antioxidant Features of Some Chilean Edible Berry Extracts. *Molecules* **2014**, *19*, 10936–10955.

## 1. Introduction

Fruits and vegetables are considered highly protective for human health, particularly against ageing and various oxidative-stress related diseases, due to their content of healthy phytochemicals [1]. Several epidemiological studies have highlighted the association between the consumption of foods with high contents of phytochemicals, mainly flavonols, phenolic acids and anthocyanins, and the prevention of degenerative diseases such as cardiovascular diseases, ageing, cancer and other degenerative disorders [2,3]. Anthocyanins are a group of red, purple, violet and blue water soluble polyphenolic pigments widely distributed in berry fruits which can act as antioxidants or free radical scavengers, thus preventing oxidative stress [4]. The term berry fruit generally refers to some small fruit that lacks big seeds and can be eaten whole. Berry fruits are often the richest source of antioxidant phytochemicals among fruits and vegetables [5], thus the chemical study of native berry fruits is of great economic significance since it can support the consumption and commercial activities of gatherers, growers, micro-companies and industries associated with the use of native plants. Chilean fruits such as arrayán, chequen, calafate, meli, maqui and murta (Figure 1) are small pigmented native berries which were collected since pre-Colombian times by South American Amerindians as a food source. At present, there is still some regional consumption of the small berries from trees and shrubs belonging to the Myrtaceae (Chilean myrtle, murta, arrayán, chequén, luma and meli), Berberidaceae (michay and calafate) as well as Eleaocarpaceae (maqui) occuring in southern Chile and Argentina. In Chile, "murta" or "murtilla" (*Myrtus ugni* Molina or *Ugni molinae* Turczaninov), a wild perennial shrub also commonly known as Chilean guava, is the best-known of the native Myrtaceae plants, where the people have long appreciated its red edible berries for its unique aroma. Infusions of the leaves of this species are anti-inflammatory and analgesic [6] and the fruits contain several volatile compounds responsible for the aroma [7].

Arrayán (*Luma apiculata* (DC.) Burret is an evergreen Myrtaceae tree occurring in southern Chile and Argentina of about 10 m in height with orange-red trunk and edible purple black berries, 1–1.5 cm in diameter, that ripen in early autumn and are half the size, with more intense color, but similar aspect and consistence as the worldwide commercialized blueberries (*Vaccinium corymbosum*). Murillo [8] describes the medicinal properties of *Eugenia apiculata* D.C. (a synonym for *L. apiculata*, also known as *Myrceugenella apiculata* (DC.) Kausel [9]). The traditional use indications include aromatic, slightly astringent, balsamic and anti-inflammatory uses. The fruits were used to prepare liquor. This information is in agreement with the aromatic flavor that is attractive for local producers of alcoholic beverages. The fruits of *Luma chequén* (Molina) A. Gray, syn: *Myrceugenella chequen* (Mol.) Kaus are edible small berries with similar size than those of arrayán and murta. de Mösbach [10] refers to uses of *L. chequen* in infusions and syrups as an astringent. The traditional use indications in traditional medicine can be related to the tannin content of the plant which is also recommended as a wound wash and to treat dysentery. Both *L. apiculata* and *L. chequen* fruits were used to prepare "chicha", a South American native fermented beverage [9]. Calafate or Magellan barberry (*Berberis microphylla* G. Forst, sin. *Berberis buxifolia*, and *Berberis heterophylla*) is another Patagonian shrub with edible dark small berries that can grow in a great variety of areas [11]. The production of calafate is concentrated in small gardens in the Regions of Aysén and Magallanes for local production of jams and juices [11]. This fruit contains several anthocyanins [12] and high content of cinnamic acids [13]. Maqui (*Aristotelia chilensis*) fruit is now one of the most famous dark colored Chilean berries because of its high content of anthocyanins [14]. Calafate, maqui and murta are antioxidant berries considered superfruits due to their high content of phenolic compounds, including several anthocyanins [6,12,15]. Several edible Myrtaceae fruits known worldwide present free radical scavenging constituents including anthocyanins [16], while Chilean Myrtaceae with high anthocyanin contents have been assessed for antioxidant activity and showed good antioxidant features [17–19]. Mass°spectrometry has undergone tremendous *technological improvements* in the last years, especially with the development of ionization methods such as electrospray (ESI), atmospheric pressure chemical ionization (APCI) and high resolution mass detectors such as time of flight (TOF). Indeed, several antioxidant phenolics in edible plants [20]; fruits [21–23]; nuts [24] and food byproducts [25] were analyzed using HPLC hyphenated with accurate high resolution time of flight analyzers (HPLC-PDA-ToF-MS). However, the chemical analysis regarding anthocyanins or metabolomics present in wild Chilean berries including arrayán, chequén, murta, and calafate was performed using low resolution methods (ESI-ion trap-MS) [12,15,19], while the phenolic constituents of *A. meli* have not beenreported to the best of our knowledge.

**Figure 1.** Pictures of (**a**) chequén, (*Luma chequén*) (**b**) murta, (*Ugni molinae*) (**c**) arrayán, (*Luma apiculata*), (**d**) blueberries, (*Vaccinium corymbosum*) (**e**) meli, (Amomyrtus meli and (**f**) calafate (*Berberis microphylla*) growing in the VIII region of Chile.

The aim of the present work was the analysis by high resolution mass spectrometry (HR-MS) of some important native berries from Chile, and the comparison of the antioxidant properties and total phenolics. In the present work the anthocyanin fingerprints and polyphenolic content of six small Chilean berries (arrayán, chequén, murta, calafate, meli and Chilean blueberry var. Brigitta, Figure 1) from the VIII region of Chile were compared and correlated with the antioxidant capacities measured by the DPPH radical bleaching, ferric reducing antioxidant power (FRAP), and the superoxide anion scavenging activity (SA) assays. The anthocyanins in berries were identified for the first time with the help of PDA analysis and high resolution time of flight mass spectrometry (HPLC-ESI-ToF-MS) plus comparison with authentic standards.

## 2. Results and Discussion

### 2.1. Accurate MS-PDA Identification of Anthocyanins in Six Small Berry Fruits from Southern Chile

Anthocyanins in berry fruits were accurately detected and identified using HPLC with UV-visible detection (PDA, Figure 2, Table 1) and high resolution time of flight mass spectrometry (HR-ToF-MS, Table 1). The 31 anthocyanins identified in the six berries (Figure 3) were mainly 3-O-glycoside conjugates and their derivatives.

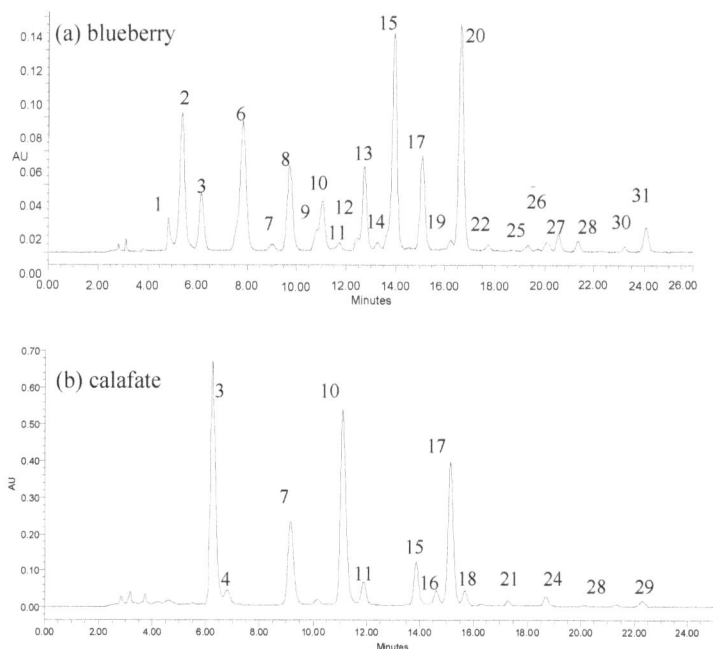

**Figure 1.** *Cont.*

19

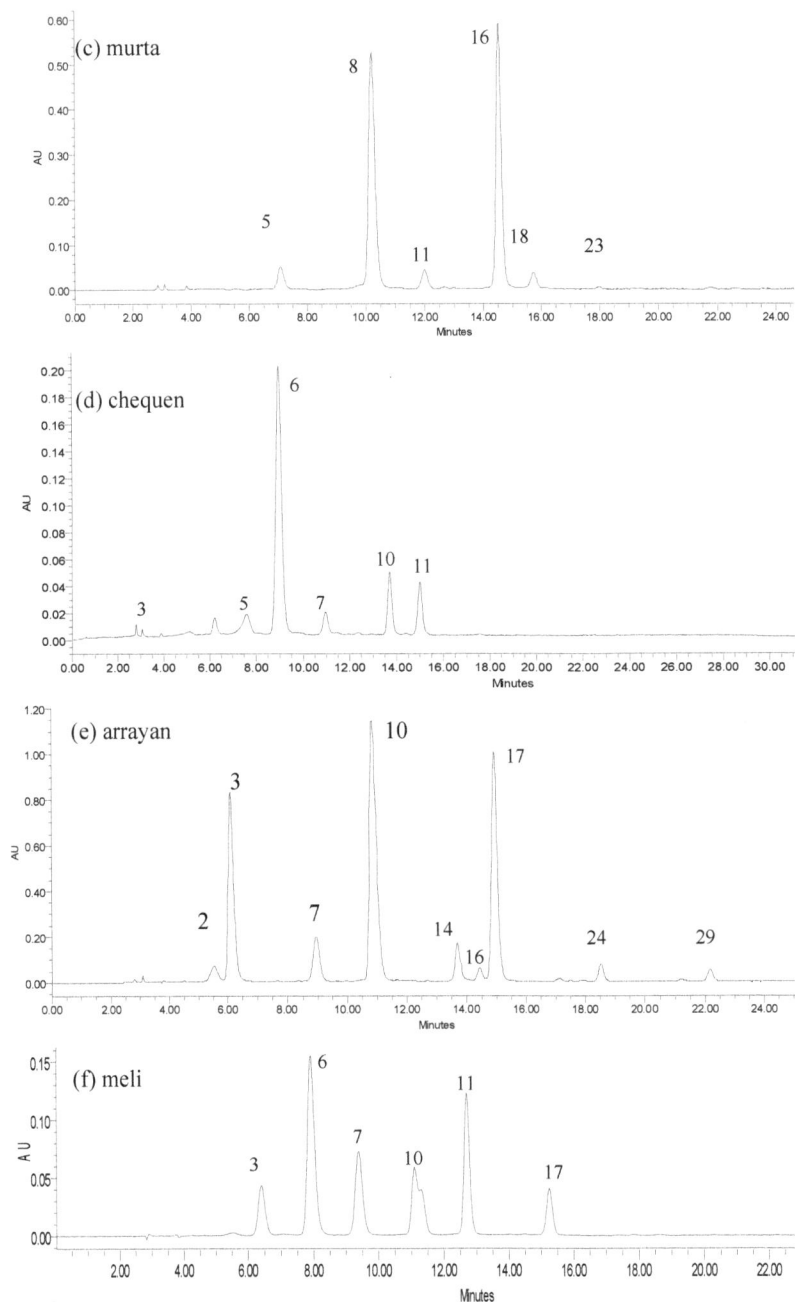

**Figure 2.** HPLC-PDA chromatograms of six berries from the VIII region of Chile.
(a) *Vaccinium corymbosum*, (b) *Berberis microphylla*, (c) *Ugni molinae*, (d) *Luma chequén*,
(e) *Luma apiculata*, and (f) Amomyrtus meli monitored at 520 nm. Peaks numbers
refer to those indicated in Table 1.

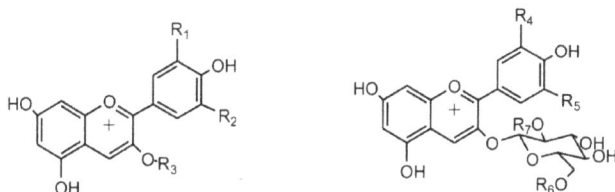

| Peak | $R_1$ | $R_2$ | $R_3$ | | Peak | $R_4$ | $R_5$ | $R_6$ | $R_7$ |
|------|-------|-------|-------|--|------|-------|-------|-------|-------|
| 3* | OH | OH | Gal | | 1 | OH | OCH₃ | Glu | H |
| 4 | OH | OH | Glu | | 2 | OH | OCH₃ | Glu | H |
| 6* | OH | H | Gal | | 5 | OH | H | Rha | H |
| 7* | OH | H | Glu | | 8 | OH | OCH₃ | Rha | H |
| 9 | OCH₃ | OCH₃ | Rha | | 9 | OCH₃ | OCH₃ | Rha | H |
| 10* | OH | OCH₃ | Glu | | 12 | H | OCH₃ | Rha | H |
| 11* | OH | OCH₃ | Gal | | 21 | OCH₃ | OCH₃ | Cou | H |
| 13* | H | OCH₃ | Gal | | 23 | OH | H | Succ | H |
| 14 | OCH₃ | OCH₃ | Gal | | 24 | H | OCH₃ | Glu | H |
| 15 | OH | OH | Ara | | 25 | OH | OH | Cou-rha | Glu |
| 16* | H | OCH₃ | Glu | | 26 | OH | OH | Ac | H |
| 17* | OCH₃ | OCH₃ | Glu | | 27 | OH | H | Ac | H |
| 18 | H | OCH₃ | Ara | | 28 | OH | OCH₃ | Ac | H |
| | | | | | 29 | OCH₃ | OCH₃ | Ac | H |
| | | | | | 30 | OH | OH | Caff | H |
| | | | | | 31 | OCH₃ | OCH₃ | Ac | H |

**Figure 3.** Structures of the anthocyanins identified in six berries from the VIII region of Chile. * Identified using standard compounds. Gal: Galactose; Glu: Glucose; Ara: Arabinose; Rha: Rhamnose; Cou: Coumaric; Succ: Succinic acid; Ac: Acetyl group; Caff: Caffeic acid.

Twenty three compounds were detected in blueberry (peaks **1–3**, **6–15**, **17**, **19**, **20**, **22**, **25–28**, **30** and **31**, Table 1) fourteen in calafate (peaks **3**, **4**, **7**, **8**, **10**, **11**, **15**, **16–18**, **21**, **24**, **28** and **29**), nine in arrayán (peaks **2**, **3**, **7**, **10**, **14**, **16**, **17**, **24** and **29**), and six in meli (peaks **3**, **6**, **7**, **10**, **11** and **17**), chequén (peaks **3**, **5**, **6**, **7**, **10** and **11**) and murta (peaks **5**, **8**, **11**, **16**, **18** and **23**). Figure S2 and S3 (Supplementary Material) show as examples full scan ToF-MS spectra of peaks **3**, **8**, **9**, **10**, **16**, **17**, **21**, **22** and **28**). Peaks **3**, **6**, **7**, **10**, **11**, **13**, **16** and **17** were identified by spiking experiments with authentic standards as delphinidin 3-$O$-galactoside (HR-MS ion at $m/z$ 465.1043, $\lambda_{max}$: 276–523), cyanidin-3-$O$-galactoside (HR-MS ion at $m/z$ 449.1052, $\lambda_{max}$: 280–511), cyanidin-3-$O$-glucoside (HR-MS ion at $m/z$ 449.1099, $\lambda_{max}$: 280–517), petunidin-3-$O$-glucoside (HR-MS ion at $m/z$ 479.1233, $\lambda_{max}$: 276–526), petunidin-3-$O$-galactoside (HR-MS ion at $m/z$ 479.1233, $\lambda_{max}$: 276–523), peonidin-3-$O$-galactoside (HR-MS ion at $m/z$ 463.1234, $\lambda_{max}$: 279–520), peonidin-3-$O$-glucoside (HR-MS ion at $m/z$ 463.1258, $\lambda_{max}$: 279–523), and malvidin-3-$O$-glucoside (HR-MS ion at $m/z$ 493.1252, $\lambda_{max}$: 276–527), (Table 1), respectively.

**Table 1.** Identification of phenolic compounds in chilean berries by LC-PDA-HR-ToF-ESI-MS data.

| Peak Number | Retention Time (min) | Uv max | HR-M + ion (ppm) | Other ions (Aglycon moiety) | Formula | Identification | Fruit |
|---|---|---|---|---|---|---|---|
| 1 | 4.8 | 276–523 | 641.1687 (−4.8) | 317.0618 (Petunidin) | $C_{28}H_{33}O_{17}$ | Petunidin-3-di-hexoside | blue |
| 2 | 5.9 | 280–517 | 611.1614 (0.3) | 449.1709 (Cyanidin-3-O-hexoside) | $C_{27}H_{31}O_{16}$ | Cyanidin- 3-O-di-hexoside * | blue, arr |
| 3 | 6.3 | 276–523 | 465.1040 (1.8) | 303.0500 (Delphinidin) | $C_{21}H_{21}O_{12}$ | Delphinidin 3-O-galactoside * | blue, cal, che, arr, lu |
| 4 | 6.8 | 276–525 | 465.1038 (1.1) | 303.0495(Delphinidin) | $C_{21}H_{21}O_{12}$ | Delphinidin-3-O-glucoside * | cal |
| 5 | 7.1 | 280–517 | 595.1478 (−31.0) | 449.1089 (Cyanidin-3-O-glucoside) | $C_{27}H_{31}O_{15}$ | Cyanidin 3-O-rutinose | mu |
| 6 | 7.8 | 280–511 | 449.1052 (−7.1) | 287.0675 (Cyanidin) | $C_{21}H_{21}O_{11}$ | Cyanidin-3-O-galactoside * | blue, che, lu |
| 7 | 9.1 | 280–517 | 449.1099 (3.3) | 287. 0507 (Cyanidin) | $C_{21}H_{21}O_{11}$ | Cyanidin-3-O-glucoside * | blue, che, lu |
| 8 | 9.8 | 276–526 | 625.1789 (3.2) | 479.1198 (Petunidin-3-O-glucoside) | $C_{28}H_{33}O_{16}$ | Petunidin-3-O-rutinoside | blue, cal, mu |
| 9 | 10.7 | 276–526 | 639.1911 (−2.2) | 493.1136 (Malvidin-3-O-glucoside) | $C_{29}H_{35}O_{16}$ | Malvidin-3-O-rutinoside | blue |
| 10 | 11.2 | 276–526 | 479.1233 (9.0) | 317.0672 (Petunidin) | $C_{22}H_{23}O_{12}$ | Petunidin-3-O-glucoside * | blue, che, arr, lu |
| 11 | 11.9 | 276–523 | 479.1224 (7.1) | 317.0646 (Petunidin) | $C_{22}H_{23}O_{12}$ | Petunidin-3-O-galactoside * | blue, cal, mu, che, lu |
| 12 | 12.5 | 276–525 | 609.1825 (0.8) | 301.0829 (Peonidin) | $C_{28}H_{33}O_{15}$ | Peonidin 3-O-rutinose | blue |
| 13 | 12.7 | 279–520 | 463.1234 (−1.3) | 301.0689 (Peonidin) | $C_{22}H_{23}O_{11}$ | Peonidin-3-O-galactoside * | blue |
| 14 | 13.4 | 276–527 | 493.1361 (3.0) | 331.0832 (Malvidin) | $C_{23}H_{25}O_{12}$ | Malvidin-3-O-galactoside * | blue, arr |
| 15 | 14.0 | 276–523 | 435.0936 (2.1) | 303.0472 (Delphinidin) | $C_{20}H_{19}O_{11}$ | Delphinidin-3-O-arabinoside | blue, cal |
| 16 | 14.7 | 276–527 | 463.1258 (3.9) | 301.1257 (Peonidin) | $C_{22}H_{23}O_{11}$ | Peonidin-3-O-glucoside | cal, mu, arr |
| 17 | 15.3 | 276–527 | 493.1252 (−19.0) | 331.0789 (Malvidin) | $C_{23}H_{25}O_{12}$ | Malvidin-3-O-glucoside * | blue, cal, arr, lu |
| 18 | 15.6 | 279–527 | 433.1131 (−0.92) | 301.0709 (Peonidin) | $C_{21}H_{21}O_{10}$ | Peonidin-3-O-arabinoside | cal, mu |
| 19 | 16.2 | 276–526 | 449.1066 (−4.0) | 317.1969 (Petunidin) | $C_{21}H_{21}O_{11}$ | Petunidin-3-O-arabinoside | blue |
| 20 | 16.7 | 280–517 | 419.0978 (−1.9) | 287. 0696 (Cyanidin) | $C_{20}H_{19}O_{10}$ | Cyanidin-3-O-arabinoside * | blue |
| 21 | 17.3 | 276–311–527 | 639.1933 (34.2) | 493.1382 (Malvidin-3-O-glucoside) | $C_{32}H_{31}O_{14}$ | Malvidin 3-O-(6'' coumaroyl) glucoside | cal |
| 22 | 17.8 | 276–527 | 463.1284 (9.5) | 330.1706 (Malvidin) | $C_{22}H_{23}O_{11}$ | Malvidin-3-O-arabinose * | blue |
| 23 | 18.0 | 280–517 | 549.1639 (7.1) | 449.1082 (Cyanidin-3-O-glucose) | $C_{25}H_{25}O_{14}$ | Cyanidin-3-O-(6'' succinoyl)-glucose | mu |
| 24 | 18.6 | 279–523 | 625.1820 (8.2) | 463.0905 (Peonidin-3-O- hexoside) | $C_{28}H_{33}O_{16}$ | Peonidin 3-O-di hexoside | cal |
| 25 | 19.4 | 276–311–523 | 919.4460 (2.1) | 303.0504 (Delphinidin) | $C_{42}H_{47}O_{23}$ | Delphinidin-3-O-rutinose (6''-p-coumaroyl)-2''-O-glucose | blue |
| 26 | 20.0 | 276–523 | 507.1135 (−0.4) | 303.0495 (Delphinidin) | $C_{23}H_{23}O_{13}$ | Delphinidin 3-O-(6'' acetyl) glucoside | blue |
| 27 | 20.6 | 280–517 | 491.1206 (3.6) | 287.1232 (Cyanidin) | $C_{23}H_{23}O_{12}$ | Cyanidin 3-O-(6'' acetyl) glucoside | blue |
| 28 | 21.4 | 276–526 | 521.1293 (−0.4) | 317.0676 (Petunidin) | $C_{24}H_{25}O_{13}$ | Petunidin 3-O-(6'' acetyl) glucoside | blue, cal |
| 29 | 22.3 | 276–527 | 535.1451 (−0.2) | 331.0789 (Malvidin) | $C_{25}H_{27}O_{13}$ | Malvidin 3-O-(6'' acetyl) galactoside | cal, arr |
| 30 | 23.2 | 276–321–523 | 627.1393 (−6.8) | 287.0743 (Cyanidin) | $C_{30}H_{27}O_{15}$ | Delphinidin-3-O-(caffeoyl)-glucose | blue |
| 31 | 24.0 | 276–527 | 535.1463 (1.5) | 331.0673 (Malvidin) | $C_{25}H_{27}O_{13}$ | Malvidin 3-O-(6'' acetyl) glucoside | blue |

Abbreviations: blue: Blueberry, cal: Calafate, mu: Murta, che: chequén, arr: Arrayán, me: Meli. * Identified by spiking experiments with authentic compounds.

22

Peaks **4** and **14** were identified as the monoglucosides delphinidin 3-*O*-glucoside and malvidin-3-*O*-galactoside (HR-MS ions at *m/z* 493.1361 and 465.1038, respectively [19,26,27]. Peaks **1**, **2** and **24** showing HR-MS molecular ions at *m/z* 611.1614, 641.1687 and 625.1820 coincident with the formulas $C_{27}H_{31}O_{16}$ (0.3), $C_{28}H_{33}O_{17}$ (−4.8) and $C_{28}H_{33}O_{16}$ (8.2) were identified as petunidin ($\lambda_{max}$: 276–523), cyanidin ($\lambda_{max}$: 280–517), and peonidin ($\lambda_{max}$: 279–523), dihexosides [12,28]. In a similar manner, peaks **5** (HR-MS at *m/z* 595.1478, $C_{27}H_{31}O_{15}$, −31.0), **8** (HR-MS at *m/z* 625.1789, $C_{28}H_{33}O_{16}$, 3.2), **9** (HR-MS at *m/z* 639.1911, $C_{29}H_{35}O_{16}$, −2.2) and **12** (HR-MS at *m/z* 609.1825, $C_{28}H_{33}O_{15}$, 0.8) were assigned as cyanidin, petunidin, malvidin and peonidin rutinosides [12,26,29,30]. Peaks **15**, **18–20** and **22** (Figure 2) with HR-MS molecular ions at *m/z* 435.0936 ($C_{20}H_{19}O_{11}$, 2.1), 433.1131 ($C_{21}H_{21}O_{10}$, −0.92), 449.1066 ($C_{21}H_{21}O_{11}$, −4.0), 419.0978 ($C_{20}H_{19}O_{10}$, −1.9) and 463.1284 ($C_{22}H_{23}O_{11}$) were identified as delphinidin ($\lambda_{max}$: 276–523), peonidin ($\lambda_{max}$: 276–527), petunidin ($\lambda_{max}$: 276–523), cyanidin ($\lambda_{max}$: 280–517) and malvidin ($\lambda_{max}$: 276–527) arabinosides, respectively [26,31], While peaks **21** (HR-MS at *m/z* 639.1933, $C_{32}H_{31}O_{14}$) and **23** (HR-MS at *m/z* 549.1639, $C_{25}H_{25}O_{14}$) were identified as malvidin 3-*O*-(6″ coumaroyl) glucoside and cyanidin-3-*O*-(6″ succinoyl)-glucose [28,30]. Peak **25** with a molecular ion at *m/z* 919.4460 ($C_{42}H_{47}O_{23}$) present in blueberries was identified as the complex anthocyanin: delphinidin-3-*O*-rutinose (4‴-*O*-p-coumaroyl)-2″-*O*-glucose [27,32]. Peaks **26–28** and **31** with HR-MS peaks at *m/z* 507.1135 ($C_{23}H_{23}O_{13}$), 491.1206 ($C_{23}H_{23}O_{12}$), 521.1293 ($C_{24}H_{25}O_{13}$), and 535.1463 ($C_{24}H_{25}O_{13}$), were identified as delphinidin, cyanidin, petunidin, and malvidin 3-*O*-(6″ acetyl) glucosides as reported [27,31], while peak **30** (HR molecular ion at *m/z* 627.1393 coincident with a formula of $C_{30}H_{27}O_{15}$ (−6.8) was identified as delphinidin-3-*O*-(6″ caffeoyl)-glucose [29]. An isomer of peak **31** (peak **29**, HR-MS ion at *m/z* 535.1451 ($C_{25}H_{27}O_{13}$, −0.2), was identified as malvidin 3-*O*-(6″ acetyl) galactoside [27,31].

## 2.2. Identification of Phenolic Acids and Flavonols

Other minor phenolic compounds [12,15,33] were present in all six blueberries analyzed which were accurately identified (Figure 4). The phenolic acids: feruloyl-quinic acid (HR-ToF-MS: 369.1105, MF: $C_{17}H_{21}O_9$, −0.3), chlorogenic acid (HR-ToF-MS: 355.1061, MF: $C_{16}H_{19}O_9$, 9.0) and neochlorogenic acid (HR-ToF-MS: 355.1038, molecular formula: $C_{16}H_{19}O_9$, 2.5), the flavonols quercetin (HR-ToF-MS: 303.0489, MF: $C_{15}H_{11}O_7$, error −5.3), myricetin (HR-ToF-MS: 319.0459, molecular formula: $C_{15}H_{11}O_8$, −1.6) rutin (HR-ToF-MS: 611.1614, MF: $C_{27}H_{31}O_{16}$, 0.3) hyperoside (HR-ToF-MS: 465.1043, MF: $C_{21}H_{21}O_{12}$, 2.2) isoquercitrin (HR-ToF-MS: 465.1032, MF: $C_{21}H_{21}O_{12}$, −0.2) and isorhamnetin (HR-ToF-MS: 317.0670, MF: $C_{16}H_{13}O_7$, 2.8; this last flavonoid was only present in chequén fruits).

## 2.3. Total Phenolics, Flavonoids and Anthocyanin Contents

The total phenolic content (TPC) varied from 5.11 ± 0.18 for chequén to 65.53 ± 1.35 μM Trolox equivalents/g DW for calafate fruits, and showed linear correlation with the antioxidant assays ($R^2$ = 0.8755 and $R^2$ = 0.9143 for TPC/DPPH and TPC/FRAP assays, respectively, Table 2) the TPC of our sample of calafate showed values two times higher than a Chilean sample from Mañihuales [11] but was close to that reported for a Chilean sample from Faro San Isidro [12]. The total anthocyanin content (TAC) ranged from 1.54 ± 0.05 for chequén to 51.62 ± 1.78 mg cyanidin-3-glucoside/g DW for calafate and showed strong linear correlation with the antioxidant assays ($R^2$ = 0.7044 and $R^2$ = 0.9914 for TAC/DPPH and TAC/FRAP assays, respectively, Table 2). The total flavonoid content (TFC) showed similar trend, varying from 2.57 ± 0.11 for *L. chequén* to 45.72 ± 2.68 mg quercetin/g DW for *Berberis microphylla*. The TFC showed linear correlation with the antioxidant assays ($R^2$ = 0.678 for TFC/DPPH and $R^2$ = 0.9856 for TFC/FRAP assays, respectively. The total anthocyanin content for our sample of calafate was close to the values reported for Chilean samples collected in La Junta and Darwin (16.76 mmol/g fresh weigh) and Faro San Isidro (15.44 mmol/g fresh weigh) taking into account conversion factors and 85% water loss (approximately 50.11 and 46.21 mg/g dry weight, respectively) [12]. The levels of anthocyanins in the fruits can explain the different intensity in the color especially for murta, which is red-rose, in comparison with calafate which is purple and blueberry and arrayán which are black (Figure 1).

## 2.4. Quantification of Individual Anthocyanins

The major anthocyanins were quantified in the six edible berries, for some of the species for the first time. The order for the sum of the major anthocyanins was: calafate > blueberries > arrayan > meli > murtilla > chequen (Table 3) which is coincident with the trend found for the total anthocyanin content (TAC) (Table 2) measured by a colorimetric method. The HPLC quantification method showed good performance, baseline was good (Figure 2), and the correlation coefficients for the standard curves of the glycosilated standard anthocyanins varied from 0.998 to 0.999. The limits of detection for three representative compounds were 0.08 to 0.12 μg/mL and the limits of quantification were 0.24 to 0.35 μg/mL (Table 4). Repeatability for retention time and peak area was good, relative standard deviations were below 2.00% [34]. As seen in Table 4 all recovery results varied from 97.93 ± 0.33 to 99.72 ± 1.34 and were within the usually required recovery range of 100% ± 5% [34]. However, the anthocyanin concentration in our Chilean blueberries sample is quite different from those published for blueberries from other locations [31,35] being the major anthocyanins found peonidin-3-*O*-arabinoside and delphinidin-3-*O*-arabinoside (37.43 ± 4.76 and 34.43 ± 3.28 mg/100 g fresh weight, respectively) followed by malvidin-3-*O*-glucoside and petunidin-3-*O*-rutinoside (Table 3). In the case of calafate (*Berberis microphylla*) the major anthocyanins were delphinidin 3-*O*-galactoside, petunidin-3-*O*-glucoside and malvidin-3-*O*-glucoside (60.42 ± 1.28, 51.39 ± 1.65 and 42.94 ± 1.25, mg/100 g fresh weight, respectively). We found as the major anthocyanin in this species delphinidin 3-*O*-galactoside, but Ruiz *et al* [15] reported delphinidin 3-*O*-glucoside as the major constituent (8.83 ± 1.53 µmol/g fresh weight), followed by petunidin-3-glucoside (4.71 ± 1.08 µmol/g fresh weight). For chequén (*Luma chequen*) the main anthocyanins were cyanidin-3-*O*-galactoside, petunidin-3-*O*-glucoside and petunidin-3-*O*-galactoside (43.46 ± 1.39, 12.83 ± 1.65 and 9.55 ± 1.02 mg/100 g fresh weight, respectively), and for arrayán (*Luma apiculata*) were petunidin-3-*O*-glucoside, malvidin-3-*O*-glucoside, delphinidin 3-*O*-galactoside and cyanidin-3-*O*-glucoside (48.21 ± 2.2, 44.75 ± 3.31, 34.43 ± 2.12 and 9.45 ± 0.15 mg/100 g fresh weight, respectively). Our sample of murtilla (*Ugni molinae*) showed two main anthocyanins (petunidin-3-*O*-rutinoside and peonidin-3-*O*-glucoside, Figure 2, Tables 1 and 3) and meli (*Amomyrtus meli*) showed six main glycosilated anthocyanins including cyanidin-3-*O*-galactoside and petunidin-3-*O*-galactoside as major ones (Tables 1 and 3). These compounds were quantified in these *Luma* species for the first time.

Table 2. Scavenging of the 1,1-diphenyl-2-picrylhydrazyl Radical (DPPH), Ferric Reducing Antioxidant Power (FRAP), Superoxide Anion scavenging activity (SA), Total Phenolic Content (TPC), Total Flavonoid Content (TFC), Total Anthocyanin Content (TAC), and Extraction Yields of Six Edible Berry Fruits From the VIII Region of Chile.

| Species | DPPH α | FRAP β | SA ° | TPC δ | TFC ψ | TAC x | Extraction Yields (%) μ |
|---|---|---|---|---|---|---|---|
| Vaccinium corymbosum | 3.32 ± 0.18 a | 96.15 ± 5.39 df | 72.61 ± 1.91 r | 45.86 ± 3.46 | 18.50 ± 3.75 p | 21.41 ± 1.65 | 6.72 |
| Berberis microphylla | 2.33 ± 0.21 ab | 124.46 ± 6.54 | 81.31 ± 2.95 s | 65.53 ± 1.35 | 45.72 ± 2.68 | 51.62 ± 1.78 | 4.99 |
| Luma chequén | 12.92 ± 0.30 | 76.22 ± 3.45 e | 43.79 ± 2.91 t | 5.11 ± 0.18 k | 2.57 ± 0.11 m | 1.54 ± 0.05 | 7.39 |
| Luma apiculata | 5.88 ± 0.21 | 93.4 ± 4.68 dg | 64.22 ± 3.46 | 27.61 ± 1.61 | 12.80 ± 2.43 np | 15.24 ± 1.49 l | 6.34 |
| Ugni molinae | 10.94 ± 0.32 c | 81.10 ± 4.58 ehj | 52.22 ± 1.81 t | 9.24 ± 0.28 k | 5.54 ± 0.91 mo | 6.85 ± 0.10 | 5.21 |
| Amomyrtus meli | 7.46 ± 0.10 b | 88.29 ± 6.34 fghi | 56.44 ± 2.32 | 17.52 ± 0.66 | 11.76 ± 2.04 no | 13.33 ± 2.69 l | 4.89 |
| Gallic acid Φ | 1.36 ± 0.22 (7.99 ± 1.29 µM) | 148.1 ± 8.35 | 94.39 ± 1.98 | - | - | - | - |
| Cyanidin 3-*O*-glucoside Φ | 8.47 ± 1.23 c (17.47 ± 2.53 µM) | 95.48 ± 6.72 ij | 76.85 ± 1.71 rs | - | - | - | - |

α Antiradical DPPH activities are expressed as $IC_{50}$ in µg/mL for extracts and compounds. β Expressed as µM trolox equivalents/g dry weight. ° Expressed in percentage scavenging of superoxide anion at 100 µg/mL. δ Total phenolic content (TPC) expressed as mg gallic acid/g dry weight. ψ Total flavonoid content (TFC) expressed as mg quercetin/g dry weight. x Total Anthocyanin content (TAC) expressed as mg cyanidin 3-*O*-glucoside/g dry weight. μ Extraction yields expressed in percent W/W extraction on the basis of freeze dried material. Φ Used as standard antioxidants. Values in the same column marked with the same letter are not significantly different (at $p < 0.05$).

## 2.5. Antioxidant Features

The order of the antioxidant activity measured by the bleaching of the radical DPPH and the ferric reducing antioxidant power (FRAP) showed by the six fruits was calafate > blueberry > arrayán > meli > murta > chequén which is also the order found for the sum of the individual major anthocyanins measured by HPLC. A similar trend was observed for superoxide anion scavenging activity (Table 2, Figure S1, Supplementary Material). Calafate showed the highest antioxidant activity (2.33 ± 0.21 μg/mL and 124.46 ± 6.54 μM TE/g dry weight in the DPPH and FRAP assays, respectively, Table 2), followed by blueberry (3.32 ± 0.18 μg/mL and 96.15 ± 5.39 μM TE/g DW), and arrayán (5.88 ± 0.21 and 93.4 ± 4.68 μM TE/g DW, Table 2). The bleaching of the radical DPPH for calafate was close to that shown by the standards gallic acid and cyanidin-3-glucoside (1.36 ± 0.22 and 8.47 ± 1.23 μg/mL, respectively). The antioxidant activities showed positive correlation with polyphenolic content assays ($0.67 \geqslant R^2 \geqslant 0.9856$). It is reported that fruits antioxidant activities and composition of phenolics are dependent of genetic differences among different species and environmental conditions and harvest and/or ripeness within the same species [11,36] which can explain the differences in phenolic composition and antioxidant capacities found between the species under study and among other reports of antioxidant activities and phenolic composition of the same species from other zones of Chile [11,12,15].

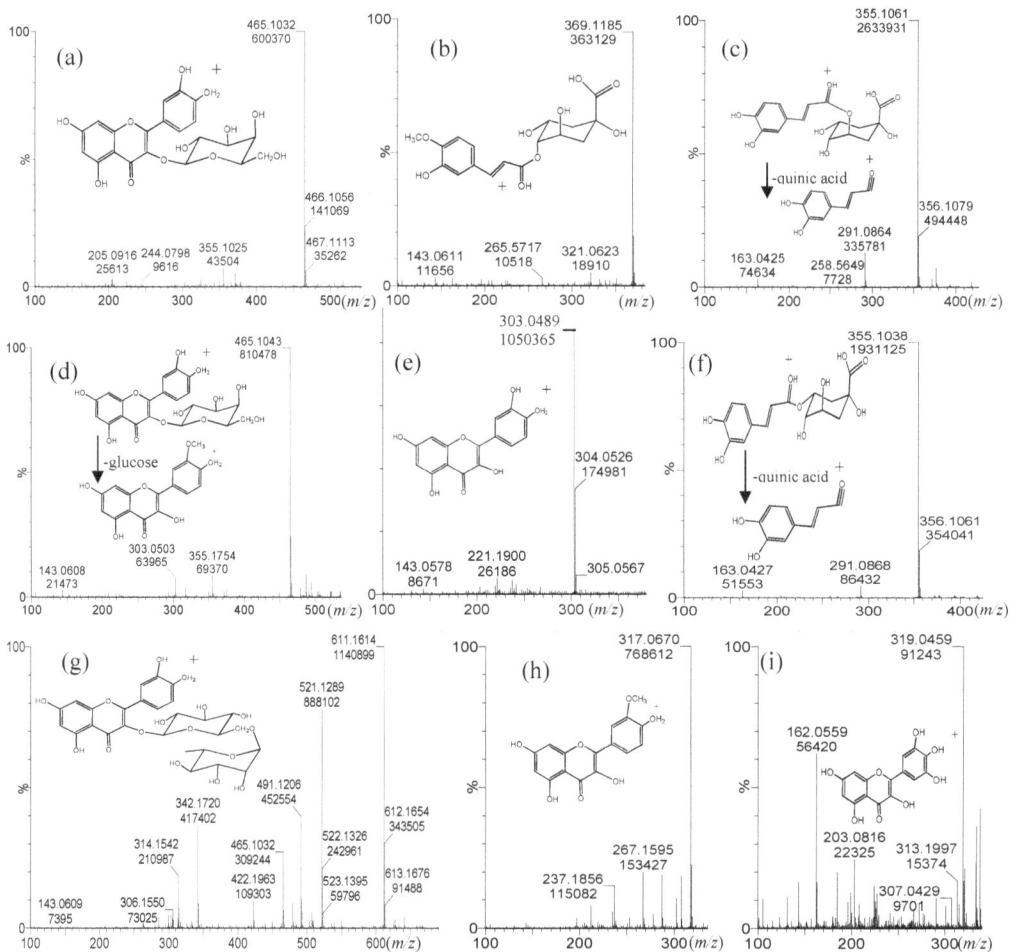

**Figure 4.** Full scan ToF MS spectra and structures of minor phenolic compounds detected in six berries from the VIII region of Chile. (**a**) Hyperoside, (**b**) feruloyl-quinic acid, (**c**) chlorogenic acid (**d**) isoquercitrin (**e**) quercetin, (**f**) neochlorogenic acid (**g**) rutin (**h**) isorhamnetin and (**i**) myricetin.

Table 3. Major anthocyanins quantified by HPLC-DAD in six edible berry fruits from the VIII Region of Chile.

| Berry Species | Anthocyanin (mg/100 g)[a] | | | | | | | | | | | | | Total |
|---|---|---|---|---|---|---|---|---|---|---|---|---|---|---|
| | 2 | 3 | 6 | 7 | 8 | 10 | 11 | 13 | 15 | 16 | 17 | 18 | 20 | |
| Vaccinium corymbosum | 19.23 ± 3.18 | 8.51 ± 3.29 a | 21.17 ± 0.32 | 0.96 ± 0.12 | 14.29 ± 2.15 | 9.27 ± 2.22 d | 1.78 ± 0.01 | 14.28 ± 0.98 | 34.43 ± 3.28 | nd | 16.42 ± 1.45 | nd | 37.43 ± 4.76 | 177.77 |
| Berberis microphylla | nd | 60.42 ± 1.28 | nd | 21.89 ± 2.74 b | nd | 51.39 ± 1.65 e | 6.45 ± 0.89 | nd | 9.28 ± 0.01 | 3.96 ± 0.02 | 42.94 ± 1.25 f | 3.84 ± 0.02 | nd | 200.17 |
| Luma chequen | nd | 3.72 ± 0.02 | 43.46 ± 1.39 | 5.29 ± 0.23 | nd | 12.83 ± 1.65 | 9.55 ± 1.02 | nd | nd | nd | nd | nd | nd | 74.85 |
| Luma apiculata | 2.6 ± 0.01 | 34.43 ± 2.12 | nd | 9.45 ± 0.15 | nd | 48.21 ± 2.2 e | nd | nd | nd | nd | 44.75 ± 3.31 f | nd | nd | 139.44 |
| Ugni molinae | nd | nd | nd | nd | 51.37 ± 0.28 | nd | 4.87 ± 0.02 | nd | nd | 61.48 ± 2.42 | nd | 4.43 ± 0.04 | nd | 122.15 |
| Amomyrtus meli | nd | 8.87 ± 1.76a | 48.39 ± 2.23 | 20.43 ± 2.39 b | nd | 13.54 ± 2.46 d | 27.12 ± 1.25 | nd | nd | nd | 8.45 ±1.13 | nd | nd | 126.80 |

[a] Expressed as mg/100 g fresh weight, measurements are expressed as mean ± SD of five parallel determinations. (Values in the same row marked with the same letter are not significantly different at $p < 0.05$). nd: not detected/determined.

Table 4. Inter-day and Intra-day accuracy and precision (as RSD%), limits of detection (LOD) and quantification (LOQ) and recovery of three major anthocyanins (compounds 3, 7 and 10).

| Compound | Nominal concentration (μg/mL) | Inter day | | | Intra day | | | LOD-LOQ (μg/mL) | Sample-Recovery (mean ± RSD%) |
|---|---|---|---|---|---|---|---|---|---|
| | | Observed concentration (μg/mL) | Accuracy (%) | RSD% | Observed concentration (μg/mL) | Accuracy (%) | RSD% | | |
| 3 | 10 | 10.94 ± 0.05 | 109.45 | 0.071 | 11.83 ± 0.05 | 118.33 | 0.04 | 0.08-0.24 | Calafate 97.93 ± 0.33 |
| 3 | 20 | 20.91 ± 0.62 | 104.55 | 0.80 | 21.73 ± 0.37 | 108.66 | 0.34 | | Arrayán 98.73 ± 1.50 |
| 3 | 40 | 40.49 ± 0.70 | 101.23 | 0.69 | 41.2 ± 0.72 | 103.00 | 0.70 | | Blueberry 98.57 ± 0.33 |
| 7 | 10 | 12.0 ± 0.08 | 120.00 | 0.08 | 12.13 ± 0.15 | 121.33 | 0.12 | 0.12-0.35 | Calafate 99.72 ± 1.34 |
| 7 | 20 | 19.81 ± 0.27 | 99.08 | 0.28 | 20.36 ± 0.76 | 101.80 | 0.75 | | Arrayán 98.97 ± 1.98 |
| 7 | 40 | 40.77 ± 0.37 | 101.93 | 0.36 | 40.46 ± 0.67 | 101.15 | 0.66 | | Chequén 99.84 ± 0.16 |
| 10 | 10 | 10.56 ± 0.81 | 105.66 | 0.77 | 9.99 ± 0.12 | 99.96 | 0.12 | 0.09-0.30 | Calafate 99.31 ± 0.35 |
| 10 | 20 | 20.20 ± 0.59 | 101.03 | 0.58 | 20.61 ± 0.33 | 103.05 | 0.32 | | Arrayán 98.59 ± 0.38 |
| 10 | 40 | 41.04 ± 0.41 | 102.64 | 0.40 | 40.8 ± 0.36 | 40.8 | 0.79 | | Chequén 98.19 ± 0.76 |

## 3. Experimental

### 3.1. Chemicals and Plant Material

Folin–Ciocalteu phenol reagent (2 N), reagent grade $Na_2CO_3$, $AlCl_3$, HCl, $FeCl_3$, $NaNO_2$, NaOH, quercetin, trichloroacetic acid, sodium acetate, HPLC-grade water, HPLC-grade acetonitrile, reagent grade MeOH and formic acid were obtained from Merck (Darmstadt, Germany) Cyanidin, delphinidin 3-$O$-galactoside, cyanidin-3-$O$-galactoside, cyanidin-3-$O$-glucoside, petunidin-3-$O$-glucoside, petunidin-3-$O$-galactoside, peonidin-3-$O$-galactoside, peonidin-3-$O$-glucoside and malvidin-3-$O$-glucoside (all standards with purity higher than 95% by HPLC) were purchased either from ChromaDex (Santa Ana, CA, USA), Extrasynthèse (Genay, France) or Wuxi Apptec Co. Ltd. (Shangai, China). Gallic acid, TPTZ (2,4,6- tri(2-pyridyl)-s-triazine), Trolox, *tert*-butylhydroperoxide, nitro blue tetrazolium, xanthine oxidase and DPPH (1,1-diphenyl-2-picrylhydrazyl radical) were purchased from Sigma-Aldrich Chemical Co. (St. Louis, MO, USA). All ripe fruits for this study (aprox. 500 g each) were collected at Región del Bio-Bio, Chile. Sampling was performed using sterile disposable gloves and rigid plastic sample containers and each sample was submitted individually by overnight courier to our laboratory in Antofagasta to prevent deterioration. This sampling methodology was previously used for other edible fruits [19,23,33]. Random healthy ripe fruits, representative of the lot, were collected from various specimens (at least 10 fruits per specimen) and different locations (at least 3) in each growing area. Ripe fruits of arrayán *(L. apiculata* (DC.) burret, chequén *(L. chequén* (Molina) A. Gray), and murta *(U. molinae* Turcz) were collected in Re-Re, Chile in May 2011. Meli (A. meli (Phil.) D. Legrand & Kausel and calafate (*B. microphylla* G. Forst) were collected in the Andean woods of Santa Bárbara, in May 2011. Blueberries (*V. corymbosum*) variety highbush Brigitta were collected in April 2011 in the area of Chillán. Voucher herbarium specimens including samples of fruits were deposited at the Laboratorio de Productos Naturales, Universidad de Antofagasta, Antofagasta, Chile, with the numbers La-111505-1, Lc-111505-2, Um-111505-1, Am-111805-1, Bm-111805-1 and Vc-110704-1, respectively.

### 3.2. Sample Preparation

Fresh fruits (Figure S4–S9, supplementary material) were carefully washed, separately homogenized in a blender and freeze-dried (Labconco Freezone 4.5 L, Kansas, MO, USA). Ten grams of each lyophilized fruit was finally pulverized in a mortar, defatted thrice with 100 mL of n-hexane and then extracted with 100 mL of 0.1% HCl in MeOH in the dark in an ultrasonic bath for one hour each time, The extracts were combined, filtered and evaporated *in vacuo* in the dark (40 °C). The extracts were suspended in 20 mL ultrapure water and loaded onto an XAD-7 (100 g) column. The column was rinsed with water (100 mL) and phenolic compounds were eluted with 100 mL of MeOH acidified with 0.1% HCl. This methodology was previously used for other edible fruits [19,23,33]. The solutions were combined and evaporated to dryness under reduced pressure (40 °C) to give 634.20, 739.20, 499.93, 672.24, 489.93 and 521.38 mg of *L. apiculata*, *L. chequén*, *B. microphylla*, *V. corymbosum*, *A. meli* and *U. molinae* fruits, respectively.

## 3.3. Liquid Chromatography Analysis

A portion of each extract (approximately 2 mg) obtained as explained above was dissolved in 2 mL 0.1% HCl in MeOH, filtered through a 0.45 μm micropore membrane (PTFE, Waters, Milford, MA, USA) before use and was injected into the HPLC-PDA and ESI-ToF-MS equipment. Qualitative HPLC-PDA analysis of the extracts was performed using a Waters Alliance 2695 system equipped with 2695 separation module unit and 2996 PDA detector and a 250 × 4.6 mm, 5 μm, 100 Å, Luna C-18 column (Phenomenex, Torrance, CA, USA), with a linear gradient solvent system of 0.1% aqueous formic acid (solvent A) and acetonitrile 0.1% formic acid (solvent B) as follows: 90% solvent A until 4 min, followed by 90%–75% solvent A over 25 min, then 75%–10% A over 35 min, then going back to 90% solvent A until 45 min. and finally reconditioning the column with 90% solvent A isocratic for 15 min. The flow rate and the injection volume were 0.5 mL/min and 20 μL, respectively. The compounds were monitored using a wavelength range of 210–800 nm.

## 3.4. Validation of the HPLC Method

Quantification was done by external standardization, using the respective standard anthocyanins, at the wavelengths of maximum absorption of the compounds. For the validation of the analytical method based on HPLC factors, linearity, precision, detection limits and accuracy were evaluated following [34]. Stock solutions of all seven standard compounds (**3, 4, 6, 7, 10, 11**, and **17**) were prepared by dissolving one milligram of each anthocyanin in methanol-formic acid 1% (1 mg/mL). Several calibration levels were prepared by diluting the stock solutions with methanol-formic acid 1% yielding concentrations of 15.65, 31.25, 62.5, 125, 250 and 500 μg/mL. The calibration curves ($R^2 > 0.098$) were obtained by plotting peak areas versus concentrations. Compound **15** was quantified using the calibration curve obtained for **3**, compounds **15–18** and **20** with the calibration curve of **11** and compound **2** with the calibration curve of compound **7**. Limits of detection (LOD) and quantitation (LOQ) were measured for three representative compouns (**3, 7** and **10**, Table 4) and are reported as the concentrations that gave signal-to-noise ratios of 3 and 10, respectively, from three replicate injections. Accuracy was determined by spiking three standard anthocyanins (**3, 7** and **10**, Table 4) at three concentration levels (10: low, 20: medium, and 40 μg/mL: high spike) in one gram of each fresh fruits, which was then extracted and assayed as described before. Mean percentage recovery in relation to the theoretically present amounts (% recovery = amount detected $\times$ 100/theoretical amount) were used as a measure of accuracy (Table 4). The relative standard deviation (RSD%) within the measurements was considered as a measure of precision and repeatability. The samples were prepared and analyzed for anthocyanin concentration on the same day and on three consecutive days (n = 5) for intra- and interday precision respectively.

## 3.5. Mass Spectrometric Conditions

Hyphenated PDA with high-resolution electrospray ionization-time of flight-mass spectrometry (HR-ESI-ToF-MS) analysis was performed using a LCT premier XE ToF mass spectrometer (Waters) equipped with an ESI interface and controlled by MassLynx V4.1 software, using the chromatographic conditions as stated above. The compounds were monitored using PDA with a wavelength range of 210–800 nm, while mass spectra were acquired with electrospray ionization and the ToF mass analyzer in both positive and negative modes over the range $m/z$: 100–1000. The capillary voltages were set at 3000 V (positive mode) and 2800 V (negative mode), respectively, and the cone voltage was 20 V. Nitrogen was used as the nebulizer and desolvation gas. The desolvation and cone gas flow rates were 300 and 20 L/h, respectively. The desolvation temperature was 400 °C, and the source temperature was 120 °C. For the dynamic range enhancement (DRE) lockmass, a solution of leucine enkephalin (Sigma–Aldrich, Steinheim, Germany) was infused by a secondary reference probe at 200 pg/mL in $CH_3CN$/water (1:1) containing 0.1% formic acid with the help of a second LC pump (Waters 515 HPLC pump). The reference mass was scanned once every five scans for each positive and negative data collection. Both positive and negative ESI data were collected using a scan time of 0.2 s, with an interscan time of 0.01 s, and a polarity switch time of 0.3 s. The full chromatograms were recorded at two different aperture voltages. The most intense fragmental ions and molecular ions could be obtained, when the aperture voltage were set at 60 V and 0 V, respectively. V-optics mode was used for increased intensity.

## 3.6. Antioxidant Assays

### 3.6.1. Free Radical Scavenging Capacity

The free radical scavenging capacity of the extracts was determined by the DPPH· assay as previously described [37], with some modifications. DPPH radical absorbs at 517 nm, but upon reduction by an antioxidant compound its absorption decreases. Briefly, 50 µL of processed SPE MeOH extract or pure compound prepared at different concentrations was added to 2 mL of fresh 0.1 mM solution of DPPH in methanol and allowed to react at 37 °C in the dark. After thirty minutes the absorbance was measured at 517 nm. The DPPH scavenging ability as percentage was calculated as: DPPH scavenging ability = $(A_{control} - A_{sample}/A_{control}) \times 100$. Afterwards, a curve of % DPPH bleaching activity versus concentration was plotted and $IC_{50}$ values were calculated. $IC_{50}$ denotes the concentration of sample required to scavenge 50% of DPPH free radicals. The lower the $IC_{50}$ value the more powerful the antioxidant activity. Gallic acid (from 1.0 to 125.0 μg/mL, $R^2$ = 0.991) and cyanidin 3-O-glucoside (from 1.0 to 125.0 μg/mL, $R^2$ = 0.997) were used as standard antioxidant compounds.

### 3.6.2. Ferric Reducing Antioxidant Power

The determination of ferric reducing antioxidant power or ferric reducing ability (FRAP assay) of the extracts was performed as described by [38] with some modifications. The stock solutions prepared were 300 mM acetate buffer pH 3.6, 10 mM TPTZ (2,4,6-tri(2-pyridyl)-s-triazine) solution in 40 mM HCl, and 20 mM $FeCl_3 \cdot 6H_2O$ solution. Plant extracts or standard methanolic Trolox solutions (150 μL) were incubated at 37 °C with 2 mL of the FRAP solution (prepared by mixing 25 mL acetate buffer, 5 mL TPTZ solution, and 10 mL $FeCl_3 \cdot 6H_2O$ solution) for 30 min in the dark. Absorbance of the blue ferrous tripyridyltriazine complex formed was then read at 593 nm. Quantification was performed using a standard calibration curve of the antioxidant Trolox (from 0.2 to 2.5 μmol/mL, $R^2$: 0.995). Samples were analyzed in triplicate and results are expressed in μmol TE/gram dry mass.

### 3.6.3. Superoxide Anion Scavenging Activity

The enzyme xanthine oxidase is able to generate superoxide anion radical $(O_2 \cdot^-)$ "in vivo" by oxidation of reduced products from intracellular ATP metabolism. The superoxide anion generated in this reaction sequence reduces the nitro blue tetrazolium dye (NBT), leading to a chromophore with a maximum of absorption at 560 nm. Superoxide anion scavengers reduce the speed of generation of the chromophore. The superoxide anion scavenging activities of isolated compounds and fractions were measured spectrophotometrically in a microplate reader as reported previously [23]. All compounds, and berry extracts were evaluated at 100 μg/mL. Values are presented as mean ± standard deviation of three determinations.

### 3.6.4. Polyphenol, Flavonoids and Anthocyanin Contents

The total polyphenolic contents (TPC) of *Luma* fruits and leaves were determined by the Folin-Ciocalteau method [19,33,39] with some modifications. An aliquot of each processed SPE extract (200 μL, approx. 2 mg/mL) was added to the Folin–Ciocalteau reagent (2 mL, 1:10 v/v in purified water) and after 5 min of reaction at room temperature (25 °C), 2 mL of a 100 g/l solution of $Na_2CO_3$ was added. Sixty minutes later the absorbance was measured at 710 nm. The calibration curve was performed with gallic acid (concentrations ranging from 16 to 500 µg/mL, $R^2$ = 0.999) and the results were expressed as mg gallic acid equivalents/g dry mass. Determination of total flavonoid content (TFC) of the methanolic extracts was performed as reported previously [40] using the $AlCl_3$ colorimetric method. Quantification was expressed by reporting the absorbance in the calibration graph of quercetin, which was used as a standard (from 0.1 to 65.0 µg/mL, $R^2$ = 0.994). Results are expressed as mg quercetin equivalents/g dry weight. The assessment of total anthocyanin content (TAC) was carried out by the pH differential method according to AOAC as described by [38,41]. Absorbance was measured at 510 and 700 nm in buffers at pH 1.0 and 4.5. Pigment concentration is expressed as mg cyanidin 3-glucoside equivalents/g dry mass and calculated using the formula:

$$TA(mg/g) = \frac{A \times MW \times DF \times 10^5}{\varepsilon \times 1}$$

where A = (A510 nm − A700 nm) pH 1.0 − (A510 nm − A700 nm) pH 4.5; MW (molecular weight) = 449.2 g/mol; DF = dilution factor; 1 = cuvette pathlength in cm; $\varepsilon$ = 26,900 L/mol.cm, molar extinction coefficient for cyanidin 3-$O$-$\beta$-D-glucoside. $10^3$: factor to convert g to mg. All spectrometric measurements were performed using a Unico 2800 UV-Vis spectrophotometer (Unico Instruments Co. Ltd., Shanghai, China).

### 3.7. Statistical Analysis

The statistical analysis was carried out using the originPro 9.0 software packages (Originlab Corporation, Northampton, MA, USA). The determination was repeated at least three times for each sample solution. Analysis of variance was performed using ANOVA. Significant differences between means were determined by Tukey comparison test ($p$ values < 0.05 were regarded as significant).

## 4. Conclusions

Thirty one anthocyanins, three phenolic acids (feruloylquinic acid, chlorogenic and neochlorogenic acid) and six flavonols (rutin, quercetin, myricetin, hyperoside, isoquercitrin and isorhamnetin) were identified for the first time in six edible berries from the VIII region of Chile using ToF-MS. Among the 31 anthocyanins identified in the six berries under study, twenty three compounds were detected in blueberry, fourteen in calafate, nine in arrayán and six were present in meli, chequén and murta. The anthocyanins detected were mainly branched 3-$O$-glycoconjugates of malvidin, delphinidin, peonidin, petunidin and cyanidin. However, significant differences in the amount of anthocyanins, (which were measured individually by HPLC for the major ones and by TAC colorimetric method) were found for the six berries, which presented also different antioxidant capacities. Blueberry fruits showed the most complex anthocyanin profile, while the fruits of chequen and murta showed a simpler pattern with only six anthocyanins, whereas arrayán and chequén showed a more complex pattern. However, the fruits of calafate (*B. microphylla*) presented the highest antioxidant features and polyphenolic content followed by the fruits of Chilean blueberries (*V. corymbosum*), arrayán (*L. apiculata*) and meli (*A. meli*), which makes calafate, arrayán and meli the better candidates for industrial crop production and potential use in functional foods and nutraceuticals.

**Supplementary Materials:** Supplementary materials can be accessed at: http://www.mdpi.com/1420-3049/19/8/10936/s1.

**Acknowledgments:** This work was financially supported by the National Fund of Scientific and Technological Development of Chile (Fondecyt No. 1140178).

**Author Contributions:** EK, CA, BS and MS designed research; AB, MS and BS performed research and analyzed the data; MS, CA and EK wrote the paper. All authors read and approved the final manuscript.

**Conflicts of Interest:** The authors declare no conflict of interest.

## References

1.  Pennington, J.A.T.; Fisher, R.A. Food component profiles for fruit and vegetable subgroups. *J. Food Compos. Anal.* **2010**, *23*, 411–418.
2.  Zamora-Ros, R.; Fedirko, V.; Trichopoulou, A.; Gonzalez, C.A.; Bamia, C.; Trepo, E.; Nothlings, U.; Duarte-Salles, T.; Serafini, M.; Bredsdorff, L.; *et al.* Dietary flavonoid, lignan and antioxidant capacity and risk of hepatocellular carcinoma in the European prospective investigation into cancer and nutrition study. *Int. J. Cancer* **2013**, *133*, 2429–2443.
3.  Manach, C.; Scalbert, A.; Morand, C.; Remesy, C.; Jimenez, L. Polyphenols: Food sources and bioavailability. *Am. J. Clin. Nutr.* **2004**, *79*, 727–747.
4.  Pojer, E.; Mattivi, F.; Johnson, D.; Stockley, C.S. The case for anthocyanin consumption to promote human health: A review. *Compr. Rev. Food Sci. F.* **2013**, *12*, 483–508.
5.  Lachman, J.; Orsák, M.; Pivec, V. Antioxidant contents and composition in some vegetables and their role in human nutrition. *Hortic. Sci.* **2000**, *27*, 65–78.
6.  Delporte, C.; Backhouse, N.; Inostroza, V.; Aguirre, M.C.; Peredo, N.; Silva, X.; Negrete, R.; Miranda, H.F. Analgesic activity of *Ugni molinae* (murtilla) in mice models of acute pain. *J. Ethnopharmacol.* **2007**, *112*, 162–165.
7.  Scheuermann, E.; Seguel, I.; Montenegro, A.; Bustos, R.; Hormazabal, E.; Quiroz, A. Evolution of aroma compounds of murtilla fruits (Ugni molinale Turcz) during storage. *J. Sci. Food Agric.* **2008**, *88*, 485–492.
8.  Murillo, A. *Plantes médicinales du Chili*; Exposition Universelle de Paris Section Chilienne; Roger y Chernoviz: Paris, France, 1889; Volum 80, p. 234.
9.  Hoffmann, A.E. *Flora Silvestre de Chile*, 3rd ed.; Gay, C., Ed.; Ediciones Fundacion Claudio Gay: Santiago, Chile, 1995; Volume 1, p. 258.

10. De Mösbach, E.W. Botánica indígena de Chile. In *Museo Chileno de Arte Precolombino, Fundación Andes y Editorial Andrés Bello*; Aldunate, C., Villagrán, C., Eds.; Editorial Andrés Bello: Santiago, Chile, 1991; pp. 95–96.

11. Mariangel, E.; Reyes-Diaz, M.; Lobos, W.; Bensch, E.; Schalchli, H.; Ibarra, P. The antioxidant properties of calafate (*Berberis microphylla*) fruits from four different locations in southern Chile. *Cienc. Investig. Agrar.* **2013**, *40*, 161–170.

12. Ruiz, A.; Hermosín-Gutiérrez, I.; Mardones, C.; Vergara, C.; Herlitz, E.; Vega, M.; Dorau, C.; Winterhalter, P.; von Baer, D. Polyphenols and antioxidant activity of Calafate (*Berberis microphylla*) fruits and other native berries from Southern Chile. *J. Agric. Food Chem.* **2010**, *58*, 6081–6089.

13. Ruiz, A.; Mardones, C.; Vergara, C.; Hermosín-Gutiérrez, I.; von Baer, D.; Hinrichsen, P.; Rodriguez, R.; Arribillaga, D.; Dominguez, E. Analysis of hydroxycinnamic acids derivatives in calafate (Berberis microphylla G. Forst) berries by liquid chromatography with photodiode array and mass spectrometry detection. *J. Chromatogr. A* **2013**, *1281*, 38–45.

14. Escribano-Bailón, M.T.; Alcalde-Eon, C.; Muñoz, O.; Rivas-Gonzalo, J.C.; Santos-Buelga, C. Anthocyanins in berries of Maqui (*Aristotelia chilensis* (Mol.) Stuntz). *Phytochem. Anal.* **2006**, *17*, 8–14.

15. Ruiz, A.; Hermosín-Gutiérrez, I.; Vergara, C.; von Baer, D.; Zapata, M.; Hitschfeld, A.; Obando, L.; Mardones, C. Anthocyanin profiles in south Patagonian wild berries by HPLC-DAD-ESI-MS/MS. *Food Res. Int.* **2013**, *51*, 706–713.

16. Reynertson, K.A.; Yang, H.; Jiang, B.; Basile, M.J.; Kennelly, E.J. Quantitative analysis of antiradical phenolic constituents from fourteen edible Myrtaceae fruits. *Food Chem.* **2008**, *109*, 883–890.

17. Theoduloz, C.; Franco, L.; Ferro, E.; Schmeda Hirschmann, G. Xanthine oxidase inhibitory activity of Paraguayan Myrtaceae. *J. Ethnopharmacol.* **1988**, *24*, 179–183.

18. Theoduloz, C.; Pacheco, P.; Schmeda Hirschmann, G. Xanthine oxidase inhibitory activity of Chilean Myrtaceae. *J. Ethnopharmacol.* **1991**, *33*, 253–255.

19. Simirgiotis, M.J.; Borquez, J.; Schmeda-Hirschmann, G. Antioxidant capacity, polyphenolic content and tandem HPLC-DAD-ESI/MS profiling of phenolic compounds from the South American berries *Luma apiculata* and *L. chequen*. *Food Chem.* **2013**, *139*, 289–299.

20. Bórquez, J.; Kennelly, E.J.; Simirgiotis, M.J. Activity guided isolation of isoflavones and hyphenated HPLC-PDA-ESI-ToF-MS metabolome profiling of *Azorella madreporica* Clos. from northern Chile. *Food Res. Int.* **2013**, *52*, 288–297.

21. Wu, S.-B.; Wu, J.; Yin, Z.; Zhang, J.; Long, C.; Kennelly, E.J.; Zheng, S. Bioactive and Marker Compounds from Two Edible Dark-Colored Myrciaria Fruits and the Synthesis of Jaboticabin. *J. Agric. Food Chem.* **2013**, *61*, 4035–4043.

22. Wu, S.-B.; Dastmalchi, K.; Long, C.; Kennelly, E.J. Metabolite Profiling of Jaboticaba (*Myrciaria cauliflora*) and Other Dark-Colored Fruit Juices. *J. Agric. Food Chem.* **2012**, *60*, 7513–7525.

23. Simirgiotis, M.J.; Ramirez, J.E.; Schmeda Hirschmann, G.; Kennelly, E.J. Bioactive coumarins and HPLC-PDA-ESI-ToF-MS metabolic profiling of edible queule fruits (*Gomortega keule*), an endangered endemic Chilean species. *Food Res. Int.* **2013**, *54*, 532–543.

24. Verardo, V.; Arráez-Román, D.; Segura-Carretero, A.; Marconi, E.; Fernández-Gutiérrez, A.; Caboni, M.F. Identification of buckwheat phenolic compounds by reverse phase high performance liquid chromatography–electrospray ionization-time of flight-mass spectrometry (RP-HPLC–ESI-TOF-MS). *J.Cereal Sci.* **2010**, *52*, 170–176.

25. Qiu, J.; Chen, L.; Zhu, Q.; Wang, D.; Wang, W.; Sun, X.; Liu, X.; Du, F. Screening natural antioxidants in peanut shell using DPPH–HPLC–DAD–TOF/MS methods. *Food Chem.* **2012**, *135*, 2366–2371.

26. Lätti, A.K.; Riihinen, K.R.; Jaakola, L. Phenolic compounds in berries and flowers of a natural hybrid between bilberry and lingonberry (*Vaccinium intermedium* Ruthe). *Phytochemistry* **2011**, *72*, 810–815.

27. Li, H.; Deng, Z.; Zhu, H.; Hu, C.; Liu, R.; Young, J.C.; Tsao, R. Highly pigmented vegetables: Anthocyanin compositions and their role in antioxidant activities. *Food Res. Int.* **2012**, *46*, 250–259.

28. Aguirre, M.J.; Isaacs, M.; Matsuhiro, B.; Mendoza, L.; Santos, L.S.; Torres, S. Anthocyanin composition in aged Chilean Cabernet Sauvignon red wines. *Food Chem.* **2011**, *129*, 514–519.

29. Wu, X.; Prior, R.L. Identification and characterization of anthocyanins by high-performance liquid chromatography–electrospray Ionization–tandem mass spectrometry in common foods in the United States: vegetables, nuts, and grains. *J. Agric. Food Chem.* **2005**, *53*, 3101–3113.

30. Abdel-Aal, E.-S.M.; Young, J.C.; Rabalski, I. Anthocyanin composition in black, blue, pink, purple, and red cereal grains. *J. Agric. Food Chem.* **2006**, *54*, 4696–4704.

31. Yousef, G.G.; Brown, A.F.; Funakoshi, Y.; Mbeunkui, F.; Grace, M.H.; Ballington, J.R.; Loraine, A.; Lila, M.A. Efficient quantification of the health-relevant anthocyanin and phenolic acid profiles in commercial cultivars and breeding selections of Blueberries (*Vaccinium* spp.). *J. Agric. Food Chem.* **2013**, *61*, 4806–4815.

32. Zheng, J.; Ding, C.; Wang, L.; Li, G.; Shi, J.; Li, H.; Wang, H.; Suo, Y. Anthocyanins composition and antioxidant activity of wild *Lycium ruthenicum* Murr. from Qinghai-Tibet plateau. *Food Chem.* **2011**, *126*, 859–865.

33. Simirgiotis, M.J. Antioxidant Capacity and HPLC-DAD-MS Profiling of Chilean Peumo (*Cryptocarya alba*) Fruits and Comparison with German Peumo (*Crataegus monogyna*) from Southern Chile. *Molecules* **2013**, *18*, 2061–2080.

34. Schierle, J.; Pietsch, B.; Ceresa, A.; Fizet, C. Method for the determination of β-carotene in supplements and raw materials by reversed-phase liquid chromatography: Single laboratory validation. *J. AOAC Int.* **2004**, *87*, 1070–1082.

35. Bunea, A.; Rugina, D.; Sconta, Z.; Pop, R.M.; Pintea, A.; Socaciu, C.; Tabaran, F.; Grootaert, C.; Struijs, K.; VanCamp, J. Anthocyanin determination in blueberry extracts from various cultivars and their antiproliferative and apoptotic properties in B16-F10 metastatic murine melanoma cells. *Phytochemistry* **2013**, *95*, 436–444.

36. Carbone, K.; Giannini, B.; Picchi, V.; Lo Scalzo, R.; Cecchini, F. Phenolic composition and free radical scavenging activity of different apple varieties in relation to the cultivar, tissue type and storage. *Food Chem.* **2011**, *127*, 493–500.

37. Simirgiotis, M.J.; Schmeda-Hirschmann, G. Determination of phenolic composition and antioxidant activity in fruits, rhizomes and leaves of the white strawberry (*Fragaria chiloensis* spp. chiloensis form chiloensis) using HPLC-DAD-ESI-MS and free radical quenching techniques. *J. Food Compos. Anal.* **2010**, *23*, 545–553.

38. Benzie, I.F.F.; Strain, J.J. The ferric reducing ability of plasma (FRAP) as a measure of "Antioxidant Power": The FRAP assay. *Anal.Biochem.* **1996**, *239*, 70–76.

39. Simirgiotis, M.J.; Caligari, P.D.S.; Schmeda-Hirschmann, G. Identification of phenolic compounds from the fruits of the mountain papaya Vasconcellea pubescens A. DC. grown in Chile by liquid chromatography-UV detection-mass spectrometry. *Food Chem.* **2009**, *115*, 775–784.

40. Simirgiotis, M.J.; Adachi, S.; To, S.; Yang, H.; Reynertson, K.A.; Basile, M.J.; Gil, R.R.; Weinstein, I.B.; Kennelly, E.J. Cytotoxic chalcones and antioxidants from the fruits of *Syzygium samarangense* (Wax Jambu). *Food Chem.* **2008**, *107*, 813–819.

41. Lee, J.; Durst, R.W.; Wrolstad, R.E. Determination of total monomeric anthocyanin pigment content of fruit juices, beverages, natural colorants, and wines by the pH differential method: Collaborative study. *J. AOAC Int.* **2005**, *88*, 1269–1278.

**Sample Availability:** *Sample Availability*: Samples of the compounds and extract of the berries are available from authors.

# Composition and Antioxidant Activity of the Anthocyanins of the Fruit of *Berberis heteropoda* Schrenk

Li-Li Sun, Wan Gao, Meng-Meng Zhang, Cheng Li, Ai-Guo Wang, Ya-Lun Su and Teng-Fei Ji

**Abstract:** In present study, the anthocyanin composition and content of the fruit of *B. heteropoda* Schrenk were determined for the first time. The total anthocyanins were extracted from the fruit of *B. heteropoda* Schrenk using 0.5% HCl in 80% methanol and were then purified using an AB-8 macroporous resin column. The purified anthocyanin extract (PAE) was evaluated by high-performance liquid chromatography with a diode array detector (HPLC-DAD) and HPLC-high resolution-electrospray ionization-mass spectrometry (HPLC-HR-ESI-MS) under the same experimental conditions. The results revealed the presence of seven different anthocyanins. The major anthocyanins purified by preparative HPLC were confirmed to be delphinidin-3-O-glucopyranoside (30.3%), cyanidin-3-O-glucopyranoside (33.5%), petunidin-3-O-glucopyranoside (10.5%), peonidin-3-O-glucopyranoside (8.5%) and malvidin-3-O-glucopyranoside (13.8%) using HPLC-HR-ESI-MS and NMR spectroscopy. The total anthocyanin content was $2036.6 \pm 2.2$ mg/100 g of the fresh weight of *B. heteropoda* Schrenk fruit. In terms of its total reducing capacity assay, DPPH radical-scavenging activity assay, ferric-reducing antioxidant power (FRAP) assay and ABTS radical cation-scavenging activity assay, the PAE also showed potent antioxidant activity. The results are valuable for illuminating anthocyanins composition of *B. heteropoda* Schrenk and for further utilising them as a promising anthocyanin pigment source. This research enriched the chemical information of *B. heteropoda* Schrenk.

Reprinted from *Molecules*. Cite as: Sun, L.-L.; Gao, W.; Zhang, M.-M.; Li, C.; Wang, A.-G.; Su, Y.-L.; Ji, T.-F. Composition and Antioxidant Activity of the Anthocyanins of the Fruit of *Berberis heteropoda* Schrenk. *Molecules* **2014**, *19*, 19078–19096.

## 1. Introduction

*Berberis heteropoda* Schrenk, a type of Berberidaceae deciduous shrub that is native to the Xinjiang Uygur Autonomous Region of China, is used in both medicine and food [1]. Its ripe fruit contains glucose, fructose, malic acid, carotene, pigments and other substances. It was reported that *B. heteropoda* Schrenk is a rich source of anthocyanin compounds and that its beneficial effects are remarkable [2]. Its ripe fruits had been used for treatment of dysentery, enteritis, pharyngitis, stomatitis, eczema and hypertension [3]. Moreover, the residents of Kazakh and Uighur have drunk tea made from it for a long time. They also made it into jams. Thus, the value of the *B. heteropoda* Schrenk fruit has great potential for development. To date, only limited literature about this fruit is available, and most of it has focused on extraction methods, optimization of technical conditions, determination of the total flavonoid/berberine content, stability analyses or antitumor effects. However, the major anthocyanin constituents and the antioxidant activity of the fruit of *B. heteropoda* Schrenk have not yet been systematically studied, and in terms of large-scale applications, the potential of plant is now basically untapped [2]. This situation has largely restricted the research and development of the fruit of *B. heteropoda* Schrenk.

Anthocyanins are the most common pigmented flavonoids and are widespread in the plant kingdom. These compounds are responsible for most of the brilliant colors (orange, red, pink, purple and blue) observed in most fruits, flowers, leaves and cereal grains [4]. The naturally occurring anthocyanins of plants are cyanidin, delphinidin, peonidin, petunidin, malvidin, and pelargonidin [4,5]. From a chemical point of view, the anthocyanin molecule is composed of a flavylium nucleus bearing one or more sugar residues. The most prevalent of these sugars, including D-glucose, D-galactose, L-rhamnose, D-xylose, and D-arabinose, are 3-glycosides or 3,5-di-glycosides [6,7]. In addition, these sugars may be esterified by aliphatic or aromatic organic acids, which can greatly help to stabilize the anthocyanin structure. However, anthocyanins are not stable and are prone to degradation. The stability of anthocyanins is affected by several factors, such as their chemical structure, their concentration, oxygen, temperature, pH, light, enzymes, metal ions and various storage conditions [8].

In addition to their natural colorants role, anthocyanins have attracted considerable global interest, mainly due to their health-promoting benefits, such as reducing the risk of coronary heart disease and preventing several chronic diseases, which are associated with their antioxidant properties. The likely mechanism postulated for their effects is that anthocyanins act as potent antioxidants by scavenging free radical species such as reactive oxygen species (ROS) or reactive nitrogen species (RNS), inhibiting certain enzymes or chelating trace metals involved in free-radical production and upregulating or protecting antioxidant defenses, breaking the free-radical chain reaction [9,10]. Because of their beneficial health effects as dietary antioxidant, dietitians have proposed that adding certain amounts of exogenous anthocyanins derived from natural sources to the daily diet will delay or prevent many degenerative diseases.

Identifying anthocyanins using HPLC-DAD is complicated by the fact that some anthocyanins show similar retention times and spectroscopic characteristics [11]. In the past several decades, a good number of technological advances in HPLC and the introduction of the ultra-performance liquid chromatography (UPLC) have achieved significant improvements in both speed and separation of anthocyanins. In addition, advancements in mass spectrometry (MS), such as tandem MS, high-resolution MS (HR-MS) and sequential collision MS (MSn) allow fast structural elucidation of anthocyanins that play a important role in food anthocyanin research [12]. Due to their "soft" ionization properties, leading to the production of intact molecular ions and the corresponding anthocyanidin fragments, ESI-MS techniques have been shown to be highly suitable for anthocyanin characterization. Now, HPLC assisted by MS (HPLC-MS) has been frequently used as an excellent tool for the simultaneous chemical separation and identification of anthocyanin compounds. To our knowledge, the anthocyanin composition of the fruit of *B. heteropoda* Schrenk has never been described. One of the objectives of this study was to identify and characterize the anthocyanins of the fruit of *B. heteropoda* Schrenk using HPLC-HR-ESI-MS/MS. Another objective was to evaluate the antioxidant activity of its extracts according to the total reducing capacity, the DPPH-radical scavenging activity, the ferric-reducing antioxidant power (FRAP) and ABTS radical-cation scavenging activity.

## 2. Results and Discussion

### 2.1. Total Anthocyanin Content (TAC)

The amount of total anthocyanin in the PAE that was determined using the pH differential method was 2036.6 ± 2.2 mg/100 g of fresh weight of *B. Heteropoda* Schrenk fruit, expressed as cyanidin-3-O-glucoside equivalents and calculated as the mean value of three measurements and the standard deviation. This value is considerably higher than that of other anthocyanin-rich fruits and vegetables. For instance, the anthocyanin content of *Berberis boliviano* Lecher fruit is 1,500 mg/100 g of fresh fruit weight [13], that of black chokeberries is 560 mg/100 g of fresh fruit weight [13], that of wild *Lycium ruthenicum* Murr. from Dulan is 520 mg/100 g of fresh fruit weight, that of wild *Lycium ruthenicum* Murr. from Gomud is 470 mg/100 g of fresh fruit weight, and that of wild *Lycium ruthenicum* Murr. from Delingha is 475 mg/100 g of fresh fruit weight [4]. The anthocyanin content of each of the above fruits was lower than that of the fresh fruit of *B. heteropoda* Schrenk.

### 2.2. HPLC Analysis

The HPLC-DAD chromatogram of the PAE of *B. heteropoda* Schrenk fruit that was obtained at 530 nm is displayed in Figure 1. The method utilized provided repeatable and satisfactory separation of the components of the PAE. Although several other components of the PAE were detected in the present study, five major anthocyanins were discovered as shown in the HPLC chromatogram. The major anthocyanins represented approximately 97% of the total peak area, and the relative amounts of anthocyanins 1–5 were approximately 30.3%, 33.5%, 10.5%, 8.5% and 13.8%, respectively. However, two minor peaks (Compounds 6 and 7) were also detected which were identified as peonidin and malvidin by HPLC and HPLC-HR-ESI-MS/MS (Table 1).

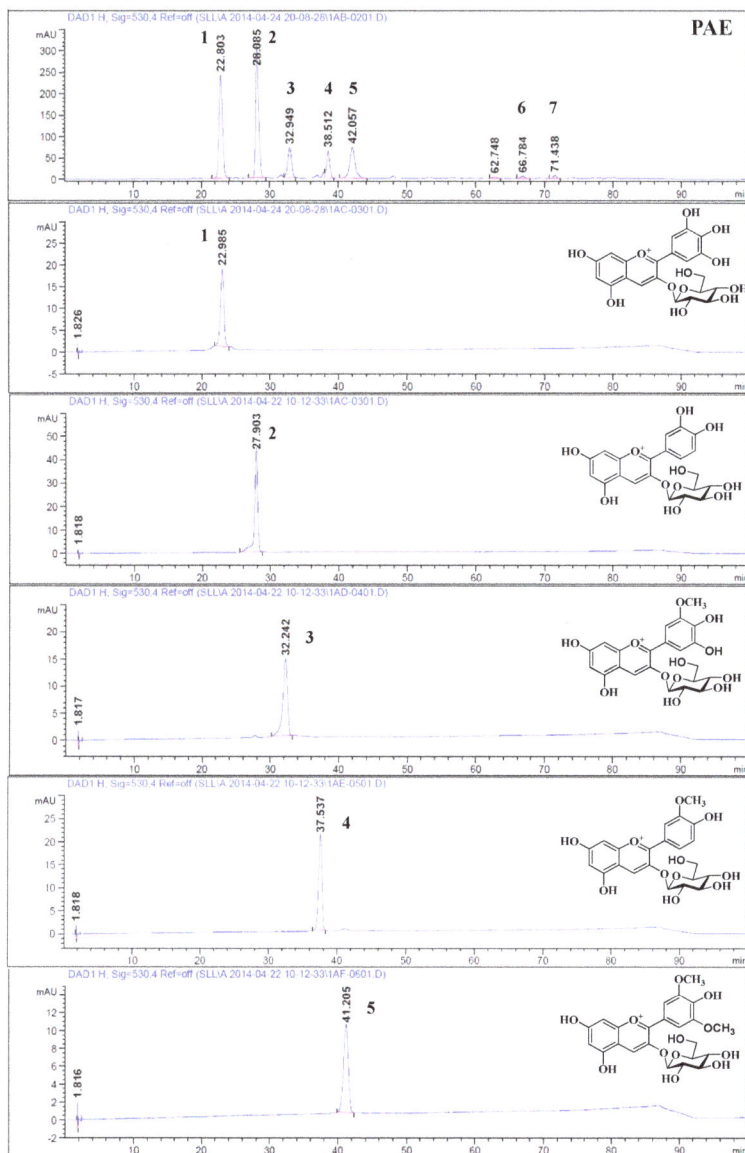

**Figure 1.** HPLC chromatogram of the five major anthocyanins of the fruit of *B. heteropoda* Schrenk that were detected at 530 nm. The peaks were numbered in order of their elution.

41

## 2.3. Structural Elucidation

Five major anthocyanins were successfully isolated for the first time from the fruit of *B. heteropoda* Schrenk using preparative HPLC techniques. Compounds 1–5 were identified by HPLC-HR-ESI-MS/MS and NMR spectroscopic analysis. Figures 1 and 2 shows the structures of these five anthocyanins.

Compound 1 was obtained as an amorphous red power. Its molecular formula of $C_{21}H_{21}O_{12}^+$ (calculated $m/z$ 465.1034), was established based on the molecular ion at $m/z$ 465.1027 $[M+H]^+$ observed in the positive-ion HPLC-HR-ESI-MS/MS spectrum. A major fragmentation that occurred at $m/z$ 303.0486 $[M+H-162]^+$ was in accordance with the presence of a delphinidin aglycone and the loss of a hexose moiety. The $^1$H-NMR spectrum revealed a symmetrical aromatic proton signal at $\delta$ 7.71 (H-2'/H-6', s), two *meta*-coupled doublet protons on the A-ring at $\delta$ 6.63 (H-6, $J$ = 1.5 Hz) and $\delta$ 6.82 (H-8, $J$ = 1.5 Hz) and a singlet proton at $\delta$ 8.91 (H-4), suggesting the presence of a delphinidin nucleus. The proton signals observed at $\delta$ 5.30–3.50 (H-1-Glu-H-6-Glu) indicated the presence of a glucopyranose moiety. The anomeric proton signal at $\delta$ 5.30 (H-1-Glu, d, $J$ = 7.5 Hz) in addition to the C-3 carbon signal at $\delta$ 144.9 ppm indicated that a sugar moiety with a $\beta$-configuration was attached to the C-3 position. From the above spectral studies, the structure of compound 1 was determined to be that of delphinidin-3-O-$\beta$-glucopyranoside, which corresponded with data in the literature [14].

Compound 2 was isolated as an amorphous red power. Its molecular formula was established as $C_{21}H_{21}O_{11}^+$ (calculated $m/z$ 449.1082), with twelve degrees of unsaturation, based on its positive-ion HPLC-HR-ESI-MS/MS analyses. A molecular ion at $m/z$ 449.1075 $[M+H]^+$ and a major fragmentation occurring at $m/z$ 287.0538 $[M+H-162]^+$ corresponded to those of a cyanidin aglycone and a hexose unit. Similarly, the observation of a set of ABX-type aromatic proton signals at $\delta$ 7.02 (H-5', d, $J$ = 8.5 Hz), $\delta$ 8.05(H-2', d, $J$ = 2.0 Hz) and $\delta$ 8.23 (H-6', dd, $J$ = 9.0, 2.5 Hz), two *meta*-coupled doublet protons on the A-ring at $\delta$ 6.65 (H-6, $J$ = 1.5 Hz) and $\delta$ 6.87 (H-8, $J$ = 1.0 Hz) and a singlet proton at $\delta$ 9.00 (H-4) in the $^1$H-NMR spectrum also implied the presence of a cyanidin nucleus. From the appearance of the anomeric proton signal at $\delta$ 5.28 (H-1-Glu, d, $J$ = 7.5 Hz) and the C-3 carbon signal at $\delta$ 145.6 ppm indicated that a $\beta$-glucopyranose moiety was attached to C-3 position. Thus, compound 2 was elucidated as cyanidin-3-O-$\beta$-glucopyranoside by comparison of the spectral data with the literature [15].

Compound 3, which also formed an amorphous red power, showed a molecular ion at $m/z$ 479.1185 $[M+H]^+$ and a major fragment ion at $m/z$ 317.0664 $[M+H-162]^+$ in the positive-ion HPLC-HR-ESI-MS/MS spectrum, which was in agreement with the molecular formula of $C_{22}H_{23}O_{12}^+$ (calculated $m/z$ 479.1192), with twelve degrees of unsaturation. These data indicated the presence of a petunidin aglycon and a hexose moiety. Comparison of the $^1$H- and $^{13}$C-NMR spectra of compound 3 with those of compound **1** demonstrated that the structure of compound **3** was very similar to that of compound **1**, except for the presence of a methyl group [$\delta_H$ 3.98 (3H, s); $\delta_C$ 57.3]. Compound **3** was thus identified as petunidin-3-O-$\beta$-glucopyranoside by comparing these spectral data with reference [16,17].

Compound **4** was also presented as an amorphous red power. The positive-ion HPLC-HR-ESI-MS/MS spectrum showed its molecular ion at $m/z$ 463.1223 [M+H]$^+$ and a major fragment at $m/z$ 301.0697 [M+H−162]$^+$, corresponding to the molecular formula of $C_{22}H_{23}O_{11}{}^+$ (calculated $m/z$ 463.1229), with twelve degrees of unsaturation. These data corresponded to those of a peonidin aglycone and a hexose moiety. Further support for this structure was obtained by comparing the $^1$H and $^{13}$C-NMR spectra of compound **4** with those of compound **2**. The spectral features of compound **4** were almost identical with those of compound **2**, except for the presence of a methyl group [$\delta_H$ 3.99 (3H, s); $\delta_C$ 56.7]. Based on the above-described observations, compound **4** was assigned as peonidin-3-O-$\beta$-glucopyranoside, which in accordance with the relevant published data [18].

Compound **5** was also obtained as an amorphous red power and had the molecular formula of $C_{23}H_{25}O_{12}{}^+$ (calculated $m/z$ 493.1339), with twelve degrees of unsaturation, as determined using positive-ion HPLC-HR-ESI-MS/MS. A molecular ion at $m/z$ 493.1332 [M+H]$^+$ and a major fragment at $m/z$ 331.0762 [M+H−162]$^+$ were consistent with a structure consisting of a malvidin aglycone and a hexose moiety. By comparing their $^1$H and $^{13}$C-NMR spectra, the structure of compound **5** was found to differ from that of compound **1** mainly in the substitution pattern in ring B. The existence of a methyl group at $\delta_H$ 3.99 (6H, s, OCH$_3$-3'/OCH$_3$-5') in addition to a carbon signal at $\delta_C$ 56.7 ppm and a symmetrical aromatic proton signal at $\delta$ 7.97 (H-2'/H-6', s) revealed two methoxyl groups, one located at C-3 and the other at C-5. Based on the above-described evidence, compound **5** was identified as malvidin-3-O-$\beta$-glucopyranoside by comparison with previous data [19].

The detailed data obtained using HPLC-HR-ESI-MS/MS are presented in Table 1 and Figure 2, and the $^1$H- and $^{13}$C-NMR spectroscopic data are shown in Table 2. This is the first time that all of the major anthocyanins in *B. heteropoda* Schrenk fruits were systematically isolated and characterized.

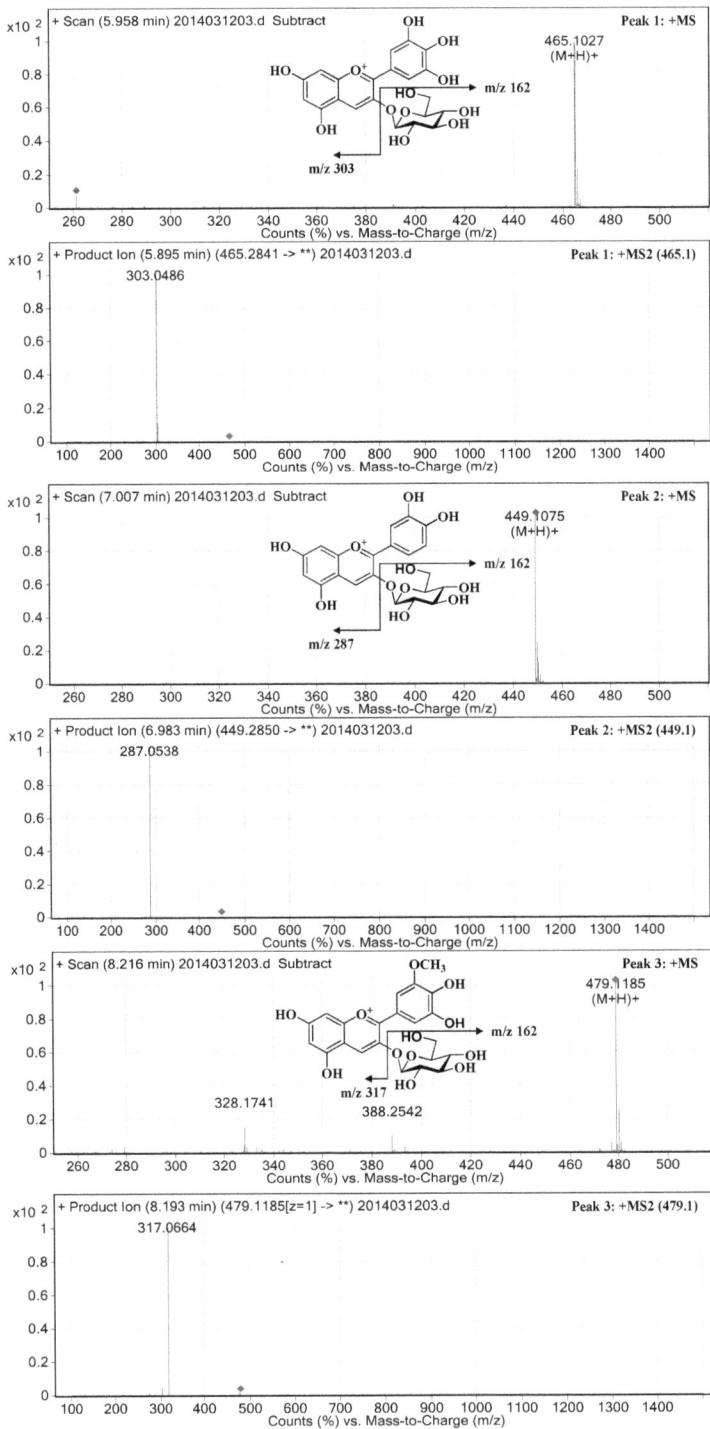

**Figure 2.** *Cont.*

44

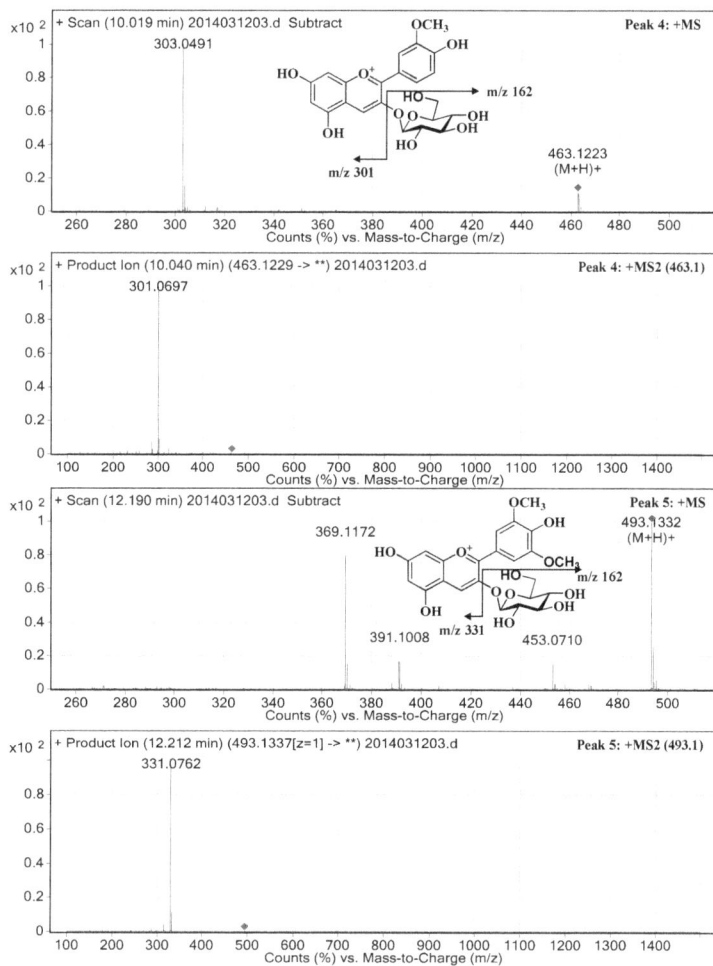

**Figure 2.** High-resolution electrospray mass spectrum of identified anthocyanins. Peak 1: delphinidin-3-O-glucopyranoside; Peak 2: cyanidin-3-O-glucopyranoside; Peak 3: petunidin-3-O-glucopyranoside; Peak 4: peonidin-3-O-glucopyranoside; and Peak 5: malvidin-3-O-glucopyranoside.

**Table 1.** Chromatographic and spectroscopic characteristics of the major anthocyanins in the fruits of *B. heteropoda* Schrenk, determined using HPLC-DAD and HPLC-HR-ESI-MS/MS.

| Peak [a] | t$_R$ (min) HPLC | (%) HPLC | λ$_{max}$ (nm) | t$_R$ (min) HPLC-HR-ESI-MS/MS | M$^+$ (m/z) | Fragment Ions (m/z) | Compound | Contents [b] (mg/100 g Fresh Fruits) |
|---|---|---|---|---|---|---|---|---|
| 1 | 22.38 | 30.34 | 524/278 | 5.96 | 465.1027 | 303.0486 | Delphinidin-3-O-glucopyranoside | 617.84 ± 0.98 |
| 2 | 27.54 | 33.47 | 516/280 | 7.01 | 449.1075 | 287.0538 | Cyanidin-3-O-glucopyranoside | 681.58 ± 1.13 |
| 3 | 34.42 | 10.55 | 527/278 | 8.22 | 479.1185 | 317.0664 | Petunidin-3-O-glucopyranoside | 212.59 ± 1.79 |
| 4 | 37.88 | 8.53 | 520/280 | 10.02 | 463.1223 | 301.0697 | Peonidin-3-O-glucopyranoside | 173.65 ± 0.66 |
| 5 | 41.48 | 13.77 | 528/278 | 12.20 | 493.1332 | 331.0762 | Malvidin-3-O-glucopyranoside | 279.83 ± 0.60 |
| 6 | 78.83 | 0.16 | 520/280 | 26.12 | 301.0704 | - | Peonidin | 3.29 ± 0.05 |
| 7 | 79.52 | 0.22 | 530/278 | 27.38 | 331.0803 | - | Malvidin | 4.49 ± 0.06 |

Notes: [a] Numbered according to the order of elution; [b] mean value ± SD (*n* = 3).

46

**Table 2.** $^1$H-NMR and $^{13}$C-NMR spectroscopic data for compounds **1–5** in CD$_3$OD/TFA-d (7:1, v/v) ($\delta$ in ppm. $J$ in Hz).

| Position | 1 $\delta_{\mathrm{H}}$[a] | 1 $\delta_{\mathrm{C}}$[c] | 2 $\delta_{\mathrm{H}}$[a] | 2 $\delta_{\mathrm{C}}$[d] | 3 $\delta_{\mathrm{H}}$[b] | 3 $\delta_{\mathrm{C}}$[d] | 4 $\delta_{\mathrm{H}}$[a] | 4 $\delta_{\mathrm{C}}$[c] | 5 $\delta_{\mathrm{H}}$[a] | 5 $\delta_{\mathrm{C}}$[d] |
|---|---|---|---|---|---|---|---|---|---|---|
| 2 | | 164.1 | | 164.4 | | 164.0 | | 164.1 | | 163.0 |
| 3 | 8.91 (s) | 144.9 | 9.00 (s) | 145.6 | 8.98 (s) | 145.1 | 9.02 (s) | 145.5 | 9.02 (s) | 145.3 |
| 4 | | 136.4 | | 137.1 | | 136.9 | | 137.4 | | 137.3 |
| 5 | | 159.6 | | 159.6 | | 159.0 | | 158.8 | | 158.9 |
| 6 | 6.63 (d, 1.5) | 103.9 | 6.65 (d, 1.5) | 103.8 | 6.67 (d, 1.2) | 103.6 | 6.66 (d, 1.5) | 103.5 | 6.67 (d, 0.5) | 103.1 |
| 7 | | 170.8 | | 170.7 | | 170.7 | | 170.9 | | 170.9 |
| 8 | 6.82 (d, 1.5) | 95.4 | 6.87 (d, 1.0) | 95.2 | 6.88 (d, 0.6) | 95.2 | 6.89 (d, 1.5) | 95.3 | 6.93 (d, 0.5) | 95.4 |
| 9 | | 157.9 | | 157.8 | | 157.8 | | 157.9 | | 157.9 |
| 10 | | 112.9 | | 113.8 | | 113.6 | | 113.6 | | 112.9 |
| 1' | | 120.3 | | 121.3 | | 120.1 | | 121.1 | | 119.9 |
| 2' | 7.71 (s) | 113.6 | 8.05 (d, 2.0) | 118.6 | 7.90 (d, 1.8) | 109.2 | 8.19 (d, 1.5) | 115.2 | 7.97 (s) | 110.6 |
| 3' | | 147.7 | | 147.4 | | 149.9 | | 149.6 | | 149.8 |
| 4' | | 146.1 | | 155.7 | | 145.7 | | 156.5 | | 164.4 |
| 5' | | 147.7 | 7.02 (d, 8.5) | 117.3 | | 147.4 | 7.05 (d, 9.0) | 117.6 | | 149.8 |
| 6' | 7.71 (s) | 113.6 | 8.23 (dd, 9.0, 2.5) | 128.1 | 7.81 (d, 1.8) | 112.9 | 8.23 (dd, 8.5, 2.0) | 128.9 | 7.97 (s) | 110.6 |
| 3'-OMe | | | | | 3.98 (s) | 57.3 | 3.99 (s) | 56.7 | 3.99 (s) | 56.7 |
| 5'-OMe | | | | | | | | | 3.99 (s) | 56.7 |
| 1" | 5.30 (d, 7.5) | 103.7 | 5.28 (d, 7.5) | 104.1 | 5.32 (d, 7.8) | 103.8 | 5.29 (overlap) | 103.9 | 5.33 (overlap) | 103.9 |
| 2" | 3.77 (m) | 75.0 | 3.73 (m) | 74.8 | 3.70 (m) | 74.9 | 3.67 (m) | 74.9 | 3.71 (m) | 74.9 |
| 3" | 3.61 (m) | 78.3 | 3.56 (m) | 78.1 | 3.57 (m) | 78.2 | 3.60 (m) | 78.2 | 3.60 (m) | 78.1 |
| 4" | 3.50 (m) | 71.3 | 3.46 (m) | 71.1 | 3.47 (m) | 71.1 | 3.47 (m) | 71.2 | 3.47 (m) | 71.3 |
| 5" | 3.61 (m) | 79.0 | 3.56 (m) | 78.7 | 3.57 (m) | 78.8 | 3.60 (m) | 78.8 | 3.60 (m) | 77.8 |
| 6"a | 3.94 (dd, 12.0, 1.5) | 62.6 | 3.92 (dd, 12.0, 2.0) | 62.4 | 3.94 (dd, 9.0, 3.0) | 62.3 | 3.93 (dd, 10.0, 2.0) | 62.4 | 3.92 (dd, 9.5, 1.5) | 62.3 |
| 6"b | 3.77 (dd, 12.5, 5.0) | | 3.73 (dd, 12.0, 4.5) | | 3.74 (dd, 12.0, 6.0) | | 3.72 (dd, 12.5, 6.0) | | 3.77 (dd,12.0, 6.0) | |

Notes: [a] The $^1$H-NMR data were obtained at 500 MHz; [b] The $^1$H-NMR data were obtained at 600 MHz; [c] The $^{13}$C-NMR data were obtained at 125 MHz; [d] The $^{13}$C-NMR data were obtained at 150 MHz.

## 2.4. Antioxidant Activity

### 2.4.1. Total Reducing Capacity of the PAE

The total reducing power of the anthocyanins can be used as an index of their antioxidative electron-donating activity. The total reducing capacity of the PAE and the ascorbic-acid control are shown in Table 3 and Figure 3A, respectively. The results showed that the reducing power of the PAE was approximately one-third that of ascorbic acid. The $IC_{50}$ value, which was the concentration that raised the absorbance at 700 nm to 0.5, of the reducing power of the PAE was 139.65 $\mu$g/mL. Furthermore, both PAE and ascorbic acid exhibited a dose-dependent reducing power. Good linearity was obtained using the PAE at concentrations of 90 to 190 $\mu$g/mL and ascorbic acid at concentrations of 30 to 62 $\mu$g/mL. The calibration curve of the PAE and ascorbic acid were Y (absorbance) = 0.0033X (concentration) + 0.0615 (r = 0.999), Y (absorbance) = 0.0099X (concentration) + 0.0434 (r = 0.995), respectively.

A

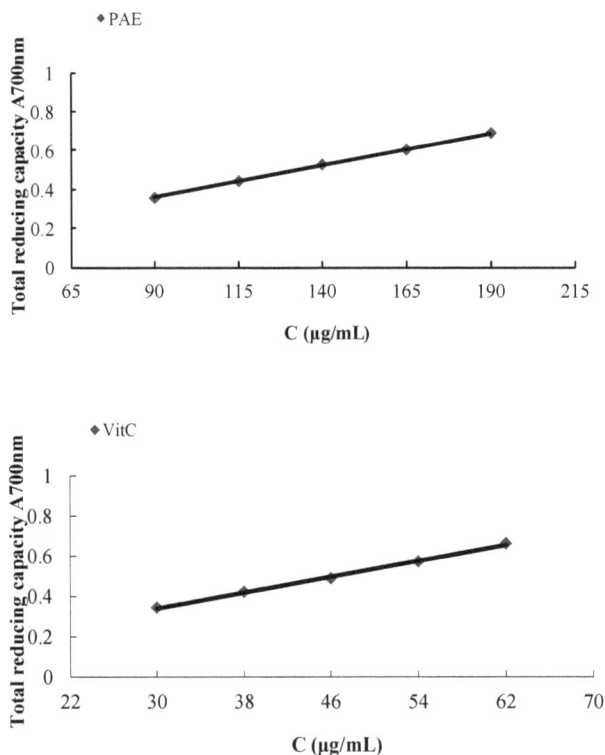

Figure 3. Cont.

**B**

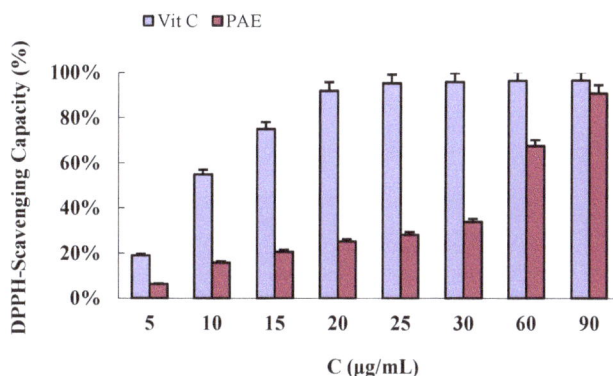

**Figure 3.** Antioxidant activities of the PAE of *B. heteropoda* Schrenk fruit determined using the total reducing capacity assay (**A**) and the DPPH free radical-scavenging assay (**B**). The data represented by different-colored bars are significantly different ($p < 0.05$).

**Table 3.** Antioxidant activities of PAE and ascorbic acid in terms of the total reducing capacity assay, DPPH radical scavenging activity assay, ferric-reducing antioxidant power (FRAP) assay and ABTS radical cation scavenging activity assay.

| Samples | $A_{700\ nm}$ = 0.5/Total Reducing Power (μg/mL) [a] | $IC_{50}$/DPPH (μg/mL) [b] |
|---|---|---|
| PAE | 139.65 ± 0.01 | 47.16 ± 0.35 |
| Ascorbic acid | 46.12 ± 0.04 | 9.30 ± 0.21 |
| Samples | FRAP value (mmol/g) | TEAC/ABTS |
| PAE | 4.32 ± 0.03 | 2.250 ± 0.17 |
| Trolox | 7.007 ± 0.14 | - |

Notes: All the trials were performed in triplicate and all the data represent the means ± standard deviation ($n = 3$). Data in the same column with different letters are significantly different ($p < 0.05$); [a] The antioxidant activity was evaluated as the concentration of the test sample required to raise the absorbance at 700 nm to 0.5; [b] The antioxidant activity was calculated as the concentration of the test sample required to decrease the absorbance at 517 nm by 50%.

## 2.4.2. DPPH Radical-Scavenging Activity of the PAE

The effect of an antioxidant on DPPH radical scavenging is generally ascribed to its hydrogen-donating ability. The decrease in absorbance at 517 nm reflects the DPPH-radical reduction potential of samples. In the present study, the concentration of the samples required to scavenge 50% of the DPPH radicals ($IC_{50}$) was used as an indicator for comparing their antioxidant activities. The results are shown in Table 3 and Figure 3B. As shown in Figure 3B, the DPPH radical-scavenging activity of the PAE and ascorbic acid increased as the concentrations increased. At a concentration of 90 μg/mL, the PAE scavenged almost all of the DPPH radicals that were present (>95%). The $IC_{50}$ value for the PAE was 47.16 μg/mL, however, the $IC_{50}$ value for the control, ascorbic acid, was 9.30 μg/mL. In terms of the $IC_{50}$ value, DPPH radical-scavenging activity of the PAE was approximately one-fifth that of ascorbic acid.

### 2.4.3. Ferric-Reducing Antioxidant Power (FRAP) of the PAE

A simple and reliable assay was adopted to evaluate the reducing capacity of antioxidants in the present study. This assay is based on the reaction of an antioxidant with the TPTZ-Fe(III) complex to generate TPTZ-Fe(II). The absorbance of TPTZ-Fe(II) at 593 nm was measured to evaluate the reducing power of the tested sample. The corresponding $FeSO_4$ value provides quantification of the antioxidant activity of the extract. A higher $FeSO_4$ value demonstrates a higher ferric-reducing capacity. In this assay, the PAE exhibited a high antioxidant activity, with a $FeSO_4$ value of 4.32 mmol/g, whereas the $FeSO_4$ value of the Trolox control was determined to be 7.007 mmol/g (Table 3). As far as the $FeSO_4$ value be concerned, Ferric-reducing capacity of PAE was approximately three-fifths that of Trolox.

### 2.4.4. ABTS$^+$ Radical-Scavenging Activity of the PAE

The ABTS$^+$ radical-scavenging assay is considered an excellent tool for investigating the antioxidant activity of hydrogen-donating antioxidants and chain-breaking antioxidants [20]. This assay is based on the inhibitory effect of the antioxidant on the absorbance of ABTS$^+$, which reflects the antioxidant capacity of the tested sample. The TEAC values were calculated using standard curves, with a higher TEAC value indicating a higher antioxidant activity. The PAE exhibited a good antioxidative activity in this test, with a TEAC value of 2.25 mmol/g (Table 3), which was comparable to that of the extract of pigeon-pea leaves (1.095 mmol/g) [21]. Therefore, *B. heteropoda* Schrenk fruit is an excellent candidate antioxidant.

## 3. Experimental Section

### 3.1. Plant Material

Fresh ripe fruit (3 kg) of *B. heteropoda* Schrenk was manually harvested in Daxigou (Latitude 44°26′ N, Longitude 80°46′ E, Altitude 1000 m), Huocheng County, Yili Kazakh Autonomous Prefecture, Xinjiang Uygur Autonomous Region of China during September 2012. The fruit was maintained at −18 °C from immediately after collection until they were used. Samples were also preserved in a refrigerator (−18 °C) for later analysis in the laboratory.

## 3.2. Reagents and Apparatus

2,2-Diphenyl-1-picrylhydrazyl (DPPH), potassium ferricyanide and ascorbic acid (vitamin C) were purchased from the Sigma-Aldrich Chemical Co. (Shanghai, China); The FRAP and ABTS assay kits were obtained from the Beyotime Institute of Biotechnology (Haimen, Jiangsu Province, China). The methanol and acetonitrile for the HPLC analysis were of chromatographic grade and were purchased from J.T. Baker (Phillipsburg, NJ, USA). All other reagents and solvents were of analytical grade.

The ultrasonicator was purchased from Tian Pong Electricity New Technology Co. Ltd. (Beijing, China) The filter papers were produced by Hangzhou Special Paper Industry (Hangzhou, Zhejiang Province, China). The 0.45-μm reinforced nylon membrane filter was purchased from ANPEL (Shanghai, China). The freeze dryer used was an EYELA FDU-1100 model (Tokyo, Japan). The L5S spectrophotometer was purchased from the Jingke Industrial Co. Ltd. (Shanghai, China). The multimode reader was a PerkinElmer EnSpire model (Waltham, MA, USA).

## 3.3. Extraction and Primary Purification of the Anthocyanins

The fresh fruit of *B. heteropoda* Schrenk (100 g) were crushed using a pulverizer, and then ultrasonically extracted for 30 min using 1 L of 0.5% HCL (v/v) in 80% methanol in the dark at room temperature. This extraction process was repeated three times to ensure exhaustive extraction. The supernatants were combined and were filtered through filter paper to remove the fruit residues, proteins and the polysaccharide-containing sediment. The filtrate was concentrated using a rotary evaporator at less than 35 °C, and then was lyophilized to produce the crude anthocyanin extract (22.666 g).

The crude anthocyanin extract was purified using an AB-8 macroporous resin column (2.5 cm × 45 cm) to remove sugars, acids and other water-soluble substances. The successive eluting agents were 0.5% aqueous HCl (1 L), 0.5% HCl in 95% ethanol (1 L), 95% ethanol (1 L). The eluent obtained using 95% ethanol (0.5% HCl) was evaporated under reduced pressure at less than 35 °C and was freeze-dried to yield the PAE (10.956 g).

## 3.4. Isolation and Identification of the Main Anthocyanins in the PAE

### 3.4.1. Isolation

The PAE (5.0 g) was further fractionated using a medium-pressure chromatographic column (4.9 × 46 cm) using a gradient elution of 0%, 20%, 25%, 30%, 35%, 40%, 50%, 70% $CH_3OH/H_2O$ (containing 0.1% HCl; 2 L of each eluent), resulting in 37 fractions. The 37 fractions were concentrated separately under reduced pressure at less than 35 °C and then were analyzed using HPLC. Fractions 12 (121 mg), 15 (209 mg) and 21 (201 mg) were further refined by preparative HPLC (Shimadzu

LC-6AD, Kyoto, Japan) using a YMC-Pack ODS column (20 × 250 mm, 10 μM, YMC Co. Ltd., Kyoto, Japan) at room temperature with a flow rate of 4.0 mL/min and detection at 530 nm. The eluting solvents were $CH_3CN/H_2O/TFA$ at a ratio of 12:87.9:0.1, 11:88.9:0.1, 13:86.9:0.1, which yielded compounds **1** (21.0 mg), **2** (21.0 mg) and **3** (3.4 mg), **4** (3.0 mg) and **5** (5.0 mg).

### 3.4.2. NMR Identification

The structures of the isolated anthocyanins were elucidated by spectroscopic analysis. $^1H$ (500 MHz) and $^{13}C$ (125 MHz) NMR spectra were obtained using an Inova 500 spectrometer (Agilent Technologies, Inc., Santa Clara, CA, USA) in $CD_3OD/CF_3COOD$ (7:1, v/v) containing TMS as an internal standard. The chemical shift values were expressed in $\delta$ (ppm) and the coupling constant ($J$) values were presented in Hertz.

### 3.5. Total Anthocyanin Content

The TAC was determined using the pH differential method [22,23]. The absorbance at 530 nm and 700 nm was measured in pH 1.0 buffer (potassium chloride-hydrochloric acid) and pH 4.5 buffer (sodium acetate-acetic acid). The results were expressed as mg of cyanidin-3-glycoside equivalents/100 g of fresh fruit weight. The TAC was calculated using the following equation:

$$A = [(A_{530} - A_{700})_{pH\ 1.0} - (A_{530} - A_{700})_{pH\ 4.5}] \tag{1}$$

$$TAC\ (mg/100\ g) = A \times MW \times DF \times \frac{1}{\mathit{"} \times L} \times \frac{V}{M} \times 100 \tag{2}$$

where MW is the molecular weight of cyanidin-3-glucoside (449.2 g/mol), DF is the dilution factor, $\varepsilon$ is the molar extinction coefficient of cyanidin-3-glucoside (26,900 $L \cdot cm^{-1} \cdot mol^{-1}$), $L$ is the cell-path length (1 cm), $V$ is the extract volume (mL) and $M$ is the fresh fruit weight (g). The data were expressed as the mean values $\pm$ SD ($n = 3$).

### 3.6. HPLC-DAD Analysis

The anthocyanins were separated using an analytical HPLC system (Agilent 1290) equipped with a G4220A 1290 Bin Pump, a G4226A 1290 Sampler, a G1316C 1290 TCC and a G4212A 1290 DAD. The analytical column used was an Agilent ZORBAX Eclipse XDB C18 column (4.6 × 150 mm, 5 μm, Agilent). The PAE (10 mg/mL) was filtered through a 0.45-μm reinforced nylon membrane filter before injection. An aliquot of 5 μL of solution was injected. The anthocyanin chromatograms were obtained in the visible spectral region (530 nm), and the spectroscopic data from 200 to 600 nm were recorded throughout the entire run.

A gradient program was applied for analysis of the anthocyanins. This program followed a previously reported technique, with a slight modification [4]. The mobile phases were as follows: A, a 3% formic-acid aqueous solution and B, 15% methanol in acetonitrile. The applied gradient conditions were as follows: 0–40 min, linear gradient from 3% to 11.5% B; 40–50 min, 11.5% B; 50–60 min, linear gradient from 11.5% to 13.5% B; 60–70 min, linear gradient from 13.5% to 15.5% B; 70–85 min, linear gradient from 15.5% to 23% B; and 85–90 min, linear gradient from 23% to 3% B. The flow rate was 0.8 mL/min, and temperature was 35 °C. The peaks of the anthocyanins from *B. heteropoda* Schrenk fruit were numbered in order of their elution.

### 3.7. HPLC-DAD-HR-ESI-MS/MS

HPLC-DAD-HR-ESI-MS/MS was performed using an Agilent 6520 Accurate-Mass Q-TOF LC/MS. The chromatographic separation was conducted using an Agilent ZORBAX Eclipse XDB C18 column (4.6 × 150 mm, 5 μm, Agilent, Wilmintton, DE, USA). The separation conditions were the same as those used for HPLC/DAD analysis, described above. The concentration of the PAE was 1.0 mg/mL. The MS parameters were as follows: capillary voltage, 4000 V; gas ($N_2$) temperature, 350 °C; flow rate, 8 L/min; and nebulizer pressure, 35 psi. The instrument was operated in the positive-ion mode with scanning from $m/z$ 0 to 1500.

### 3.8. Antioxidant Activity

3.8.1. Total Reducing Capacity Assay

The total reducing capacity of the PAE was evaluated as described by Cui *et al.*, with a slight modification [22]. For this analysis, the PAE was dissolved in distilled water for the preparation of solutions of various concentrations (90, 115, 140, 165, and 190 μg/mL). One milliliter of a sample solution was mixed with 2.5 mL of sodium phosphate buffer (0.2 M, pH 6.6) and 2.5 mL of 1% (w/v) potassium ferricyanide. The mixtures were incubated at 50 °C for 20 min in the dark and then was centrifuged at 1000 g for 10 min after adding 2.5 mL of 10% trichloroacetic acid. The supernatants (2.5 mL) were collected and mixed in a test tube with 2.5 mL of distilled water and 1.0 mL of 0.1% (w/v) ferric chloride. After a 10-min incubation at room temperature in the dark, the absorbance of the resulting solution at 700 nm was recorded using a spectrophotometer. An equivalent volume of distilled water, instead of the sample, was used as a control. The reducing power of ascorbic acid (30, 38, 46, 54, 62 μg/mL) was also determined for comparison. The increased absorbance of the reaction mixture at 700 nm indicated an increased reducing capacity. The concentration of the test sample that was required to raise the level of absorbance to 0.5 at 700 nm was calculated.

### 3.8.2. DPPH Radical-Scavenging Activity Assay

A modified method was employed to evaluate the DPPH radical-scavenging activity [24]. Briefly, 2.0 mL of an ethanolic solution of PAE at various concentrations (5, 10, 15, 20, 25, 30, 60 and 90 µg/mL) was added to 2.0 mL of 0.2 mM DPPH that was dissolved in ethanol. The mixture was then shaken vigorously and was maintained for 30 min at room temperature in the dark. The absorbance of the mixed solution at 517 nm was determined. Ascorbic acid (vitamin C) was used as the reference compound. All of the samples were analyzed in triplicate. The scavenging activity of each sample was calculated according to the following formula:

$$\text{Scavenging activity} (\%) = [1 - (A_{\text{DPPH sample}} - A_{\text{sample control}})/A_{\text{DPPH blank}}] \times 100 \qquad (3)$$

where $A_{\text{DPPH sample}}$ = absorbance of 2 mL of the sample solution + 2 mL of DPPH solution; $A_{\text{sample control}}$ = absorbance of 2 mL of the sample solution + 2 mL of ethanol; and $A_{\text{DPPH blank}}$ = absorbance of 2 mL of ethanol + 2 mL of DPPH solution. The concentration of sample required to cause 50% inhibition ($IC_{50}$ value) was determined.

### 3.8.3. Ferric-Reducing Antioxidant Power (FRAP) Assay

The FRAP of the PAE was determined using a total antioxidant-capacity assay kit following the FRAP method (Beyotime Institute of Biotechnology, Haimen, China), using a previous report as a reference [25]. The stock solutions included a TPTZ (2,4,6-tripyridyl-s-triazine) solution, a TPTZ diluent, the detection buffer, 1.0 mL of a 100 mM $FeSO_4$ solution and 0.1 mL of a 10 mM Trolox solution. A working solution was freshly prepared by mixing the TPTZ diluent, the TPTZ solution and the detection buffer in a 10:1:1 (v/v) ratio, respectively. The working solution was maintained at 37 °C before use. An aliquot of 5 µL of the PAE solution was allowed to react with 180 µL of the FRAP working solution for 3–5 min at 37 °C. Then, the absorbance of the mixture at 593 nm was measured using a multimode reader. Five microliters of distilled water, instead of a sample, was used for the blank. The standard curve was obtained using $FeSO_4$ in the concentration range of 0.15–1.5 mM. The results were expressed as $FeSO_4$ values, which were calculated using on the standard curve.

### 3.8.4. ABTS+ Radical-Scavenging Capacity Assay

The ABTS+ radical-scavenging capacity was determined according to the instruction of the Beyotime Institute of Biotechnology and methods described previously [21]. This capacity is based on the ability of different substances to scavenge the ABTS+ radical in comparison with that of a standard (Trolox). The stock solutions included an ABTS solution and an oxidant solution. The green ABTS+

solution was prepared by reacting ABTS with an equal quantity of an oxidant in and allowing the mixture to stand for 12–16 h at room temperature in the dark before use. The resulting solution was diluted with 80% ethanol to obtain an absorbance of $0.70 \pm 0.05$ at 734 nm. The solution was prepared freshly for each assay. An aliquot of 10 µL of the PAE solutions was mixed with 200 µL of the diluted $ABTS^+$ solution and maintained for 2–6 min at room temperature in the dark. Then, the absorbance of the mixture at 734 nm was recorded. A calibration curve was prepared using Trolox (a water-soluble analogue of vitamin E) in a range of concentrations from 0.15 to 1.5 mM.

To ensure a correct measurement, the test sample was diluted with water to the concentration at which the percentage of inhibition was 20%–80%. The result was calculated based on the Trolox standard curve and was expressed as the Trolox-equivalent antioxidant capacity (TEAC), which was defined as the mmol of Trolox for which the antioxidant activity was equivalent to the activity of 1 g of the sample.

*3.9. Statistical Analysis*

All of the assays of the antioxidant capacities (total reducing capacity, DPPH radical-scavenging activity, FRAP and ABTS radical-scavenging capacity) were performed in triplicate. The analytical data were expressed as the mean values from assays conducted in triplicates and the standard deviation (SD).

## 4. Conclusions

Our research has provided the chemical basis for and evidence of high levels of antioxidant activity of the fruit of *B. heteropoda* Schrenk. The high anthocyanins content indicated that the fruit of *B. heteropoda* Schrenk can be considered as an excellent source of natural colorants and a functional food that benefits human health However, we have only taken a glimpse into the antioxidant activity of the fruit of *B. heteropoda* Schrenk. Further investigations must be conducted to elucidate its other biological effects. In conclusion, the composition of the major anthocyanins and the antioxidant activities of the fruit of *B. heteropoda* Schrenk were systematically investigated for the first time. The results of this research are important for the further development of applications of the fruit of *B. heteropoda* Schrenk.

**Supplementary Materials:** Supplementary materials can be accessed at: http://www.mdpi.com/1420-3049/19/11/19078/s1.

**Acknowledgments:** The authors gratefully acknowledge Jianbei Li for his help with the HPLC-HR-ESI-MS/MS analysis.

**Author Contributions:** Tengfei Ji conceived and designed the experiments; Lili Sun performed the experiments and analyzed the data; Wan Gao, Mengmeng Zhang, Cheng Li, Aiguo Wang and Yalun Su contributed materials and analysis tools; Lili Sun and Tengfei Ji wrote the paper.

**Conflicts of Interest:** The authors declare no conflict of interest.

## References

1.  Xu, X., Baharguli, H., Eds.; *Kazak Medicine Blog*, 1st ed.; The Ethnic Publishing House: Beijing, China, 2009; Volume 1, p. 156.
2.  Teng, Y.; Zhang, G.Q.; Peng, Z.M. Research on *Berberis heteropoda* Schrenk Red Pigment Extraction and Its Stability. *Food Sci.* **2007**, *28*, 67–70.
3.  Ministry of Health of Forces Logistics of Xinjiang. *Handbook of Chinese Herbal of Xinjiang*; People's Publishing House: Xinjiang, China, 1970; pp. 153–283.
4.  Zheng, J.; Ding, C.X.; Wang, L.S.; Li, G.L.; Shi, J.Y.; Li, H.; Wang, H.L.; Suo, Y.R. Anthocyanins composition and antioxidant activity of wild *Lycium ruthenicum* Murr. from Qinghai-Tibet Plateau. *Food Chem.* **2011**, *126*, 859–865.
5.  Liliana, S.; José, G.C.; Ovidio, A.; Coralia, O. Anthocyanin Composition of Wild Colombian Fruits and Antioxidant Capacity Measurement by Electron Paramagnetic Resonance Spectroscopy. *J. Agric. Food Chem.* **2012**, *60*, 1397–1404.
6.  Clifford, M. Anthocyanins: Nature, occurrence and dietary burden. *J. Sci. Food Agric.* **2000**, *80*, 1063–1072.
7.  Cooney, J.M.; Dwayne, J.J.; McGhie, T. LC-MS identification of anthocyanins in boysenberry extract and anthocyanin metabolites in human urine following dosing. *J. Sci. Food Agric.* **2004**, *84*, 237–245.
8.  Sari, P.; Wijaya, C.H.; Sajuthi, D.; Supratman, U. Colour properties, stability, and free radical scavenging activity of jambolan (*Syzygium cumini*) fruit anthocyanins in a beverage model system: Natural and copigmented anthocyanins. *Food Chem.* **2012**, *132*, 1908–1914.
9.  Meltem, T.; Seref, T.; Ufuk, D.; Mehmet, O. Effects of various pressing programs and yields on the antioxidant activity, antimicrobial activity, phenolic content and colour of pomegranate juices. *Food Chem.* **2013**, *138*, 1810–1818.
10. Rice-Evans, A.C.; Miller, N.J.; Paganga, G. Structure-antioxidant activity relationships of flavonoids and phenolic acids. *Free Radic. Biol. Med.* **1996**, *20*, 933–956.
11. Chirinos, R.; Campos, D.; Betalleluz, I.; Giusti, M.M.; Schwartz, S.J.; Tian, Q.; Pedreschi, R.; Larondelle, Y. High-performance liquid chromatography with photodiode array detection (HPLC-DAD)/HPLC-Mass Spectrometry (MS) profiling of anthocyanins from Andean Mashua Tubers (Tropaeolum tuberosum Ruiz and Pavon) and their contribution to the overall antioxidant activity. *J. Agric. Food Chem.* **2006**, *54*, 7089–7097.
12. Sun, J.H.; Lin, L.Z.; Chen, P. Recent Applications for HPLC-MS Analysis of Anthocyanins in Food Materials. *Curr. Anal. Chem.* **2013**, *9*, 397.
13. Carla, D.C.J.; Carls, F.; He, J.; Tian, Q.Q.; Steven, J.S.; Giusti, M.M. Characterisation and preliminary bioactivity determination of *Berberis boliviana* Lechler fruit anthocyanins. *Food Chem.* **2011**, *128*, 717–724.
14. Choung, M.G.; Baek, I.Y.; Kang, S.T.; Han, W.Y.; Shin, D.C.; Moon, H.P.; Kang, K.H. Isolation and determination of anthocyanins in seed coats of black soybean (*Glycine max* (L.) Merr.). *J. Agric. Food Chem.* **2001**, *49*, 5848–5851.

15. Lee, J.H.; Cho, K.M. Changes occurring in compositional components of black soybeans maintained at room temperature for different storage periods. *Food Chem.* **2012**, *131*, 161–169.

16. Lee, J.H.; Choung, M.G. Identification and characterisation of anthocyanins in the antioxidant activity-containing fraction of *Liriope platyphylla* fruits. *Food Chem.* **2011**, *127*, 1686–1693.

17. Lee, J.H.; Kang, N.S.; Shin, S.O.; Shin, S.H.; Lim, S.G.; Suh, D.Y.; Baek, I.Y.; Park, K.Y.; Ha, T.J. Characterisation of anthocyanins in the black soybean (*Glycine max* L.) by HPLC–DAD–ESI/MS analysis. *Food Chem.* **2009**, *112*, 226–231.

18. Fossen, T.; Slimestad, R.; Øvstedal, D.O.; Andersen, Ø.M. Anthocyanins of grasses. *Biochem. Syst. Ecol.* **2002**, *30*, 855–864.

19. Alcalde-Eon, C.; Escribano-Bailón, M.T.; Santos-Buelga, C.; Rivas-Gonzalo, J.C. Changes in the detailed pigment composition of red wine during maturity and ageing. A comprehensive study. *Anal. Chim. Acta* **2006**, *563*, 238–254.

20. Leong, L.P.; Shui, G. An investigation of antioxidant capacity of fruits in Singapore markets. *Food Chem.* **2002**, *76*, 69–75.

21. Kong, Y.; Wei, Z.F.; Fu, Y.J.; Gu, C.B.; Zhao, C.J.; Yao, X.H.; Thomas, E. Negative-pressure cavitation extraction of cajaninstilbene acid and pinostrobin from pigeon pea [*Cajanus cajan* (L.) Millsp.] leaves and evaluation of antioxidant Activity. *Food Chem.* **2011**, *128*, 596–605.

22. Cui, C.; Zhang, S.M.; You, L.J.; Ren, J.Y.; Luo, W.; Chen, W.F.; Zhao, M.M. Antioxidant capacity of anthocyanins from *Rhodomyrtus tomentosa* (Ait.) and identification of the major anthocyanins. *Food Chem.* **2013**, *139*, 1–8.

23. Meng, J.F.; Fang, Y.L.; Qin, M.Y.; Zhuang, X.F.; Zhang, Z.W. Varietal differences among the phenolic profiles and antioxidant properties of four cultivars of spine grape (*Vitis davidii* Foex) in Chongyi County (China). *Food Chem.* **2012**, *134*, 2049–2056.

24. Jing, L.H.; Shen, X.J.; Toshihiko, S.; Tomomasa, K.; Zhou, J.C.; Zhao, L.M. Characterization and activity of anthocyanins in Zijuan Tea (*Camellia sinensis* var. *kitamura*). *J. Agric. Food Chem.* **2013**, *61*, 3306–3310.

25. Luo, J.G.; Li, L.; Kong, L.Y. Preparative separation of phenylpropenoid glycerides from the bulbs of *Lilium lancifolium* by high-speed counter-current chromatography and evaluation of their antioxidant activities. *Food Chem.* **2012**, *131*, 1056–1062.

**Sample Availability:** *Sample Availability:* Samples of the compounds 1 and 2 are available from the authors.

# Obtaining Ready-to-Eat Blue Corn Expanded Snacks with Anthocyanins Using an Extrusion Process and Response Surface Methodology

Anayansi Escalante-Aburto, Benjamín Ramírez-Wong,
Patricia Isabel Torres-Chávez, Jaime López-Cervantes,
Juan de Dios Figueroa-Cárdenas, Jesús Manuel Barrón-Hoyos,
Ignacio Morales-Rosas, Néstor Ponce-García and Roberto Gutiérrez-Dorado

**Abstract:** Extrusion is an alternative technology for the production of nixtamalized products. The aim of this study was to obtain an expanded nixtamalized snack with whole blue corn and using the extrusion process, to preserve the highest possible total anthocyanin content, intense blue/purple coloration (color $b$) and the highest expansion index. A central composite experimental design was used. The extrusion process factors were: feed moisture (FM, 15%–23%), calcium hydroxide concentration (CHC, 0%–0.25%) and final extruder temperature (T, 110–150 °C). The chemical and physical properties evaluated in the extrudates were moisture content (MC, %), total anthocyanins (TA, mg·kg$^{-1}$), pH, color ($L$, $a$, $b$) and expansion index (EI). ANOVA and surface response methodology were applied to evaluate the effects of the extrusion factors. FM and T significantly affected the response variables. An optimization step was performed by overlaying three contour plots to predict the best combination region. The extrudates were obtained under the following optimum factors: FM (%) = 16.94, CHC (%) = 0.095 and T (°C) = 141.89. The predicted extrusion processing factors were highly accurate, yielding an expanded nixtamalized snack with 158.87 mg·kg$^{-1}$ TA (estimated: 160 mg·kg$^{-1}$), an EI of 3.19 (estimated: 2.66), and color parameter $b$ of $-0.44$ (estimated: 0.10).

Reprinted from *Molecules*. Cite as: Escalante-Aburto, A.; Ramírez-Wong, B.; Torres-Chávez, P.I.; López-Cervantes, J.; de Dios Figueroa-Cárdenas, J.; Barrón-Hoyos, J.M.; Morales-Rosas, I.; Ponce-García, N.; Gutiérrez-Dorado, R. Obtaining Ready-to-Eat Blue Corn Expanded Snacks with Anthocyanins Using an Extrusion Process and Response Surface Methodology. *Molecules* **2014**, *19*, 21066–21084.

## 1. Introduction

The energy density of ingested foods has increased globally; populations have become more urban, and the high consumption of products with large amounts of sugar, carbohydrates, dyes and saturated fats has been reported [1]. There is also an increase in the worldwide consumption of cereal-based foods, especially snack

58

products. This market is expanding rapidly and will continue growing in the coming years. Snack foods are considered high energy density products, and they are directly related to promoting weight gain and to causing certain illnesses such as obesity and other related diseases (metabolic syndrome, cardiovascular events, hypertension, cancer) [1].

Nixtamalized corn snacks are consumed mostly by the American and Latin-American population. On the other hand, the extrusion process (EP) has been used as a rapid and efficient technology to obtain a large variety of products, including nixtamalized products. The EP has some advantages compared to the traditional nixtamalization process, including requiring less time and energy input and no production of water effluents (nejayote).

Producing extruded snacks with whole grains, such as corn, improves the nutritional quality of the final product. The germ of the grain is composed of polyunsaturated acids, the pericarp contains dietary fiber, and phytochemicals (bioactive ingredients) such as anthocyanins are present in the outer part of the endosperm (aleurone layer). These compounds have been isolated from pigmented corn grains and have shown health benefits such as anti-radical activity, which plays an important role in preventing chronic-degenerative illnesses [2].

The chemical structure of anthocyanins is based on the aglycon molecule (the flavylium ion or 2-phenylbenzopirilium) with various substitutions of methoxyl and hydroxyl groups in different positions. Generally, these compounds are linked to one or more glycosidic molecules (hexoses and pentoses) and they also can be joined to different organic acids (cinnamic and aliphatic acids). Nevertheless, anthocyanins are unstable compounds at high pH levels and temperatures [3].

Anthocyanins are soluble compounds responsible for the blue/purple coloration in pigmented corns. As the chemical forms of anhocyanins change, their color also changes. The modification of certain process conditions such as pH and temperature, can produce the formation of a bluish quinoidal base or a colorless carbinol pseudobase [4]. These aspects affect the color of the products made with pigmented corns, modifying their appearance and acceptance.

Several investigations have focused on products obtained from pigmented corns produced by traditional, ecological and extrusion nixtamalization processes [5–10]. Nixtamalization by extrusion diminishes the anthocyanin losses in the end products by up to 50%–60% (tortillas and expanded extrudates), and ecological nixtamalization reduced the pH levels and the formation of compounds with proteins and carbohydrates that positively affected the retention of nutraceutical compounds [11]. All of those investigations are focused in improving the anthocyanins retention, in order to develop a healthy product with acceptable sensorial characteristics, which is the goal of food technologists.

It has been established that there are advantages to using whole grains with additional nutraceutical value (phytochemicals) to diminish the prevalence of chronic degenerative diseases. Innovations in snack products that already are on the market, lead to finding new ways of processing that result in improved products with low amounts of carbohydrates, dyes, saturated fats and low energy density.

The aim of this study was to produce expanded nixtamalized expanded snacks using extrusion process with whole blue corn, and to apply response surface methodology to obtain a product with a high total anthocyanin content, an intense purple/blue coloration, and a high expansion index.

## 2. Results and Discussion

### 2.1. Effects of FM, CHC and T on Extrudates Moisture Content (MC)

The second order equations coefficients, analysis of variance (ANOVA) and determination coefficients of feed moisture, final extruder temperature and calcium hydroxide concentration effect on the chemical and physical properties of the extrudates, are presented in Table 1.

The ANOVA showed that the FM was the processing variable that most affected the MC in linear terms ($p < 0.0001$). The quadratic terms $(FM)^2$ and $(T)^2$ presented a significant and a very significant effect on MC ($p < 0.0152$ and $p < 0.0001$, respectively).

The model fitting for MC in terms of the actual factors is presented in Equation (1):

$$Y_{MC} = -160.23 + 3.57(FM) + 2.07(T) + 0.009\,(FM)(T) - 0.11(FM)^2 - 0.008(T)^2 \quad (1)$$

The extrudates MC varied from 8.1% to 14.7% (Table 2). Figure 1a shows that higher values of FM yielded the highest MC in the extrudates at a T of 130 °C. In Figure 1b, the interaction of FM*CHC shows that at low FM content, the CHC had no effects the on extrudate MC. The interaction effect of T*CHC is presented in Figure 1c, where T showed an interesting effect; at 130 °C, the extrudates reached the highest MC regarding the CHC level.

According to the results presented in Table 2, the extrudates produced at higher FM and T had greater losses of moisture content. The evaporation of water when the product emerged from the die provoked proportional losses of moisture from 11% up to 43% when compared with the original FM. It has been reported that after the extrudate is released from the die and reaches the maximum expansion, the product starts to contract under elastic recoil. The sudden drop of temperature diminishes the viscosity, causing the evaporation of 8–10 g of moisture per 100 g of fluid [12].

Table 1. Coefficients of the second order equations (prediction models), analysis of variance and determination coefficients, showing the relationship among the processing factors and the chemical and physical properties of the extrudates.

| Coefficients | MC [a] | TA | pH | $L$ | $a$ | $b$ | EI |
|---|---|---|---|---|---|---|---|
| Intercept | | | | | | | |
| $\beta$ | 13.45 | 147.63 | 6.85 | 21.61 | 3.87 | −0.27 | 2.13 |
| Lineal | | | | | | | |
| $\beta_1$ | 1.51 *** | 5.55ns | 0.021ns | −2.54 *** | −0.35 *** | −0.23 *** | −0.45 *** |
| $\beta_2$ | −0.33ns | 1.93ns | −0.0009ns | 0.46ns | 0.18 *** | −0.016ns | 0.002ns |
| $\beta_3$ | 0.050ns | 0.39ns | −0.013ns | 0.29ns | 0.011ns | −0.004ns | −0.015ns |
| Quadratic | | | | | | | |
| $\beta_{11}$ | −0.66 ** | 2.86ns | −0.008ns | 1.11 *** | 0.17 *** | 0.13 ** | 0.11 ** |
| $\beta_{22}$ | −1.29 *** | 7.39 * | −0.035ns | 0.31ns | 0.23 *** | 0.030ns | −0.009ns |
| $\beta_{33}$ | −0.40ns | 2.05ns | −0.001ns | 0.38ns | −0.0007ns | 0.063ns | 0.001ns |
| Interaction | | | | | | | |
| $\beta_{12}$ | 0.27ns | −4.94ns | −0.006ns | −0.62ns | −0.15 ** | −0.005ns | −0.005ns |
| $\beta_{13}$ | 0.25ns | 3.26ns | 0.0003ns | −0.016ns | −0.16 ** | 0.020ns | −0.035ns |
| $\beta_{23}$ | −0.005ns | −1.66ns | 0.020ns | 0.53ns | 0.033ns | 0.010ns | 0.021ns |
| $R^2$ | 0.89 | 0.44 | 0.17 | 0.89 | 0.92 | 0.72 | 0.95 |

[a] MC = moisture content; TA = total anthocyanins; $L$ = white (100) to black (0), $a$ = red (+) to green (−), $b$ = yellow (+) to blue (−), EI = Expansion index; $\beta_1$, feed moisture; $\beta_2$, final extruder temperature; $\beta_3$, calcium hydroxide concentration. ns = not significant ($p > 0.1$); * $p < 0.1$; ** Significant ($p < 0.05$); *** Very significant ($p < 0.01$).

Table 2. Experiment design used to obtain different combinations of extrusion feed moisture/calcium hydroxide concentration/temperature for production of expanded nixtamalized blue corn extrudates.

| Tr [c] | Process Factors [a] | | | Response Variables [b] | | | | | | |
|---|---|---|---|---|---|---|---|---|---|---|
| | FM | T | CHC | MC | TA | pH | $L$ | $a$ | $b$ | EI |
| | $X_1$ | $X_2$ | $X_3$ | $Y_1$ | $Y_2$ | $Y_3$ | $Y_4$ | $Y_5$ | $Y_6$ | $Y_7$ |
| 1 | 19 (0) [d] | 130 (0) | 0.13 (0) | 13.6 | 150.7 | 6.8 | 22.43 | 4.01 | −0.33 | 2.13 |
| 2 | 15 (−1.682) | 130 (0) | 0.13 (0) | 10.0 | 141.1 | 6.8 | 28.89 | 5.01 | 0.35 | 3.26 |
| 3 | 19 (0) | 130 (0) | 0.13 (0) | 14.0 | 154.7 | 6.8 | 21.4 | 3.83 | −0.35 | 2.05 |
| 4 | 19 (0) | 130 (0) | 0.13 (0) | 13.3 | 130.4 | 6.8 | 22.38 | 4.00 | −0.04 | 2.19 |
| 5 | 21.38 (1) | 141.89 (1) | 0.2 (1) | 12.3 | 174 | 6.9 | 21.47 | 3.69 | −0.38 | 1.78 |
| 6 | 19 (0) | 130 (0) | 0.13 (0) | 13.2 | 153.8 | 6.9 | 20.96 | 3.76 | −0.12 | 2.19 |
| 7 | 19 (0) | 130 (0) | 0.25 (1.682) | 13.1 | 143.5 | 6.7 | 22.49 | 4.17 | 0.11 | 2.02 |
| 8 | 21.38 (1) | 118.11 (−1) | 0.05 (−1) | 11.9 | 165.5 | 6.8 | 19.44 | 4.13 | −0.17 | 1.72 |
| 9 | 19 (0) | 130 (0) | 0 (−1.682) | 13.1 | 142.6 | 7.1 | 23.44 | 3.73 | −0.40 | 2.33 |
| 10 | 21.38 (1) | 141.89 (1) | 0.05 (−1) | 11.6 | 176.9 | 6.7 | 20.17 | 4.05 | −0.22 | 1.80 |
| 11 | 16.62 (−1) | 141.89 (1) | 0.05 (−1) | 8.5 | 170.8 | 6.7 | 25.72 | 4.77 | 0.32 | 2.54 |
| 12 | 16.62 (−1) | 118.11 (−1) | 0.2 (1) | 9.4 | 144.5 | 6.8 | 23.96 | 4.33 | 0.12 | 2.62 |
| 13 | 23 (1.682) | 130 (0) | 0.13 (0) | 14.7 | 149.5 | 7.0 | 21.18 | 3.83 | −0.28 | 1.71 |
| 14 | 21.38 (1) | 118.11 (−1) | 0.2 (1) | 12.4 | 183.4 | 6.8 | 20.83 | 3.58 | −0.32 | 1.76 |
| 15 | 19 (0) | 110 (−1.682) | 0.13 (0) | 11.0 | 163.3 | 6.8 | 21.21 | 4.25 | −0.23 | 2.21 |
| 16 | 16.62 (−1) | 141.89 (1) | 0.2 (1) | 8.0 | 169.1 | 6.8 | 29.28 | 4.97 | 0.14 | 2.80 |
| 17 | 19 (0) | 150 (1.682) | 0.13 (0) | 10.1 | 153 | 6.8 | 22.35 | 4.94 | −0.25 | 2.08 |
| 18 | 16.62 (−1) | 118.11 (−1) | 0.05 (−1) | 9.7 | 153.8 | 6.7 | 24.7 | 4.2 | 0.41 | 2.58 |
| 19 | 19 (0) | 130 (0) | 0.13 (0) | 13.0 | 142.3 | 6.9 | 20.73 | 3.87 | −0.44 | 2.14 |
| 20 | 19 (0) | 130 (0) | 0.13 (0) | 13.1 | 157.4 | 6.9 | 21.63 | 3.73 | −0.33 | 2.06 |

[a] FM = Feed moisture (%), T = Final extruder temperature (°C), CHC = Calcium hydroxide concentration (%); [b] MC = Moisture content (%), TA = Total anthocyanins (mg· kg$^{-1}$), $L$ = White (100) to black (0), $a$ = Red (+) to green (−), $b$ = Yellow (+) to blue (−), EI = Expansion index; [c] Tr = Treatment; [d] Numbers in parentheses corresponded to coded values.

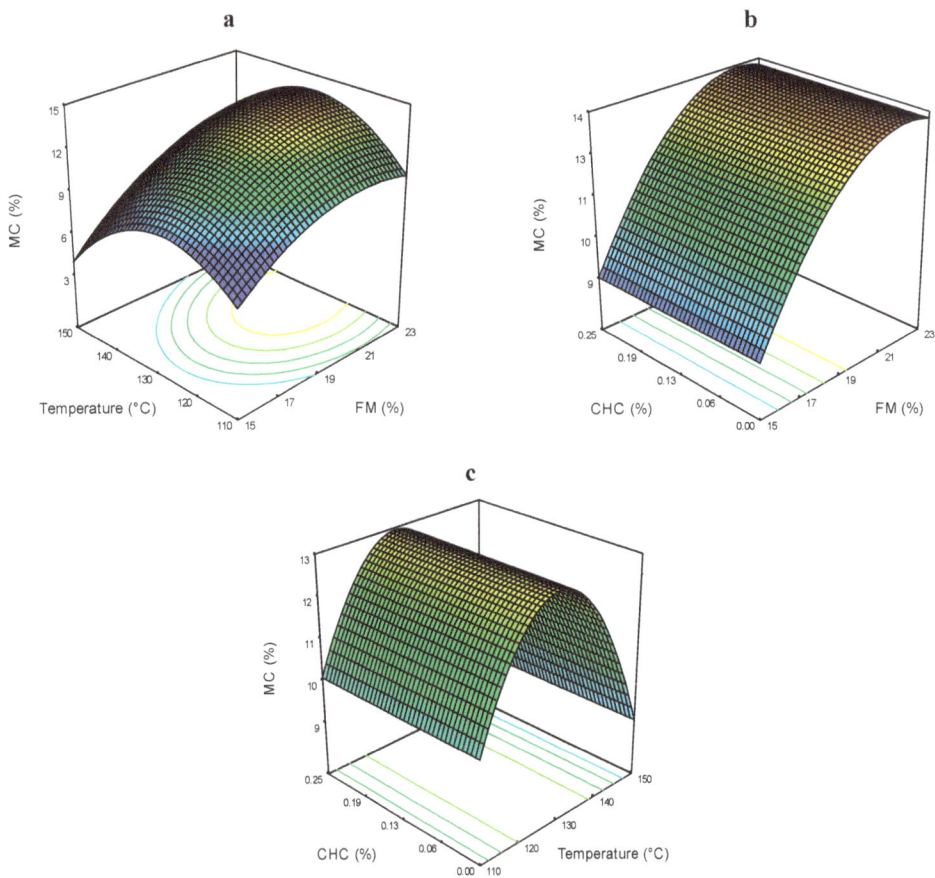

**Figure 1.** Moisture content (MC) of expanded nixtamalized blue corn extrudates, as a function of: (**a**) feed moisture (FM) and temperature (T); (**b**) feed moisture (FM) and calcium hydroxide concentration (CHC); and (**c**) temperature (T) and calcium hydroxide concentration (CHC).

## 2.2. Effects of FM, CHC and T on Total Anthocyanins (TA)

The ANOVA (Table 1) showed that the quadratic term $(T)^2$ was the processing factor that had the most significant effect ($p < 0.0459$). The model fitting for TA in terms of the actual factors is presented in Equation (2):

$$Y_{TA} = 960.61 - 12.61(T) + 0.049(T)^2 \tag{2}$$

The values obtained from this analysis ranged between 130.4 and 183.4 mg· kg$^{-1}$ (Table 2). The interaction effects of FM*T on the TA are shown in Figure 2a; at higher levels of these process factors, the TA levels were the highest. At lower FM and T levels, the TA was the lowest. These results are inconsistent with other studies where it was demonstrated that at higher temperatures, the destruction of anthocyanins increased. Temperature has a notable effect on anthocyanin structures; these compounds lose their color at elevated temperatures, and they become paler because the equilibrium among the anthocyanin species shifts towards other chemical forms [4].

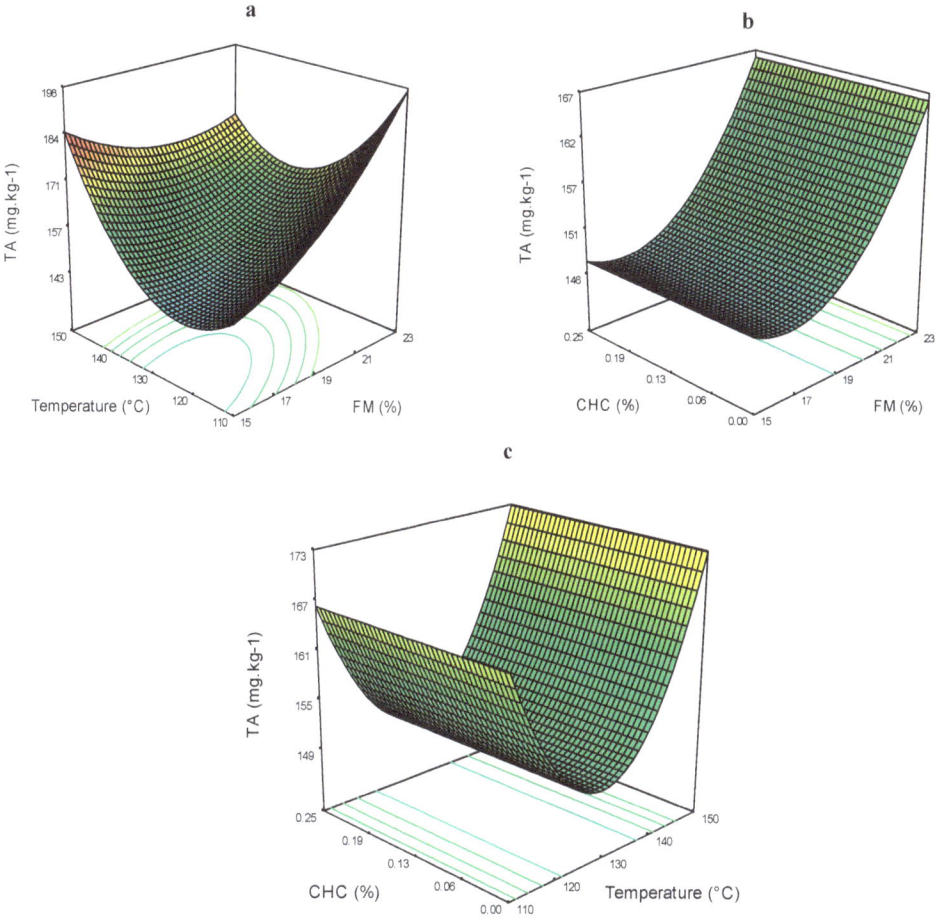

**Figure 2.** Total anthocyanins (TA) of expanded nixtamalized blue corn extrudates, as a function of: (**a**) feed moisture (FM) and temperature (T); (**b**) feed moisture (FM) and calcium hydroxide concentration (CHC); and (**c**) temperature (T) and calcium hydroxide concentration (CHC).

However, the results obtained in our research are consistent with those from a previous study [7], where it was demonstrated that even when the total anthocyanin content decreased due to higher processing temperatures, there was an increment of 11.3% in the cyanidin 3-glucoside (major anthocyanin in blue corn) content. Other authors [13] showed that purified anthocyanins degraded at a faster rate than anthocyanins in unpurified extracts obtained from the food material. These findings lead to the presumption that some anatomical parts of the grain (like the pericarp) could be acting as protectors of the anthocyanins, reducing their degradation by high processing temperatures.

In Figure 2b, it can be observed that there was no effect of CHC on AT content; rather FM had a greater effect such that at high levels of this process factor, the TA was slightly higher at any given CHC level. The interaction effect of T*CHC is presented in Figure 2c, where TA contents were higher at temperatures of 141.89 and 118.11 °C.

Nevertheless, it has been reported that many factors affect the stability of these compounds, such as extraction procedures and glycosylated substituents, which could affect the total anthocyanin content in the extrudates samples [14].

## 2.3. Effects of FM, CHC and T on pH

There were no significant effects of the extrusion processing factors on pH values (Table 1). This was probably due to the minimum calcium hydroxide concentrations (range 0%–0.25%) used in the treatments to obtain the extrudates. These results differ from those obtained by Zazueta et al. [15], who reported that the CHC had an effect on physical and chemical properties of blue maize extrudates elaborated by extrusion. In the case of the pH, alkaline media induced starch swelling and gelatinization, which expose reactive sites of the starch. There are formation of starch-calcium complexes that changes conformational and structural characteristics of the starch based foods. The pH ranged from 6.7 to 6.1 in all the treatments (Table 2). It can be seen that at the CHC used in this study, the anthocyanin content could not be affected by pH, allowing the manipulation of other extrusion parameters.

## 2.4. Effects of FM, CHC and T on Color Parameters (L, a, b)

Luminosity (L) was evaluated on a scale of 100 (white) to 0 (black). Extrudates with lower values of L appeared darker (purple) to the unaided eye. The ANOVA (Table 1) showed that the FM affected the color parameter L very significantly in linear ($p < 0.0001$) and quadratic ($p < 0.0040$) terms. The values of this parameter ranged between 19.44 and 28.89 (Table 2). Figure 3a,b show that at low FM levels and high CHC and T, the L parameter reached the highest values. The interaction T*CHC had no significant effect on L values (Figure 3c).

The model fitting for L in terms of the actual factors is presented in Equation (3):

$$Y_L = 109.27 - 8.10(FM) + 0.185(FM)^2 \tag{3}$$

When anthocyanins are exposed to higher temperatures, some degradation reactions occur, forming colorless structures that fade the original red-blue coloration. It has been reported that in corn extrudates the parameter $L$ was highly affected by the FM of the extrusion process. It can be seen that at higher processing temperatures, the lightness ($L$) increased in the blue corn extrudates (Figure 3a,c). Some authors [16] have reported an increase in the lightness of anthocyanin extracts containing cyanidin and pelargonidin with mono- and diglucoside moieties. This effect was attributed to the transition of the colored flavylium cation into colorless and yellowish carbinol and chalcone forms, respectively. It is possible that the same anthocyanin degradation mechanism occurred in the extrudate samples obtained in this study.

The positive and negative values of the color parameter $a$ indicate red and green shades, respectively. The ANOVA (Table 1) showed that the linear terms of FM ($p < 0.0001$) and T ($p < 0.0012$) presented a highly significant effect. The quadratic terms of these processing variables also had very significant effects: $FM^2$ ($p < 0.0013$) and $T^2$ ($p < 0.0001$). The interactions FM*T and FM*CHC showed significant effects ($p < 0.0191$ and $p < 0.0140$, respectively).

The model fitting of $a$ in terms of the actual factors is presented in Equation (4):

$$Ya = 27.92 - 0.48(\text{FM}) - 0.30(\text{T}) + 16.96(\text{CHC}) - 0.0052(\text{FM})(\text{T}) - 0.88(\text{FM})(\text{CHC}) + 0.029(\text{FM})^2 + 0.0016(\text{T})^2 \tag{4}$$

The color parameter $a$ of the extrudates ranged between 3.58 and 5.01 (Table 2). The effects of the extrusion processing conditions are presented in Figure 4a–c. It can be seen that the three processing factors affected this evaluation. The FM*T interaction shows that at low levels of FM and high T, the color parameter $a$, is the highest (Figure 4a). In Figure 4b, the FM*CHC interaction, shows that as the FM diminishes and CHC increases, the values for color parameter $a$ are higher. The stability of anthocyanins is highly dependent on pH at levels of 6–7 or higher. The coloration of these compounds changes to blue-purplish and the quinoidal forms are degraded rapidly by the air oxidation, thereby decreasing the positive $a$ values.

Figure 4c shows the effects of T*CHC on the color parameter $a$, and it can be seen that at higher T, the $a$ values increases by increasing CHC. It is probably that the release of acylated anthocyanins increases the relative proportion of the stable red flavylium cation, thus protecting the red coloration at higher pH [3]. The effect of CHC in the color parameter $a$ of pigmented nixtamalized products was evaluated by other authors [17], concluding that extrusion process improves the retention of anthocyanins and produced more retention of blue-red colorations in the obtained products.

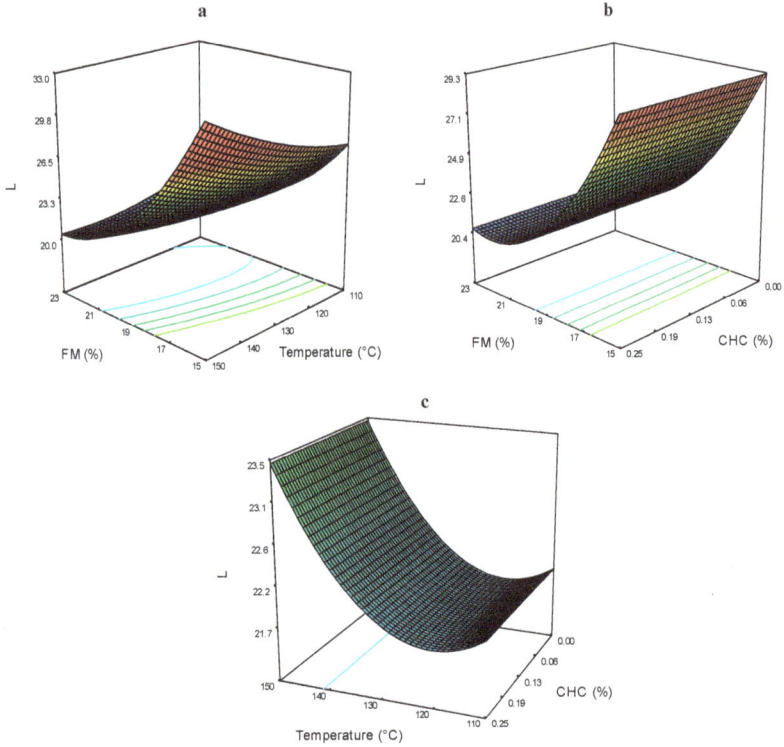

**Figure 3.** Color parameter $L$ of expanded nixtamalized blue corn extrudates, as a function of: (**a**) feed moisture (FM) and temperature (T); (**b**) feed moisture (FM) and calcium hydroxide concentration (CHC); and (**c**) temperature (T) and calcium hydroxide concentration (CHC).

In the color spectrum, positive and negative $b$ values indicate yellow and blue shades, respectively. The ANOVA (Table 1) results showed that the FM had a highly significant effect on this parameter in its linear term ($p < 0.0008$) and a significant in its quadratic term ($p < 0.0325$).

The model fitting of $b$ in terms of the actual factors is presented in Equation (5):

$$Y_b = 9.16 - 0.89(\text{FM}) + 0.0208(\text{FM})^2 \tag{5}$$

The values of $b$ obtained in the extrudates ranged from $-0.44$ to $0.41$ (Table 2). Figure 5a shows the effects of FM*T on the color parameter $b$. The FM had a significant effect: as the FM increased, the $b$ values became positive, and practically no effects of T. The effect of FM*CHC on color parameter $b$ is presented in Figure 5b, which shows the same trend as that in the FM*T interaction. As the FM decreased, the parameter $b$ increased regarding the CHC. Due to that anthocyanins are pH-dependent compounds, changes in this parameter were expected. However, the presence of one or more acyl groups in the molecule prevents the hydrolysis of the flavylium form and promotes the synthesis of quinoidal structures with blue shades, making these types of compounds more stable and less sensitive to pH changes [4]. The effects of T*CHC on color parameter $b$, are presented in Figure 5c. It can be observed that $b$ remained constant (non-significant effect) regarding the CHC and T. It can be assumed that the acylated anthocyanins contained in the blue corn extrudates, were not affected by these processing factors.

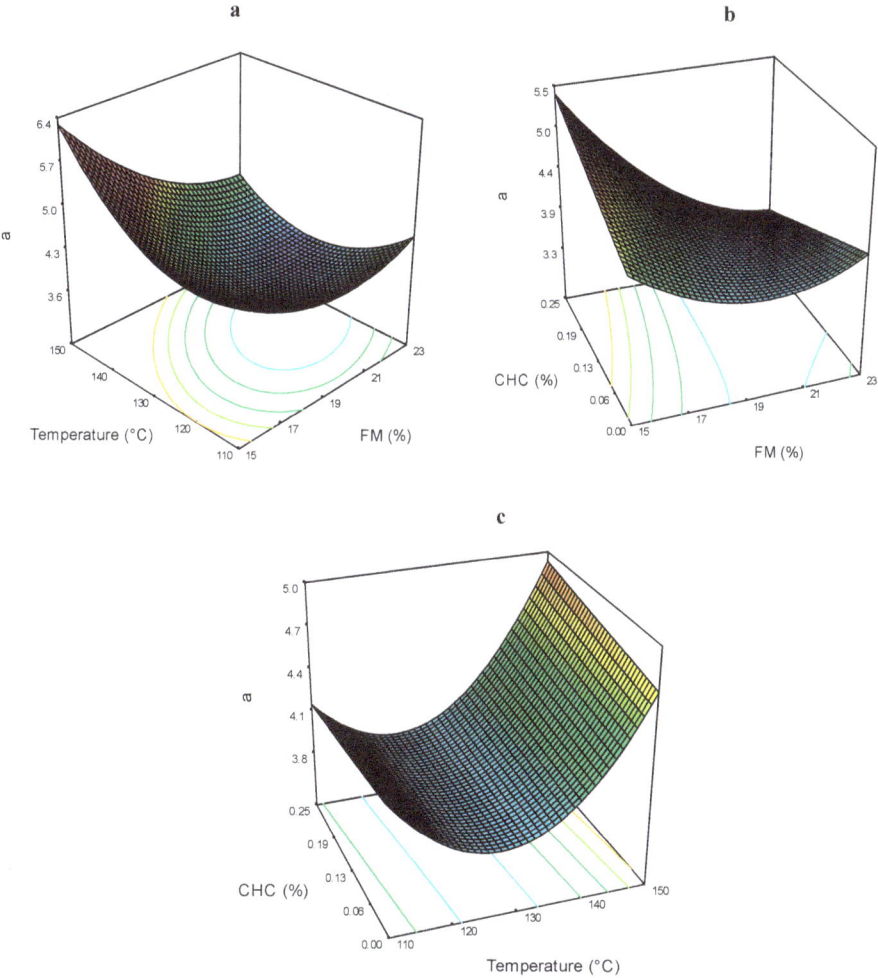

Figure 4. Color parameter $a$ of expanded nixtamalized blue corn extrudates, as a function of: (a) feed moisture (FM) and temperature (T); (b) feed moisture (FM) and calcium hydroxide concentration (CHC); and (c) temperature (T) and calcium hydroxide concentration (CHC).

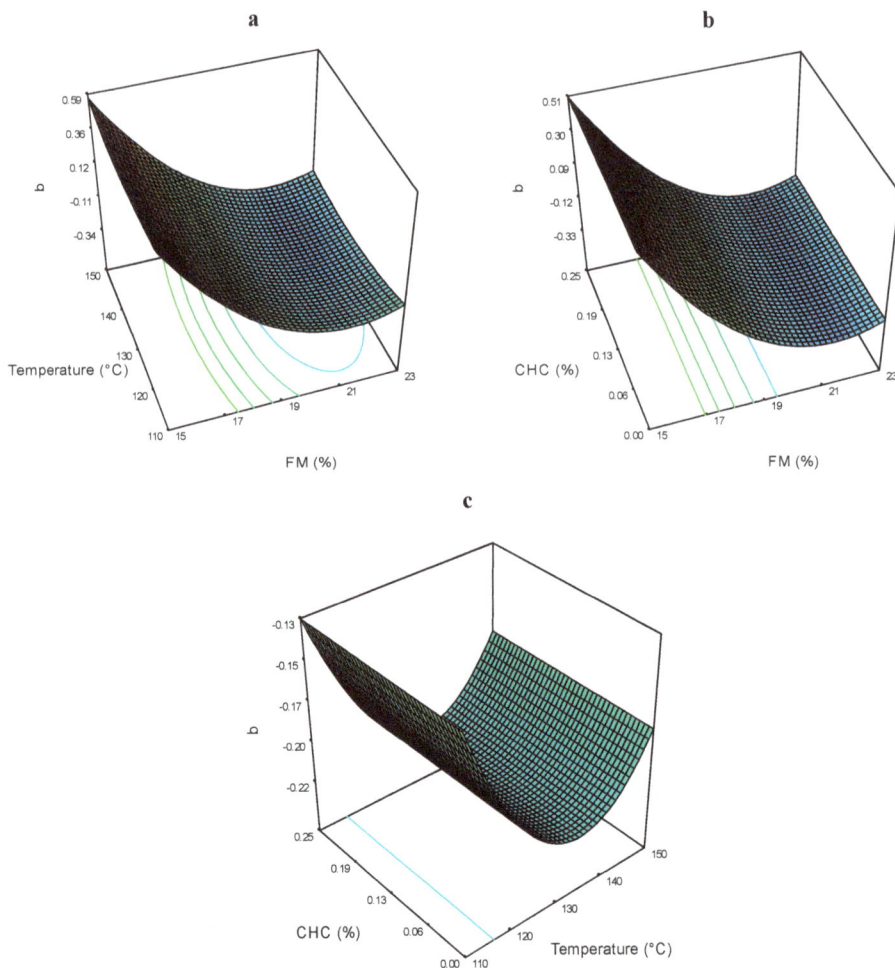

**Figure 5.** Color parameter *b* of expanded nixtamalized blue corn extrudates, as a function of: (**a**) feed moisture (FM) and temperature (T); (**b**) feed moisture (FM) and calcium hydroxide concentration (CHC); and (**c**) temperature (T) and calcium hydroxide concentration (CHC).

## 2.5. Effects of FM, CHC and T on Expansion Index (EI)

The ANOVA (Table 1) showed that the linear and quadratic terms of the FM had a very significant effect in the EI values ($p < 0.0001$ and $p < 0.0037$, respectively). The model fitting of IE in terms of the actual factors is presented in Equation (6):

$$Y_{EI} = 12.75 - 0.931(\text{FM}) + 0.019(\text{FM})^2 \qquad (6)$$

The EI of the extrudates ranged from 1.71 to 3.26 (Table 2). The effects of FM*T on the EI are presented in Figure 6a. At any T, the FM had a significant effect, demonstrating that the higher the FM, the lower the EI. The same trend was observed for FM in the FM*CHC interaction (Figure 6b), where CHC had no significant effect on the EI; in addition the T*CHC interaction had no effect on the EI (Figure 6c).

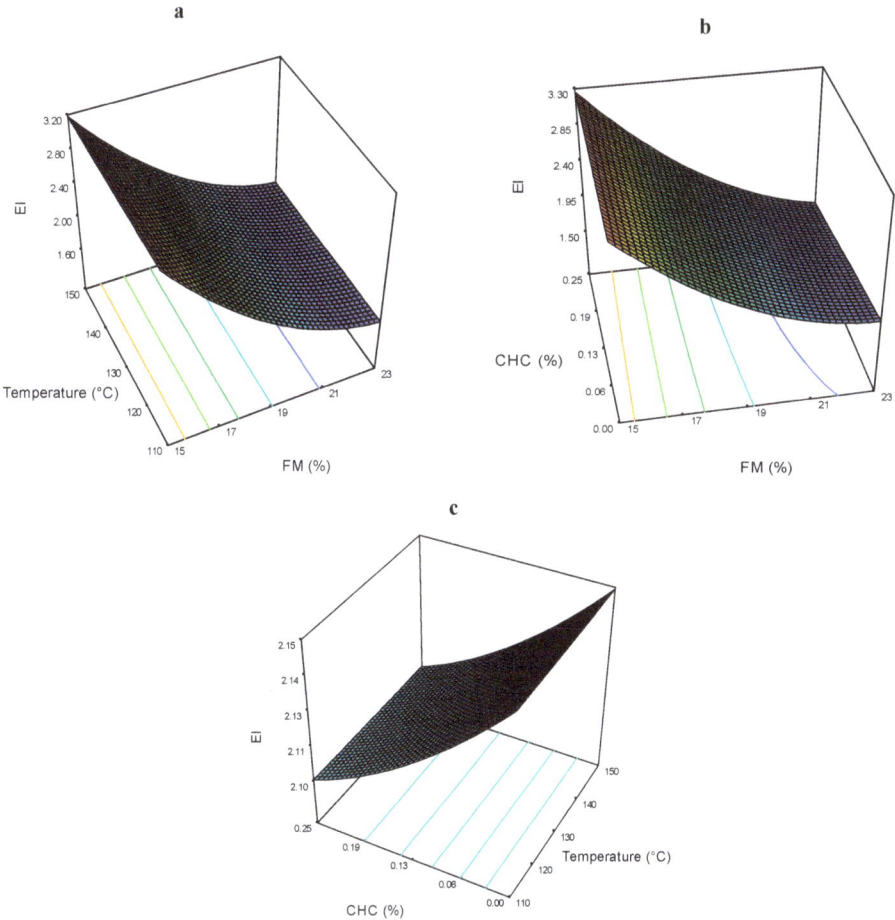

**Figure 6.** Expansion index (EI) of expanded nixtamalized blue corn extrudates, as a function of: (**a**) feed moisture (FM) and temperature (T); (**b**) feed moisture (FM) and calcium hydroxide concentration (CHC); and (**c**) temperature (T) and calcium hydroxide concentration (CHC).

The effects of FM on the extrudate expansion have been widely studied and its role affecting this parameter was established because of the sudden drop of pressure (from the inside of the extruder barrel to the atmospheric pressure). The pressure drop causes an extensive expulsion of water vapor from the melt, which emerges in the form of bubbles and allows the expansion of the molten extrudate [18]. The results obtained in our study were in according with those obtained by Thymi *et al.* [19], who reported that the extrudate expansion is most dependent on the material moisture content, as higher FM decreased the expansion of corn extrudates. The effects of temperature on the expansion of starch-based extrudates have also been reported; however, in the expanded extrudates, the effects of this processing factor were not statistically significant, possibly because the temperature range (110–150 °C) was not sufficiently large.

## 2.6. Optimization and Model Prediction Performance

In this part of the study, optimization was defined as the processing conditions that provided an optimum value as a function (maximum or minimum) of certain variables subject to constraints that were previously imposed [20]. Cereal snack products containing natural compounds with additional health benefits (such as antioxidants) are highly accepted by consumers. In addition, texture and appearance (attractive color) are two of the most important sensorial characteristics for these products.

The goals for optimization in this study were to maximize the total anthocyanin content and the expansion index in order to obtain extrudates with the highest amount of antioxidants (anthocyanins) and acceptable crunchiness, assuming that the higher the expansion index, the crunchier the extrudate. In the case of color parameter $b$, the goal was to minimize the value so that the blue/purplish coloration was more intense, making the appearance of the extrudates more attractive.

Figures 7–9 show the central points of the best combination regions (optimum) corresponding to the following processing factors: FM(%) = 16.62/T(°C) = 141.89, FM(%) = 17.27/CHC(%) = 0.11, and T(°C) = 141.89/CHC(%) = 0.08. From these three set of values, an average one was computed for each processing factor, resulting in the following processing factors: FM(%) = 16.94, T(°C) = 141.89 (fourth zone of the extruder) and CHC(%) = 0.095. These conditions estimated the production of expanded nixtamalized blue corn extrudates with a total anthocyanin content of 160 mg· kg$^{-1}$, expansion index of 2.66, and color parameter $b$ of 0.10.

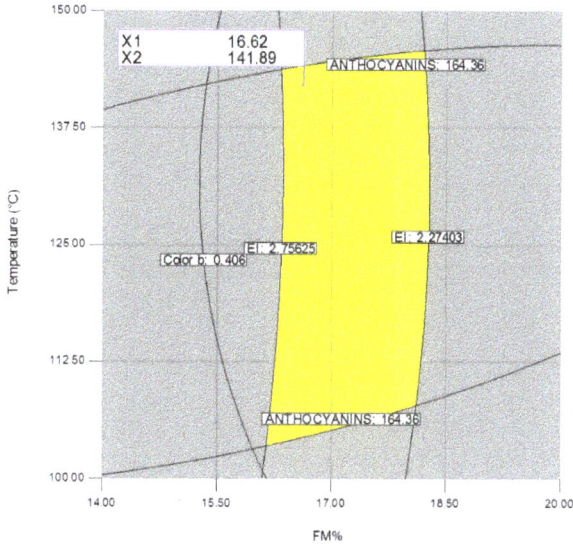

**Figure 7.** Overlay plot and optimized region (best combination) of the extrusion processing factors feed moisture (FM) and temperature (T), for producing expanded nixtamalized blue corn extrudates.

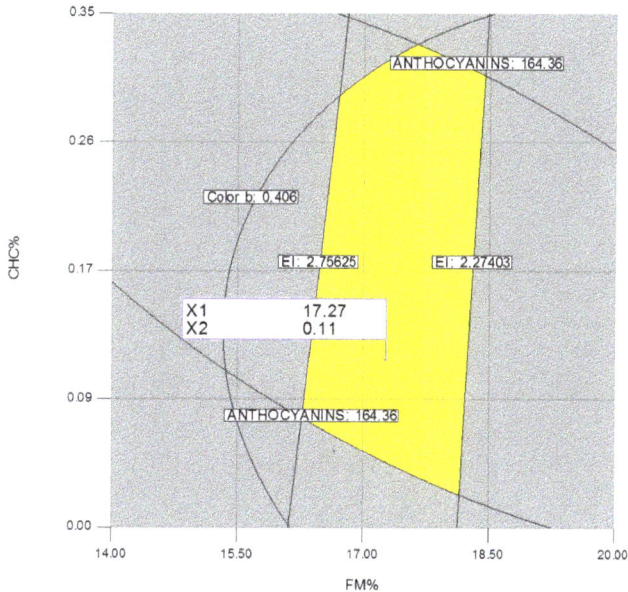

**Figure 8.** Overlay plot and optimized region (best combination) of the extrusion processing factors feed moisture (FM) and calcium hydroxide concentration (CHC), for producing expanded nixtamalized blue corn extrudates.

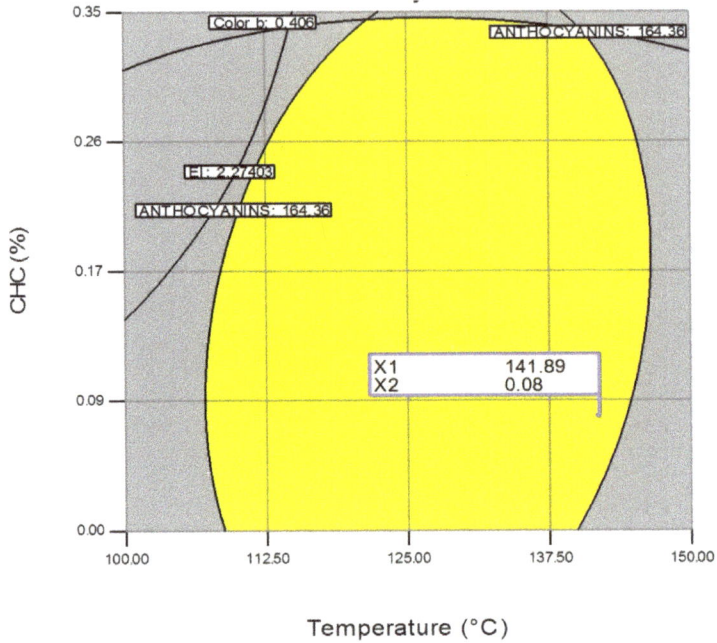

**Figure 9.** Overlay plot and optimized region (best combination) of the extrusion processing factors temperature (T) and calcium hydroxide concentration (CHC), for producing expanded nixtamalized blue corn extrudates.

## Evaluations of the Expanded Nixtamalized Blue Corn Snacks Produced with the Optimum Extrusion Conditions

The experimental validation of the processing factors was performed in the extruder with the conditions obtained by the overlay plots. Results of the chemical and physical characteristics evaluated in the nixtamalized blue corn expanded extrudates are presented in Table 3.

**Table 3.** Predicted values and experimental results of chemical and physical evaluations on the expanded nixtamalized blue corn extrudates obtained from the optimum processing conditions.

| Response Variable | Predicted Value | Experimental Value |
|---|---|---|
| TA (mg· kg$^{-1}$) [a] | 160 | 158.87 ± 2.26 [c] |
| Color $b$ [b] | 0.10 | −0.45 ± 0.08 [c] |
| Expansion Index | 2.66 | 3.19 ± 0.11 [d] |

[a] Total anthocyanins; [b] $b$ = Yellow (+) to blue (−); [c] Average ± standard deviation, $n = 4$; [d] Average ± standard deviation, $n = 40$.

Despite the use of biological material (whole blue corn), the model's prediction led to highly accurate results. The experimental value for TA in the extrudates was $158.87\ mg \cdot kg^{-1}$ (estimated, $160\ mg \cdot kg^{-1}$), meaning 99.2% fitting. The experimental result for the expansion index was 3.19. This parameter was higher than expected (estimated, 2.66), representing 83.3% fitting and more expansion (crunchiness) in the extrudate. Finally, the experimental result for the color parameter $b$ was $-0.45$. This result was better than expected (estimated, 0.10), meaning that the negative values indicated an increased intensity of the blue/purple shades.

## 3. Experimental Section

### 3.1. Raw Material

Creole soft blue corn was obtained in Toluca, Mexico (2010 crop). The grains (10 kg) were cleaned (Clipper BLOUNT/Ferrell-Ross, Model M2BC; Bluffton Inc., Bluffton, IN, USA) and ground in a six blade mill with rubbed shell (Pulvex SA de CV, Model 200, serial 1030401, Mexico, DF, Mexico) and passed through a 0.8 mm mesh. The ground corn was stored in sealed polyethylene bags and kept in the dark (to avoid anthocyanin degradation) at 5 °C until use. Commercial lime (calcium hydroxide) (Calhidra de Sonora, SA de CV, Hermosillo, Son., Mexico) and distilled water were used.

### 3.2. Extrusion Process

Samples (300 g each) of ground corn were mixed in a laboratory blender (Kitchen Aid, Model MK45SSWH, St. Joseph, MI, USA), with different concentrations of calcium hydroxide (0%–0.25%) and distilled water (15%–23%) for 5 min. Each ground corn sample was conditioned with calcium hydroxide and water according to the experiment design (Table 2). The conditioned ground corn samples were kept in sealed polyethylene bags at 5 °C in dark conditions for 12 h before extrusion.

A single-screw extruder (Brabender Instruments, Model E19/25 D, OHG, Duisburg, Germany) with four heat/cool zones was used. The temperatures inside the barrel in the first, second and third zones were 60, 80, 110 °C, respectively, and the fourth zone temperature was set according to the experiment design (110–141.89 °C). The conditioned ground corn was fed into the extruder under the following conditions: screw number 3 (nominal compression ratio 3:1 and diameter 19 mm); screw speed of 120 rpm; hopper feed rate of 50 rpm; and 3 mm die opening diameter. The obtained extrudates were cooled at room temperature (25 °C), dried at 60 °C for 30 min in a tunnel dryer, and then stored at 5 °C in sealed polyethylene bags in the dark until analysis.

## 3.3. Extrudate Evaluations

### 3.3.1. Moisture Content (MC)

The AACCI Approved Method 44–15.02 [21] was used, and three replicates of the analysis were performed for each treatment

### 3.3.2. Total Anthocyanins (TA)

The analysis was assessed according to Abdel-Aal and Hucl [22]. Samples (3 g) of ground extrudates were weighed in a 50 mL centrifuge tube, and acidified ethanol (ethanol with 1 N HCl, 85:15 v/v, 24 mL) was added. The solutions were adjusted at pH 1 with 4N HCl and then shaken and centrifuged (Thermo Scientific, Model Heraes Biofuge Primo R., Dreieich, Germany) at $27,200\times g$ for 15 min; this step was performed twice. The supernatant was separated into a centrifuge tube, and the volume was adjusted to 50 mL with acidified ethanol. The absorbance was measured at 535 nm (against a blank) in a UV-visible spectrophotometer (Varian Australia PT LTD, Cary 50 CONC, Victoria, Australia). The analysis was made in triplicate.

### 3.3.3. pH

This determination was measured according the AACCI Approved Method 02–52 [23]. Three replicates were performed.

### 3.3.4. Color

The parameters $L$, $a$, and $b$ for each treatment were measured using a Hunter Lab Miniscan XE Plus (Hunter Association Laboratories, Reston, VA, USA). The color value $L$ indicates lightness on a scale of 100 (white) to black (0), positive and negative $a$ color values indicate red and green shades, respectively, and positive and negative $b$ color values indicated yellow and blue shades, respectively. These measurements were made in triplicate for each treatment.

### 3.3.5. Expansion Index (EI)

The expansion index was measured using a Digital Caliper (Mitutoyo Corp., Model CD–6 CS, Kanagawa, Japan) and was calculated as the ratio of the extrudate diameter to the diameter of the extruder die (3 mm). Forty replicates of each determination were performed.

## 3.4. Experimental Design and Statistical Analysis

A central composite experimental design of three factors and five levels was used (Table 2). The independent variables were: feed moisture, FM ($X_1$, 15%–23%); calcium hydroxide concentration, CHC ($X_2$, 0%–0.25%) and fourth zone extruder temperature,

T ($X_3$, 110–150 °C), coded with levels of −1.682, −1, 0, +1 and +1.682. The response variables in the extrudates were: moisture content (%), total anthocyanins (mg· $kg^{-1}$), pH, color parameters ($L$, $a$, $b$) and expansion index (EI). The empirical model representing the interaction between the independent and response variables is presented in Figure 10.

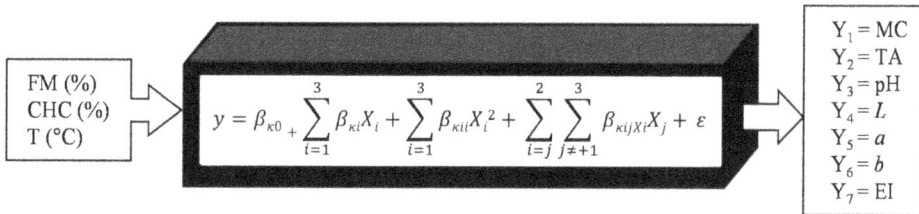

$$y = \beta_{k0} + \sum_{i=1}^{3} \beta_{ki}X_i + \sum_{i=1}^{3} \beta_{kii}X_i^2 + \sum_{i=j}^{2} \sum_{j\neq+1}^{3} \beta_{kij}X_iX_j + \varepsilon$$

FM (%)   CHC (%)   T (°C)

$Y_1 = MC$
$Y_2 = TA$
$Y_3 = pH$
$Y_4 = L$
$Y_5 = a$
$Y_6 = b$
$Y_7 = EI$

**Figure 10.** Empirical model of the interaction between processing and response variables.

This mathematical expression is used to model the response variables ($Y_1$–$Y_7$) where the $k$ value changes from 1 to 7; where $\beta_{k0}$ represents a constant, $\beta_{ki}$ the linear coefficient, $\beta_{kii}$ the quadratic coefficient, $\beta_{kji}$ the interaction effect of the response variables, and $\varepsilon$ the experimental error.

All data obtained from the response variables were recorded, and an analysis of variance (ANOVA) was performed with a confidence level of 95%. Besides, a backward regression analysis was applied, and non-significant factors ($p > 0.1$) were eliminated from the second-order polynomial equation. Then a new equation was recalculated to achieve the final predictive model for each response variable. Response surface methodology (RSM) was used [24]. Contour plots for each determination were obtained using Design Expert Software V.7.0.0 (Stat-Ease, Minneapolis, MN, USA).

### 3.5. Optimization of the Extrusion Process

To obtain the best combinations of factors for the extrusion process (FM, CHC, T), response surface methodology was used. The response variables to optimize the process were: total anthocyanins (maximize), color $b$ (minimize) and expansion index (maximize). Once the graphical contour plots were obtained, superposition surface methodology was applied to achieve the optimization technique [17]. Three contour plots were made, and the optimum combination of the processing variables were selected. Validation of the optimization model was performed according to the estimated processing values.

## 4. Conclusions

The FM, in linear and quadratic terms, was the factor that most significantly affected all of the evaluations performed in the extrudates, except for the TA determinations, where the quadratic term of T showed the most significant effect. The color parameter $a$ was the response variable that was most significantly affected by the three processing factors and their interactions. Certainly during the process, there are complex transformations in anthocyanins showed by the effect of the extrusion factors in the color parameter $a$. The TA content was not affected either by FM or CHC, leaving the processing temperature as the most important factor influencing the retention of anthocyanins during the elaboration of this kind of products.

According to the overlay plots, the optimum extrusion conditions to obtain expanded nixtamalized blue corn extrudates were: FM(%) = 16.94, CHC(%) = 0.095 and T($^\circ$C) = 141.89, and the estimated values of TA, EI and color parameter $b$ were: 160 mg·kg$^{-1}$, 2.66 and 0.10, respectively. The expanded nixtamalized blue corn extrudates obtained experimentally under those conditions were: TA 158.87 mg·kg$^{-1}$; EI 3.19; and color parameter $b$, $-0.44$ (intense blue/purple coloration).

The surface response methodology was a useful tool to obtain an expanded nixtamalized blue corn snack with anthocyanins and acceptable texture and color. Extrusion process and the anthocyanin content of expanded nixtamalized snacks can be optimized to obtain functional products.

**Acknowledgments:** Escalante-Aburto is grateful to CONACyT for the doctoral scholarship provided. Additional thanks go to Abril Z. Graciano Verdugo and Q.B. César B. Otero León for their technical assistance and for providing the color evaluation equipment. The authors thank SEP-PROMEP for financial support for the Project "Application of Physical, Rheological and Biological Methods in Processing Corn" through the network: Conventional and Alternative Technologies for Cereals Processing.

**Author Contributions:** A.E.-A., and B.R-W., designed research; P.I.T.-C., J.L-C., J.D.F.-C., J.M.B.-H., R.G.-D., performed research and analyzed the data; A.E.-A., I.M.-R. and N.P.-G., performed the extrusion experiments. A.E.-A, and B.R.-W. wrote the paper. All the authors read and approved the final manuscript.

**Conflicts of Interest:** The authors declare no conflict of interest.

## References

1.  Swinburn, B.A.; Caterson, I.; Seidell, J.C.; James, W.P.T. Diet, nutrition and the prevention of excess weight gain and obesity. *Public Health Nutr.* **2004**, *7*, 123–146.
2.  Mendoza-Díaz, S.; Ortíz-Valerio, M.C.; Castaño-Tostado, E.; Figueroa-Cárdenas, J.D.; Reynoso-Camacho, R.; Ramos-Gómez, M.; Campos-Vega, R.; Loarca-Piña, G.F. Antioxidant capacity and antimutagenic activity of anthocyanin and carotenoid extracts from nixtamalized pigmented creole maize races (*Zea mays* L.). *Plant Food Hum. Nutr.* **2012**, *67*, 442–449.

3.  De Pascual-Teresa, S.; Santos-Buelga, C.; Rivas-Gonzalo, J.C. LC-MS analysis of anthocyanins from purple corn cob. *J. Sci. Food Agric.* **2002**, *82*, 1003–1006.

4.  Bridle, P.; Timberlake, C.F. Anthocyanins as natural food colours—Selected aspects. *Food Chem.* **1997**, *58*, 103–109.

5.  Salinas, M.Y.; Martínez, B.F.; Soto, H.M.; Ortega, P.R.; Arellano, V.J.L. Efecto de la Nixtamalización sobre las Antocianinas del Grano de Maíces Pigmentados. *Agrociencia* **2003**, *37*, 617–628.

6.  Aguayo-Rojas, J.; Mora-Rochín, S.; Cuevas-Rodríguez, E.; Serna-Saldivar, S.; Gutierrez-Uribe, J.; Reyes-Moreno, C.; Milán-Carrillo, J. Phytochemicals and antioxidant capacity of tortillas obtained after lime-cooking extrusion process of Whole Pigmented Mexican Maize. *Plant Food Hum. Nutr.* **2012**, *67*, 178–185.

7.  Escalante-Aburto, A.; Ramírez-Wong, B.; Torres-Chávez, P.I.; Figueroa-Cárdenas, J.D.; López-Cervantes, J.; Barrón-Hoyos, J.M.; Morales-Rosas, I. Effect of extrusion processing parameters on anthocyanin content and physicochemical properties of nixtamalized blue corn expanded extrudates. *CyTA—J. Food* **2013**, *11*, 29–37.

8.  Camacho-Hernández, I.L.; Zazueta-Morales, J.J.; Gallegos-Infante, J.A.; Aguilar-Palazuelos, E.; Rocha Guzmán, N.E.; Navarro-Cortez, R.O.; Jacobo-Valenzuela, N.; Gómez-Aldapa, C.A. Effect of extrusion conditions on physicochemical characteristics and anthocyanin content of blue corn third generations snacks. *CyTA—J. Food* **2014**, *12*, 320–330.

9.  Cortés-Gómez, A.; San Martín-Martínez, E.; Martínez-Bustos, F.; Vázquez-Carrillo, G.M. Tortillas of blue maize (*Zea mays* L.) prepared by a fractionated process of nixtamalization: Analysis using response surface mehtodology. *J. Food Eng.* **2005**, *66*, 273–281.

10. Mora-Rochid, S.; Gutiérrez-Uribe, J.A.; Serna-Saldívar, S.O.; Sánchez-Peña, P.; Reyes-Moreno, C.; Milán-Carrillo, J. Phenolic content and antioxidant activity of tortillas produced from pigmented maize processed by conventional nixtamalization or extrusion cooking. *J. Cereal Sci.* **2010**, *52*, 502–508.

11. Méndez, L.I.R.; Cárdenas, J.D.F.; Gómez, M.R.; Lagunas, L.L.M. Nutraceutical properties of flour and tortillas made with an ecological nixtamalization process. *J. Food Sci.* **2013**, *78*, C1529–C1534.

12. Guy, R. Snack foods. In *Extrusion Cooking. Technologies and Applications*, 1st ed.; Guy, R., Ed.; CRC: Boca Raton, FL, USA, 2001; pp. 161–181.

13. Nayak, B.; Berrios, J.D.J.; Powers, J.R.; Tang, J. Thermal degradation of anthocyanins from purple potato (Cv. Purple Majesty) and impact on antioxidant capacity. *J. Agric. Food Chem.* **2011**, *59*, 11040–11049.

14. Jing, P.; Giusti, M.M. Effects of extraction conditions on improving the field and quality of an anthocyanins-rich purple corn (*Zea mays* L.) color extract. *J. Food Sci.* **2007**, *72*, C363–C368.

15. Zazueta-Morales, J.J.; Martínez-Bustos, F.; Jacobo-Valenzuela, N.; Ordorica-Falomir, C.; Paredes-López, O. Effect of the addition of calcium hydroxide on some characteristics of extruded products from blue maize (*Zea mays* L.) using response surface methodology. *J. Sci. Food Agric.* **2001**, *81*, 1379–1386.

16. Sadilova, E.; Stintzing, F.C.; Carle, R. Thermal degradation of acylated and nonacylated anthocyanins. *J. Food Sci.* **2006**, *71*, C504–C512.

17. Sánchez-Madrigal, M.A.; Quintero-Ramos, A.; Martínez-Bustos, F.; Meléndez-Pizarro, C.O.; Ruíz-Gutiérrez, M.G.; Camácho-Dávila, A.; Torres-Chávez, P.I.; Ramírez-Wong, B. Effect of different calcium sources on the bioactive compounds stability of extruded and nixtamalized blue maize flours. *J. Food Sci. Technol.* **2014**.

18. Arhaliass, A.; Bouvier, J.M.; Legrand, J. Melt growth and shrinkage at the exit of the die in the extrusion-cooking process. *J. Food Eng.* **2003**, *60*, 185–192.

19. Thymi, S.; Krokida, M.K.; Pappa, A.; Maroulis, Z.B. Structural properties of extruded corn starch. *J. Food Eng.* **2005**, *68*, 519–526.

20. Altan, A.; McCarthy, K.L.; Maskan, M. Extrusion cooking of barley flour and process parameter optimization by using response surface methodology. *J. Sci. Food Agric.* **2008**, *88*, 1648–1659.

21. AACC International. Moisture-air-oven methods (Approved October 30, 1975; Reapproved November 3, 1999). In *Approved Methods of Analysis*, 11th ed.; AACC International: St. Paul, MN, USA, 1995; Method 44-15.02.

22. Abdel-Aal, E.S.M.; Hucl, P. A rapid method for quantifying total anthocyanins in blue aleurone and purple pericarp wheats. *Cereal Chem.* **1999**, *76*, 350–354.

23. AACC International. Hydrogen-Ion Activity (pH) (Approved October 30, 1975; Reapproved November 3, 1999). In *Approved Methods of Analysis*, 11th ed.; AACC International: St. Paul, MN, USA, 1995; Mthod 02-52.01.

24. Myers, R.H.; Montgomery, D.C.; Anderson-Cook, C.M. The analysis of second-order response surfaces. In *Response Surface Methodology. Process and Product Optimization Using Designed Experiments*, 3rd ed.; Myers, R.H., Montgomery, D.C., Anderson-Cook, C.M., Eds.; Wiley: Hoboken, NJ, USA, 2009; pp. 219–264.

**Sample Availability:** *Sample Availability*: Samples of the compounds are not available from the authors.

# Decreasing pH Results in a Reduction of Anthocyanin Coprecipitation during Cold Stabilization of Purple Grape Juice

David C. Manns, Passaporn Siricururatana, Olga I. Padilla-Zakour and Gavin L. Sacks

**Abstract:** Anthocyanin pigments in grape juice can coprecipitate with potassium bitartrate (KHT) crystals during cold stabilization, but factors that reduce these adsorptive losses are not well understood. We hypothesized that coprecipitation on a % w/w basis should be decreased at lower pH. In initial experiments, model juice solutions containing an anthocyanin monoglucoside extract and varying pH values were subjected to cold-storage to induce KHT crystallization, and anthocyanins in the resulting precipitant were characterized by HPLC. The pH of the model juice was directly correlated with the % w/w concentration of anthocyanins in the KHT crystals, with a maximum observed at pH 3.40 (0.20% w/w) and a minimum at pH 2.35 (0.01% w/w). A pH dependency was also observed for anthocyanin-KHT coprecipitation in purple Concord grape juice, although the effect was smaller. Coprecipitation was significantly greater for anthocyanin monoglucosides and acylated anthocyanins as compared to anthocyanin diglucosides at pH > 3.05, but coprecipitation of mono- and acylated forms declined more sharply at lower pH values.

Reprinted from *Molecules*. Cite as: Manns, D.C.; Siricururatana, P.; Padilla-Zakour, O.I.; Sacks, G.L. Decreasing pH Results in a Reduction of Anthocyanin Coprecipitation during Cold Stabilization of Purple Grape Juice. *Molecules* **2015**, *20*, 21066–21084.

## 1. Introduction

While a wide range of phenolic compounds have been detected in juices from Concord grape juices and related grape cultivars, including hydroxycinnamates, flavan-3-ols, flavonols, and stilbenes (e.g., resveratrol) [1,2], the major species are the anthocyanins [1]. These anthocyanins are the major compounds responsible for the pigmentation of red and purple grapes and are critical to consumer acceptance of grape-derived products like juices and wines [3]. Additionally, the anthocyanins along with other polyphenols and their metabolites have been implicated as important phytonutrients capable of reducing the incidence of chronic disease [4,5]. Due to their overall importance to the acceptability of fruit juices and related products, several publications have considered the impact of production practices on anthocyanin stability, particularly acid-catalyzed hydrolysis and polymerization reactions [6–8].

In grape juice and wines, an additional source of anthocyanin losses during production is coprecipitation of anthocyanins with potassium bitartrate [9–12]. Grapes are uniquely high in tartaric acid compared to other fruits, 2–14 g/kg, or 0.01–0.07 M [13], and also contain high concentrations of potassium, 0.01–0.06 M [14,15]. These concentrations are at or above the solubility of potassium bitartrate (KHT) in pure water at $0\,^\circ C$ (0.01 M), although the apparent solubility of KHT in real juices is higher due to the presence of polyphenols, polysaccharides, and other constituents that can inhibit crystallization [16,17]. To prevent formation of KHT crystals in finished products, a cold-stabilization step is usually performed on grape juices and wines prior to bottling [18].

The factors affecting the kinetics and thermodynamics of KHT precipitation are well studied [19–21]. Coprecipitation of anthocyanins with KHT resulted in a 20%–40% loss of total anthocyanins from Concord juice in one report [9], with similar losses reported elsewhere [10,12]. The loss of anthocyanins during cold stabilization is comparable to gains achieved by widely used juice processing treatments such as the use of pectolytic enzymes [22] or thermal treatments [23]. Thus, elimination of anthocyanin coprecipitation could be considered an unexploited route to increasing final anthocyanin concentrations in grape juice. The coprecipitation process also results in enrichment of the anthocyanins in the KHT precipitate compared to the remaining solution by about an order of magnitude [10,12]. Coprecipitation of tannins, hydroxycinnamic acids, flavonols, and other organic compounds are also reported to occur, with preferential loss of less polar species [12].

The mechanism for the loss of anthocyanins during cold stabilization is not well understood. Occlusion of anthocyanins within the crystal is unlikely to occur, as the proportions of coprecipitating compounds are different than their proportions in solution [16]. Rather, the interaction of anthocyanins and KHT appears to be adsorptive in nature [11], a process which also inhibits crystal growth, changes crystal morphology, and increases the apparent solubility of KHT in grape products *vs.* pure water [16]. The interactions between the KHT crystal face and phenolics are variously proposed to be ionic, hydrogen-bonding, or charge-transfer in nature [24,25]. X-ray crystallography data indicates that the {010} face is populated by the bitartrate species, and it was hypothesized that this would result in a positive surface charge on this face created by excess potassium ions, and consequentially the adsorption of Lewis bases, e.g. the neutral forms of anthocyanins [25]. In contradiction, Alongi *et al.* [10] observed that anthocyanin species which favored the flavylium cation form (lower $pK_h$ value) were more likely to be lost via coprecipitation, indicating that the anthocyanins may interact directly with bitartrate at the surface.

Regardless of the mechanism, the loss of anthocyanins and other polyphenolics during KHT precipitation is undesirable to the wine and grape industries, but strategies to reduce these losses are largely unknown. A study by our group showed that anthocyanin coprecipitation is significantly less in juice concentrate (59 Brix) as compared to single strength juice [10]. Cold-stabilization of single-strength Concord juice prior to concentration resulted in moderate losses (~20%) of anthocyanins, similar to previous reports, while concentration prior to cold-stabilization (so-called "direct to concentrate") resulted in no significant loss of anthocyanins. Compositional analysis of KHT crystals yielded similar results—although comparable losses of KHT occurred in both systems, the precipitate from the direct to concentrate had lower anthocyanin content (0.13% $vs.$ 0.80% w/w). The improved anthocyanin stability achieved in concentrate did not appear to result from increased co-pigmentation. Because anthocyanin species that existed more in charged forms (higher $pK_h$ values) were more likely to coprecipitate, it was hypothesized that the reduction in coprecipitation in concentrate could be credited to the lower pH of concentrate. The pH of concentrate (2.5) is lower than single-strength juice (pH = 3.1), which should result in a neutralization of the surface charge of the KHT surface [24]. However, because concentrate differs from juice in many other respects (greater ionic strength, lower water activity, $etc.$), this was not conclusive.

In this work, we investigated if a pH decrease could induce changes in the degree of coprecipiation of anthocyanins with potassium bitartrate during grape juice cold-stabilization. The current study investigated the effects of pH on coprecipitation of anthocyanins with KHT, in both a model juice system and a purple Concord grape juice.

## 2. Results and Discussion

### 2.1. Anthocyanin-Bitartrate Coprecipitation in a Model Juice

As an initial evaluation of this hypothesis, a model juice containing a blackcurrant anthocyanin extract was prepared with potassium (0.02–0.04 M) and tartaric acid (0.02–0.04 M) concentrations within the range ordinarily encountered in grape juice: 0.01–0.06 M for potassium [14,15] and 0.01–0.07 M for tartaric acid [13]. Initial concentrations of the four primary anthocyanins present in the model blackcurrant juice were 46.35, 161.1, 32.38, and 197.2 μg/mL malvidin-3-glucoside equivalents for delphinidin-3-glucoside (D-3-G), delphinidin-3-rutinoside (D-3-R), cyanidin-3-glucoside (C-3-G), and cyanidin-3-rutinoside (C-3-R), respectively. The pH range, 2.35–3.40, was selected to bracket the range typically observed in single strength grape juice (pH = 3.0–3.5) as well as in 59 Brix juice concentrate (pH = 2.5). We observed negligible KHT precipitation in most treatments with 0.02 M tartaric acid and/or potassium (data not shown), and characterization of anthocyanin content in the resulting crystals was not feasible. As a result, only data for 0.04 M K$^+$ and 0.04 M tartaric acid with varying pH are reported. During cold storage of model juices we observed precipitation of KHT (Figure 1A), with pH 2.95–3.40 model juices precipitated 1.62–1.76 g (8.63–9.33 mmoles) of KHT, significantly greater than the mass precipitated at pH 2.35 (0.790, or 4.2 mmoles) and pH 2.70 (1.31, or 6.98 mmoles). Anthocyanins in KHT crystals were quantified by HPLC following redissolution, and the concentration of anthocyanin in KHT crystals (%, w/w) as a function of pH is shown in Figure 1B. The highest anthocyanin concentration was 0.19% w/w at pH 3.4, and decreased to a low of 0.01% w/w at pH 2.35.

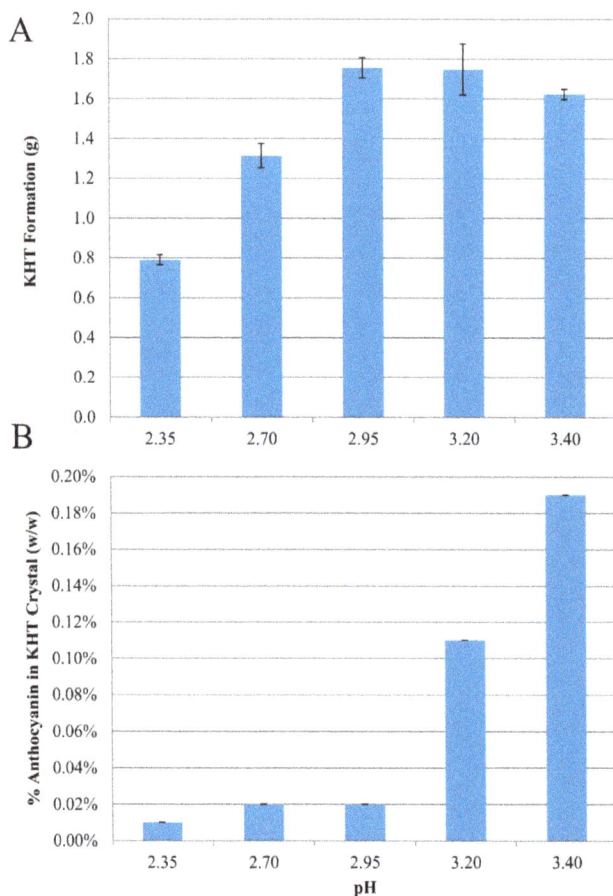

**Figure 1.** Effects of pH on (**A**) mass (g) of KHT crystals recovered by filtration after cold stabilization of blackcurrant model juices, and (**B**) total anthocyanin (% w/w) coprecipitating with KHT crystals recovered from cold stabilized blackcurrant model juice.

Decreasing pH resulted in decreased coprecipitation for all four black currant anthocyanins (Figure 2A), and significant differences ($p < 0.05$ by ANOVA) in the relative losses of each species were also observed (Figure 2B). The delphinidin- and cyanidin-3-glucoside were enriched by a factor of 1.5 to 3, while the corresponding rutinosides were depleted by 10%–50%.

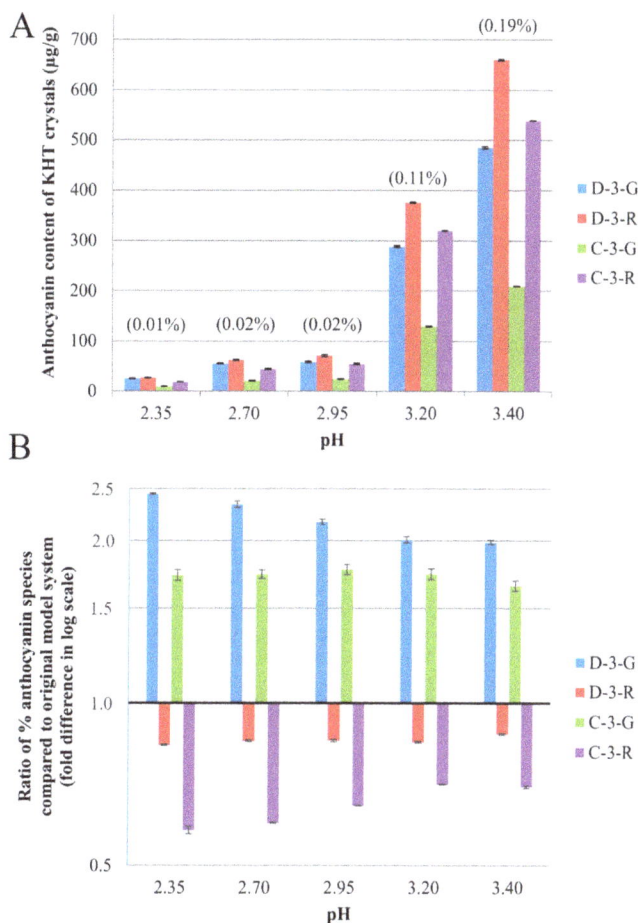

**Figure 2.** (**A**) Anthocyanin concentration (µg/g) in KHT crystals as a function of pH recovered from model juices containing 0.04 M tartaric acid and 0.04 M KCl. The number in parenthesis represented the total anthocyanin in KHT crystals (% w/w); (**B**) Selectivity of coprecipitation, calculated as the ratio of an anthocyanin species concentration in KHT crystals (normalized to total anthocyanin in KHT crystal) to the anthocyanin species concentration in original model juice (normalized to total anthocyanin in original juice). D-3-G: delphinidin-3-glucoside; D-3-R: delphinidin-3-rutinoside; C-3-G: cyanidin-3-glucoside; C-3-R: cyanidin-3-rutinoside.

## 2.2. Anthocyanin-Bitartrate Coprecipitation in a Purple Concord Grape Juice

The model juice study was duplicated with a real Concord juice adjusted to one of six pH values prior to cold-stabilization. The anthocyanin composition, organic acid composition, and basic juice chemistry of the original juice are reported in Table 1.

**Table 1.** Initial anthocyanin concentrations and basic juice parameters in Concord juice.

| Analyte | Concentration [a] |
|---|---|
| *Anthocyanin Diglucosides* | |
| Del-3,5-Di | 26.0 (0.38) |
| Cy-3,5-Di | 32.0 (0.63) |
| Pet-3,5-Di | 16.4 (0.65) |
| Peo-3,5-Di | 33.9 (0.68) |
| Mvn-3,5-Di | 30.2 (0.72) |
| *Anthocyanin Monoglucosides* | |
| Del-3-Glu | 94.8 (1.0) |
| Cy-3-Glu | 74.4 (0.82) |
| Pet-3-Glu | 28.4 (0.65) |
| Peo-3-Glu | 13.7 (0.37) |
| Mvn-3-Glu | 19.2 (0.59) |
| *Acylated Anthocyanins* | 122.6 (1.3) |
| **Total Anthocyanins** | 491.6 (0.84) |
| *Organic Acids* | |
| Citric Acid | 0.23 (0.01) |
| Tartaric Acid | 11.5 (0.07) |
| Malic Acid | 2.38 (0.03) |
| Total Soluble Solids | 20.0 (0.1) |
| Initial pH | 3.05 |

[a]: All anthocyanin units are in mg/L. The diglucosides are expressed in malvidin-3,5-diglucoside equivalents. The monoglucosides and modified anthocyanins are expressed in malvidin-3-glucoside equivalents. Organic acids are expressed as g/L. The Total Soluble Solids measurement is expected to be >90% sugars in grapes [26] and is expressed in units of Brix. Parenthetical values are standard deviations.

Although the KHT crystals had uniform appearance at high pH, the crystals formed at lower pH were heterogeneous in appearance, with some nearly transparent and others deeply pigmented (Figure 3).

As with the model juice, the total mass of KHT precipitate formed decreased with decreasing pH (Figure 4A). For most anthocyanin species, the total mass of each anthocyanin species (in grams) in the final juice and in the precipitate was 90%–110% of the initial mass (Supplementary Figure S1), except for two anthocyanins with 80%–90% recovery (petunidin-3,5-diglucoside and delphinidin-3,5-diglucoside). These two exceptions were challenging to quantify by HPLC due to their low concentrations.

**Figure 3.** KHT precipitates from Concord juices as a function of pH.

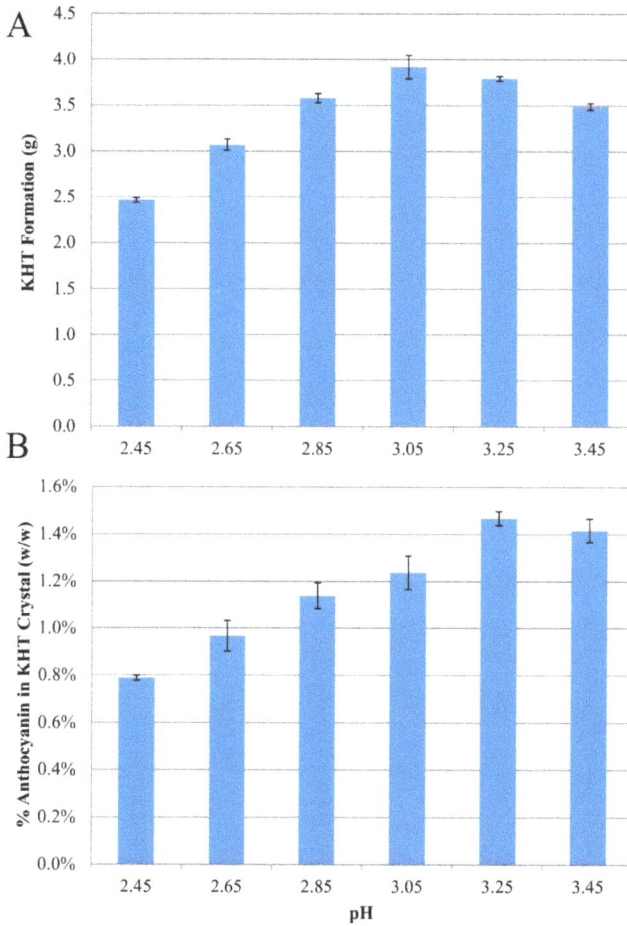

**Figure 4.** Effects of pH on (**A**) mass (g) of KHT crystals recovered by filtration after cold stabilization of Concord juices, and (**B**) total anthocyanin content (% w/w) of KHT crystals recovered from cold stabilized Concord juice.

The concentration of anthocyanin in the KHT precipitate was at a maximum at pH 3.25–3.45 (1.4% w/w, Figure 4B), and decreased significantly with decreasing pH below 3.05 to a minimum at pH 2.45 (0.8% w/w). Similar to what was observed with model juice, decreasing the pH to 3.05 or lower resulted in a significant decrease in total anthocyanin coprecipitation, with the lowest content observed at the lowest pH (Figure 4B).

The mass fraction of each anthocyanin (%w/w) in KHT crystal (Figure 5A), the percent loss of each anthocyanin from juice (Figure 5B), and the percent loss of each anthocyanin class (Figure 5C) were determined by HPLC.

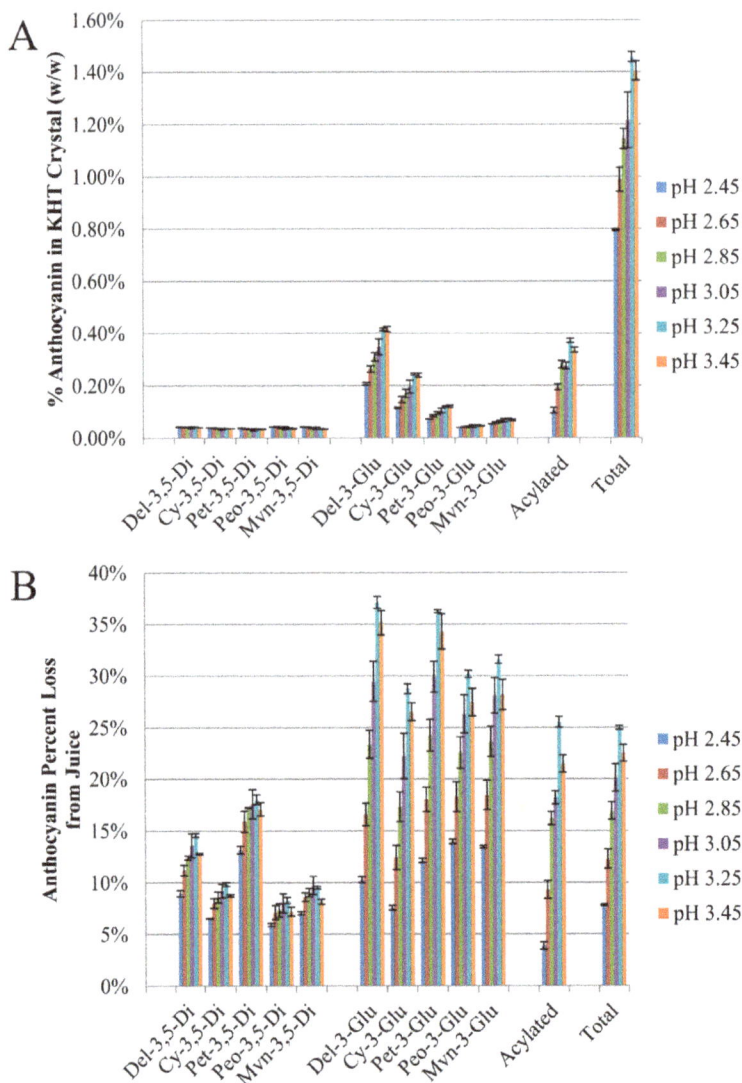

Figure 5. *Cont.*

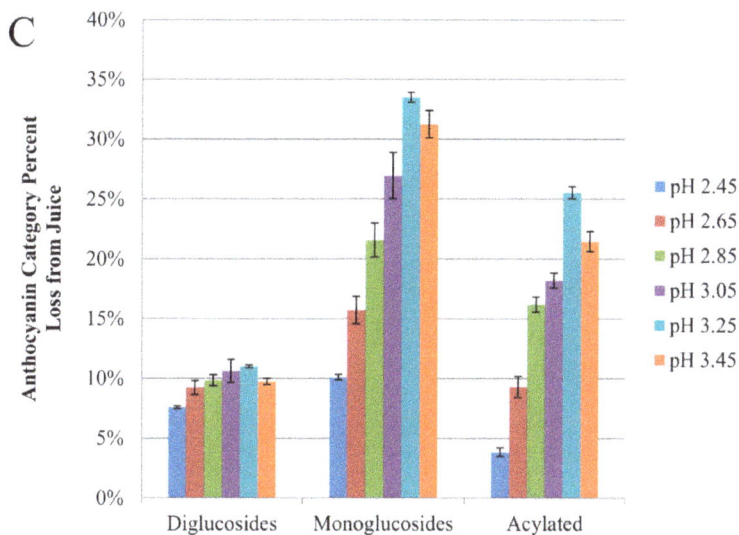

**Figure 5.** Effects of pH on (**A**) the mass fraction (%w/w) of individual anthocyanins in KHT crystals; (**B**) the percent loss ((Original−Final)/Original ×100%) of individual anthocyanin species lost from Concord juice; and (**C**) the percent loss of anthocyanin classes lost from Concord juice.

Monoglucosides accounted for the majority of anthocyanin coprecipitate, with delphinidin-3-glucoside accounting for nearly one-third of the total (Figure 5A). However, when normalized against the original anthocyanin content of each species in the juice, the percent decreases was roughly comparable across all monoglucosides as well as acylated species, with losses in the range of 25%–35% for these species (Figure 5B). Less coprecipitation losses were seen for all monoglucosides at low pH, with losses in the range of 7%–15% at pH 2.45. Small but significant differences in the effect of pH on anthocyanin monoglucoside loss were observed among aglycones. Specifically, the difference in this change was inversely correlated to HPLC retention time on a non-polar column ($p < 0.05$). For example, delphinidin-3-glucoside coprecipitated to a greater extent than malvidin-3-glucoside at high pH, but this trend was reversed at the lowest pH value (Figure 5B). A greater selectivity effect was observed as a result of the glycoside. At pH 3.25 or greater, anthocyanin monoglucosides and acylated forms were enriched in the KHT precipitate as compared to diglucosides (Figure 5C). As the pH decreased below 3, the concentration of anthocyanin monoglucosides and acylated anthocyanins in KHT precipitate also decreased, with maximum differences of 3- to 5-fold at pH 2.45 *vs.* pH 3.25. By comparison, diglucosides showed no significant decrease in anthocyanin content of KHT crystals until pH 2.45, and the difference was less than a factor of 2.

## 2.3. Discussion

Previous work has shown that concentration of juice prior to cold-stabilization decreases the mass fraction of anthocyanins lost to coprecipitation with potassium bitartrate by over four-fold [10]. Because the loss of individual anthocyanin species was correlated with the hydration constant ($pK_h$) of the species, it was hypothesized that the decrease in coprecipitation was due to the lower pH of concentrate and resulting neutralization of the bitartrate crystal faces [7]. In exploratory studies, we attempted to use commercial grape anthocyanin extracts in our model juice systems. However, we observed very little coprecipitation of anthocyanins during these preliminary experiments. A possible explanation is that the commercial anthocyanin extract had already undergone cold-stabilization, resulting in the loss of the species most prone to coprecipitation. As an alternative, we chose to use commercially available black currant extract as our source of anthocyanins. Black currants contain four dominant anthocyanins [27] which account for >99% of the total anthocyanin content, which simplifies the chromatographic separation. Two of these anthocyanins are major anthocyanin species in grapes (cyanidin-3-glucoside and delphinidin-3-glucoside), while the other two (cyanidin-3-rutinoside and delphinidin-3-rutinoside) are not observed in grapes.

The low concentrations of the KHT crystal formed at pH < 3 (Figure 1A) were expected due to the low concentration of bitartrate species at pH < 3.0, the $pK_a$ of tartaric acid. Interestingly, although the concentration of the bitartrate species is predicted to be at a maximum at pH 3.65 [26], the pH 3.40 model juice did not produce the most KHT precipitate. Instead, the model juices followed the order pH 2.95 > pH 3.2 > pH 3.4 > pH 2.7 > pH 2.35. This is likely due to increasing adsorption of anthocyanins to the growing crystal faces at higher pH. This coprecipitation effect is well known to limit crystal growth and the extent of precipitation [11,16,25].

The maximum anthocyanin concentration observed to coprecipitate (0.19% w/w at pH 3.40) was less than the 0.8% w/w content of KHT recovered from cold stabilization of single strength Concord grape juice [10]. The difference between our model system and Concord may be due to the higher concentrations of both acylated and monoglucoside anthocyanins in Concord, which are more likely to coprecipitate [9,10]. pH had a significant and dramatic effect on anthocyanin loss, with a sharp, order of magnitude decrease in anthocyanin content observed at pH ⩽ 2.95 as compared to pH 3.4 (Figure 1B). The observation that a decreased pH results in decreased coprecipitation supports the hypothesis previously advanced by Alongi *et al.* [10] to explain differences in anthocyanin coprecipitation in concentrate and single strength juice. The authors observed that KHT precipitation from juice concentrate resulted in negligible losses as compared to precipitation from single strength juice. The higher pH of juice (3.05) could result in a negative surface charge of KHT and thus increase interactions between the flavylium form of anthocyanins and the deprotonated sites of the bitartrate crystals, while at the low pH of concentrate (2.5) the KHT surface would be neutralized and interactions would diminish. This suggests that the transition from neutral to charged surface occurs around pH 3.0 for KHT. This is below the pH range usually observed for red grape juices and wines, which may explain why this phenomenon had not been previously reported. According to Celotti *et al.* [24], a transition in the surface charge appears to occur between pH 2.8 and 3.0, as measured by streaming potential experiments. Surprisingly, the authors report that the surface charge became more negative with decreasing pH, an observation at odds with what would be expected to occur to surface charge with decreasing pH, and thus may be an error in sign in the earlier publication.

While pH had the major effect on the mass fraction of anthocyanin coprecipitation (over an order of a magnitude), sugar moiety and aglycone also had a smaller but still significant effect. Glucosides were preferentially lost over corresponding rutinosides, and delphinidins were preferentially lost as compared to cyanidins. The reason for the preferential loss of glucosides is unclear, but may be because the rutinosides are sterically hindered as disaccharides and thus less able to adsorb to the KHT crystal surface. The observation that delphinidins preferentially coprecipitated as compared to cyanidins is in contradiction to our previous work [10], where we observed a greater loss of cyanidin-3-glucoside from Concord grape juice (15% decrease) than delphinidin-3-glucoside (3%) from single strength juice during cold stabilization. This unexpected result is discussed in more detail later.

### 2.3.1. Anthocyanin-Bitartrate Coprecipitation in a Purple Concord Juice

As with the model juice system, decreasing pH resulted in a decrease in total anthocyanin coprecipitation (Figure 5). The decrease in anthocyanin loss did appear to be uniform among KHT crystals, however, and the heterogenous appearance crystals at low pH (Figure 3) may indicate that adsorption of anthocyanins onto the KHT face at pH < 3 involves cooperative binding by anthocyanin species, in which initial adsorption is slow and subsequent binding is faster [28].

The maximum concentration of anthocyanin in the KHT precipitate was at a maximum at pH 3.25 (1.4% w/w, Figure 4B), higher than that observed in previous work on cold stabilization of single strength Concord juice (0.8% w/w, [10]). The higher degree of coprecipitation may be due to the higher concentration of anthocyanins in the current study as opposed to the previous work (492 mg/L *vs.* 278 mg/L as malvidin-3-glucoside equivalents).

Similar to what was observed with model juice, decreasing the pH to < 3.05 resulted in a significant decrease in total anthocyanin coprecipitation, with the lowest content observed at the lowest pH (Figure 4B).

The effect of pH on coprecipitation varied among anthocyanin classes. Anthocyanin monoglucosides and acylated forms were enriched in the KHT precipitate as compared to monoglucosides at higher pH values, pH > 3 (Figure 5A,B). This was in agreement with a previous study which observed the trend acylated > mono- > di- for % w/w anthocyanin in KHT precipitate from unadjusted Concord juice (pH = 3.1) [10], which follows an order of increasing polarity. Similarly, we observed that the loss of monoglucoside anthocyanin species at high pH was inversely correlated to HPLC retention time ($p < 0.05$), suggesting that the likelihood a species would coprecipitate at high pH was correlated to its hydrophobicity and solubility in grape juice. In other words, less soluble species are more likely to coprecipitate with KHT.

An alternate explanation for why monoglucoside and acylated forms of anthocyanins coprecipitate more readily than diglucosides at pH > 3 may be due to changes in KHT surface charge. As with model juice, the sharp decrease in anthocyanin coprecipitation below pH 3 in both model and real juice may relate to the neutralization of KHT surface charge below pH 3.1 [10]. Because diglucosides generally have lower $pK_h$ values than their corresponding monoglucosides [29], they will exist in a neutral form to a greater extent and thus be less likely to coprecipitate with the negatively charged KHT surface at high pH. Following KHT surface neutralization at pH < 3, this mechanism is of less importance and under these low pH conditions, coprecipitation would require interaction of neutral anthocyanin species with the neutral KHT crystal face. This would explain why there would be a much larger decrease in monoglucoside and acylated losses (higher $pK_h$) with decreasing pH than in diglucosides (lower $pK_h$), as shown in Figure 5C. However, in contradiction to this hypothesis, the effect of pH on losses of delphinidin-3-glucoside ($pK_h$ = 2.35) and cyanidin-3-glucoside ($pK_h$ = 3.01) was comparable (Figure 5B).

The observed decrease in coprecipitation with decreasing pH may partially explain the decrease in coprecipitation observed with juice concentration prior to cold-stabilization. In previous work with 59 Brix concentrate, a 6-fold decrease in co-precipitation was observed (0.8% w/w vs. 0.13% w/w) as compared to juice [10]. This is somewhat higher than the ~2-fold decrease observed in the current work. Because diglucoside coprecipitation is not affected by pH, the stronger effect of pH on total anthocyanin loss observed in the earlier work may be partially due to the low concentration of diglucosides (<2% as malvidin-3-glucoside equivalents) as compared to ~25% in the current work. However, other differences exist in juice concentrate (higher ionic strength, higher concentration of $K^+$, $HT^-$, and anthocyanins) which could also account for the observed effects.

## 2.3.2. Potential Implications for Juice Processing and Retention or Isolation of Anthocyanins

Our study demonstrates that a lower pH reduces anthocyanin coprecipitation, particularly for acylated and monoglucoside species. The 20%–40% losses of anthocyanins observed during cold stabilization are comparable to gains achieved by more common juice processing treatments like the use of pectolytic enzymes [22], and thus may represent an interesting target for purple grape juice processors to improve color. As a caveat, changes in anthocyanin concentration may not translate into perceivable differences to consumers, since juice color will also depend on other factors like copigmentation and pH [21]. Potentially, coprecipitation losses to anthocyanins during cold-stabilization of grape juice and wine could be eliminated by intentionally reducing the pH below 3 prior to production. Reducing the pH much below 3 is likely undesirable, at least in single strength juice, as insufficient KHT precipitation would occur. A reduction of pH can be achieved by concentration prior to cold-stabilization, as previously demonstrated [10], but this is a complex process, and would not be appropriate for wine or for juices that are intended to be bottled without concentration. Alternatively, the pH could be reduced chemically (*i.e.*, by addition of tartaric acid) or by physical means (*i.e.*, electrodialysis) prior to cold stabilization. Following cold stabilization, the pH could be raised by analogous chemical or physical processes. Cation-exchange resins could also be used to reduce pH, but these resins are well known to adsorb anthocyanins [30], so an improvement to anthocyanin content probably would not be realized.

Alternatively, coprecipitation with KHT could be exploited to selectively enrich and isolate anthocyanins. Coprecipitation via adsorption is a classic analytical strategy for enriching trace analytes, although the strategy has been used primarily for enriching trace metal cations [31]. In the case of anthocyanins, commercial products are generally sold as crude preparations due to the cost and difficulty of purifying these compounds from complex natural sources [32,33]. As shown here, the concentration of anthocyanins in KHT crystals can exceed 1%, comparable to the loadings achievable with reversed phase resins, but with the advantage that KHT is a fraction of the cost of commercial resins.

## 3. Experimental Section

### 3.1. Chemicals

Black currant powder containing 20% w/w anthocyanin as cyanidin-3-glucoside equivalents was used as an anthocyanin source (Artemis International Inc., Fort Wayne, IN, USA). Malvidin-3-glucose was purchased from Sigma-Aldrich, Inc. (St. Louis, MO, USA). Citric acid monohydrate, malic acid, and anhydrous sodium hydrogen phosphate were purchased from J.T. Baker Chemical Co. (Phillipsburg, NJ,

USA). D-Glucose, D-fructose, L-(+)-tartaric acid, potassium chloride, sodium chloride, and 0.01 M hydrochloric acid were purchased from Thermo Fisher Scientific Inc. (Fair Lawn, NJ, USA). Water from a Nanopure water purifier (Barnstead Thermolyne, Boston, MA, USA) was used throughout the study.

## 3.2. Preparation and Cold-Stabilization Treatments of Model Juices

A full factorial design was used to produce model juice systems with varying pH values, $K^+$ concentrations, and tartaric acid concentrations. All model juices contained 80 g/kg glucose, 80 g/kg fructose, and 250 mg/L anthocyanin as cyanidin-3-glucoside equivalents (similar to red grape juice). Five pH values were used: 2.35, 2.70, 2.95, 3.20, and 3.40, and were prepared by the appropriate combination of 0.1 M citric acid and 0.2 M sodium hydrogen phosphate buffer solutions. Two $K^+$ concentrations were used: 0.02 M and 0.04 M, added in the form of KCl. Two tartaric acid concentrations were used: 0.02 M and 0.04 M. The total number of model juice systems investigated was 5 pH × 2 $K^+$ × 2 tartaric = 20 systems. Each juice system was prepared in duplicate. Cold stabilization was performed by storing all model juices at −3 °C for 7 weeks without any bitartrate crystal seeding. The pH of model juices was measured before and after cold stabilization using a pH meter model Orion 3 Star Series pH Benchtop (Thermo Electron Corp., Beverly, MA, USA).

## 3.3. Processing, Preparation, and Cold-Stabilization Treatments of Concord Juice

Frozen deseeded Concord grape mash was obtained from the New York State Agricultural Experiment Station collected from grapes harvested from the Lake Erie region (New York, NY, USA) during the 2013 growing season. The frozen juice was quickly defrosted during the initial stages of kettle pasteurization and diluted to 20.0 Brix, determined by digital refractometry (Misco model #PA203X; Misco Refractometer, Solon, OH, USA). The mash was depectinized with 30 mL of Adex-G (DSM Enzymes, Heerlen, The Netherlands) per 19 L of mash. After bulk pasteurization at 60 °C (140°F) using a 85 L tilting steam-jacketed kettle (model 20CD 1979; Lee Industries, Philipsburg, PA, USA), the juice was immediately pre-filtered through several layers of cheesecloth before being passed through two plate filter beds packed with 0.75% (w/w) Celite 503 (Imerys Filtration Minerals, San Jose, CA, USA) using a size 7 Shriver plate and frame filter press (FLSmidth, Salt Lake City, UT, USA).

Six equal aliquots of juice were pH adjusted from the base pH of 3.05 using either 1 M HCl or 1 M NaOH as required to form the pH 2.45, 2.65, 2.85, 3.05, 3.25, and 3.45 sample series. Following adjustment, and prior to cold stabilization, samples were analyzed for soluble solids by refractometry, organic acids by HPLC [34], and anthocyanins. Triplicate 500 mL portions of each pH-adjusted juice were transferred to autoclave-sterilized 500 mL Pyrex storage bottles (Thermo Fisher

Scientific, Waltham, MA, USA) and cold stabilized for three weeks at $-3\,°C$, shielded from light.

### 3.4. Characterization of Anthocyanins in KHT Crystals

KHT crystals were collected by filtration on a glass fiber filter (Type A/E, PALL Corp, Ann Arbor, MI, USA), followed by a washing step with cold 95% ethanol to remove any loosely adhering material on the crystal surface. The crystals were dried to constant weight in an oven at $60\,°C$ and weighed prior to anthocyanin analysis.

### 3.4.1. KHT Solubilization

For the blackcurrant model juice samples, 50 mg of KHT crystals were dissolved at room temperature in 3 mL of 1M NaCl acidified with HCl (0.01 M). When less than 50 mg of precipitate was formed, a proportionally reduced volume of the acidified NaCl solution was used for dissolution. After dissolution (approximately 15 min) each sample was immediately filtered through a 0.2 μm regenerated cellulose membrane (Sigma-Aldrich, St. Louis, MO, USA) in preparation for HPLC analysis. The KHT crystals recovered from the Concord juice sample set were prepared in a similar fashion with the exception that the entire recovered portion of KHT crystal was dissolved in 500 mL of the NaCl/HCl solution.

### 3.4.2. HPLC Analysis of Anthocyanins in Model Juice Samples

For the blackcurrant samples, resuspended anthocyanins were analyzed using an Agilent 1100 series HPLC system with inline degasser, autosampler and diode array detector (Agilent Technologies, Santa Clara, CA, USA). A 250 mm × 4.6 mm Varian LiChrospher RP-18 endcapped column (particle size 5 μm, pore size 100 Å; Varian, Inc., Palo Alto, CA, USA) was maintained at $30\,°C$ by an Eppendorf CH-30 external column heater. A 50 μL aliquot of each sample was injected on to the HPLC system. Mobile phase A consisted of water/phosphoric acid (99.5:0.5) and mobile phase B consisted of acetonitrile/water/phosphoric acid (50:49.5:0.5). Analytes of interest were resolved over a 38 min gradient elution profile starting at 0% B for 2 min, increasing to 20% B over 5 min, increasing to 36% B over 15 min, increasing to 100% B over 6 min, holding at 100% B for 2 min, followed by an 8 min return to starting conditions. The flow rate was 1 mL/min. The eluent was monitored at 520 nm. Analytes were identified based on comparison of relative retention times to those previously reported for anthocyanins in blackcurrant juice [27]. Quantification of each anthocyanin was based on a malvidin-3-glucose standard curve and thus reported in units of malvidin-3-glucoside equivalents, as is common for anthocyanin analyses [35]. The total anthocyanin content was calculated as the sum of all major anthocyanins identified by HPLC analysis and expressed in units of malvidin-3-glucoside equivalents. Two analytical replicates were performed on each sample.

### 3.4.3. HPLC Analysis of Anthocyanins in Real Juice Samples

Due to the presence of diglucosidic forms, anthocyanins from the Concord juice sample set were resolved using a 100 mm × 2.1 mm ID (2.6 μm particle size) Kinetex pentafluorophenyl (PFP) column (Phenomenex, Torrance, CA, USA) attached to an Agilent 1260 HPLC system equipped with an inline degasser, autosampler, thermostated column compartment, and diode array detector and analyzed using a previously published method [36]. Retention times had been previously demonstrated to not vary significantly among juice samples and standards. Monoglucosides and modified anthocyanins (acylated, methylated, *etc.*) were quantified using a malvidin-3-glucoside standard while diglucosides were quantified using a malvidin-3,5-diglucoside standard. The stability of the anthocyanins in the acidified NaCl solution preparation was evaluated by repeatedly analyzing one sample from each pH solution at 0, 24, 48, 72 h. No significant differences were observed.

### 3.5. Statistical Analysis

Results were reported in mean ± standard deviation. Data were subjected to analysis of variance (ANOVA). For treatments with a significant effect, means were compared with Tukey-Kramer HSD at 95% confidence interval using the JMP® 8.0 statistical software package (SAS institute Inc., Cary, NC, USA).

### 4. Conclusions

At pH ≤ 3, potassium bitartrate removal by cold stabilization of both real and model juices results in significantly less coprecipitation of anthocyanins with KHT crystals. This is likely due to changes in the charged of the KHT crystal surface, and may explain a previous observation that cold-stabilization of concentrate results in negligible coprecipitation. Potentially, pH adjustments could be made to juice by chemical or physical processes prior to cold stabilization to reduce losses of anthocyanins. Acylated and monoglucosides coprecipitate to a much greater extent than diglucosides at pH > 3, but this effect was minimized at low pH. Other minor selectivity effects were observed due to the aglycone.

**Supplementary Materials:** Supplemental Figure S1: Mass balance of anthocyanin species before and after cold stabilization. Supplementary materials can be accessed at: http://www.mdpi.com/1420-3049/20/01/0556/s1.

**Acknowledgments:** This work was supported by the New York State Agricultural Experiment Station (NYSAES) and the New York Wine and Grape Foundation Total Quality Focus & Sustainability Program.

**Author Contributions:** G.L.S. and O.I.P.-Z. conceived and designed the experiments; P.S. and D.M. performed the experiments; G.L.S., P.S. and D.M. analyzed the data; G.L.S., O.I.P.-Z. and D.M wrote the paper.

**Conflicts of Interest:** The authors declare no conflict of interest.

## References

1. Oszmianski, J.; Lee, C.Y. Isolation and HPLC determination of phenolic compounds in red grapes. *Am. J. Enol. Vitic.* **1990**, *41*, 204–206.
2. Wang, Y.; Catana, F.; Yang, Y.; Roderick, R.; van Breemen, R.B. An LC-MS method for analyzing total resveratrol in grape juice, cranberry juice, and in wine. *J. Agric. Food Chem.* **2002**, *50*, 431–435.
3. Morris, J.R.; Striegler, R.K. Grape juice: Factors that influence quality, processing technology, and economics. In *Processing Fruits: Science and Technology,*, 2nd ed.; Barrett, D.M., Somogyi, L.P., Ramaswamy, H.S., Eds.; CRC: Boca Raton, FL, USA, 2005; pp. 585–616.
4. Wallace, T.C. Anthocyanins in cardiovascular disease. *Adv. Nutr.* **2011**, *2*, 1–7.
5. Rodriguez-Mateos, A.; Heiss, C.; Borges, G.; Crozier, A. Berry (poly)phenols and cardiovascular health. *J. Agric. Food Chem.* **2014**, *62*, 3842–3851.
6. Harbourne, N.; Marete, E.; Jacquier, J.C.; O'Riordan, D. Stability of phytochemicals as sources of anti-inflammatory nutraceuticals in beverages—A review. *Food Res. Int.* **2013**, *50*, 480–486.
7. Sacchi, K.L.; Bisson, L.F.; Adams, D.O. A review of the effect of winemaking techniques on phenolic extraction in red wines. *Am. J. Enol. Vitic.* **2005**, *56*, 197–206.
8. Delgado-Vargas, F.; Jiménez, A.R.; Paredes-López, O. Natural pigments: Carotenoids, anthocyanins, and betalains—Characteristics, biosynthesis, processing, and stability. *Crit. Rev. Food Sci. Nutr.* **2000**, *40*, 173–289.
9. Ingalsbe, D.; Neubert, A.; Carter, G. Concord grape pigments. *J. Agric. Food Chem.* **1963**, *11*, 263–268.
10. Alongi, K.S.; Padilla-Zakour, O.I.; Sacks, G.L. Effects of concentration prior to cold-stabilization on anthocyanin stability in Concord grape juice. *J. Agric. Food Chem.* **2010**, *58*, 11325–11332.
11. Correa-Gorospe, I.; Polo, M.; Hernandez, T. Characterization of the proteic and the phenolic fraction in tartaric sediments from wines. *Food Chem.* **1991**, *41*, 135–146.
12. Vernhet, A.; Dupre, K.; Boulange-Petermann, L.; Cheynier, V.; Pellerin, P.; Moutounet, M. Composition of tartrate precipitates deposited on stainless steel tanks during the cold stabilization of wines. Part II. Red wines. *Am. J. Enol. Vitic.* **1999**, *50*, 398–403.
13. Amerine, M.A.; Ough, C.S. *Methods for Analysis of Musts and Wines*; Wiley: New York, NY, USA, 1980.
14. Mattick, L.; Shaulis, N.; Moyer, J. The effect of potassium fertilization on the acid content of Concord grape juice. *Am. J. Enol. Vitic.* **1972**, *23*, 26–30.
15. Zoecklein, B.W.; Fugelsang, K.C.; Gump, B.H.; Nury, F.S. *Wine Analysis and Production*; Kluwer Academic/Plenum Publishers: New York, NY, USA, 1999.

16. Balakian, S.; Berg, H. The role of polyphenols in the behavior of potassium bitartrate in red wines. *Am. J. Enol. Vitic.* **1968**, *19*, 91–100.

17. Boulange-Petermann, L.; Vernhet, A.; Dupre, K.; Moutounet, M. Kht cold stabilization: A scanning electron microscopy study of the formation of surface deposits on stainless steel in model wines. *Vitis* **1999**, *38*, 43–45.

18. Konja, G.; Lovric, T. Berry fruit juices. In *Fruit Juice Processing Technology*; Nagy, S., Chen, C.S., Shaw, P.E., Eds.; Agscience: Auburndale, FL, USA, 1993; pp. 436–514.

19. Dunsford, P.; Boulton, R. The kinetics of potassium bitartrate crystallization from table wines. I. Effect of particle size, particle surface area and agitation. *Am. J. Enol. Vitic.* **1981**, *32*, 100–105.

20. Dunsford, P.; Boulton, R. The kinetics of potassium bitartrate crystallization from table wines. II. Effect of temperature and cultivar. *Am. J. Enol. Vitic.* **1981**, *32*, 106–110.

21. Gerbaud, V.; Gabas, N.; Blouin, J.; Laguerie, C. Nucleation studies of potassium hydrogen tartrate in model solutions and wines. *J. Cryst. Growth* **1996**, *166*, 172–178.

22. Buchert, J.; Koponen, J.M.; Suutarinen, M.; Mustranta, A.; Lille, M.; Törrönen, R.; Poutanen, K. Effect of enzyme-aided pressing on anthocyanin yield and profiles in bilberry and blackcurrant juices. *J. Sci. Food Agric.* **2005**, *85*, 2548–2556.

23. Montgomery, M.W.; Reyes, F.G.R.; Cornwell, C.; Beavers, D.V. Sugars and acid analysis and effect of heating on color stability of northwest Concord grape juice. *J. Food Sci.* **1982**, *47*, 1883–1885.

24. Celotti, E.; Bornia, L.; Zoccolan, E. Evaluation of the electrical properties of some products used in the tartaric stabilization of wines. *Am. J. Enol. Vitic.* **1999**, *50*, 343–350.

25. Rodriguez-Clemente, R.; Correa-Gorospe, I. Structural, morphological, and kinetic aspects of potassium hydrogen tartrate precipitation from wines and ethanolic solutions. *Am. J. Enol. Vitic.* **1988**, *39*, 169–179.

26. Boulton, R.B.; Singleton, V.L.; Bisson, L.F.; Kunkee, R.E. *Principles and Practices of Winemaking*; Kluwer Academic/Plenum Publishers: New York, NY, USA, 1999.

27. Slimestad, R.; Solheim, H. Anthocyanins from black currants (*Ribes nigrum* L.). *J. Agric. Food Chem.* **2002**, *50*, 3228–3231.

28. Evans, J.W. Random and cooperative sequential adsorption. *Rev. Mod. Phys.* **1993**, *65*, 1281–1329.

29. Wrolstad, R.E. Symposium 12: Interaction of natural colors with other ingredients—Anthocyanin pigments—Bioactivity and coloring properties. *J. Food Sci.* **2004**, *69*, C419–C421.

30. Ohta, H.; Tonohara, K.; Yoza, K.I.; Nogata, Y. Tartar stabilization of Concord grape juice by means of ion-exchange resins—(Studies on quality of grape juice.12). *J. Jpn. Soc. Food Sci. Technol.-Nippon Shokuhin Kagaku Kogaku Kaishi* **1992**, *39*, 1105–1111.

31. Harvey, D. *Modern Analytical Chemistry*; McGraw-Hill: Boston, MA, USA, 2000.

32. Côté, J.; Caillet, S.; Doyon, G.; Sylvain, J.F.; Lacroix, M. Analyzing cranberry bioactive compounds. *Crit. Rev. Food Sci. Nutr.* **2010**, *50*, 872–888.

33. Kraemer Schafhalter, A.; Fuchs, H.; Pfannhauser, W. Solid phase extraction (SPE)—A comparison of 16 materials for the purification of anthocyanins from Aronia melanocarpa var nero. *J. Sci. Food Agric.* **1998**, *78*, 435–440.
34. Castellari, M.; Versari, A.; Spinabelli, U.; Galassi, S.; Amati, A. An improved hplc method for the analysis of organic acids, carbohydrates, and alcohols in grape musts and wines. *J. Liquid Chromatogr. Rel. Technol.* **2000**, *23*, 2047–2056.
35. Waterhouse, A.L.; Price, S.F.; McCord, J.D. Reversed-phase high-performance liquid chromatography methods for analysis of wine polyphenols. *Methods Enzymol.* **1999**, *299*, 113–121.
36. Manns, D.C.; Mansfield, A.K. A core-shell column approach to a comprehensive high-performance liquid chromatography phenolic analysis of *Vitis vinifera* L. and interspecific hybrid grape juices, wines, and other matrices following either solid phase extraction or direct injection. *J. Chromatogr. A* **2012**, *1251*, 111–121.

**Sample Availability:** *Sample Availability*: Samples of the compounds are not available from the authors.

# The Encapsulation of Anthocyanins from Berry-Type Fruits. Trends in Foods

Paz Robert and Carolina Fredes

**Abstract:** During the last decade, many berry-type fruits have been recognised as good sources of anthocyanins. Nevertheless, the use of anthocyanins in the development of food colourants and healthy and/or functional ingredients has been limited because of their low stability under given environmental conditions and interaction with other compounds in the food matrix. This review compiles information about the encapsulation of anthocyanins from twelve different berry-type fruit species as a technology for improving the stability and/or bioavailability of anthocyanins. Encapsulation by spray drying has been the primary method used to encapsulate anthocyanins, and some studies attempt to keep anthocyanin microparticles stable during storage. Nevertheless, more studies are needed to determine the stability of anthocyanin microparticles in food matrices over the product shelf life in the development of food colourants. Studies about encapsulated anthocyanins in simulated gastrointestinal models have primarily been conducted on the release of anthocyanins from microparticles to evaluate their bioavailability. However, adding anthocyanin microparticles to a food vehicle must guarantee the health properties attributed to the specific anthocyanins present in berry-type fruits.

Reprinted from *Molecules*. Cite as: Robert, P.; Fredes, C. The Encapsulation of Anthocyanins from Berry-Type Fruits. Trends in Foods. *Molecules* **2015**, *20*, 5875–5888.

## 1. Introduction

Berry-type fruits have long been regarded as having considerable health benefits because of their nutritional attributes, particularly their total antioxidant activity against cellular oxidation reactions [1]. These benefits have stimulated research to investigate the phenolic status and antioxidant activity of distinct berry fruit species and new varieties in different countries. There is a great variety of species from diverse botanical families (e.g., blueberries (*Vaccinium corymbosum* L., Ericaceae), strawberries (*Fragaria ananassa* Duch., Rosaceae), red raspberries (*Rubus idaeus* L., Rosaceae), and blackberries (*Rubus* sp., Rosaceae)) that produce the small purple or red fruits that are denoted as berries. In botanical terms, a berry is a fruit with many seeds and mesocarp flesh that evolves from a flower with a superior ovary [2]. Therefore, in strictly botanical terms, none of the above mentioned fruits are true berries. Nevertheless, these fruit species share a red-blue colour and a high polyphenol content and antioxidant activity. Because of the various botanical families and fruits to which berry-type fruits belong, a wide range of phenolic contents and corresponding antioxidant activities can be expected according to the specific compounds that are present in these species [3,4].

In recent years, other berry-type fruits from worldwide: bilberries (*Vaccinium myrtillus* L., Ericaceae), blackcurrant (*Ribes nigrum* L., Grossulariaceae), pomegranate (*Punica granatum* Linn., Punicaceae) and açaí (*Euterpe oleracea* Mart, Arecaceae), have also gained increased interest mainly due to the potential health benefits and the consumer demand of novel fruits [1].

Anthocyanins are responsible for the colour of berry-type fruits [5–10]. In addition to their colourant properties, anthocyanins have been associated with a wide range of biological, pharmacological, anti-inflammatory, antioxidant, and chemoprotective properties [11]. Anthocyanins are also beneficial against many chronic diseases [12,13]. New evidence highlights the potential use of berry-type fruit species as a source of bioactive compounds, primarily anthocyanins, in food and nutraceutical industries. However, the use of anthocyanin-rich extracts as food colourants and healthy foods is limited because of the low stability of anthocyanins under the environmental conditions (heat, oxygen, and light among others) experienced during processing and/or storage [14]. Anthocyanins are water-soluble pigments that correspond to the glycoside or acyl-glycoside of anthocyanidins and are stored in the plant cell vacuole [15,16]. The stability of anthocyanins is affected by pH, temperature, the presence of light, metal ions, oxygen enzymes, ascorbic acid, sugars and their degradation products, proteins and sulphur dioxide [17]. Additionally, anthocyanin bioavailability is low because of its sensitivity to pH changes. In general, anthocyanins are generally stable at pH values of 3.5 and below and they degrade at higher pH values [11]. Because the high instability of isolated anthocyanins from berry-type fruits has a direct impact on their colour stability and potential health benefits, encapsulation technology can be used to improve the stability and/or bioavailability of anthocyanins [18]. The objectives of this paper are to review the evidence regarding the encapsulation of anthocyanins from berry-type fruits by spray drying and to compile the new applications of anthocyanin microparticles in foods to propose future perspectives.

## 2. The Encapsulation of Anthocyanins from Berry-Type Fruits

Encapsulation is a technique by which active solid, liquid or gas compounds are introduced into a matrix or a polymeric wall system to protect the "actives" from environmental conditions, their interactions with other food components or to control their release (for a specific place and/or time) [19]. The polymers used in microencapsulation are called encapsulating agents (EA). The resulting microparticles are vesicles or small particles in which the size can vary from sub-microns to several millimetres.

During the last four years, several studies on the encapsulation of anthocyanins from different berry-type fruits have been reported (Table 1). In all of these, anthocyanin encapsulation has primarily been focused on providing protection from environmental conditions (light, oxygen, temperature and water), avoiding oxidation and increasing the shelf life of active compounds. Currently, anthocyanin encapsulation is focused on studying the anthocyanins released in simulated gastrointestinal tracts [20–25].

In the encapsulation of anthocyanins, fruit pulps, juices and extracts are selected from commonly consumed fruit species or exotic ones, and Andes berry (*Rubus glaucus* Benth., Rosaceae), bayberry (*Myrica* gole L., Myricaceae), black mulberry (*Morus nigra*, Moraceae), corozo (*Bactris guineensis*, Arecaceae), jaboticaba (*Myrciaria jaboticaba* (Vell.) O. Berg, Myrtaceae) and *kokum* (*Garcinia indica* Choisy, Guttiferae) have been used as raw materials. Pomace extract from bilberry [23,26], blackcurrant [27] and blueberry [24,25] and peel extract from jaboticaba [28,29] have also been used because anthocyanins are accumulated primarily in the fruit epicarp (peel) [30]. The anthocyanins from berry-type fruit have been encapsulated primarily by the oil dispersed phase, double emulsion (w/o/w), extrusion, emulsification/heat gelation, microgel synthesis, freeze drying, supercritical $CO_2$, spray drying and ionic gelification (Table 1), but the most common method used to encapsulate anthocyanins from berry-type fruits is spray drying (Figure 1). Compared to the other encapsulation methods, spray drying allows one to obtain berry-type microparticle powders in a one step process. In addition, spray dryers are equipment commonly available in food and pharmaceutical industries.

## 2.1. Raw Materials

Several methods for treating fruits as raw materials for anthocyanin encapsulation have been described. The simplest method is to employ fruit pulp [31] or fruit juice [32–36] without any additional solvent; a filtration process is followed to eliminate solids before the preparation of a feed drying solution for the encapsulation process. The feed drying solution is obtained by homogenising the pulp or juice with one or more EAs. More steps are involved when either fruit pomace (PO) or peel (PE) are used as raw materials. Bilberry PO [23,37] commonly comes from pomace extraction with methanol, following filtration, evaporation and lyophilisation processes performed by providers. Bilberry PO also contains other polyphenols, tannins, carbohydrates and roughage [23,37]. Jaboticaba PE [28] comes from a peel extraction with ethanol (70%) that is acidified (pH 2.0) with hydrochloric acid, following filtration and evaporation processes. The feed drying solution is obtained in a similar way as that of the fruit extracts described above.

**Table 1.** The encapsulation of anthocyanins from different berry-type fruits.

| Raw Material | Encapsulation Method | Encapsulating Agent | References |
|---|---|---|---|
| Andes berry fruit extract | Spray drying | maltodextrin, gum Arabic, corn starch, yucca starch, Capsul® TA (Ingredion Incorporated, Westchester, IL, USA), Hi-CA™100 (Ingredion Incorporated, Westchester, IL, USA) | [38] |
| Bayberry fruit extract | Oil dispersed phase—spray drying | ethyl cellulose | [20] |
| Bayberry fruit juice | Spray drying | maltodextrin | [33] |
| Bayberry fruit juice | Spray drying | whey protein isolate or maltodextrin | [34] |
| Bilberry fruit extract | Emulsion | whey protein isolate | [21] |
| Bilberry fruit extract | Double emulsion (w/o/w) | pectin (calcium chloride), PGPR | [22] |
| Bilberry pomace extract | (a) extrusion; (b) emulsification/heat gelation; (c) spray drying | (a) amidated pectin; (b) whey protein isolate; (c) maltodextrin + pectin | [37] |
| Bilberry pomace extract | (a) emulsification/heat gelation; (b) extrusion | (a) whey protein isolate; (b) amidated pectin | [26] |
| Bilberry pomace extract | (a) extrusion; (b) emulsification/heat gelation; (c) spray drying | (a) amidated pectin; (b) whey protein isolate; (c) maltodextrin + pectin | [23] |
| Blackcurrant pomace | Spray drying | maltodextrin or inulin | [27] |
| Blackberry pulp | Spray drying | maltodextrin | [39] |
| Blackmulberry fruit juice | Spray drying | maltodextrin or gum Arabic | [35] |
| Blueberry fruit/pomace extract | Spray drying | whey protein isolate or gum Arabic | [24] |
| Blueberry fruit extract | Spray drying | mesquite gum | [40] |
| Blueberry pomace extract | Microgel synthesis | oxidized potato starch + sodium trimetaphosphate (STMP) | [25] |
| Corozo fruit extract | Spray drying | maltodextrin | [41] |
| Grape fruit juice | Freeze drying | maltodextrin + gum Arabic | [42] |
| Jaboticaba peel extract | Spray drying | maltodextrin, gum Arabic + maltodextrin or Capsul™ + maltodextrin | [28] |
| Jaboticaba peel extract | (a) Rapid Extraction of Supercritical Solution (RESS); (b) ionic gelification | (a) polyethyleneglycol (PEG); (b) Ca-alginate | [29] |
| Kokum fruit extract | Spray drying | maltodextrin, gum Acacia or tricalcium phosphate | [36] |
| Pomegranate fruit juice or fruit extract | Spray drying | maltodextrin or soybean protein isolate | [32] |

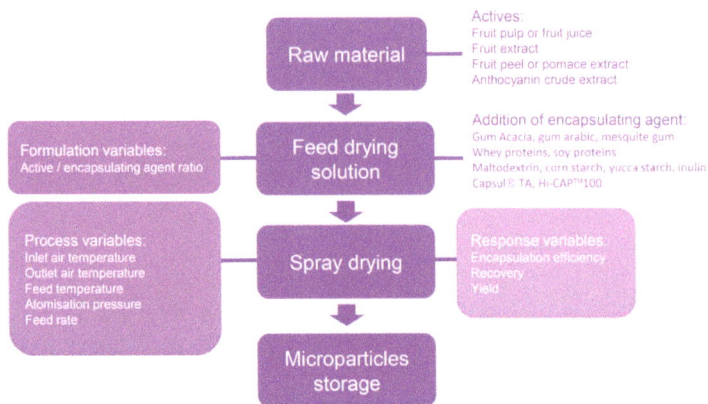

Figure 1. Microencapsulation of anthocyanins from berry-type fruits by spray drying and variables that must be considered in the feed formulation and process.

Some studies on anthocyanin encapsulation have described different methods of anthocyanin extraction from fruits before the preparation of the feed drying solution that involved more steps in the encapsulation process. Zheng *et al.* [20] have described the use of a microwave-assisted extraction method for bayberry fruits by using ethanol (80%) as a solvent and a rotary evaporation at 55 °C to remove excess solvent and some other low boiling point impurities. Similarly, other authors described the use of ethanol/water (1:1) for pomegranate arils [32], ethanol (96%) for blueberry fruits [40] and ethanol (80%) and citric acid (0.5%) for blackcurrant pomace [27]. Amberlite XAD16N column has been used to follow a solid-liquid fractionation of anthocyanins from whole blueberries and blueberry pomace with three different solvent systems (acetonic, ethanolic and methanolic) [24] and from corozo fruits without seeds with methanol/acetic acid [41], then the resulting extracts were freeze-dried.

*2.2. Encapsulating Agents (EA)*

A variety of EAs have been studied for anthocyanin encapsulation. Natural gums such as gum arabic [24,28,35,38], mesquite gum [40] and gum Acacia [36], proteins such as whey proteins [24,34] and soy proteins [32], polysaccharides such as maltodextrins of different dextrose equivalents [23,27,28,32–39], inulin [27], corn starch [38] and yucca starch [38], and modified polysaccharides such as Capsul® TA [28,38] and Hi-CAP™100 [38] have been successfully used in spray-drying. The choice of EA is very important for the proper encapsulation efficiency (EE), the stability of the active compounds in the microparticles during storage and the release properties in foods and the gastrointestinal tract. In the microencapsulation of anthocyanins, maltodextrin has been shown to be essential for preserving the

102

integrity of anthocyanins for encapsulation [28]. Nevertheless, the use of this EA is limited by its solubility in water, and the subsequent release of anthocyanins in liquid media, which does not allow colour stability when the microparticles are applied in liquid foods such as dairy products [32].

## 2.3. The Encapsulation of Anthocyanins from Berry-Type Fruits by Spray Drying

Spray drying is widely used in the food industry to encapsulate active compounds and protect materials in an economic, simple and continuous way [19]. However, it is considered to be an immobilisation technology rather than a true encapsulation technology because some active compounds may be exposed superficially on the microparticles [43].

By using this technique, the feed or dispersion solution is sprayed (with a nozzle or a rotating disc), in the form of fine drops in a hot air flow. When the liquid droplets come into contact with the hot air, a resulting powder is instantaneously produced by the rapid evaporation of water [44]. In addition to the simplicity of spray drying, another advantage of this technique is that it is useful for encapsulating heat sensitive materials because the time of exposure to elevated temperatures is very short (5–30 s) [45]. Using this method, it is possible to obtain powder microparticles with low water activity that facilitate the transportation, handling, and storage of the product and ensure microbiological quality [44].

It is known that optimum drying conditions should be used to obtain a high encapsulation efficiency, which leads to the use of experimental design. The feed temperature, inlet air temperature and air outlet temperature (process variables) and active/encapsulating agent ratio (formulation variable) [44] have been reported as important variables in encapsulation efficiency, recovery, yield, and antioxidant activity (Figure 1). Inlet air temperature has been associated with anthocyanin oxidation and/or degradation reactions induced by heat. Different encapsulating agents have different optimum parameters because encapsulating agent features, such as solubility and viscosity, affect the formation rate of a crust on the particle surface [44,46]. The response surface methodology is applied to optimise the response variables that fit to some regression model, generally a second-order regression. For multiple optimisations, a desirable function is used.

A laboratory spray dryer (BÜCHI Labortechnik AG, Postfach, Switzerland) is generally used for anthocyanin encapsulation experiments and, in this instrument, the inlet air temperature can be regulated but the outlet air temperature varies according to the inlet temperature.

## 2.4. The Characterisation of Berry-Type Fruit Microparticles

The EEs represent the anthocyanin-polymer interaction from electrostatic interactions, or hydrogen bonding. Encapsulation efficiency is obtained by the quantification of the superficial and total anthocyanins [33]. Anthocyanin quantification is primarily achieved either colourimetrically or by high-performance liquid chromatography (HPLC). The principal method is the pH differential method by spectrophotometry, which is widely used in industry because it is a rapid and easy procedure to perform [47]. However, this method cannot provide any information regarding the compositional profile of anthocyanins and the relative amounts of these compounds.

The EE of anthocyanins from berry-type fruits has commonly been determined by quantifying the total anthocyanins by spectrometry [22–25,28,29,32,33,36,39,40,42] or HPLC-DAD [21,23,26]. This profile indicates that the EEs of specific anthocyanins have not been reported in those studies. The EE of each anthocyanin may have special relevance when a fruit species has a wide profile of anthocyanins. For example, fifteen and thirteen anthocyanins have been identified in different bilberry [48] and blueberry [49] genotypes, respectively. Therefore, a wide range of EEs could be expected because of the different structural features of the anthocyanins identified in these fruits. However, there are not any studies on the structure-encapsulation efficiency relation. Additionally, the EE of specific anthocyanins can be crucial when the primary bioactivity and/or specific health effect of some fruits may be attributed to certain anthocyanins with specific structural features. Fang and Bhandari [33], using HPLC-DAD-ESIMS, have analysed the individual phenolics (gallic acid, cyanidin 3-glucoside, quercetin 3-galactoside, quercetin 3-glucoside and quercetin deoxyhexoside [tentatively identified]) of a bayberry juice solution before spray drying, the powders obtained immediately after spray drying, and the powders after 6 months of storage. The five phenolic compounds were found in the bayberry powders, with a high retention percentage (~93%–97%). Nevertheless, at the end of six months of storage, the phenolic contents declined at different rates, depending on the storage conditions. Cyanidin 3-glucoside was not detected when the powder was stored at 40 °C for 6 months, suggesting that the cyanidin 3-glucoside in bayberry powder is less stable than gallic acid and flavonols under the conditions of the study [33]. Different EAs could allow different EEs. Robert *et al.* [32] reported that the EE reached higher values for total anthocyanin than total polyphenol contents performed by spectrometry, showing the ability of maltodextrin MD (DE 12–20) and soybean protein isolate to bind anthocyanins. Thus, the flavylium cation could be related to the better polymer-anthocyanin interaction. This finding was consistent with Ersus and Yurdagel [50] who obtained the greatest pigment retention in the microencapsulation of anthocyanins from black carrot (*Daucus carota* L.) by spray drying with maltodextrin and DE 20–23.

The recovery from spray drying was reportedly influenced by the encapsulating agent properties (viscosity and solubility), anthocyanin/encapsulating agent ratio, and the inlet air temperature. A high recovery of anthocyanins could be attributed to short drying times or the rapid formation of a dry crust, which allows for water diffusion but retains the active properties.

Some studies on the stability of anthocyanin microparticles during storage have been reported [27,32,33]. Pseudo-first order kinetics for encapsulated anthocyanins from pomegranate juice and pomegranate ethanolic extract with maltodextrin (DE 12-20) and soybean protein isolate have been reported [31]. The storage conditions of anthocyanin microparticles, such as the polymer nature, temperature, and water activity (aw), among others influence their stability. The effect of the polymer's nature was observed in encapsulated blackcurrant with different maltodextrins (DE 11, DE 18 and DE 21) and inulin during storage at 8 °C and 25 °C [27]. Anthocyanins were significantly more stable in inulin, and no effect from the DE in maltodextrins was found [27]. In encapsulated pomegranate juice with maltodextrin (DE 12-20) and soybean protein isolate stored at 60 °C, maltodextrin showed the best anthocyanin protective effect, and it had the lowest degradation rate constant [32]. The effect of the aw was studied in encapsulated bayberry juice, with maltodextrin DE10 stored at different temperatures (4 °C, 25 °C and 40 °C) and aw amounts (0.11, 0.22, 0.33 and 0.44) [33]. At a higher aw during storage, the degradation of anthocyanins increases [33]. The storage temperature plays an important role in the stability of anthocyanins, which is associated with the glass transition temperature of a spray-dried powder (rubber-glassy transition).

Other studies have been conducted with spray-dried juice using carrier agents although the encapsulation parameters were not evaluated [31]. In this context, spray-dried açai juice with maltodextrin (DE10 and DE20), Arabic gum and tapioca starch with aw of 0.33 and 0.53 were stored at 25 °C and 35 °C. The maltodextrin DE10 and the lower aw powders showed the best anthocyanin protection [31].

## 3. Applications in Foods

### 3.1. Anthocyanins as Colourants

Stability is a crucial factor to consider when anthocyanin pigments are used as food colourants. This issue has special relevance when natural colourants (based on anthocyanins) are compared with synthetic ones. Arocas et al. [51] compared the colour stability of three natural red colourants (cochineal (E120), enocyanin (E163) and dark carrot (E163)) and three artificial colours (allura red (E129), Carmoisine (E122) and Ponceau 4R (E124)), determining major differences in the natural colours' responses to changes in pH (3–8) and temperature (80 °C for 30 min), thus giving them a lower stability in relation to artificial colourants. Fracassetti et al. [52] evaluated the storage effects on the total and individual anthocyanin content of a lyophilised powder (used as a colourant) from wild blueberry (*Vaccinium angustifolium*) at 25 °C, 42 °C, 60 °C, and 80 °C for 49 days. The storage reduced the total and individual anthocyanin contents, and the reduction was slower at 25 °C (after 2 weeks), whereas it was more rapid at 60 °C and 80 °C after 3 days. Based on this result, the half-lives were determined to be 139, 39, and 12 days at 25 °C, 42 °C, and 60 °C, respectively. This result is less than five months at 25 °C.

The interaction of anthocyanins with ascorbic acid (AA) is particularly noteworthy because the effect of this interaction is largely negative. West and Mauer [53] have studied the colour and chemical stability of six anthocyanins (highly purified and present in semi-purified extracts) from grape pomace, purple corn, and black rice, in combination with ascorbic acid solutions at different pH values (3.0 to 4.0) and temperatures (6–40 °C), and in lyophilised powders at different relative humidity values (43%–98% RH). The results indicated that in liquids, stability was negatively correlated with increased pH and temperature, and that for powders, stability was inversely proportional to the relative humidity, also confirming the mutual destruction of the anthocyanins and ascorbic acid in solution. This finding is particularly important for beverages because they form a product category in which anthocyanins and AA are part of the same formulation.

The first antecedents of anthocyanin application in yoghurt were based on the addition of açai juice [54] and lyophilised fruit from *Berberis boliviana* [55] to commercial yoghurt. Nevertheless, these successful examples differ from the results of studies about anthocyanin addition during the process of yoghurt formulation. Karaaslan *et al.* [56] have evaluated the incorporation of wine grape ethanolic extracts before the yoghurt production process, noting a significant degradation of anthocyanins during storage. Similar results were reported by Sun-Waterhouse *et al.* [57], who indicated that fermentation affects the anthocyanin content of blackcurrant extract during the yoghurt formulation process. Additionally, Scibisz *et al.* [58] determined that certain probiotic cultures significantly affect the stability of anthocyanins from cranberry fruit when added to yoghurt.

Microencapsulation has been used as a strategy to address the stability of anthocyanins that are used as natural colourants. Robert *et al.* [32] compared the degradation kinetics of encapsulated and un-encapsulated anthocyanins from pomegranate juice, obtaining similar degradation constants when they are applied to yoghurt, showing the loss of the protection after addition to the yoghurt. This finding may be explained by the solubility of the EA in hydrophilic matrices. Therefore, the selection of an EA that allows microparticle dispersion may be an alternative to designing a dairy product such as yoghurt with natural colourant based on a berry-type fruit.

*3.2. Anthocyanins in Healthy Foods*

In recent years, encapsulation technology has increased in importance in the food industry, particularly in the development of functional and/or healthy foods.

The use of encapsulated anthocyanins as an ingredient in healthy foods should allow for the protection of the anthocyanins (the preservation of their nutritional properties) until they are consumed within the food vehicle [43]. However, to the best of our knowledge there are not studies of anthocyanin release from microparticles in food matrices. To find the kinetic release of anthocyanins from release curves in food, the data are fitted to mathematical models (Peppas [59], Higuchi [60] and Hixson-Crowel [61]), allowing researchers to obtain the release rate constants and to explain the release mechanism. When the anthocyanin microparticles are soluble in the food matrix, the anthocyanins are quickly released. Without this protection, the anthocyanins would otherwise be exposed to adverse conditions in the food (pH, enzymes, and other food components). In this case, the microparticles would be best suited to functional dry-mixes or instant food. Contrary, when microparticles are insoluble in food matrices, the release is very slow and they could be used for the formulation of liquid functional-foods. Therefore, the solubility of encapsulating agent determines the applicability of the microparticles in foods.

Some studies have been conducted on the effect of process variables of the spray drying (inlet air temperature, outlet temperature, feeding rate, and polymer nature (EA)) on the antioxidant capacity (measured by ABTS or DPPH) of berry-type fruit powders [23,27,40]. In all of them, a reduction of anthocyanin content (mainly due to the inlet air (140–205 °C) and outlet temperature) led to a decrease in antioxidant capacity, the last being used as a response variable [23,27,40]. In other work, the effect of storage temperature and aw on the antioxidant capacity of bayberry powders was undertaken [33].

The information about EA was evaluated for anthocyanin controlled release in simulated gastrointestinal digestion models [20–25], and they indicated the use of ethyl cellulose [20], whey protein isolate [21,23,24], pectin, PGPR [22], amidated pectin [23], maltodextrin + pectin [23], gum arabic [24], and potato starch/STMP [25].

Kropat et al. [26] have determined the anthocyanin release in cell culture from two applied microencapsulation systems (Table 1). Fifteen different anthocyanins, composed of five aglycons called delphinidin (del), cyanidin (cy), petunidin (pt), malvidin (ma), and peonidin (peo) and three glycons called glucose (glc), galactose (gal), and arabinose (arab), were identified in bilberry extract by HPLC-DAD, where del-glycosides (34.3% wt) and cy-glycosides (27.6% wt) predominated. Although the anthocyanins were stabilised by microencapsulation, the rate of the decrease in the anthocyanin content appeared to be substantially affected by their respective aglycon. Del-glycosides disappeared completely within 30 min, whereas cy-, pt-, ma-, and peo-glycosides were relatively stable.

## 4. Future Perspectives

During recent years, the official rules of the European Union (EU) and the United States have restricted the use of synthetic colourants (especially red ones) as food additives because of their potential adverse health effects. These effects are related to their carcinogenic effects in experimental models [62,63], possible allergenic effects [64], and hyperactivity in children (3 and 8–9 years) [65]. McCann *et al.* [65] have evaluated the mixture of synthetic colourants with sodium benzoate (frequently used in soft drinks and other foods), considering among the synthetic red colourants azorubine red (Carmoisine (E-122), Ponceau 4R (E-124) and Allura red AC (E-129)). Their findings indicated that childhood hyperactivity could be exacerbated by the use of synthetic colourants. Since the establishment of this evidence, the European Food Safety Authority has decreased the allowed daily intake levels of these synthetic colourants. Additionally, Chilean Food Health Regulations do not allow the use of colourants in infant foods (until 3 years) [66]. Therefore, international regulation trends in addition to informed consumers are a market opportunity for natural food colourants based on anthocyanin pigments in which microencapsulation is a key technology for the development of natural colourants that are stable in food matrices.

Studies about encapsulated anthocyanins in simulated gastrointestinal models have primarily been conducted on the release of anthocyanins from microparticles without using a food vehicle [22–25]. The use of encapsulated anthocyanins as ingredients in healthy foods should allow for the protection of anthocyanins until reaching the gastrointestinal site where anthocyanin release is desired.

**Acknowledgments:** Anillo ACT 1105 (CONICYT, Chile) and FONDECYT N° 3150342.

**Author Contributions:** Both authors have contributed equally to this work.

**Conflicts of Interest:** The authors declare no conflict of interest.

## References

1. Seeram, N.P. Berry fruits: Compositional elements, biochemical activities, and the impact of their intake on human health, performance, and disease. *J. Agric. Food Chem.* **2008**, *56*, 627–629.
2. Bowling, B. *The Berry Grower's Companion*; Timber Press, Inc.: Portland, OR, USA, 2000.
3. Fredes, C.; Montenegro, G.; Zoffoli, J.; Santander, F.; Robert, P. Comparison of total phenolic, total anthocyanin and antioxidant activity of polyphenol-rich fruits grown in Chile. *Cienc. Investig. Agrar.* **2014**, *41*, 49–60.
4. Speisky, H.; López-Alarcón, C.; Gómez, M.; Fuentes, J.; Sandoval-Vicuña, C. First web-based database on total phenolics and oxygen radical absorbance capacity (ORAC) of fruits produced and consumed within the South Andes Region of South America. *J. Agric. Food Chem.* **2012**, *60*, 8851–8859.

5.  Capocasa, F.; Scalzo, J.; Mezzetti, B.; Battino, M. Combining quality and antioxidant attributes in the strawberry: The role of genotype. *Food Chem.* **2008**, *111*, 872–878.

6.  Koca, I.; Karadeniz, B. Antioxidant properties of blackberry and blueberry fruits grown in the Black Sea Region of Turkey. *Sci. Hortic.* **2009**, *121*, 447–450.

7.  Caliskan, O.; Bayazit, S. Phytochemical and antioxidant attributes of autochthonous Turkish pomegranates. *Sci. Hortic.* **2012**, *147*, 81–88.

8.  Josuttis, M.; Carlen, C.; Crespo, P.; Nestby, R.; Toldam-Andersen, T.B.; Dietrich, H.; Krüger, E. A comparison of bioactive compounds of strawberry fruit from Europe affected by genotype and latitude. *J. Berry Res.* **2012**, *2*, 73–95.

9.  Skrede, G.; Martinsen, B.K.; Wold, A.-B.; Birkeland, S.-E.; Aaby, K. Variation in quality parameters between and within 14 Nordic tree fruit and berry species. *Acta Agric. Scand. Sect. B—Soil Plant Sci.* **2012**, *62*, 193–208.

10. Wang, S.Y.; Chen, H.; Camp, M.J.; Ehlenfeldt, M.K. Genotype and growing season influence blueberry antioxidant capacity and other quality attributes. *Int. J. Food Sci. Technol.* **2012**, *47*, 1540–1549.

11. De Pascual-Teresa, S.; Sanchez-Ballesta, M.T. Anthocyanins: From plant to health. *Phytochem. Rev.* **2007**, *7*, 281–299.

12. Wallace, T.C. Anthocyanins in cardiovascular disease. *Adv. Nutr.* **2011**, *2*, 1–7.

13. Huang, W.-Y.; Davidge, S.T.; Wu, J. Bioactive natural constituents from food sources-potential use in hypertension prevention and treatment. *Crit. Rev. Food Sci. Nutr.* **2013**, *53*, 615–630.

14. Frank, K.; Köhler, K.; Schuchmann, H.P. Stability of anthocyanins in high pressure homogenisation. *Food Chem.* **2012**, *130*, 716–719.

15. Routray, W.; Orsat, V. Blueberries and their anthocyanins: Factors affecting biosynthesis and properties. *Compr. Rev. Food Sci. Food Saf.* **2011**, *10*, 303–320.

16. Clifford, M.N. Review Anthocyanins—Nature, occurrence and dietary burden. *J. Sci. Food Agric.* **2000**, *80*, 1063–1072.

17. Fernandez-Lopez, J.A.; Angosto, J.; Gimenez, P.; Leon, G. Thermal stability of selected natural red extracts used as food colorants. *Plant Foods Hum. Nutr.* **2013**, *68*, 11–17.

18. Mahdavi, S.A.; Jafari, S.M.; Ghorbani, M.; Assadpoor, E. Spray-drying microencapsulation of anthocyanins by natural biopolymers: A review. *Dry Technol.* **2014**, *32*, 509–518.

19. Desai, K.G.H.; Jin Park, H. Recent developments in microencapsulation of food ingredients. *Dry Technol.* **2005**, *23*, 1361–1394.

20. Zheng, L.; Ding, Z.; Zhang, M.; Sun, J. Microencapsulation of bayberry polyphenols by ethyl cellulose: Preparation and characterization. *J. Food Eng.* **2011**, *104*, 89–95.

21. Betz, M.; Steiner, B.; Schantz, M.; Oidtmann, J.; Mäder, K.; Richling, E.; Kulozik, U. Antioxidant capacity of bilberry extract microencapsulated in whey protein hydrogels. *Food Res. Int.* **2012**, *47*, 51–57.

22. Frank, K.; Walz, E.; Gräf, V.; Greiner, R.; Köhler, K.; Schuchmann, H.P. Stability of anthocyanin-rich w/o/w-emulsions designed for intestinal release in gastrointestinal environment. *J. Food Sci.* **2012**, *77*, N50–N57.

23. Oidtmann, J.; Schantz, M.; Mäder, K.; Baum, M.; Berg, S.; Betz, M.; Kulozik, U.; Leick, S.; Rehage, H.; Schwarz, K.; *et al.* Preparation and comparative release characteristics of three anthocyanin encapsulation systems. *J. Agric. Food Chem.* **2012**, *60*, 844–851.

24. Flores, F.P.; Singh, R.K.; Kerr, W.L.; Pegg, R.B.; Kong, F. Total phenolics content and antioxidant capacities of microencapsulated blueberry anthocyanins during *in vitro* digestion. *Food Chem.* **2014**, *153*, 272–278.

25. Wang, Z.; Li, Y.; Chen, L.; Xin, X.; Yuan, Q. A study of controlled uptake and release of anthocyanins by oxidized starch microgels. *J. Agric. Food Chem.* **2013**, *61*, 5880–5887.

26. Kropat, C.; Betz, M.; Kulozik, U.; Leick, S.; Rehage, H.; Boettler, U.; Teller, N.; Marko, D. Effect of microformulation on the bioactivity of an anthocyanin-rich bilberry pomace extract (*Vaccinium myrtillus* L.) in vitro. *J. Agric. Food Chem.* **2013**, *61*, 4873–4881.

27. Bakowska-Barczak, A.M.; Kolodziejczyk, P.P. Black currant polyphenols: Their storage stability and microencapsulation. *Ind. Crops Prod.* **2011**, *34*, 1301–1309.

28. Silva, P.I.; Stringheta, P.C.; Teófilo, R.F.; de Oliveira, I.R.N. Parameter optimization for spray-drying microencapsulation of jaboticaba (*Myrciaria. jaboticaba*) peel extracts using simultaneous analysis of responses. *J. Food Eng.* **2013**, *117*, 538–544.

29. Santos, D.T.; Albarelli, J.Q.; Beppu, M.M.; Meireles, M.A. Stabilization of anthocyanin extract from jabuticaba skins by encapsulation using supercritical $CO_2$ as solvent. *Food Res. Int.* **2013**, *50*, 617–624.

30. Wang, S.Y.; Camp, M.J.; Ehlenfeldt, M.K. Antioxidant capacity and α-glucosidase inhibitory activity in peel and flesh of blueberry (*Vaccinium* spp.) cultivars. *Food Chem.* **2012**, *132*, 1759–1768.

31. Tonon, R.V.; Brabet, C.; Hubinger, M.D. Anthocyanin stability and antioxidant activity of spray-dried açai (*Euterpe oleracea* Mart.) juice produced with different carrier agents. *Food Res. Int.* **2010**, *43*, 907–914.

32. Robert, P.; Gorena, T.; Romero, N.; Sepulveda, E.; Chavez, J.; Saenz, C. Encapsulation of polyphenols and anthocyanins from pomegranate (*Punica granatum*) by spray drying. *Int. J. Food Sci. Technol.* **2010**, *45*, 1386–1394.

33. Fang, Z.; Bhandari, B. Effect of spray drying and storage on the stability of bayberry polyphenols. *Food Chem.* **2011**, *129*, 1139–1147.

34. Fang, Z.; Bhandari, B. Comparing the efficiency of protein and maltodextrin on spray drying of bayberry juice. *Food Res. Int.* **2012**, *48*, 478–483.

35. Fazaeli, M.; Emam-Djomeh, Z.; Ashtari, A.K.; Omid, M. Food and Bioproducts Processing Effect of spray drying conditions and feed composition on the physical properties of black mulberry juice powder. *Food Bioprod. Process.* **2012**, *90*, 667–675.

36. Nayak, C.A; Rastogi, N.K. Effect of selected additives on microencapsulation of anthocyanin by spray drying. *Dry Technol.* **2010**, *28*, 1396–1404.

37. Baum, M.; Schantz, M.; Leick, S.; Berg, S.; Betz, M.; Frank, K.; Rehage, H.; Schwarz, K.; Kulozik, U.; Schuchmann, H.; *et al.* Is the antioxidative effectiveness of a bilberry extract influenced by encapsulation? *J. Sci. Food Agric.* **2014**, *94*, 2301–2307.

38. Villacrez, J.L.; Carriazo, J.G.; Osorio, C. Microencapsulation of Andes Berry (*Rubus glaucus* Benth.) aqueous extract by spray drying. *Food Bioprocess Technol.* **2014**, *7*, 1445–1456.

39. Ferrari, C.C.; Pimentel, S.; Germer, M.; de Aguirre, J.M. Drying Technology: An international journal effects of spray-drying conditions on the physicochemical properties of blackberry powder effects of spray-drying conditions on the physicochemical properties of blackberry powder. *Dry Technol.* **2012**, *30*, 154–163.

40. Jiménez-Aguilar, D.M.; Ortega-Regules, A.E.; Lozada-Ramírez, J.D.; Pérez-Pérez, M.C.I.; Vernon-Carter, E.J.; Welti-Chanes, J. Color and chemical stability of spray-dried blueberry extract using mesquite gum as wall material. *J. Food Compos. Anal.* **2011**, *24*, 889–894.

41. Osorio, C.; Acevedo, B.; Hillebrand, S.; Carriazo, J.; Winterhalter, P.; Morales, A.L. Microencapsulation by spray-drying of anthocyanin pigments from corozo (*Bactris guineensis*) fruit. *J. Agric. Food Chem.* **2010**, *58*, 6977–6985.

42. Gurak, P.D.; Correa Cabral, L.M.; Rocha-Leão, M.H. Production of grape juice powder obtained by freeze- drying after concentration by reverse osmosis. *Braz. Arch. Biol. Technol.* **2013**, *56*, 1011–1017.

43. Voos, P.; Faas, M.; Spasojevic, M.; Sikkema, J. Encapsulation for preservation of functionality and targeted delivery of bioactive food components. *Int. Dairy J.* **2010**, *20*, 293–302.

44. Gharsallaoui, A.; Roudaut, G.; Chambin, O.; Voilley, A.; Saurel, R. Applications of spray-drying in microencapsulation of food ingredients: An overview. *Food Res. Int.* **2007**, *40*, 1107–1121.

45. Gouin, S. Microencapsulation: Industrial appraisal of existing technologies and trends. *Trends Food Sci. Technol.* **2004**, *15*, 330–347.

46. Kenyon, M. Modified starch, maltodextrin and corn syrup solids as wall material for food encapsulation. In *Encapsulation and Controlled Release of Food Ingredients*; Risch, S., Reineccius, G., Eds.; American Chemical Society: Washington, DC, USA, 1995; pp. 42–50.

47. Lee, J.; Durst, R.W.; Wrolstad, R.E. Determination of total monomeric anthocyanin pigment content of fruit juices, beverages, natural colorants, and wines by the pH differential method: Collaborative study. *J. AOAC Int.* **2005**, *88*, 1269–1278.

48. Cassinese, C.; de Combarieu, E.; Falzoni, M.; Fuzzati, N.; Pace, R.; Sardone, N. New liquid chromatography method with ultraviolet detection for analysis of anthocyanins and anthocyanidins in *Vaccinium myrtillus* fruit dry extracts and commercial preparations. *J. AOAC Int.* **2007**, *90*, 911–920.

49. Lohachoompol, V.; Mulholland, M.; Srzednicki, G.; Craske, J. Determination of anthocyanins in various cultivars of highbush and rabbiteye blueberries. *Food Chem.* **2008**, *111*, 249–254.

50. Ersus, S.; Yurdagel, U. Microencapsulation of anthocyanin pigments of black carrot (*Daucus carota* L.) by spray drier. *J. Food Eng.* **2007**, *80*, 805–812.

51. Arocas, A.; Varela, P.; González-Miret, M.L.; Salvador, A.; Heredia, F.; Fiszman, S. Differences in colour gamut obtained with three synthetic red food colourants compared with three natural ones: pH and heat stability. *Int. J. Food Prop.* **2013**, *16*, 766–777.

52. Fracassetti, D.; Del Bo, C.; Simonetti, P.; Gardana, C.; Klimis-Zacas, D.; Ciappellano, S. Effect of time and storage temperature on anthocyanin decay and antioxidant activity in wild blueberry (*Vaccinium angustifolium*) powder. *J. Agric. Food Chem.* **2013**, *61*, 2999–3005.

53. West, M.E.; Mauer, L.J. Color and chemical stability of a variety of anthocyanins and ascorbic acid in solution and powder forms. *J. Agric. Food Chem.* **2013**, *61*, 4169–4179.

54. Coïsson, J.D.; Travaglia, F.; Piana, G.; Capasso, M.; Arlorio, M. *Euterpe oleracea* juice as a functional pigment for yogurt. *Food Res. Int.* **2005**, *38*, 893–897.

55. Wallace, T.C.; Giusti, M.M. Determination of color, pigment, and phenolic stability in yogurt systems colored with nonacylated anthocyanins from *Berberis boliviana* L. as compared to other natural/synthetic colorants. *J. Food Sci.* **2008**, *73*, C241–C248.

56. Karaaslan, M.; Ozden, M.; Vardin, H.; Turkoglu, H. Phenolic fortification of yogurt using grape and callus extracts. *LWT—Food Sci. Technol.* **2011**, *44*, 1065–1072.

57. Sun-Waterhouse, D.; Zhou, J.; Wadhwa, S.S. Drinking yoghurts with berry polyphenols added before and after fermentation. *Food Control.* **2013**, *32*, 450–460.

58. Ścibisz, I.; Ziarno, M.; Mitek, M.; Zaręba, D. Effect of probiotic cultures on the stability of anthocyanins in blueberry yoghurts. *LWT—Food Sci. Technol.* **2012**, *49*, 208–212.

59. Peppas, N.A.; Sahlin, J.J. A simple equation for the description of solute release. III. Coupling of diffusion and relaxation. *Int. J. Pharm.* **1986**, *57*, 69–172.

60. Higuchi, T. Mechanism of sustained-action medication. Theoretical analysis of rate of release of solid drugs dispersed in solid matrices. *J. Pharm. Sci.* **1963**, *52*, 1145–1149.

61. Hixson, A.W.; Crowell, J.H. Dependence of reaction velocity upon surface and agitation. *Ind. Eng. Chem.* **1931**, *23*, 923–931.

62. Osman, M.; Sharaf, I.; Osman, H.; El-Khouly, Z.; Ahmed, E. Synthetic organic food colouring agents and their degraded products: Effects on human and rat cholinesterases. *Br. J. Biomed. Sci.* **2004**, *61*, 128–132.

63. Li, H.; Xiong, Z.; Dai, X.; Zeng, Q. The effect of perspiration on photo-induced chemical reaction of azo dyes and the determination of aromatic amine products. *Dyes Pigm.* **2012**, *94*, 55–59.

64. Eigenmann, P.; Haenggeli, C. Food colourings and preservatives—Allergy and hyperactivity. *Lancet* **2004**, *364*, 823–824.

65. McCann, D.; Barrett, A.; Cooper, A.; Crumpler, D.; Dalen, L.; Grimshaw, K.; Kitchin, E.; Lok, K.; Porteous, L.; Prince, E.; *et al.* Food additives and hyperactive behaviour in 3-year-old and 8/9-year-old children in the community: A randomised, double-blinded, placebo-controlled trial. *Lancet* **2007**, *370*, 1560–1567.

66. Ministerio de Salud. Reglamento sanitario de los alimentos [Internet]. 2011, p. 173. Available online: http://web.minsal.cl/portal/url/item/d61a26b0e9043de4e0400 101650149c0.pdf (accessed on 10 October 2014).

# Section 2:
# Grape Anthocyanins and Wine Quality

# Effect of Two Anti-Fungal Treatments (Metrafenone and Boscalid Plus Kresoxim-methyl) Applied to Vines on the Color and Phenol Profile of Different Red Wines

Noelia Briz-Cid, María Figueiredo-González, Raquel Rial-Otero,
Beatriz Cancho-Grande and Jesús Simal-Gándara

**Abstract:** The effect of two anti-fungal treatments (metrafenone and boscalid + kresoxim-methyl) on the color and phenolic profile of Tempranillo and Graciano red wines has been studied. To evaluate possible modifications in color and phenolic composition of wines, control and wines elaborated with treated grapes under good agricultural practices were analyzed. Color was assessed by Glories and CIELab parameters. Color changes were observed for treated wines with boscalid + kresoxim-methyl, leading to the production of wines with less color vividness. Phenolic profile was characterized by HPLC analysis. Boscalid + kresoxim-methyl treatment promoted the greatest decrease on the phenolic content in wines.

Reprinted from *Molecules*.  Cite as:  Briz-Cid, N.; Figueiredo-González, M.; Rial-Otero, R.; Cancho-Grande, B.; Simal-Gándara, J. Effect of Two Anti-Fungal Treatments (Metrafenone and Boscalid Plus Kresoxim-methyl) Applied to Vines on the Color and Phenol Profile of Different Red Wines. *Molecules* **2014**, *19*, 8093–8111.

## 1. Introduction

The main difficulty in growing grapes for wine is the fight against fungal diseases caused by fungi such as grey mold (*Botrytis cinerea*), powdery mildew (*Uncinula necator*) and downy mildew (*Plasmopara viticola*). Although practicing different traditional techniques it can be possible to minimize the incidence of these fungi during cultivation of the grape, the most effective means to combat them is the application of fungicides. With time, fungi can develop resistance to the most frequently applied fungicides, making it necessary to replace the traditionally used fungicides by others that include new generation active substances or new fungicides [1]. Several studies report that fungicides applied to vine may persist at trace levels in the grapes and thus be transferred to grape juice and ultimately to the wine [2–4], modifying the sensory quality of the final wine by causing changes in the fermentation kinetics and in the aromatic profile [2,5–7].

115

However, the effect of the presence of fungicide residues in grapes on the extraction of polyphenolic compounds during the winemaking process or their evolution during wine storage or aging remains almost completely unexplored. Some studies have confirmed that the phenolic composition of Monastrell red wines was altered by the presence of fungicide residues [4,8]. In fact, reductions in the anthocyanin content were found in wines obtained from grapes treated with famoxadone, fenhexamid and trifloxystrobin, while the hydroxycinnamic acid content decreased in the case of treatments with famoxadone, fluquinconazole, kresoxim-methyl and trifloxystrobin [4]. Thereby, the extraction of phenolic compounds during fermentation could be affected as a consequence of the presence of fungicide residues and this could originate problems in the stabilization of the wine color characteristics during storage.

The main objective of the present study was to evaluate the effect of some new fungicides (metrafenone, boscalid and kresoxim-methyl) on the accumulation of the major phenolic compounds in Tempranillo and Graciano red wines produced in La Rioja (N.E. Spain). For this, vineyards were treated with these fungicides under good agricultural practices.

## 2. Results and Discussion

Tempranillo (T) and Graciano (G) are the most distinctive red grapes of La Rioja (N.E. Spain). The first one is the most characteristic grape variety of this region and is able to produce wines with long aging, very balanced in alcohol content, color and acidity [9]. On the other hand, Graciano is often used as a blending partner of Tempranillo-based wines due to its contribution to improve the color of T wines and to add aroma, tannins and acidity [10]. Although it is less usual, some wineries also produce monovarietal G wines [10]. Graciano presents a certain resistance to diseases such as downy mildew and powdery mildew, has low fertility, is late maturing and produces wines with considerable acidity and polyphenolic content, ideal for aging, with a very intense aroma. The color and phenolic profile of both varieties as well as the effect of fungicide treatments on these parameters were established in this study.

### 2.1. Influence of New Generation Fungicides Residues on the Color

The color of a red wine is closely related with the grape variety, degree of ripeness, time and temperature of maceration process. The color of a red wine can be established by using colorimetric indexes and the CIELab parameters. Regarding colorimetric indexes, if we compare T and G control wines (without fungicide treatments), T-Control showed a higher yellow color contribution (38%) and a higher tonality (77) in comparison to G-Control (% yellow and tonality of 35 and 62, respectively). However, T-Control showed a lower red color contribution (49%) and color intensity (0.5) than G-Control (% red and color intensity of 56 and 0.6, respectively). Regarding to CIELab space, T-Control wines showed lower chroma ($C_{ab}^* = 29$), higher lightness ($L^* = 70$) and hue angle ($h_{ab} = 4$) than G-Control wines (whose values were 39, 67 and 1, respectively). All these results indicate that Graciano variety provides darker and colorful wines than the Tempranillo variety. These results are in agreement with those of other authors [9].

In addition, the effect of tested fungicides on the color of T and G wines has been analyzed. The comparison between the colorimetric indexes in treated and control wines for T and G varieties can be observed in Figure 1a. Independent of the variety and fungicide treatment, the yellow, red and blue percentages in all treated wines varied less than $\pm 10\%$ with respect to the control. Nevertheless, the treatment with boscalid + kresoxim-methyl caused an increment in the tonality of wines of both varieties (87 in T and 74 in G wines) respect to their respective control wines (77 and 62). The comparison between CIELab coordinates ($h_{ab}$, $C_{ab}^*$ and $L^*$) of treated wines respect to the control wines can be observed in the Figure 1b. Some marked differences (higher than 20%) were registered again in CIELab coordinates for those wines obtain from grapes treated with boscalid + kresoxim-methyl. These wines showed lower chroma and higher hue than their respective control wines and as a consequence, this fungicide treatment could promote wines with higher tonality and lower color vividness.

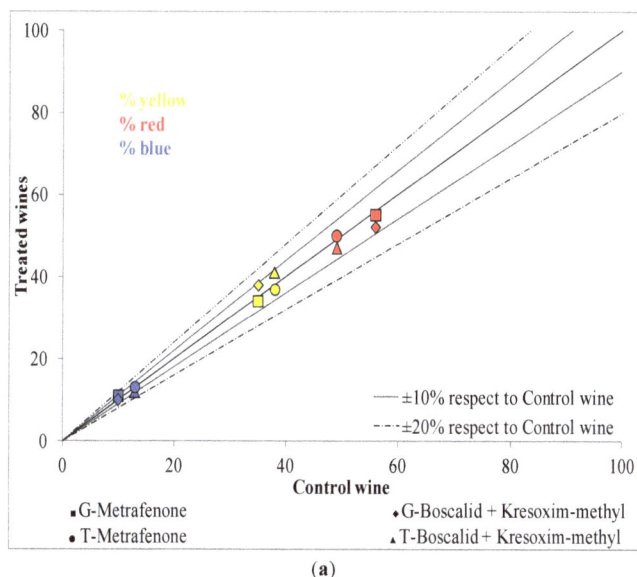

(a)

**Figure 1.** *Cont.*

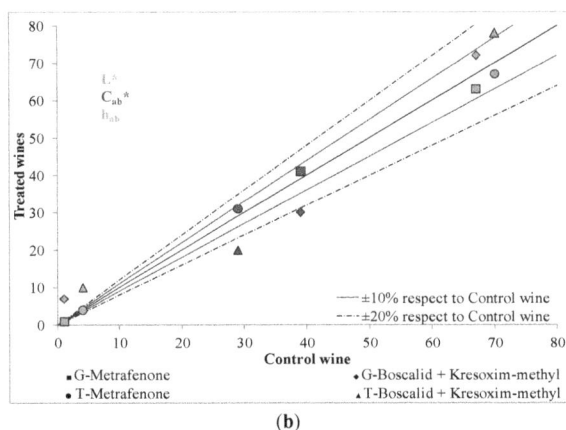

**Figure 1.** Colorimetric indexes (**a**) and CIELab data (**b**) of treated wines (Y axis) and control wines (X axis) for Graciano (G) and Tempranillo (T) wines.

In order to know if the differences observed in the CIELab parameters represent chromatic changes that can be perceived by the human eye, the $\Delta E^*_{ab}$ parameter (difference in color between treated and control samples in the CIELab space, calculated as the Euclidean distance between their location defined by L*, a* and b*) was calculated [9]. Changes in the CIELab parameters were more pronounced for boscalid + kresoxim-methyl treatments since $\Delta E^*_{ab}$ parameter ranged from 10.7 to 11.5, while in wines treated with metrafenone the value was around 4.

### 2.2. Influence of New Generation Fungicides Residues on the Phenolic Profile of Wines

#### 2.2.1. Anthocyanins

Anthocyanins are the principal compounds responsible of the color of red wines. They are transferred from the skin of grapes to the wine during the early days of winemaking. Total anthocyanins of T and G wines and their distribution (% monomeric, % copigmented and % polymeric forms), determined by UV-Vis spectroscopy according to Boulton [11], can be seen in Table 1. The monomeric forms decreased in the first stages because they participate in the copigmentation process with other non- pigmented phenolic compounds, being flavonols the most effective copigments [12]. On the other hand, acetaldehyde, tannins and other phenolic compounds (catechins, proanthocyanidins), are involved in processes of condensation-polymerization with monomeric anthocyanins, leading to the formation of more stable polymeric pigments than the free monomeric forms [13–15]. For sample comparisons, a variability greater than 10% in the results determined in treated wines respect to the control was considered as a difference statistically significant ($p < 0.05$) according to the statistical treatment applied ($t$-Student test).

**Table 1.** Anthocyanin composition and quantitative data (mg·L$^{-1}$ ± SD) of monomeric anthocyanins in Tempranillo and Graciano wines.

| Wines | T-Control | T-Metrafenone | T-Boscalid-Kresoxim-methyl | G-Control | G-Metrafenone | G-Boscalid-Kresoxim-methyl |
|---|---|---|---|---|---|---|
| **Anthocyanins by UV/Vis** | | | | | | |
| Monomeric (%) | 34.68 ± 0.949 | 38.91 ± 1.753 | 28.83 * ± 1.689 | 29.58 ± 1.244 | 38.20 * ± 2.645 | 34.36 * ± 1.060 |
| Copigmented (%) | 41.87 ± 0.678 | 37.37 ± 1.594 | 53.66 * ± 0.740 | 43.05 ± 1.054 | 39.48 ± 1.983 | 46.37 ± 0.994 |
| Polymeric (%) | 23.46 ± 0.271 | 23.71 ± 0.159 | 17.51 * ± 0.949 | 27.37 ± 0.190 | 22.33 * ± 0.662 | 19.27 * ± 0.066 |
| TOTAL Anthocyanins (absorbance units) | 3.905 ± 0.021 | 3.880 ± 0.014 | 4.100 * ± 0.057 | 5.180 ± 0.028 | 4.345 * ± 0.021 | 4.270 * ± ≤0.001 |
| **Monomeric anthocyanins by HPLC** | | | | | | |
| **Malvidin derivatives** | | | | | | |
| malvidin-3-O-glucoside | 161.59 ± 4.204 | 152.12 ± 2.326 | 158.57 ± 4.068 | 163.75 ± 1.684 | 128.79 * ± 3.121 | 121.94 * ± 4.149 |
| malvidin-3-O-(6-O-p-coumaroyl)glucoside | 13.25 ± 0.886 | 11.26 ± 0.018 | 13.46 ± 0.445 | 8.87 ± 0.723 | 9.63 ± 0.783 | 7.82 ± 0.202 |
| malvidin-3-O-(6-O-acetyl)glucoside | 5.79 ± 0.592 | 5.05 ± 0.088 | 6.62 ± 0.058 | 11.44 ± 0.382 | 10.93 ± 1.116 | 10.30 * ± 0.300 |
| malvidin-3-O-(6-O-caffeoyl)glucoside | 0.85 ± 0.021 | 0.91 ± 0.039 | 1.55 * ± 0.023 | 1.40 ± 0.131 | 1.62 ± 0.168 | 1.85 * ± 0.100 |
| vitisin A | 0.54 ± 0.032 | 0.56 ± ≤0.001 | 0.47 ± 0.055 | 1.37 ± 0.049 | 2.10 * ± 0.204 | 1.93 * ± 0.100 |
| vitisin B | 0.25 ± 0.005 | 0.37 * ± 0.020 | 0.06 * ± 0.006 | 0.46 ± 0.043 | 0.36 ± 0.004 | 0.47 ± 0.040 |
| subTOTAL (mg·L$^{-1}$) (%) | 182.12 (78.5) | 170.27 (79.9) | 180.73 (79.6) | 187.29 (80.8) | 153.44 * (86.7) | 144.31 * (85.6) |
| **Petunidin derivatives** | | | | | | |
| petunidin-3-O-glucoside | 25.38 ± 1.947 | 22.82 ± 0.834 | 25.32 ± 0.395 | 10.59 ± 0.095 | 6.35 * ± 0.074 | 4.70 * ± 0.176 |
| petunidin-3-O-(6-O-p-coumaroyl)glucoside | 3.00 ± 0.125 | 2.34 * ± 0.144 | 2.80 ± 0.059 | 0.39 ± 0.005 | 0.03 * ± ≤0.001 | 0.41 ± 0.020 |
| petunidin-3-O-(6-O-acetyl)glucoside | 1.20 ± 0.117 | 1.12 ± 0.059 | 1.00 ± 0.054 | 0.60 ± 0.045 | 0.60 ± 0.047 | 0.27 * ± 0.016 |
| subTOTAL (mg·L$^{-1}$) (%) | 29.58 (12.7) | 26.28 (12.3) | 29.12 (12.7) | 11.58 (5.0) | 6.98 * (3.9) | 5.38 * (5.2) |
| **Delphinidin derivatives** | | | | | | |
| delphinidin-3-O-glucoside | 10.06 ± 0.735 | 8.24 * ± 0.284 | 11.06 ± 0.070 | 5.08 ± 0.071 | 1.99 * ± 0.219 | 1.36 * ± 0.064 |
| delphinidin-3-O-(6-O-p-coumaroyl)glucoside | 2.99 ± 0.111 | 0.95 * ± 0.001 | 1.20 * ± 0.007 | n.d. | n.d. | n.d. |
| delphinidin-3-O-(6-O-acetyl)glucoside | 0.66 ± 0.041 | 0.68 ± 0.017 | 0.70 ± 0.015 | 0.37 ± 0.009 | 0.32 ± 0.035 | 1.36 * ± 0.156 |
| subTOTAL (mg·L$^{-1}$) (%) | 13.71 (5.9) | 9.87 * (4.6) | 12.96 (5.6) | 5.45 (2.4) | 2.31 * (1.3) | 2.72 * (1.6) |
| **Peonidin derivatives** | | | | | | |
| peonidin-3-O-glucoside | 4.53 ± 0.123 | 3.98 * ± 0.186 | 4.69 ± 0.027 | 18.76 ± 0.385 | 8.04 * ± 0.749 | 10.39 * ± 0.290 |
| peonidin-3-O-(6-O-p-coumaroyl)glucoside | 0.82 ± 0.023 | 1.00 ± 0.010 | 0.84 ± 0.082 | 4.39 ± 0.173 | 2.90 * ± 0.259 | 2.49 * ± 0.060 |
| peonidin-3-O-(6-O-acetyl)glucoside | 0.06 ± 0.004 | 0.37 * ± 0.004 | 0.10 * ± 0.009 | 2.95 ± 0.053 | 2.17 * ± 0.232 | 2.30 * ± 0.222 |
| peonidin-3-O-(6-O-caffeoyl)glucoside | n.d. | n.d. | n.d. | 0.16 ± 0.012 | 0.17 ± 0.008 | 0.25 * ± ≤0.001 |
| subTOTAL (mg·L$^{-1}$) (%) | 5.41 (2.3) | 5.37 (2.5) | 5.63 * (2.4) | 26.25 (11.3) | 13.28 * (7.5) | 15.43 * (9.1) |
| **Cyanidin derivatives** | | | | | | |
| cyanidin-3-O-glucoside | 0.44 ± 0.003 | 0.48 ± 0.024 | 0.46 ± 0.028 | 0.43 ± 0.024 | 0.25 * ± 0.032 | 0.25 * ± 0.017 |
| cyanidin-3-O-(6-O-p-coumaroyl)glucoside | 0.56 ± 0.033 | 0.51 ± 0.002 | 0.61 ± 0.031 | 0.43 ± 0.006 | 0.41 ± 0.021 | 0.10 * ± 0.009 |
| cyanidin-3-O-(6-O-acetyl)glucoside | 0.34 ± 0.017 | 0.41 * ± 0.011 | 0.38 * ± 0.004 | 0.22 ± 0.012 | 0.34 * ± 0.001 | 0.49 * ± 0.007 |
| subTOTAL (mg·L$^{-1}$) (%) | 1.34 (0.6) | 1.40 (0.7) | 1.45 (0.6) | 1.09 (0.5) | 1.00 (0.6) | 0.84 * (0.5) |
| TOTAL *monomeric anthocyanins* (mg·L$^{-1}$) | 232.16 | 213.17 | 229.89 | 231.65 | 177.01 * | 168.68 * |

*: Statistical differences according to the *t*-student test ($p < 0.05$).

As can be seen in Table 1, the total anthocyanins content was lower in T wines than in G wines, although the percentages of monomeric, copigmented and polymeric forms were similar in both varieties (around 30%, 42% and 25%, respectively). Statistical tests confirmed a significant difference between control and wines treated with boscalid + kresoxim-methyl for both varieties; meanwhile the metrafenone treatment seems only to affect the anthocyanin percentages in G wines.

Monomeric anthocyanins content in T and G control wines, determined by chromatographic analysis, was similar (232.16 and 231.65 mg·L$^{-1}$, respectively), as can be seen in Table 1. However, fungicide treatments applied in the field affected the anthocyanin content differently, depending on the variety (as can be seen in Figure 2). Thereby, while for T wines no effects were observed, for G wines reductions of about 25% respect to the G-Control were observed for the two treatments. The most abundant anthocyanin compounds in both wines were malvidin derivatives (78.5% and 80.8% in T and G control wines, respectively), being malvidin-3-O-glucoside the most abundant anthocyanin in all wines (Table 1). Petunidin derivatives were the following most abundant derivatives for T wines (12.7%), being petunidin-3-O-glucoside the main compound in this group; in contrast, the second most abundant group for G wines was the peonidin derivatives (11.3%). Concentrations of the other minor derivative groups (ranging from 0.5% to 6%) were as follows: delphinidin > peonidin > cyanidin derivatives in T wines; meanwhile for G wines the minor groups were: petunidin > delphinidin > cyanidin derivatives.

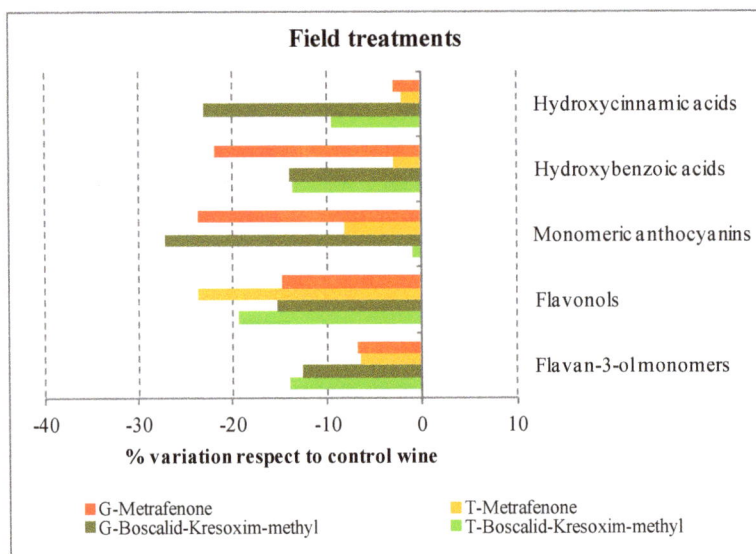

Figure 2. Effect of each antifungal treatment on the phenolic profile of Graciano (G) and Tempranillo (T) wines.

## 2.2.2. Flavan-3-ol Monomers and Proanthocyanidins

The total content of flavan-3-ol monomers was lower in T-Control than in G-Control (28.6 *versus* 49.6 mg·L$^{-1}$, respectively), as can be seen in Table 2. This is in agreement with other authors [16,17].

The main flavan-3-ol monomers in T and G wines were (+)-catechin (C), (−)-epicatechin (EC) and (−)-gallocatechin (GC) (Table 2). While the contents of C and EC in G-Control were similar, the content of EC in T-Control was about 40% of the C content. Significant differences were only observed for boscalid + kresoxim-methyl treatments, independent of the variety, with reductions of about 13% (see Figure 2).

In addition to flavan-3-ol monomers, there are also dimeric, trimeric, oligomeric and condensed procyanidins. These compounds are better extracted with longer maceration in the presence of alcohol [18]. We can find procyanidins —polymers of (epi)catechin that release cyanidin- and prodelphinidins —polymers of (epi)gallocatechin that release delphinidin [19]. This large group of compounds differs in the nature of their constitutive units, their number (degree of polymerization) and the position of linkages between them. All of the structures have not been analyzed, and only the procyanidin dimers and some of the trimers have been completely identified.

After acid-catalyzed cleavage of the polymer in presence of phloroglucinol, the mean degree of polymerization (aDP, calculated as the ratio of total units to terminal units), the average molecular weight (aMW) and the total proanthocyanidins concentration (calculated as the sum of all units) were determined [19,20]. As it can be seen in Table 2, the concentration of total proanthocyanidins in T-Control was 241.94 mg·L$^{-1}$, slightly higher than that obtained in G-Control (219.18 mg·L$^{-1}$). In general, no important effects of metrafenone residues in the total proanthocyanidins content were observed (with reductions of around 5%) for T and G wines. Nevertheless the treatments with boscalid + kresoxim-methyl produced a reduction of over 15% for T and G wines. In addition, the aDP value obtained for T and G wines was 2.6 and 1.8, respectively, indicating that the proanthocyanidins present in these wines are mostly dimers. These results are in good agreement with those reported in similar wines by other authors [21].

**Table 2.** Flavan-3-ol composition and quantitative data (mg·L$^{-1}$ ± SD) in Tempranillo and Graciano wines.

| Wines | T-Control | T-Metrafenone | T-Boscalid-Kresoxim-methyl | G-Control | G-Metrafenone | G-Boscalid-Kresoxim-methyl |
|---|---|---|---|---|---|---|
| **Flavan-3-ol monomers** | | | | | | |
| catechin (C) | 19.49 ± 0.013 | 19.26 ± 0.036 | 18.32 * ± 0.054 | 24.39 ± 0.419 | 23.39 ± 0.276 | 20.94 * ± 0.232 |
| epicatechin (EC) | 7.93 ± 0.035 | 7.43 ± 0.239 | 6.27 * ± 0.003 | 23.77 ± 0.270 | 21.70 ± 0.721 | 20.98 * ± 0.167 |
| galocatechin (GC) | 1.14 ± 0.023 | n.d. | n.d. | 1.48 ± 0.004 | 1.42 * ± 0.004 | 1.44 * ± 0.001 |
| subTOTAL (mg·L$^{-1}$) | 28.56 | 26.69 | 24.59 * | 49.64 | 46.29 | 43.36 * |
| **Proanthocyanidins** | | | | | | |
| aDP (%) | 2.6 | 2.7 | 2.6 | 1.8 | 1.8 | 1.8 |
| aMW (%) | 765.4 | 811.5 | 778.1 | 526.4 | 516.9 | 521.9 |
| procyanidins (%PC) | 59 | 60 | 68 | 79 | 81 | 81 |
| prodelphinidins (% PD) | 40 | 40 | 3 | 20 | 18 | 18 |
| galloylated (% G) | 0.5 | 0.5 | 10.6 | 1.3 | 1.2 | 0.8 |
| subTOTAL (mg·L$^{-1}$) | 241.94 ± 0.11 | 228.05 ± 1.83 | 200.80 * ± 1.11 | 219.18 ± 1.46 | 207.51 ± 0.35 | 191.40 * ± 2.23 |

*: Statistical differences according to the t-student test ($p < 0.05$).

### 2.2.3. Flavonols

Flavonols were the minor flavonoid group with respect to the other flavonoid groups described above. As it can be seen in Table 3, flavonol content in T-Control was 8.20 mg·L$^{-1}$, being this concentration higher than that obtained in G-Control (6.46 mg·L$^{-1}$). Flavonol 3-$O$-glucoside derivatives are the main group contributing to the total flavonol content for T- and G-based wines. Five 3-$O$-glucoside derivatives (myricetin, quercetin, laricitrin, isohamnetin and syringetin) were detected in G wines, while kaempherol-3-$O$-glucoside was also detected in T wines. The concentrations of these derivatives ranged from 4.74 to 6.01 mg·L$^{-1}$ in T wines, myricetin-3-$O$-glucoside being the main flavonol; in G wines they ranged from 3.62 to 4.28 mg·L$^{-1}$, syringetin-3-$O$-glucoside being the most abundant. The next group is formed by the 3-$O$-glucuronide derivatives of myricetin, quercetin and kaempherol, with concentrations ranging between 1.07 and 1.37 mg·L$^{-1}$ in T wines, and between 1.12 and 1.43 mg·L$^{-1}$ in G wines. Furthermore, while in T-Control two galactoside derivatives were identified (quercetin and kaempherol), only quercetin-3-$O$-galactoside was detected in G-Control. Finally, four aglycone forms (myricetin, quercetin, kaempherol and laricitrin) were identified in both varieties in concentrations ranging from 0.35 to 0.60 and from 0.39 to 0.72 mg·L$^{-1}$ in T and G wines, respectively.

**Table 3.** Flavonol composition and quantitative data (mg·L$^{-1}$ ± SD) in Tempranillo and Graciano wines.

| Wines | T-Control | T-Metrafenone | T-Boscalid-Kresoxim-methyl | G-Control | G-Metrafenone | G-Boscalid-Kresoxim-methyl |
|---|---|---|---|---|---|---|
| **3-O-glucoside derivatives** | | | | | | |
| myricetin-3-O-glucoside | 3.46 ± 0.279 | 2.48 * ± 0.072 | 2.50 * ± 0.215 | 0.61 ± 0.001 | 0.37 * ± ≤0.001 | 0.37 * ± 0.001 |
| quercetin-3-O-glucoside | 0.32 ± 0.021 | 0.36 * ± 0.021 | 0.26 * ± 0.012 | 0.13 ± 0.002 | n.d. | 0.13 ± 0.002 |
| laricitrin-3-O-glucoside | 1.01 ± 0.030 | 0.78 * ± 0.035 | 0.86 * ± 0.028 | 0.74 ± 0.035 | 0.61 * ± 0.002 | 0.60 * ± 0.048 |
| kaempherol-3-O-glucoside | 0.12 ± 0.001 | 0.09 * ± 0.002 | 0.14 * ± 0.009 | n.d. | n.d. | n.d. |
| isohamnetin-3-O-glucoside | 0.15 ± 0.005 | 0.13 * ± 0.004 | 0.14 ± 0.006 | 0.25 ± 0.016 | 0.24 ± 0.005 | 0.44 * ± 0.022 |
| syringetin-3-O-glucoside | 0.95 ± 0.025 | 0.90 ± 0.078 | 0.86 ± 0.057 | 2.55 ± 0.161 | 2.39 ± 0.072 | 2.36 ± 0.171 |
| **subTOTAL (mg·L$^{-1}$) (%)** | **6.01** (73.3) | **4.74** * (75.7) | **4.76** * (71.9) | **4.28** (66.2) | **3.62** * (65.8) | **3.90** (71.3) |
| **3-O-glucuronide derivatives** | | | | | | |
| myricetin-3-O-glucuronide | 0.55 ± 0.006 | 0.43 * ± 0.040 | 0.41 * ± 0.024 | 0.23 ± 0.017 | 0.21 ± 0.007 | 0.15 * ± 0.003 |
| quercetin-3-O-glucuronide | 0.63 ± 0.037 | 0.50 * ± 0.029 | 0.63 ± 0.048 | 0.93 ± 0.038 | 0.95 ± 0.011 | 0.72 * ± 0.022 |
| kaempherol-3-O-glucuronide | 0.19 ± 0.017 | 0.14 * ± 0.001 | 0.17 ± 0.001 | 0.23 ± ≤0.001 | 0.28 * ± 0.004 | 0.25 ± 0.001 |
| **subTOTAL (mg·L$^{-1}$) (%)** | **1.37** (16.7) | **1.07** * (17.1) | **1.21** (18.3) | **1.39** (21.5) | **1.43** * (26.0) | **1.12** * (20.5) |
| **3-O-galactoside derivatives** | | | | | | |
| quercetin-3-O-galactoside | 0.14 ± 0.001 | 0.10 * ± 0.001 | 0.14 ± 0.003 | 0.07 ± 0.002 | n.d. | 0.06 * ± ≤0.001 |
| kaempherol-3-O-galactoside | 0.07 ± 0.001 | n.d. | 0.08 ± 0.001 | n.d. | n.d. | n.d. |
| **subTOTAL (mg·L$^{-1}$) (%)** | **0.21** (2.6) | **0.10** * (1.6) | **0.22** (3.3) | **0.07** (1.1) | | **0.06** * (1.1) |
| **Aglycons** | | | | | | |
| myricetin | 0.27 ± 0.025 | 0.15 * ± 0.005 | 0.23 ± 0.006 | 0.24 ± 0.019 | 0.18 * ± ≤0.001 | 0.11 * ± ≤0.001 |
| quercetin | 0.21 ± 0.011 | 0.14 * ± ≤0.001 | 0.14 * ± 0.004 | 0.37 ± 0.028 | 0.27 * ± ≤0.001 | 0.17 * ± ≤0.001 |
| kaempherol | 0.06 ± ≤0.001 | n.d. | n.d. | 0.05 ± ≤0.001 | n.d. | 0.05 ± ≤0.001 |
| laricitrin | 0.07 ± ≤0.001 | 0.06 * ± 0.002 | 0.06 * ± ≤0.001 | 0.06 ± 0.001 | n.d. | 0.05 * ± ≤0.001 |
| **subTOTAL (mg·L$^{-1}$) (%)** | **0.60** * (7.3) | **0.35** * (5.6) | **0.43** * (6.5) | **0.72** (11.1) | **0.45** * (8.2) | **0.39** * (7.2) |
| **TOTAL flavonols (mg·L$^{-1}$)** | **8.20** | **6.26** * | **6.62** * | **6.46** | **5.50** * | **5.47** * |

*: Statistical differences according to the t-student test ($p < 0.05$).

124

**Table 4.** Phenolic acid composition and quantitative data ($mg \cdot L^{-1} \pm SD$) in Tempranillo and Graciano wines.

| Wines | T-Control | T-Metrafenone | T-Boscalid-Kresoxim-methyl | G-Control | G-Metrafenone | G-Boscalid-Kresoxim-methyl |
|---|---|---|---|---|---|---|
| **Hydroxybenzoic acids** | | | | | | |
| gallic acid | 17.00 ± 0.788 | 17.25 ± 0.160 | 13.88 * ± 0.754 | 10.55 ± 0.046 | 5.41 * ± 0.110 | 7.29 * ± 0.152 |
| 3,5-dihydroxibenzoic acid | 6.01 ± 0.252 | 4.93 * ± 0.226 | 5.69 ± 0.189 | 3.42 ± 0.070 | 2.42 * ± 0.120 | 2.21 * ± 0.048 |
| protocatechuic acid | 2.54 ± 0.098 | 2.45 ± 0.124 | 2.17 * ± 0.151 | 1.72 ± 0.106 | 1.58 ± 0.008 | 1.97 ± 0.003 |
| **Hydroxybenzoic acids** | | | | | | |
| vanillic acid | 2.58 ± 0.005 | 3.19 ± 0.002 | 2.21 * ± 0.047 | 6.25 ± 0.348 | 6.23 ± 0.200 | 5.93 ± 0.588 |
| syringic acid | 4.07 ± 0.007 | 3.42 * ± 0.231 | 3.86 ± 0.297 | 3.52 ± 0.355 | 4.26 ± 0.134 | 4.52 ± 0.393 |
| subTOTAL ($mg \cdot L^{-1}$) (%) | 32.20 (51.4) | 31.24 (51.2) | 27.82 * (50.2) | 25.46 (65.2) | 19.90 * (60.1) | 21.91 * (67.9) |
| **Hydroxycinnamic acids and their derivatives** | | | | | | |
| caftaric acid | 16.44 ± 1.022 | 16.48 ± 0.948 | 14.74 ± 0.478 | 7.31 ± 0.130 | 7.37 ± 0.340 | 5.77 * ± 0.487 |
| caffeic acid | 0.46 ± 0.012 | 0.50 ± ≤0.001 | 0.32 * ± 0.001 | 0.12 ± 0.007 | 0.07 * ± 0.001 | 0.01 * ± ≤0.001 |
| c-coutaric acid | 2.36 ± 0.141 | 2.14 ± 0.080 | 2.13 ± 0.132 | 1.08 ± 0.003 | 0.94 * ± 0.009 | 0.78 * ± 0.034 |
| t-coutaric acid | 10.08 ± 0.824 | 9.69 ± 0.631 | 9.40 ± 0.645 | 3.17 ± 0.017 | 3.30 ± 0.192 | 2.37 * ± 0.228 |
| p-coumaric acid | 1.10 ± 0.112 | 0.98 ± 0.060 | 0.98 ± 0.044 | 1.58 ± 0.127 | 1.20 * ± 0.017 | 1.30 * ± 0.073 |
| subTOTAL ($mg \cdot L^{-1}$) (%) | 30.43 (48.6) | 29.78 (48.8) | 27.57 (49.8) | 13.27 (34.0) | 12.88 (38.9) | 10.23 * (31.7) |
| **Stylbene** | | | | | | |
| resveratrol | n.d. | n.d. | n.d. | 0.30 ± 0.002 | 0.32 ± 0.011 | 0.10 * ± 0.002 |
| TOTAL phenolic acids ($mg \cdot L^{-1}$) | 62.63 | 61.02 | 55.39 * | 39.03 | 33.10 * | 32.24 * |

*: Statistical differences according to the $t$-student test ($p < 0.05$).

125

Statistical differences in the flavonol profiles were observed for both treatments (metrafenone and boscalid + kresoxim-methyl) and varieties (Table 3) although greater reductions (around 20%–24%) were observed for T treated wines (Figure 2).

### 2.2.4. Acids

The total phenolic acid content in T-Control was about 38% higher than that observed in G-Control. Stilbenes, hydroxybenzoic acids, hydroxycinnamic acids and their derivatives identified in T and G wines are listed in Table 4. Hydroxybenzoic acids content ranged between 27.82 and 32.20 mg·L$^{-1}$ in T wines, gallic and 3,5-dihydroxybenzoic acids being the main constituents among the five identified. Meanwhile, the content of these compounds in G wines comprised between 19.90 and 25.46 mg·L$^{-1}$, with gallic and vanillic acids being the most abundant compounds. Hydroxycinnamic acids (caffeic and coumaric acids) and their respective esters (caftaric and coutaric acid) were detected at concentrations between 27.57 and 30.43 mg·L$^{-1}$ in T wines, and between 10.23 and 13.27 mg·L$^{-1}$ in G wines, caftaric acid being the main compound in both varieties. In addition, the stilbene resveratrol was identified in G wines, but at low concentrations (between 0.10 and 0.32 mg·L$^{-1}$), representing less than 1% of total non-flavonoids. This compound was not detected in the wines obtained from T grapes.

The different phytosanitary treatments had a variable effect in the accumulation of these compounds in the wines (Figure 2). In general terms, the treatments with metrafenone had no effect on the accumulation of hydroxycinnamic and hydroxybenzoic acids, except for G wines where a reduction of 22% respect to the control wine was observed. To the contrary, treatments with boscalid + kresoxim-methyl caused significant reductions for both acids and varieties, except for hydroxycinnamic acids in T treated wines.

## 3. Experimental

### 3.1. Fungicide Experiments

Fungicide experiments were performed out in 2012 at two experimental vineyards located in Aldeanueva de Ebro, La Rioja (N.E. Spain), belonging to D.O.Ca. Rioja. The vineyards produce *Vitis vinifera* cv. Tempranillo and cv. Graciano red grapes. The experimental vineyards were 3,000 m$^2$ in area, approximately, and contained 30 rows with 40–50 vines each one; the gaps between rows and grapevines were 2.6 and 1.2 m, respectively. Each experimental vineyard was divided into three experimental plots: the first was untreated and used to produce the control wine, the other two were treated with the phytosanitary products Collis® (BASF, 20% w/v boscalid + 10% w/v kresoxim-methyl) and Vivando® (BASF, 50% w/v metrafenone), respectively, in accordance with good agricultural practices (GAP), using the doses recommended by the manufacturer and keeping the pre-harvest time in vines.

## 3.2. Winemaking Process and Wine Samples

The winemaking process was carried out in the experimental cellar located at the University of La Rioja. Identical vinifications were performed with the grapes from each experimental plot as follows: grapes were crushed, destemmed and placed in a metallic fermentation vessel (40 L) that was supplied with $SO_2$ (at 50 mg·$L^{-1}$) concentration. The temperature during alcoholic fermentation–maceration, which took 14 days, was 17–21 °C. At the end of the process, the wine was strained off, grape residues were pressed and the wine–must mixtures were transferred to a metallic vessel where it was supplied with $SO_2$ (at 30 mg·$L^{-1}$). Prior to bottling, a step of cold clarification was carried out.

## 3.3. Analytical Standards, Reagents and Materials

Malvidin-3-O-glucoside chloride, quercetin, catechin, epicatechin, resveratrol, and gallic, 3,5-dihydroxybenzoic, protocatechuic, vanillic, syringic, p-coumaric and caffeic acids were purchased from Sigma Aldrich (St. Louis, MO, USA). Individual stock solutions of each compound were prepared in methanol. Different working standards solutions were prepared by appropriate dilution in 12% ethanol in water and then stored in dark vials at −80 °C. Solvents (water, methanol, acetone and ethyl acetate) of HPLC grade and other inorganic reagents (formic, hydrochloric, acetic, trifluoroacetic and ascorbic acids, phloroglucinol, sodium acetate anhidro, and sodium bisulfite) were purchased from Sigma Aldrich. The sorbent materials used for SPE were: Oasis MCX cartridges (500 mg, 6 mL size) from Waters Corp (Milford, MA, USA); Strata-X-A 33u Polymeric Strong Anion sorbent (60 mg, 3 mL size) and Strata C18-E (2 g, 12 mL size) from Phenomenex (Torrance, CA, USA).

## 3.4. Characterization of the Color Fraction and Phenolic Content

The characterization of the color fraction was determined by spectrophotometric parameters, colorimetric indexes and CIELab parameters, using a Beckman Coulter DU 730 Life Science UV/Vis spectrophotometer. All of the measurements were carried out in duplicate, using quartz cells of 1 mm path length. A hydroalcoholic solution (12% ethanol) was used as blank in all measures.

*Colorimetric indexes.* Absorbances at 420, 520 and 620 nm were measured to assess the must color by chromatic parameters such as % red, % yellow and % blue, color intensity (CI) and tonality (T), according to Glories [22].

*CIELab coordinates.* The must color was also assessed by the CIELab space [23]. The parameters that define the CIELab space are: rectangular coordinates such as red/green color component (a*), yellow/blue color component (b*) and lightness (L*); and the cylindrical coordinates such as chroma ($C_{ab}$*) and hue angle ($h_{ab}$).

*Copigmented, monomeric, polymeric and total anthocyanins.* Each group of anthocyanins was determined according to Boulton [11]. Briefly, this method consisted of adjusting the pH of a wine to 3.6 and then filtering the wine through a 0.45 μm mesh filter. Then, the following tests were conducted:

- $A^{acet}$: 20 μL of 10% (*v/v*) acetaldehyde was added to 2 mL of prepared wine and the sample was allowed to sit for 45 min at room temperature before measuring $A_{520\ nm}$;
- $A^{20}$: to another 100 μL of prepared wine, 1,900 μL hydroalcoholic solution was added and absorbance $A_{520\ nm}$ was also measured;
- $A^{SO2}$: 160 μL of 5% (w/v) $SO_2$ was added to 2 mL of prepared wine and absorbance $A_{520\ nm}$ was measured.

From these readings, the different forms of anthocyanins were expressed in absorbance units as:

copigmented anthocyanins = $A^{acet} - A^{20}$
monomeric anthocyanins = $A^{20} - A^{SO2}$
polymeric anthocyanins = $A^{SO2}$
total anthocyanins = $A^{acet}$

The percent distribution of the various forms was calculated as:

% copigmented = $[(A^{acet} - A^{20})/A^{acet}] \times 100$
% monomeric = $[(A^{20} - A^{SO2})/A^{acet}] \times 100$
% polymeric = $[A^{SO2}/A^{acet}] \times 100$

*3.5. Determination of Phenolic Compounds*

3.5.1. Extraction Procedures

**Flavan-3-ol monomsers and proanthocyanidins.** Proanthocyanidins were extracted and characterized according to the procedure described by Kennedy and Jones [24], with minor modifications [20]. Briefly, bleaching of anthocyanins pigments is necessary prior to retained proanthocyanidins by anion exchange sorbent. After eluting with 75% acetone in water, proanthocyanidins were acid-catalyzed in presence of phloroglucinol. This process followed by HPLC analysis is a useful alternative for quantification and characterization of longer proanthocyanidins.

(a) *Flavan-3-ol monomers and proanthocyanidins extraction.* Wine (2 mL) was adjusted to pH 1.0 with a drop of concentrated hydrochloric acid, transferred to a 5 mL test tube containing sodium bisulfite (800 mg) and stirred for 20 min. Under these conditions, most of monomeric anthocyanins are combined with bisulfite to form colorless sulfonic acid adducts [25], which can be readily retained by anion exchange sorbents. This bleached wine was diluted 1:2 with ultrapure water and an aliquot (2 mL) was loaded into a Strata-X-A mixed-mode anion exchange/reversed phase SPE cartridge, previously activated with 75% acetone in water (2 mL) followed by water (4 mL). Afterwards, the cartridge was washed with water (4 mL) and flavan-3-ols and proanthocyanidins were eluted with 75% acetone in water (8 mL), whereas anthocyanins and organic acids were still retained through anion exchange interactions. This eluate was brought to dryness on a rotary evaporator at 35 °C and then reconstituted in methanol (200 μL). In order to quantify monomeric flavan-3-ols, a portion of this methanolic extract (50 μL) were diluted to 500 μL with 2.5% acetic acid in water, filtered (0.20 μm) and analyzed by HPLC/DAD–ESI/MS.

(b) *Acid-catalyzed degradation of proanthocyanidins in presence of phloroglucinol.* Proanthocyanidins were characterized following the acid-catalyzed cleavage of the polymer in the presence of phloroglucinol excess according to the procedure described by Kennedy and Jones [24] with minor modifications. In brief, a solution containing 0.2 M HCl, 50 mg·mL$^{-1}$ phloroglucinol and 10 mg·mL$^{-1}$L-ascorbic acid was prepared in methanol as the phloroglucinolysis reagent. Methanolic wine extract (100 μL) was allowed to react with phloroglucinol solution (200 μL) in a water bath for 40 min at 50 °C. Afterwards, the reaction was cooled down and quenched by the addition of 15 mM sodium acetate aqueous solution (2.7 mL). The reaction mixture was then purified by SPE using a Strata-X-A cartridge SPE previously conditioned with 75% acetone in water (2 mL) followed by water (4 mL). The cartridge was washed with water (4 mL) and the phloroglucinolysis products were eluted with 75% acetone in water (8 mL). This eluate was evaporated to dryness on a rotary evaporator at 35 °C, reconstituted in 2.5% acetic acid in water (1 mL), filtered (0.20 μm) and analyzed by HPLC/DAD–ESI/MS.

*Anthocyanins.* Wine samples were previously evaporated under a stream of nitrogen to remove ethanol and reconstituted with water. A sample of the reconstituted wine (2 mL) was loaded onto a Strata C18 cartridge, previously activated with methanol (10 mL) followed by water (10 mL). The sorbent was dried by blowing $N_2$ for 30 min. After washing with ethyl acetate (20 mL), the anthocyanin fraction was eluted with 0.1% TFA in methanol (30 mL). The eluate was evaporated to dryness (35 °C, 10 psi) and redisolved in 12% ethanol in water (1 mL). The ethanolic extract was passed through a filter of 0.45 μm pore size prior to HPLC/DAD–ESI/MS analysis.

*Phenolic acids, resveratrol and flavonols.* Wine samples were previously evaporated under a stream of nitrogen to remove ethanol and reconstituted with water. The reconstituted wine (3 mL, adjusted to pH 7) was loaded into a MCX cartridge previously activated with methanol (5 mL) followed by water (5 mL). The sorbent was washed with 0.1 M hydrochloric acid (5 mL) followed by water (5 mL). The acid and flavonol fractions were eluted with methanol (15 mL). The eluate was evaporated to dryness (35 °C, 10 psi) and redissolved in 12% ethanol in water (1 mL). The ethanolic extract was passed through a filter of 0.45 μm pore size prior to HPLC/DAD–ESI/MS analysis.

### 3.5.2. HPLC/DAD–ESI/MS Analysis

Identification of these groups of polyphenols was performed according to the HPLC/DAD-ESI/MS procedures described by Figueiredo-González *et al.* [26] and Quijada-Morín *et al.* [20]. HPLC measurements were made by using a Thermo Separation-Products (TSP, Waltham, MA, USA) system comprised of a P2000 binary pump equipped with a TSP AS1000 autosampler, and a TSP SCM1000 vacuum membrane degasser. An analytical column, Phenomenex C18 Luna (150 × 3 mm i.d., 5 μm), with a guard column, Pelliguard LC-18 (50 × 4.6 mm i.d., 40 μm; Supelco, Bellefonte, PA, USA) was used for separation of anthocyanins, phenolic acids, resveratrol and flavonols and other analytical column, Phenomenex C18 Luna (150 × 3 mm i.d., 3 μm) was used for separation of flavan-3-ol monomers and proanthocyanidins. UV–Vis spectra were scanned from 200 to 600 nm on a diode array UV6000LP DAD detector. For confirmation purposes, the HPLC–DAD system was coupled to a TSQ Quantum Discovery triple-stage quadrupole mass spectrometer from Thermo Fisher Scientific (Waltham, MA, USA). The mass spectrometer was operated in the negative electrospray ionization (ESI) mode under the following specific conditions: spray voltage 4,000, capillary temperature of 250 °C, sheath gas and auxiliary gas pressure of 30 and 10 units, collision energy 25 and tube lens offset 110. The detection was accomplished in the full-scan mode, from *m/z* 100 to 1,700, and in the MS/MS mode.

*Flavan-3-ol monomers and proanthocyanidins.* Acetic acid extract (20 μL) was injected into the column and eluted at 30 °C. Mobile phase A and B were 0.1% formic acid aqueous solution and 95% acetonitrile (in 5% mobile phase A) respectively, and the flow rate was 0.4 mL·min$^{-1}$. The linear gradient used was as follows: 0–2 min, 98% A and 2% B; 20–22 min, 90% A and 10% B; 50 min, 85% A and 15% B; 60 min, 80% A and 20% B; 70 min, 60% A and 40% B; 72–80 min, 10% A and 90% B; 82–92 min, 98% A and 2% B. DAD chromatograms were registered at 280 nm.

Due to the lack of the corresponding standards, extension subunits, *i.e.*, flavan-3-ol phloroglucinol adducts, were quantified using their molar response factors relative to catechin as reported by Kennedy and Jones [24]. In any case, the presence of the phloroglucinol adducts was confirmed by mass spectrometry. The mass spectrum of the gallocatechin phloroglucinol adduct obtained in the

ESI negative mode exhibited a $[M-H]^-$ ion at $m/z$ 429 and a $[2M-H]-$ ion at $m/z$ 859. MS/MS fragmentation of $m/z$ 429 produced a daughter ion at $m/z$ 303 $[M-H-C_6H_6O_3]^-$, which was indicative for a loss of phloroglucinol (126 Da) and the retro-Diels-Alder (RDA) product at $m/z$ 261 $[M-H-C_8H_8O_4]^-$. The MS analysis of catechin and epicatechin adducts showed a $[M-H]^-$ ion at $m/z$ 413 and a $[2M-H]-$ ion at $m/z$ 827. MS/MS fragmentation product ions of $m/z$ 413 were detected at $m/z$ 287 $[M-H-C_6H_6O_3]^-$ (loss of phloroglucinol) and at $m/z$ 261 $[M-H-C_8H_8O_3]^-$ (RDA fission).

*Anthocyanins.* Ethanolic extract (20 µL) was injected into the column and eluted at 35 °C. Mobile phase A and B were 5% formic acid aqueous solution and methanol, respectively, and the flow rate was 1 mL·min$^{-1}$. The following linear gradient was used: 0–5 min, 90% A and 10% B; 15 min, 80% A and 20% B; 30 min, 70% A and 30% B; 40–85 min, 68% A and 32% B; 87 min, 60% A and 40% B; 96 min, 50% A and 50% B; 98–108 min, 5% A and 95% B; 110–120 min, 90% A and 10% B. DAD chromatograms were registered at 520 nm.

*Phenolic acids and resveratrol.* Ethanolic extract (20 µL) was injected into the column and eluted at 35 °C. Mobile phase A and B were 0.2% formic acid aqueous solution and methanol, respectively, and the flow rate was 0.8 mL·min$^{-1}$. The following linear gradient was used: 0 min, 97% A and 3% B; 40 min, 70% A and 30% B; 50–53 min, 50% A and 50% B; 55–65 min, 5% A and 95% B; 67–77 min, 97% A and 3% B. DAD detection wavelengths of 280, 320 and 309 nm were selected for phenolic acids, hydroxycinnamic acids and resveratrol, respectively.

*Flavonols.* Ethanolic extract (20 µL) was injected into the column and eluted at 35 °C. Mobile phase A and B were 2.5% formic acid aqueous solution and methanol, respectively, and the flow rate was 1 mL·min$^{-1}$. The following linear gradient was used: 0 min, 80% A and 20% B; 10 min, 75% A and 25% B; 30 min, 65% A and 35% B; 40–42 min, 60% A and 40% B; 45 min, 50% A and 50% B, 48 min, 40% A and 60% B; 50–60 min, 5% A and 95% B; 62–72 min, 80% A and 20% B. DAD chromatograms were registered at 370 nm.

### 3.6. Statistical Analysis

For sample comparison, the data are presented as means ± standard deviation (SD) of analyses performed in triplicate. Significant differences among treated wines and control wines for each variety and compound were assessed by the *t*-student test. Data analyses were performed using the STATGRAPHICS Centurion XV 15.2.05 Software (Statpoint Technologies, Inc., Warrenton, VA, USA).

## 4. Conclusions

Results showed that the wines obtained from grapes treated under good agricultural practices with boscalid + kresoxim-methyl had lower chroma and higher hue than control wines, resulting in less colorful wines. The $\Delta E^*_{ab}$ parameter confirmed that these CIELab changes could be perceived by wine consumers. Although significant differences were observed for all determined phenolic compounds in wines, for both treatments and varieties, the results showed again that G and T wines obtained from grapes treated with boscalid + kresoxim-methyl were the most affected.

**Acknowledgments:** This work was granted by EU FEDER funds and by the Spanish Ministry of Education and Science grant (AGL2011-30378-C03-01). The authors are also very grateful to María Teresa Martínez and Jesús Sanz for their work to obtain the analyzed samples.

**Author Contributions:** Noelia Briz-Cid and María Figueiredo-González were responsible for the experimental work and the statistical treatment of the results. Raquel Rial-Otero, Beatriz Cancho-Grande and Jesús Simal-Gándara initially planned the experimental work and supervised the obtained results. Finally, all the authors contributed, in the same degree, in the data interpretation and the writing of the paper.

**Conflicts of Interest:** The authors declare no conflict of interest.

## References

1. Russell, P.E. Fungicide resistance: Occurrence and management. *J. Agric. Sci.* **1995**, *124*, 317–323.
2. Cabras, P.; Angioni, A.; Garau, V.L.; Pirisi, F.P.; Farris, G.A.; Madau, G.; Emonti, G. Pesticides in fermentative processes of wine. *J. Agric. Food Chem.* **1999**, *47*, 3854–3857.
3. González-Rodríguez, R.M.; Cancho-Grande, B.; Simal-Gándara, J. Efficacy of new commercial formulations to control downy mildew and dissipation of their active fungicides in wine after good agricultural practices. *J. Sci. Food Agric.* **2009**, *89*, 2625–2635.
4. Barba, A.; Oliva, J.; Payá, P. Influence of fungicide residues in wine quality. In *Fungicides*; Carisse, O., Ed.; InTech Europe: Rijeka, Croatia, 2010; pp. 421–440.
5. Oliva, J.; Navarro, S.; Barba, A.; Navarro, G.; Salinas, M.R. Effect of pesticide residues on the aromatic composition of red wines. *J. Agric. Food Chem.* **1999**, *47*, 2830–2836.
6. García, M.A. Influencia de los Residuos de Fungicidas en la Cinética Fermentativa y Calidad de Vinos Blancos de la D.O. Jumilla. Ph.D. Thesis, Universidad de Murcia, Facultad de Química, Jumilla, Murcia, Spain, 2002.
7. Noguerol-Pato, R.; González-Rodríguez, R.M.; González-Barreiro, C.; Cancho-Grande, B.; Simal-Gándara, J. Influence of tebuconazole residues on the aroma composition of Mencía red wines. *Food Chem.* **2011**, *124*, 1525–1532.
8. Oliva, J.; Barba, A.; San Nicolás, F.T.; Payá, P. Efectos de residuos de fungicidas en la composición fenólica de vinos tintos (var. Monastrell). *Tecnología Del Vino* **2005**, *23*, 37–40.

9. García-Marino, M.; Escudero-Gilete, M.L.; Heredia, F.J.; Escribano-Bailón, M.T.; Rivas-Gonzalo, J.C. Color-copigmentation study by tristimulus colorimetry (CIELAB) in red wines obtained from Tempranillo and Graciano varieties. *Food Res. Int.* **2013**, *51*, 123–131.

10. García-Marino, M.; Hernández-Hierro, J.M.; Santos-Buelga, C.; Rivas-Gonzalo, J.C.; Escribano-Bailón, M.T. Multivariate analysis of the polyphenol composition of Tempranillo and Graciano red wines. *Talanta* **2011**, *85*, 2060–2066.

11. Boulton, R.B. A method for the assessment of copigmentation in red wines. In Proceedings of the Forty-seventh Annual Meeting of the American Society for Enology and Viticulture; Reno, NV, USA: 26–28 June 1996.

12. Hermosín-Gutiérrez, I. Copigmentación y piranoantocianos: el papel de los flavonoles y los ácidos hidroxicinámicos en el color del vino tinto. *ACE Revista de Enología*. 2007, 81. Available online: http://www.acenologia.com/ciencia81_2.htm (accessed on 4 June 2014).

13. Mateus, N.; de Freitas, V. Evolution and stability of anthocyanin-derived pigments during port wine aging. *J. Agric. Food Chem.* **2001**, *49*, 5217–5222.

14. Tsanova-Savova, S.; Dimov, S.; Ribarova, F. Anthocyanins and color variables of Bulgarian aged red wines. *J. Food Comp. Anal.* **2002**, *15*, 647–654.

15. Alcalde-Eon, C.; Escribano-Bailon, M.T.; Santos-Buelga, C.; Rivas-Gonzalo, J.C. Identification of dimeric anthocyanins and new oligomeric pigments in red wine by means of HPLC-DAD-ESI/MS. *J. Mass Spectrom.* **2007**, *42*, 735–748.

16. Monagas, M.; Núñez, V.; Bartolomé, B.; Laureano, O.; Ricardo da Silva, J.M. Monomeric, oligomeric and polymeric flavan-3-ol composition of wines and grapes from *Vitis vinifera* L. cv. Graciano, Tempranillo and Cabernet Sauvignon. *J. Agric. Food Chem.* **2003**, *51*, 6475–6481.

17. González-Manzano, S.; Dueñas, M.; Rivas-Gonzalo, J.C.; Escribano-Bailón, M.T.; Santos-Buelga, C. Studies on the copigmentation between anthocyanins and flavan-3-ols and their influence in the colour expression of red wine. *Food Chem.* **2009**, *114*, 649–656.

18. Ribéreau-Gayon, R.; Dubourdieu, D.; Donèche, B.; Lonvaud, A. The chemistry of wine Stabilization and treatments. In *Handbook of Enology*; John Wiley & Sons Ltd.: Chichester, UK, 2006; Volume 2, pp. 141–203.

19. Cheynier, V.; Sarni-Manchado, P. Wine taste and mouthfeel. In *Managing Wine Quality, Viticulture and Wine Quality*; Reynolds, A.G., Ed.; Woodhead Publishing Limited: Cambridge, UK, 2010; Volume 1, pp. 29–58.

20. Quijada-Morín, N.; Regueiro, J.; Simal-Gándara, J.; Tomás, E.; Rivas-Gonzalo, J.C.; Escribano-Bailón, M.T. Relationship between the sensory-determined astringency and the flavanolic composition of red wines. *J. Agric. Food Chem.* **2012**, *60*, 12355–12361.

21. González-Manzano, S.; Santos-Buelga, C.; Pérez-Alonso, J.J.; Rivas-Gonzalo, J.C.; Escribano-Bailón, M.T. Characterization of the mean degree of polymerization of proanthocyanidins in red wines using liquid chromatography–mass spectrometry (LC-MS). *J. Agric. Food Chem.* **2006**, *54*, 4326–4332.

22. Glories, Y. La couleur des vins rouges. 1-ère partie. Les equilibres des anthocyanes et des tanins. *Connaissance de la Vigne et du Vin* **1984**, *18*, 195–217.

23. OIV. *Compendium of International Methods of Wine and Must Analysis*; International Organisation of Vine and Wine: Paris, France, 2000.

24. Kennedy, J.A.; Jones, G.P. Analysis of proanthocyanidin cleavage products following acid-catalysis in the presence of excess phloroglucinol. *J. Agric. Food Chem.* **2001**, *49*, 1740–1746.

25. Alcalde-Eon, C.; Escribano-Bailón, M.T.; Santos-Buelga, C.; Rivas-Gonzalo, J.C. Separation of pyranoanthocyanins from red wine by column chromatography. *Anal. Chim. Acta* **2004**, *513*, 305–318.

26. Figueiredo-González, M.; Martínez-Carballo, E.; Cancho-Grande, B.; Santiago-Blanco, J.L.; Martínez-Rodríguez, M.C.; Simal-Gándara, J. Pattern recognition of three Vitis vinifera L. red grapes varieties based on anthocyanin and flavonol fingerprints, with correlations between their biosynthesis pathways. *Food Chem.* **2012**, *130*, 9–19.

**Sample Availability:** *Sample Availability*: Not available.

# Antioxidant Activity and Acetylcholinesterase Inhibition of Grape Skin Anthocyanin (GSA)

Mehnaz Pervin, Md. Abul Hasnat, Yoon Mi Lee, Da Hye Kim, Jeong Eun Jo and Beong Ou Lim

**Abstract:** We aimed to investigate the antioxidant and acetylcholinesterase inhibitory activities of the anthocyanin rich extract of grape skin. Grape skin anthocyanin (GSA) neutralized free radicals in different test systems, such as 2,-2'-azinobis-(3-ethylbenzothiazoline-6-sulfonic acid) (ABTS) and 2,2-diphenyl-1-picrylhydrazyl (DPPH) assays, to form complexes with $Fe^{2+}$ preventing 2,2'-azobis(2-amidinopropane) dihydrochloride (AAPH)-induced erythrocyte hemolysis and oxidative DNA damage. Moreover, GSA decreased reactive oxygen species (ROS) generation in isolated mitochondria thus inhibiting 2',-7'-dichlorofluorescin (DCFH) oxidation. In an *in vivo* study, female BALB/c mice were administered GSA, at 12.5, 25, and 50 mg per kg per day orally for 30 consecutive days. Herein, we demonstrate that GSA administration significantly elevated the level of antioxidant enzymes in mice sera, livers, and brains. Furthermore, GSA inhibited acetylcholinesterase (AChE) in the *in vitro* assay with an $IC_{50}$ value of 363.61 µg/mL. Therefore, GSA could be an excellent source of antioxidants and its inhibition of cholinesterase is of interest with regard to neurodegenerative disorders such as Alzheimer's disease.

Reprinted from *Molecules*. Cite as: Pervin, M.; Hasnat, M.A.; Lee, Y.M.; Kim, D.H.; Jo, J.E.; Lim, B.O. Antioxidant Activity and Acetylcholinesterase Inhibition of Grape Skin Anthocyanin (GSA). *Molecules* **2014**, *19*, 9403–9418.

## 1. Introduction

Free radicals or reactive oxygen species (ROS) play many important roles in physiological and pathological processes. Oxidative stress is a biological phenomenon that results from a biochemical imbalance between the formation and clearance/buffering of free radicals [1]. Mitochondria are the major source of cellular ROS. The accumulation of ROS induces oxidative damage of mitochondrial DNA (mtDNA), proteins, and lipids, and has been shown to contribute to the decline in physiological function of cells resulting in a variety of diseases and accelerated aging [2].

Enzymatic systems in cells and body fluids regulate the level of ROS, which otherwise might generate a cascade of products and lead to assailing oxidants. The main classes of antioxidant enzymes in our antioxidant defense system are the superoxide dismutases (SOD), catalases (CAT), and glutathione peroxidases (GPx) [3]. Under oxidative stress conditions, antioxidant enzymes modulate the activities of these ROS and play a role in vascular function [4].

Acetylcholinesterase (AChE) is a hydrolase that plays a key role in cholinergic transmission by catalyzing the rapid hydrolysis of the neurotransmitter acetylcholine (ACh) [5]. Natural products might slow the progression of Alzheimer's disease (AD) by simultaneously protecting neurons from oxidative stress and acting as cholinesterase inhibitors [6].

Antioxidant supplementation is one plausible strategy to maintain redox homeostasis by directly quenching excessive ROS and protecting or reinforcing endogenous antioxidative defense systems against oxidative stress [7]. At present, many antioxidants are synthetically manufactured. They may possess some side effects and toxic properties to human health [8]. Therefore, studies of antioxidant substances in foods and natural products have gained increasing interest.

Anthocyanins are flavonoids, a class of secondary plant metabolites, with phenolic groups in their chemical structure that are responsible for the pigmentation in several different fruits. In grapes, they are found almost exclusively in the skins, with only a limited number of varieties showing these compounds in the pomace [9]. Recently, a great number of researchers have identified and characterized various anthocyanins found in grape skin with Liquid chromatography–mass spectrometry (LC/MS). The main *Vitis vinifera* grape anthocyanins, cyanidin, malvidin, delphinidin, petunidin, and peonidin, are present as monoglucoside, acetylmonoglucoside, and *p*-coumaroylmonoglucoside derivatives [10]. The most common anthocyanin in *V. vinifera* is malvidin-3-O-glucoside [11]. Choi *et al.* [12] showed that anthocyanins are primarily responsible for the antioxidant activity of this grape variety, which was also reported in other grape varieties. Anthocyanins are used in the treatment of coronary heart disease and are administered as dietary supplements for the prevention and treatment of metabolic disorders, in particular, obesity, diabetes, and osteoarthritis [13]. Anthocyanins also possess anticancer and antineurodegenerative properties [14–16]. It has been shown that anthocyanins have beneficial effects on memory and cognition, suggesting a clear neuroprotective role [17,18].

Therefore, the primary objective of our research was to investigate the antioxidant effects of grape skin anthocyanins using various *in vitro* and *in vivo* methods. In this study, physiologically relevant antioxidant activities of GSA, including erythrocyte membrane protection, DNA protection, and mitochondria protection were investigated for the first time. Moreover, we studied the *in vivo* antioxidant activity of GSA in mice serum, liver, and brain. Another major objective was to determine potential *in vitro* anticholinesterase activity of GSA.

## 2. Results and Discussion

### 2.1. Radical Scavenging Activity

GSA's free radical scavenging activity was evaluated on different free radical species: DPPH and ABTS. GSA was able to reduce the stable radical DPPH to the yellow-colored diphenyl picryl hydrazine. GSA exhibited a significant concentration-dependent inhibition of DPPH activity, with a 50% inhibition ($IC_{50}$) at a concentration of 95.54 µg/mL. The results are presented in Table 1. The $IC_{50}$ value of vitamin C was 71.50 µg/mL. These results indicate that GSA might act as an electron or hydrogen donator to scavenge $DPPH^\bullet$ radicals.

The ABTS radical scavenging assay is shown in Table 1: GSA showed scavenging activity in a concentration-dependent manner. The concentration for 50% inhibition of GSA and vitamin C were found to be 62.74 and 20.32 µg/mL, respectively. These results indicate that GSA has strong scavenging power for ABTS radicals and should be explored as a potential antioxidant. Previous studies have confirmed the free radical scavenging activity of red grape pomace seeds and skin extracts [19].

**Table 1.** DPPH, ABTS radical scavenging and metal chelating activities of GSA.

| Extract | DPPH Radical | | ABTS Radical | | Metal Chelating | |
|---|---|---|---|---|---|---|
| | Scavenging Activity (%) | $IC_{50}$ Value µ g/mL | Scavenging Activity (%) | $IC_{50}$ Value µ g/mL | Scavenging Activity (%) | $IC_{50}$ Value µ g/mL |
| GSA | 95.54 ± 0.43 [a] | 95.14 ± 1.13 [a] | 97.67 ± 1.009 [a] | 62.74 ± 0.43 [a] | 56.26 ± 1.67 [a] | 180.49 ± 19.40 [a] |
| Positive control | 97.75 ± 0.28 [b] (Vitamin C) | 71.50 ± 1.05 [b] | 99.78 ± 0.34 [b] (Vitamin C) | 20.32 ± 0.20 [b] | 89.82 ± 2.69 [b] (EDTA 100 µM) | 7.089 ± 0.78 [b] |

All data are expressed as mean ± SD ($n = 3$). Different letters in each column denote statistically significant difference compare to the positive control group at $p < 0.05$. Scavenging activity (%) was determined at 500 µg/mL.

### 2.2. $Fe^{2+}$ Chelation

Chelation is an important parameter because iron is required for oxygen transport, respiration, and the activities of many enzymes. However, iron is an extremely reactive metal and can catalyze oxidative changes in lipid, proteins, and other cellular components. $Fe^{2+}$ ion can trigger a Fenton reaction when it encounters $H_2O_2$, and the product of this reaction (hydroxyl radical) can cause great oxidative damage [20]. Therefore, ferrous ion-chelating activity is considered an important indicator in any oxidative stress involving ferrous ion. $Fe^{2+}$ ion is the most powerful pro-oxidant among the various species of metal ions [21]. Ferrozine can quantitatively form complexes with $Fe^{2+}$. However, in the presence of chelating agents, the complex formation is disrupted, resulting in a decrease in the red color of the complex. Measurement of color reduction therefore allows for estimation of the metal chelating activity of the co-existing chelator. The capacity of GSA to chelate $Fe^{2+}$ is shown in Table 1. We found that GSA could chelate $Fe^{2+}$ efficiently and

137

therefore reduce the production of free radicals. The $IC_{50}$ value for GSA's chelating abilities was 180.49 µg/mL. EDTA was used as reference in this assay and its $IC_{50}$ value for $Fe^{2+}$-chelation was 7.08 µg/mL.

## 2.3. Oxidative Hemolysis Inhibition Assay

The oxidative hemolysis inhibition assay system is based on the property of erythrocytes that renders them susceptible to oxidative damage and utilizes the biologically relevant radical source, AAPH-derived peroxyl radicals, to attack the erythrocyte membrane and cause erythrocyte hemolysis [22]. The rate of cell lysis can be regarded as an *in vitro* marker of oxidative damage. As shown in Figure 1, inhibition rates of erythrocyte hemolysis were 17.63%, 19.36%, 25.37%, 35.12%, 68.35%, and 68.35% for GSA, and 22.54%, 27.47%, 35.70%, 46.85%, and 72.89% for vitamin C at the tested concentrations of 31.25, 62.5, 125, 250, and 500 µg/mL, respectively. Our results are in agreement with other studies showing that polyphenols are able to protect erythrocytes from oxidative stress or increase their resistance to damage caused by oxidants [23,24].

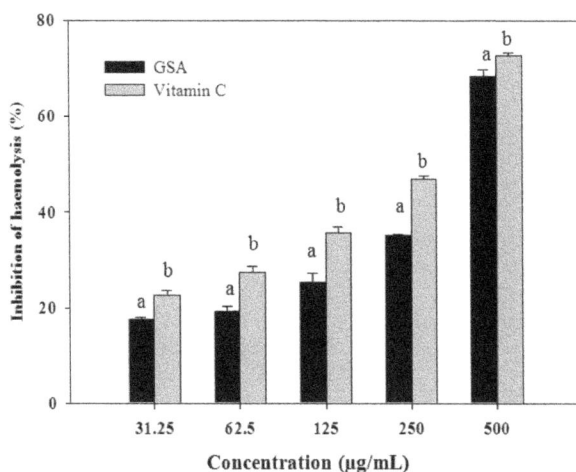

**Figure 1.** Anti- haemolytic activity of GSA on APPH-induced erythrocyte haemolysis *in vitro*. Data are expressed as mean $\pm$ SD ($n = 3$). Columns with different letters are significantly different at $p < 0.05$ level.

## 2.4. Oxidative DNA Damage Prevention

ROS, such as superoxide anion ($O_2{}^-$), hydrogen peroxide ($H_2O_2$), and hydroxyl radical ($^\bullet OH$) can cause damage to biological macromolecules leading to lipid peroxidation, protein oxidation, and DNA base modification and strand breaks [21]. Permanent modification of DNA as a result of oxidative damage is the first step in several pathological and physiological conditions such as cancer and aging, respectively. The inhibition of $H_2O_2$-induced DNA strand breakage by GSA was assessed by measuring the conversion of the supercoiled pBR322 plasmid DNA to open circular and linear forms by gel electrophoresis. Because hydroxyl radical ($^\bullet$OH) modifies and destroys DNA in a nonspecific manner, protection capacity against $^\bullet$OH-induced oxidation of DNA was also measured to evaluate an antioxidant. Figure 2, shows the inhibitory effect of GSA on pBR322 plasmid DNA cleavage caused by $H_2O_2$. Conversion of the supercoiled form of this plasmid DNA to the open-circular and further linear forms has been used as an index of DNA damage. The plasmid DNA was mainly in the supercoiled form in the absence of $Fe^{2+}$ and $H_2O_2$ (lane 1, control). After the addition of $Fe^{2+}$ and $H_2O_2$, the quantity of supercoiled DNA decreased due to conversion into the relaxed circular and linear forms. However, further fragmentation of linear form decreased in the presence of GSA (125–500 µg/mL). Both GSA and vitamin C were concentration–dependent for preventing DNA damage. The observed scission-inhibition could be due to the scavenging of hydroxyl radicals by GSA. Devasagayam *et al.* [25] studied the ability of natural antioxidants, such as carotenoids and flavonoids to protect the pBR322 plasmid DNA against ROS. In a previous study Noroozi *et al.* [26] reported that flavonoids and vitamin C were effective against hydrogen peroxide initiated oxidative DNA damage to human lymphocytes.

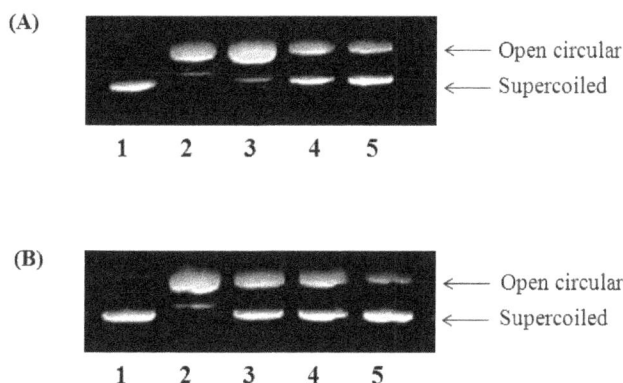

**Figure 2.** Protective effect of GSA on hydroxyl radical-mediated pBR322 DNA strand breaks. (**A**) GSA (**B**) Vitamin C. Lane 1: normal DNA control; lane 2: $FeSO_4$ + $H_2O_2$ (DNA damage control); lane: 3–5: $FeSO_4$ + $H_2O_2$ + DNA in the presence of GSA (125, 250 and 500 µg/mL, respectively).

*2.5. Evaluation of Antioxidation of GSA Using a Mitochondria-Based Assay*

Approximately 90% of cellular ROS are produced in the mitochondria [27]. ROS levels are thought to increase with age owing to the accumulation of damaged mitochondria in a self-perpetuating cycle. ROS-induced impairment of mitochondria results in increased ROS production, which in turn leads to further mitochondrial damage [28]. Measurement of ROS in living organisms has been a significant analytical challenge. Most ROS are highly reactive and short lived and therefore are difficult to detect in complex biological matrices. Additionally, ROS are often produced and/or neutralized in subcellular compartments, which requires detection methods specific to subcellular localization [29].

A physiologically relevant mitochondria-based assay was used to assess the antioxidant capability of GSA against oxidative stress in mitochondria. Ascorbic acid was used as the reference antioxidant. GSA could inhibit DCFH oxidation by scavenging ROS, thus resulting in decreased fluorescence intensity. As illustrated in Figure 3, the tested sample and ascorbic acid standard exhibited strong antioxidant capacity in a concentration-dependent manner. Inhibition of DCFH oxidation was 29.59%, 32.78%, 38.73%, 46.22%, and 65.62% for GSA, and 31.40%, 38.46%, 40.40%, 47.61%, and 68.95% for vitamin C at the tested concentrations of 31.25, 62.5, 125, 250, and 500 μg/mL, respectively (Figure 3). GSA did not show any significant difference compared to vitamin C ($p < 0.05$). The method for monitoring $H_2O_2$ generation in isolated mitochondria by DCFH-DA chemical probe was first introduced in 1983 [30]. It is known that once DCFH-DA enters the cell, the acetyl moiety is cleaved by intracellular esterases; subsequent oxidation by ROS, particularly $H_2O_2$ and hydroxyl radical, yields the fluorescent product, DCF. The principle of this method is that antioxidants can scavenge ROS generated in mitochondria, thus inhibiting DCFH oxidation.

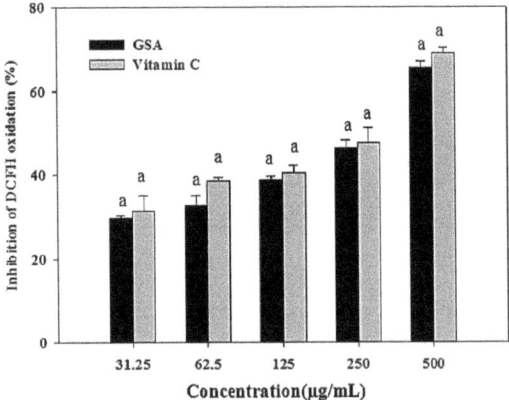

**Figure 3.** Protective efffect of GSA against oxidative damage on isolated mouse liver mitochondria. ROS generation was assayed as inhibition of DCFH oxidation. Values represent the mean ± SD ($n = 3$). Columns with different letters are significantly different at $p < 0.05$ level.

## 2.6. Antioxidant Activities in Vivo

The cooperative defense systems that protect the body from free radical damage include antioxidant nutrients and enzymes. As shown in Figure 4, after administration of GSA (50 mg/kg) SOD, CAT, and GPx activities were noticeably increased in mice serum, liver, and brain than those in the control group ($p < 0.05$). In most of the cases, compared with the control group, levels of antioxidant enzymes were not significantly elevated for the GSA extract at 12.5 and 25 mg/kg ($p < 0.05$). Administration of ascorbic acid (50 mg/kg) also showed significant increase in SOD, CAT, and GPx levels. These data suggest that GSA has significant effects on the levels of antioxidant enzymes in mice.

Regarding the *in vivo* study, evidence has shown that ethanolic extract of white button mushroom (*Agaricus bisporus*)-fed mice led to a significantly higher level of antioxidant enzymes (SOD, GsH-Px, and CAT) in mice serum, liver, and heart [31]. Grape skin anthocyanin activates the antioxidant enzymes SOD, CAT, and GPx in $H_2O_2$ treated retinal cells [32]. Puiggros *et al.* [33] provided evidence that grape seed procyanidin extract increased the Cu/Zn-SOD activity in rats and Fao cell line hepatocytes. In *in vivo* assay, numerous factors such as bioavailability, digestibility, and metabolism of the compound may influence biological potentials. Previous studies indicated that anthocyanin can rapidly reach the plasma after oral administration. [34,35]. Two previous works reported the capacity of dietary anthocyanins from grapes and berries to reach the brain [36,37]. Moreover, the results of a previous clinical study suggested that antioxidative anthocyanins are obviously absorbed from grape juice and wine [38].

## 2.7. In Vitro Cholinesterase Inhibition

Acetylcholinesterase (AChE) is a hydrolase that plays a key role in cholinergic transmission by catalyzing the rapid hydrolysis of the neurotransmitter acetylcholine (ACh) [5]. The use of acetylcholinesterase inhibitors elicits numerous responses, which mediate the symptoms of Alzheimer's disease [39]. When studied for its possible inhibitory effect in the *in vitro* assay, GSA showed AChE inhibitory activity in a dose-dependent manner. As illustrated in Figure 5, at tested concentrations of GSA (31.25–500 µg/mL), acetylcholinesterase inhibitory activities were 17.94%, 21.47%, 29.16%, 45.57%, and 55.58%, respectively ($IC_{50}$ = 363.61 µg/mL). Tacrine was used as a reference inhibitor and was more active than GSA ($p < 0.05$). At 10 µM tacrine showed a 66.30% inhibition of AChE. Several studies recently supported that different plant extracts and active compounds, including anthocyanins (pelargonidin, delphinidin and cyanidin), terpenoids, also have anticholinesterase activity [40–42]. The leaves of pomegranate and grapes exhibited considerable acetylcholinesterase inhibitory activity [43,44]. Several studies showed that grape and blueberry anthocyanin have clear neuroprotective roles [16,18,45].

**(A)**

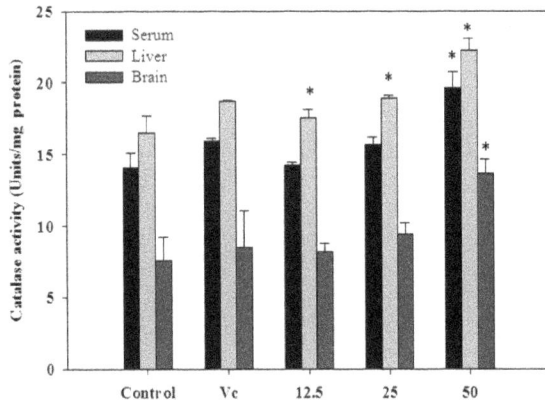

**(B)**

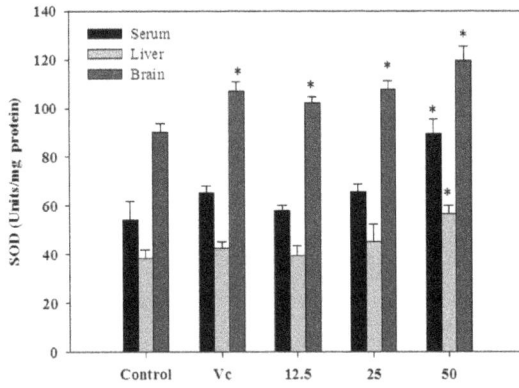

**(C)**

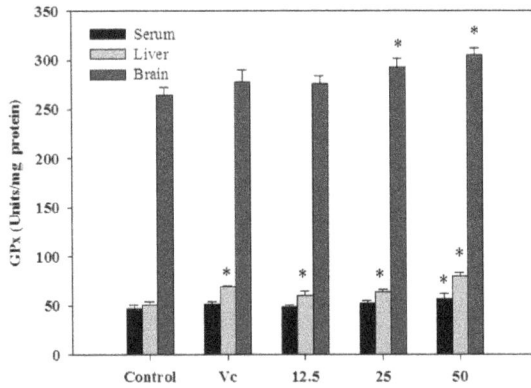

**Figure 4.** Effects of GSA on level of catalase (CAT) (**A**) SOD (**B**) and GPx (**C**) in serum, liver and brain of mice. Values are the mean $\pm$ SD ($n = 5$); $* p < 0.05$ compared with normal control group. All activities were expressed as unit per milligram of protein (U/mg protein).

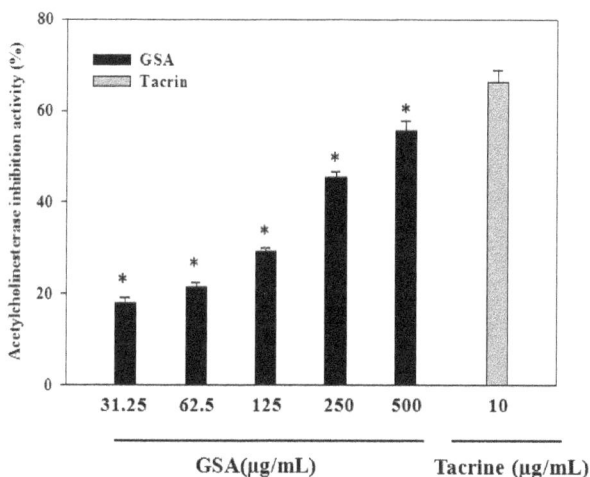

**Figure 5.** Acetylcholinesterase inhibitory activity of GSA. Values represent the mean $\pm$ SD ($n = 3$). * $p < 0.05$, compare to the positive control group.

## 3. Experimental Section

### 3.1. Samples and Chemicals

Anthocyanin rich grape skin extract was manufactured by Kitolife (Pyeong-Teak, Korea). The anthocyanin content of grape skin (*Vitis vinifera* L.) extract was standardized at 80% (w/w) by high-performance liquid chromatography (HPLC) system. These products were manufactured according to a previously described method [32]. In brief, grap skin (*V. vinifera* L. cv. Aglianico) were collected and extracted in methanol (0.75% HCl) solution for five days at room temperature. Anthocyanins extracted from skins of grape contain the following four major compounds: malvidin 3-*O*-glucoside, petunidin 3-*O*-glucoside, delphinidin 3-*O*-glucoside, and cyaniding 3-*O*-rutinoside [13,32,46].

2,2-Diphenyl-1-picrylhydrazyl (DPPH), 2,2'-azinobis-(3-ethylbenzothiazoline-6-sulphonic acid) diammonium salt (ABTS), gallic acid, sodium nitrite, Folin–Ciocalteu reagent (FC reagent), butylated hydroxyluene (BHT), ascorbic acid (AA), α-tocopherol, potassium persulphate, ferrous chloride, ammonium thiocyanate, ethylene-di-amino-tetraacetic acid (EDTA), linoleic acid, anhydrous sodium phosphate (dibasic), anhydrous sodium phosphate (monobasic), 5,5'-dimethyl-pyrroline-1-oxide (DMPO), pyrogallol and ferrous sulphate ($FeSO_4$), ethylene-bis-(oxyethylenenitrilo)-tetraacetic acid (EGTA), HEPES, glutamate, succinate, and 2',7'-dichlorofluorescin diacetate (DCFH-DA) were purchased from Sigma-Aldrich (St. Louis, MO, USA). Sodium hydroxide and ferric chloride were obtained from Wako Pure Chemical Industries Ltd. (Osaka, Japan). The catalase assay kit and SOD assay kit were purchased from Cayman Chemical Company (Ann Arbor, MI, USA). The pBR322

143

DNA and $6\times$ DNA loading dye were purchased from Fermentas Inc. (Cromwell Park, Glen Burnie, USA). All other reagents were of analytical grade.

### 3.2. 2,2-Diphenyl-1-picrylhydrazyl (DPPH) Radical Scavenging Activities

Free radical scavenging activity on DPPH by GSA was assessed by a previously described colorimetric method [20]. In brief, an 80 μL aliquot of sample solution at different concentrations (31.25–500 μg/mL) was mixed with 80 μL DPPH solution (0.3 mM in methanol). The reaction mixture was incubated for 30 min in the dark at room temperature. The absorbance of the resulting solution was measured at 517 nm with a spectrophotometer (Sunrise-Basic Tecan, Salzburg, Austria). Controls were prepared in a similar manner using the corresponding extraction solvent in place of the sample solution. The radical scavenging capacity of the tested samples was measured using the following equation:

$$\text{Scavenging activity (\%)} = (1 - \text{Absorbance of sample}/\text{Absorbance of control}) \times 100$$

### 3.3. 2,2'-Azinobis-(3-ethylbenzothiazoline-6-sulfonic acid) (ABTS) Radical-Scavenging Activity

Free radical scavenging activity on ABTS by GSA was determined using the method described by He et al. [47] with slight modifications. The ABTS$^{\bullet+}$ radical was generated by the reaction of 7 mM 2,2'-azinobis-(3-ethylbenzothiazoline-6-sulphonic acid) diammonium salt (ABTS, 5 mL) with 2.45 mM of potassium persulphate (88 μL). The mixture was left to stand for 12–16 h in the dark at room temperature. Absorbance of the reactant was later adjusted to $0.70 \pm 0.02$, at room temperature, at a wavelength of 734 nm. Different concentrations of tested extract were mixed with 0.7 mL of ABTS$^{\bullet+}$ solution and the mixture was shaken for 5 min. The reduction of the ABTS$^{\bullet+}$ radical was determined by reading the absorbance at 734 nm using a UV spectrophotometer (Pharmaspec UV-1700, Shimadzu, Kyoto, Japan). The controls contained the extraction solvent instead of the test sample. The scavenging activity of ABTS free radical was calculated as:

$$\text{Scavenging activity (\%)} = (1 - \text{Absorbance of sample}/\text{Absorbance of control}) \times 100$$

### 3.4. $Fe^{2+}$ Chelation Assay

The ferrous ion-chelating activity of GSA was estimated in accordance with the method described by Cheng et al. [20]. GSA (31.25–500 μg/mL). Each sample was incubated with 50 μL of 2 mM $FeCl_2$ for 5 min. The reaction was initiated by adding 200 μL of 5 mM ferrozine. After incubation for 5 min at room temperature, the absorbance of the mixture was measured at 562 nm against the blank, which was performed in the same way using $FeCl_2$ and water. EDTA (3.12–100 μg/mL) served as

144

the positive control, and a sample without the sample or EDTA served as the negative control. The $Fe^{2+}$-chelating activity was calculated using the equation below:

$$\text{Chelating activity (\%)} = (1 - \text{Absorbance of sample/Absorbance of control}) \times 100$$

### 3.5. Oxidative Hemolysis Inhibition Assay

Anti-hemolytic activity was assayed according to the method described by Carvalho et al. [23]. Blood was collected from female BALB/c mice (weighing $20 \pm 2$ g). RBCs were separated from plasma by centrifugation at 1500 g for 20 min. The crude RBC was then washed five times with five volumes of phosphate-buffered saline (PBS, pH 7.4). The RBC was suspended in four volumes of PBS solution for hemolysis assay. Two mL of RBC suspension were mixed with 2 mL of PBS solution containing GSA (31.25–500 μg/mL). The erythrocyte suspension was agitated gently while being incubated with APPH (final concentration, 50 mM) at 37 °C for 3 h. After incubation, 8 mL of PBS solution was added to the reaction mixture, followed by centrifugation at 1,000 g for 10 min. The absorbance of the supernatant was recorded at 540 nm in a spectrophotometer. Percentage inhibition was calculated by the following equation:

$$\text{\% Inhibition} = (1 - \text{Absorbance of sample/Absorbance of control}) \times 100$$

### 3.6. Assay for Effects of GSA on DNA Oxidative Damage

The protective effect of GSA on DNA strand breaks induced by hydroxyl radicals was measured by the conversion of pBR322 DNA to an open circular form according to the method described by Cheng et al. [20] with some minor modifications. Briefly, 1 μL of plasmid pBR322 DNA (0.5 μg/μL) was treated with 3 μL of $FeSO_4$ (0.08 mM), 4 μL of 30% $H_2O_2$ (v/v), 3 μL distilled water, and 2 μL of the tested sample at different concentrations (125–500 μg/mL). The mixture was then incubated in a water bath at 37 °C for 1 h. Then 2 μL of DNA loading dye (6×) was added to the mixture. The DNA samples were resolved on a 0.8% (w/v) agarose gel using ethidium bromide staining. Gels were scanned on a gel documentation system (Nextep, Seoul, Korea) and bands were quantified using NEXTEP analysis software. To avoid photolysis of samples, experiments were conducted in the dark.

### 3.7. Mitochondria-Based Assay

#### 3.7.1. Isolation of Mitochondria from Liver

Mitochondria were isolated from the livers of mice according to the methods described by He et al. [47], with some modifications. Briefly, livers were rinsed using cold homogenization media, and subsequently homogenized in the homogenization buffer A (1:4, w/v; sucrose 0.32 mol/L, EDTA 1 mmol/L, Tris–HCl 10 mmol/L, and bovine serum albumin (BSA) 65 mmol/L, pH 7.4). This homogenate was centrifuged for at 45 $g$ for 10 min, and the unbroken tissue, cells, and nuclei were discarded. The supernatant obtained was centrifuged at 15,000 $g$ for 10 min, and the pellet was collected and resuspended in the homogenization buffer A. This procedure was repeated until a single pellet was obtained. The pelleted mitochondria were resuspended in 30 mL of buffer B (KCl 137 mmol/L, HEPES 10 mmol/L, MgCl2 2.5 mmol/L, and EDTA 0.5 mmol/L, pH 7.2) and stored at −20 °C until use. The concentration of mitochondrial protein was determined using the Bradford protein assay with BSA as a standard.

### 3.7.2. Mitochondrial Reactive Oxygen Species (ROS) Measurements

ROS production in mitochondria was measured using 2',7'-dichlorofluorescin diacetate (DCFH-DA), a $H_2O_2$-sensitive fluorescent probe, as previously described by He *et al.* [47], with modifications. Briefly, 40 μL of appropriate dilutions of extract was added into a mixture containing 30 μL glutamate (40 mmol/L), 30 μL succinate (40 mmol/L) and 165 μL $H_2O_2$ buffer in a 96-well plate, and followed by 75 μL DCFH-DA (52 μmol/L). Then, 60 μL (1.5 mg/mL) mitochondrial suspension was added to initiate the reaction, which was incubated at 37 °C for 10 min. The change in fluorescence of the reaction mixture was recorded at 485 nm excitation and 530 nm emission in a spectrophotometer. Inhibition of DCFH oxidation was calculated by the following equation:

$$\% \, \text{Inhibition} \, = \, (1 - \text{Absorbance of sample} / \text{Absorbance of control}) \times 100$$

### 3.8. In Vivo Antioxidant Activity

### 3.8.1. Animals and Experimental Design

*In vivo* antioxidant activity was assayed according to the methods described by Liu *et al.* [31] with some modifications. Female BALB/c mice (weighing $20 \pm 2$ g, 8 weeks old) were purchased from Orient Bio Inc. (Seongnam, Gyeonggi, Korea). Animals were acclimatized under controlled conditions for 1 week before experimental feeding. Mice were housed in specific pathogen-free conditions in an animal room at Konkuk University, maintained on a 12-h light/dark cycle, and provided food and water *ad libitum*. All animal procedures were carried out according to a protocol approved by the Institutional Animal Care and Use Committee of the Konkuk University. After one week of adaptation, the mice were randomly divided into five groups of 5 animals each: Group 1 (control) received vehicle (water), Group 2 received vitamin C (positive control), Group 3, Group 4, and Group 5 received GSA at 12.5, 25, and 50 mg/kg body weight, respectively, by gavage for 30 consecutive days.

### 3.8.2. Biochemical Assay

Twenty-four hours after the last drug administration, mice were sacrificed and blood samples were collected. The blood samples were then centrifuged at 10,000 $g$ at 4 °C for 10 min to obtain blood serum, which was then stored at −80 °C for further analysis. The liver and brain were removed immediately, washed and homogenized in ice-cold physiological saline to prepare a 10% (w/v) homogenate. Then, the homogenate was centrifuged at 1,000 $g$ at 4 °C for 10 min to remove cellular debris, and the supernatant was collected for analysis.

Antioxidant enzymatic activities were determined using SOD, CAT, and GPx assay kits (Cayman Ann Arbor, MI, USA) following the manufacturer's instructions. One unit of SOD is defined as the amount of enzyme needed to exhibit 50% dismutation of the superoxide radical. Detection of catalase activity was based on the reaction of the enzyme with methanol in the presence of an optimal concentration of $H_2O_2$. The formaldehyde produced was measured spectrophotometrically with 4-amino-3-hydrazino-5-mercapto-1,2,4-triazole as the chromogen. Catalase activity was expressed as μmol of formaldehyde per min per g of protein from homogenates. GPx activity was measured on the basis of the reaction of GSH and 5,5'-dithiobis-(2-nitrobenzoic acid). The protein contents in the supernatants obtained from the liver and brain were determined by the Bradford Protein assay kit. All the above treatments were performed at 4 °C.

### 3.9. In Vitro Anticholinesterase Inhibition Assay

The AChE inhibition assay was carried out in a multi-well plate using a modified method, as described by Ellman *et al.* [48]. Electric eel acetylcholinesterase was used, while acetyl thiocholine iodide (ATCI) was used as the substrate of the reaction. 5,5-dithiobis(2-nitrobenzioc) acid (DTNB) was used for measurement of AChE activity. Briefly, 150 μL of 0.1 M sodium phosphate buffer (pH 8.0), 10 μL test compound solution, and 20 μL of enzyme solution (0.09 units/mL) were mixed and incubated for 15 min at 25 °C. 10 μL of DTNB (10 mM) was then added and reaction was initiated by the addition of substrate (10 μL of ATCI, 14 mM solution). The hydrolysis of the ATCI can be measured by the formation of the product, 5-thio-2-nitrobenzoate, a colored anion formed by the reaction of DTNB and thiocholine, which is released by enzyme hydrolysis. Absorbance was measured at 412 nm (Shimadzu, 1200, Japan) after 10 min. Tacrine, a standard AChE inhibitor, was used as positive control. The percent of acetylcholinesterase inhibition was calculated as following:

% Inhibition $= 100 - [$Absorbance of the test compound$/$Absorbance of the control$] \times 100$

## 3.10. Statistical Analysis

Data are expressed as mean $\pm$ standard deviation (SD). All analysis was carried out in at least three replicates for each sample. Results were analyzed statistically using SPSS 15.0, Sigma plot 10.0, and GraphPad Prism 5 software (San Diego, CA, USA). A value of $p < 0.05$ was considered statistically significant.

## 4. Conclusions

GSA possesses strong antioxidant activity as demonstrated by biologically relevant assays, such as the oxidative DNA damage prevention assay, hemolysis inhibition assay, liver mitochondria oxidative damage prevention assay. Administration of GSA could significantly enhance the activities of antioxidant enzymes (SOD, CAT, and GPx) in mice sera, liver, and brain. Moreover, the results suggest that GSA can inhibit cholinesterase activities. Altogether, our results show that GSA has great value for preventing oxidative stress-related disease and can be a prominent source of anticholinesterase activity.

**Acknowledgments:** This work was supported by the Basic Science Program through the National Research Foundation of Korea (NRF) funded by the Ministry of Education, Science and Technology (Grant No. 2011-0022244).

**Author Contributions:** M.P. and M.A.H. designed and conducted the research. M.P. wrote the paper. M.A.H., D.H.K., and J.E.J. helped in processing tissue and sample. Y.M.L. aided in the critical review of the manuscript. B.O.L. provided material support, technical support, and aided in study supervision, administrative, obtained funding and aided critical review of the manuscript.

**Conflicts of Interest:** The authors declare no conflict of interest.

## References

1. Cannizzo, E.S.; Clement, C.C.; Morozova, K.; Valdor, R.; Kaushik, S.; Almeida, L.N.; Follo, C.; Sahu, R.; Cuervo, A.M.; Macian, F.; *et al.* Age-related oxidative stress compromises endosomal proteostasis. *Cell Rep.* **2012**, *2*, 136–149.
2. Matsuda, T.; Kanki, T.; Tanimura, T.; Kang, D.; Matsuura, E.T. Effects of overexpression of mitochondrial transcription factor A on lifespan and oxidative stress response in Drosophila melanogaster. *Biochem. Biophys. Res. Commun.* **2013**, *430*, 717–721.
3. Sies, H. Oxidative stress: Oxidants and antioxidants. *Exp Physiol.* **1997**, *82*, 291–295.
4. Fukai, T.; Ushio-Fukai, M. Superoxide dismutases: Role in redox signaling, vascular function, and diseases. *Antioxid. Redox Signal.* **2011**, *15*, 1583–1606.
5. Gabrovska, K.; Marinov, I.; Godjevargova, T.; Portaccio, M.; Lepore, M.; Grano, V.; Dianoc, N; Mita, G.M. The influence of the support nature on the kinetics parameters, inhibition constants and reactivation of immobilized acetylcholinesterase. *Int. J. Biol. Macromol.* **2008**, *43*, 339–345.

6.  Costa, P.; Gonçalves, S.; Valentao, P.; Andrade, P.B.; Romano, A. Accumulation of phenolic compounds in *in vitro* cultures and wild plants of *Lavandula viridis* L'Hér and their antioxidant and anti-cholinesterase potential. *Food Chem. Toxicol.* **2013**, *57*, 69–74.

7.  Tang, Y.; Gao, C.; Xing, M.; Li, Y.; Zhu, L.; Wang, D.; Yang, X.; Liu, L.; Yao, P. Quercetin prevents ethanol-induced dyslipidemia and mitochondrial oxidative damage. *Food Chem. Toxicol.* **2012**, *50*, 1194–1200.

8.  Suhaj, M. Spice antioxidants isolation and their antiradical activity: A review. *J. Food Compos. Anal.* **2006**, *19*, 531–537.

9.  Liazid, A.; Guerrero, R.F.; Cantos, E.; Palma, M.; Barroso, C.G. Microwave assisted extraction of anthocyanins from grape skins. *Food Chem.* **2011**, *124*, 1238–1243.

10. Flamini, R.; Mattivi, F.; Rosso, M.D; Arapitsas, P.; Bavaresco, L. Advanced knowledge of three important classes of grape phenolics: Anthocyanins, stilbenes and flavonols. *Int. J. Mol. Sci.* **2013**, *14*, 19651–19669.

11. Kennedy, J.A.; Saucier, C.; Glories, Y. Grape and wine phenolics: History and perspective. *Am. J. Enol. Vitic.* **2006**, *57*, 239–248.

12. Choi, J.Y.; Lee, S.J.; Lee, S.J.; Park, S.; Lee, J.H.; Shim, J.H.; Abd El-Aty, A.M.; Jin, J.S.; Jeong, E.D.; Lee, W.S.; *et al.* Analysis and tentative structure elucidation of new anthocyanins in fruit peel of *Vitis coignetiae* Pulliat (meoru) using LC-MS/MS: Contribution to the overall antioxidant Activity . *J. Sep. Sci.* **2010**, *33*, 1192–1197.

13. Manfra, M.; de Nisco, M.; Bolognese, A.; Nuzzo, V.; Sofo, A.; Scopa, A.; Santi, L.; Tenore, G.C.; Novellino, E. Anthocyanin composition and extractability in berry skin and wine of *Vitis vinifera* L. cv. Aglianico. *J. Sci. Food Agric.* **2011**, *91*, 2749–2755.

14. Cai, H.; Marczylo, T.H.; Teller, N.; Brown, K.; Steward, W.P.; Marko, D.; Gescher, A.J. Anthocyanin-rich red grape extract impedes adenoma development in the ApcMin mouse: Pharmacodynamic changes and anthocyanin levels in the murine biophase. *Eur. J. Cancer* **2010**, *46*, 811–817.

15. Rahman, M.M.; Ichiyanagi, T.; Komiyama, T.; Sato, S.; Konishi, T. Effects of anthocyanins on psychological stress-induced oxidative stress and neurotransmitter status. *J. Agric. Food Chem.* **2008**, *56*, 7545–7550.

16. Gutierres, J.M; Carvalho, F.B.; Schetinger, M.R.; Marisco, P.; Agostinho, P.; Rodrigues, M.; Rubin, M.A.; Schmatz, R.; da Silva, C.R.; de P Cognato, G.; *et al.* Anthocyanins restore behavioral and biochemical changes caused by streptozotocin-induced sporadic dementia of Alzheimer's type. *Life Sci.* **2013**, *96*, 7–17.

17. Shukitt-Hale, B.; Cheng, V.; Joseph, J.A. Effects of blackberries on motor and cognitive function in aged rats. *Nutr. Neurosci.* **2009**, *12*, 135–40.

18. Krikorian, R.; Shidler, M.D.; Nash, T.A.; Kalt, W.; Vinqvist-Tymchuk, M.R.; Shukitt-Hale, B.; Joseph, J.A. Blueberry supplementation improves memory in older adults. *J. Agric. Food Chem.* **2010**, *58*, 3996–4000.

19. Rockenbach, I.I.; Gonzaga, L.V.; Rizelio, V.M.; Gonçalves, A.E.D.S.S.; Genovese, M.I.; Fett, R. Phenolic compounds and antioxidant activity of seed and skin extracts of red grape (*Vitis vinifera* and *Vitis labrusca*) pomace from Brazilian wine making. *Food Res. Int.* **2011**, *44*, 897–901.

20. Cheng, N.; Wang, Y.; Gao, H.; Yuan, J.; Feng, F.; Cao, W.; Zheng, J. Protective effect of extract of *Crataegus pinnatifida* pollen on DNA damage response to oxidative stress. *Food Chem. Toxicol.* **2013**, *59*, 709–714.

21. Halliwell, B.; Gutteridge, J.M.C. Oxygen toxicology, oxygen radicals, transition metals and disease. *J. Biochem.* **1984**, *219*, 1–4.

22. Takebayashi, J.; Chen, J.; Tai, A. A method for evaluation of antioxidant activity based on inhibition of free radical-induced erythrocyte hemolysis. *Methods Mol. Biol.* **2010**, *594*, 287–296.

23. Carvalho, M.; Ferreira, P.J.; Mendes, V.S.; Silva, R.; Pereira, J.A.; Jeronimo, C.; Silva, B.M. Human cancer cell antiproliferative and antioxidant activities of *Juglans regia* L. *Food Chem. Toxicol.* **2010**, *48*, 441–447.

24. Mendes, L.; Freitas, V.; Baptista, P.; Carvalho, M. Comparative antihemolytic and radical scavenging activities of strawberry tree (*Arbutus unedo* L.) leaf and fruit. *Food Chem. Toxicol.* **2011**, *49*, 2285–2291.

25. Devasagayam, T.P.A.; Subramanian, M.; Singh, B.B.; Ramanathan, R.; Das, N.P. Protection of plasmid pBR322 DNA by flavonoids against single strand breaks induced by singlet molecular oxygen. *J. Photochem. Photobiol. B: Biol.* **1995**, *30*, 97–103.

26. Noroozi, M.; Angerson, W.J.; Lean, M.E.J. Effects of flavonoids and vitamin C on oxidative damage to human lymphocytes. *Am. J. Clin. Nutr.* **1998**, *67*, 1210–1218.

27. Balaban, R.S.; Nemoto, S.; Finkel, T. Mitochondria, oxidants, and aging. *Cell* **2005**, *120*, 483–495.

28. Brown, K.; Xie, S.; Qiu, X.; Mohrin, M.; Shin, J.; Liu, Y.; Zhang, D.; Scadden, D.T.; Chen, D. SIRT3 reverses aging-associated degeneration. *Cell Rep.* **2013**, *3*, 319–327.

29. Ushio-Fukai, M. Compartmentalization of redox signaling through NADPH oxidase-derived ROS. *Antioxid. Redox Signal.* **2009**, *11*, 1289–1299.

30. Bass, D.; Parce, J.W.; Dechatelet, L.R.; Szejda, P.; Seeds, M.; Thomas, M. Flow cytometric studies of oxidative product formation by neutrophils: A graded response to membrane stimulation. *J. Immunol.* **1983**, *130*, 1910–1917.

31. Liu, J.; Jia, L.; Kan, J.; Jin, C.H. *In vitro* and *in vivo* antioxidant activity of ethanolic extract of white button mushroom (*Agaricus bisporus*). *Food Chem. Toxicol.* **2013**, *51*, 310–316.

32. Hwang, J.W.; Kim, E.K.; Lee, S.J.; Kim, Y.S.; Moon, S.H.; Jeon, B.T.; Sung, S.H.; Kim, E.T.; Park, P.J. Antioxidant activity and protective effect of anthocyanin oligomers on H2O2-triggered G2/M arrest in retinal cells. *J. Agric. Food Chem.* **2012**, *60*, 4282–4288.

33. Puiggros, F.; Sala, E.; Vaque, M.; Ardevol, A.; Blay, M.; Fernández-Larrea, J.; Arola, L.; Bladé, C.; Pujadas, G.; Salvadó, M.J. *In vivo*, *in vitro*, and *in silico* studies of CU/ZN-superoxide dismutase regulation by molecules in grape seed procyanidin extract. *J. Agric. Food Chem.* **2009**, *57*, 3934–3942.

34. Faria, A.; Pestana, D.; Monteiro, R.; Teixeira, D.; Azevedo, J.; Freitas, V.D. Bioavailability of Anthocyanin-pyruvic Acid Adducts in Rat. In Proceedings of the International Conference on Polyphenols and Health, Yorkshire, Leeds, UK, 7–10 December 2009; pp. 170–171.

35. Faria, A.; Pestana, D.; Azevedo, J.; Martel, F.; de Freitas, V.; Azevedo, I.; Mateus, N.; Calhau, C. Absorption of anthocyanins through intestinal epithelial cells - Putative involvement of GLUT2. *Mo.l Nutr. Food Res.* **2009**, *53*, 1430–1437.

36. Passamonti, S.; Vrhovsek, U.; Vanzo, A.; Mattivi, F. Fast access of some grape pigments to the brain. *J. Agric. Food Chem.* **2005**, *53*, 7029–7034.

37. Talave'ra, S.; Felgines, C.; Texier, O.; Besson, C.; Gil-Izquierdo, A.; Lamaison, J.L.; Rémésy, C. Anthocyanin metabolism in rats and their distribution to digestive area, kidney, and brain. *J. Agric. Food Chem.* **2005**, *53*, 3902–3908.

38. Bitsch, R.; Netzel, M.; Frank, T.; Strass, G.; Bitsch, I. Bioavailability and biokinetics of anthocyanins from red grape juice and red wine. *J. Biomed. Biotechnol.* **2004**, *5*, 293–298.

39. Tabet, N. Acetylcholinesterase inhibitors for Alzheimer's disease: Antiinflammatories in acetylcholine clothing. *Age Ageing* **2006**, *35*, 336–338.

40. Costa, P.; Grosso, C.; Gonçalves, S.; Andrade, P.B.; Valentão, P.; Bernardo-Gil, M.G.; Romano, A. Supercritical fluid extraction and hydrodistillation for the recovery of bioactive compounds from *Lavandula viridis* L'Hér. *Food Chem.* **2012**, *135*, 112–121.

41. Szwajgier, D. Anticholinesterase activities of selected polyphenols—a short report. *Pol. J. Food Nutr. Sci.* **2014**, *64*, 59–64.

42. Ryu, H.W.; Curtis-Long, M.J.; Jung, S.; Jeong, I.Y.; Kim, D.S.; Kang, K.Y.; Park, K.H. Anticholinesterase potential of flavonols from paper mulberry (*Broussonetia papyrifera*) and their kinetic studies. *Food Chem.* **2012**, *132*, 1244–1250.

43. Gupta, A.; Gupta, R. A survey of plants for presence of cholinesterase activity. *Phytochemistry* **1997**, *46*, 827–831.

44. Bekir, J.; Mars, M.; Souchard, J.P.; Bouajila, J. Assessment of antioxidant, anti-inflammatory, anti-cholinesterase and cytotoxic activities of pomegranate (*Punica granatum*) leaves. *Food Chem. Toxicol.* **2013**, *55*, 470–475.

45. Gutierres, J.M.; Carvalho, F.B.; Schetinger, M.R.; Agostinho, P.; Marisco, P.C.; Vieira, J.M.; Rosa, M.M.; Bohnert, C.; Rubin, M.A.; Morsch, V.M.; *et al.* Neuroprotective effect of anthocyanins on acetylcholinesterase activity and attenuation of scopolamine-induced amnesia in rats. *Int. J. Dev. Neurosci.* **2014**, *33*, 88–97.

46. Fournand, D.; Vicens, A.; Sidhoum, L.; Souquet, J.M.; Moutounet, M.; Cheynier, V. Accumulation and extractability of grape skin tannins and anthocyanins at different advanced physiological stages. *J. Agric. Food Chem.* **2006**, *54*, 7331–7338.

47. He, K.; Li, X.; Ye, X.; Yuan, L.; Li, X.; Chen, X.; Deng, Y. A mitochondria-based method for the determination of antioxidant activities using 2',7'-dichlorofluorescin diacetate oxidation. *Food Res. Int.* **2012**, *48*, 454–461.

48. Ellman, G.L.; Courteney, K.D.; Valentino, A.J.; Featherstone, R.M. A new and rapid colorimeteric determination of acetylcholinesterase activity. *Biochem. Pharmacol.* **1961**, *7*, 88–95.

**Sample Availability:** *Sample Availability*: Samples of the compounds are available from the authors.

# Effects of Climatic Conditions and Soil Properties on Cabernet Sauvignon Berry Growth and Anthocyanin Profiles

Guo Cheng, Yan-Nan He, Tai-Xin Yue, Jun Wang and Zhen-Wen Zhang

**Abstract:** Climatic conditions and soil type have significant influence on grape ripening and wine quality. The reported study was conducted in two "Cabernet Sauvignon (*Vitis vinifera* L.V)" vineyards located in Xinjiang, a semiarid wine-producing region of China during two vintages (2011 and 2012). The results indicate that soil and climate affected berry growth and anthocyanin profiles. These two localities were within a distance of 5 km from each other and had soils of different physical and chemical composition. For each vineyard, the differences of anthocyanin concentrations, and parameters concerning berry growth and composition between the two years could be explained by different climatic conditions. Soil effect was studied by investigation of differences in berry composition and anthocyanin profiles between the two vineyards in the same year, which could be explained mainly by the different soil properties, vine water and nitrogen status. Specifically, the soils with less water and organic matter produced looser clusters, heavier berry skins and higher TSS, which contributed to the excellent performance of grapes. Compared with 2011, the increases in anthocyanin concentrations for each vineyard in 2012 could be attributed to smaller number of extreme temperature (>35 °C) days and rainfall, lower vine water status and N level. The explanation for higher anthocyanin concentrations in grape skins from the soils with less water and organic matter could be the vine status differences, lighter berry weight and heavier skin weight at harvest. In particular, grapes from the soils with less water and organic matter had higher levels of $3'5'$-substituted, $O$-methylated and acylated anthocyanins, which represented a positive characteristic conferring more stable pigmentation to the corresponding wine in the future. The present work clarifies the effects of climate and soil on berry growth and anthocyanin profiles, thus providing guidance for production of high-quality wine grapes in different regions.

Reprinted from *Molecules*. Cite as: Cheng, G.; He, Y.-N.; Yue, T.-X.; Wang, J.; Zhang, Z.-W. Effects of Climatic Conditions and Soil Properties on Cabernet Sauvignon Berry Growth and Anthocyanin Profiles. *Molecules* **2014**, *19*, 13683–13703.

## 1. Introduction

Anthocyanins are an important group of flavonoids and the predominant pigments in red and black grape berries. After veraison, they accumulate in the grape skins via the phenylpropanoid biosynthetic pathway [1]. The anthocyanins in grapes are responsible for the colour of the corresponding wines, which is an important parameter used to evaluate wine quality [2]. Accumulation of anthocyanins in grape skins is influenced by many environmental factors such as light and temperature and nutrient supply [3]. In *Vitis vinifera* L. varieties, they occur mainly as glycosides and acylglucosides of five anthocyanidins: malvidin, petunidin, peonidin, delphinidin and cyanidin [4]. Monoglucosides exist as 3-*O*-glucosides in *V. vinifera* L. varieties, and the acylated anthocyanins are formed by the combination with coumaric or caffeic acid [5].

The concept of terroir, including the grape cultivar, always interacts with climatic conditions, soils, cultural practices and training systems, all of which influence the grape and wine quality [6]. Generally, cultivar, soil and climate are considered to be the three main components of terroir [7]. It is known that patterns of anthocyanins are controlled by strict genotype, and the anthocyanin profiles for each variety are relatively stable and their distribution varies considerably among different grape cultivars [8]. More recently, the wine industry has turned its attention to the factors of soil and climate. They point out the soil type may play an important role in determining wine characteristics and quality, resulting in different wine styles, even under the same mesoclimate [7]. In addition, other studies show that the influence of vintage on grape metabolic profiles is greater than the soil, where climatic characteristics such as the temperature and water balance are the main factors [9]. It should be noted that anthocyanin synthesis is depressed by excessively high temperatures with considering the specific circumstances [7]. Further research has revealed that the inhibition of temperature stress on berry secondary metabolites can be attributed to ABA-mediated actions [10]. Moreover, some research has determined that moderate temperatures encourage anthocyanin accumulation and alter partitioning, while others determined that high temperatures could be inhibitory to accumulation due to the differences in gene expression and chemical degradation of metabolites [7,11]. In addition to the effects of climate and soil, there is a great need to evaluate the influences of water deficits and berry exposure to sunlight on anthocyanin concentrations [12,13]. The water deficits enable anthocyanin concentrations to rise by both increasing content per berry and reducing fruit growth [14,15]. Generally, low level of light exposure could reduce flavonoids content, but increased light results in increased contents of anthocyanins and other flavonoids [16]. However, some authors have reported no change with different light treatments [17], or even that high light resulted in decreased anthocyanin concentrations [18].

Despite the fact that soil and climate are two important components of terroir, little experimental work concerning their influence on wine grape composition has been done in China. Therefore, in this work the effects of climate and soil on grape development and anthocyanin profiles were analysed. The soil effect was examined in two vineyards with distinct soil types, and climate effect was studied through the mesoclimate by inter-annual analysis. Specifically, the effects of soil components and climatic conditions were evaluated on: (1) vine water and nutrient status; (2) yield parameters and berry characteristics; (3) anthocyanin composition and concentrations. The present work was designed to study anthocyanin biosynthesis under different types of soils and climatic conditions. Special emphasis was put on the relationship between anthocyanin accumulation and environmental conditions.

## 2. Results and Discussion

### 2.1. Climatic Conditions

The wine grape-growing regions of China display unique ecological conditions either from south to north or from east to west. The two vineyards in the present study were located in northwest China in a region with a semi-arid climate, with high biologically effective day temperatures, a big temperature difference between daytime and nighttime and low annual rainfall. All the meteorological data of 2011 and 2012 were obtained from the local meteorological administration. The two studied vineyards were within a radius of 5 km with a uniform mesoclimate. With respect to annual mean temperature and average maximum temperature, 2012 was warmer than 2011 (Table 1). With regard to the study period, the accumulated heat expressed as growing degree days (GDD, calculated from daily mean temperatures, base 10 °C) and sunlight duration in 2011 was lower than in 2012. Total rainfall, calculated from April to September, was 180.1 mm in 2011 and 103.6 mm in 2012 (Table 1). Regarding the seasonal evolution of temperature and rainfall (Figure 1), rainfall in 2011 was higher than in 2012 (mainly during 60–90 DAF), although vines did not get any rainfall at harvest (90–120 DAF) in 2011. According to the seasonal pattern during berry developmental phases (Figure 1), daily maximum temperature exceeded 30 °C for most of the summer period. A higher number of days with extreme temperatures (>35 °C) were observed in 2011 than in 2012 (23 and 13 days, respectively), mainly during expanding stage and veraison (0–90 DAF). Rainfall was on the rise from veraison (60–90 DAF) to harvest in 2012, but reduced gradually in 2011.

### 2.2. Soil Chemical and Physical Properties Analysis

The selected soils from two vineyards presented very different physical and chemical properties (Table 2), although both of the soils were classified as "slit loam" according to the texture classes in the soil survey manual. The soils from Yuanyi farm were richer in sand than those of Guangdong farm, especially in topsoil (0–20 cm). Specifically, the sand percentage was 41.99% at Yuanyi farm, and only 20.03% at Guangdong farm (Table 2).

**Table 1.** Meteorological parameters for the growing season (April–October) and the ripening period (August–September) in the study area (2011 and 2012).

| Year | Mean Temperature (°C) | | | Average Maximum Temperature (°C) | | | Average Minimum Temperature (°C) | | | Growing Degree Days (°C) | | | Sunlight Duration (h) | | | Rain (mm) | | |
|---|---|---|---|---|---|---|---|---|---|---|---|---|---|---|---|---|---|---|
| | Aug-Sep | Apr-Oct | Year | Aug-Sep | Apr-Oct | Year | Aug-Sep | Apr-Oct | Year | Aug-Sep | Apr-Oct | Year | Aug-Sep | Apr-Oct | Year | Aug-Sep | Apr-Oct | Year |
| 2011 | 22.1 | 20.3 | 8.2 | 28.7 | 26.8 | 13.6 | 16.1 | 14.6 | 3.4 | 740.9 | 2290.0 | 2290.6 | 627.4 | 2173.7 | 2899.0 | 41.2 | 180.1 | 217.8 |
| 2012 | 22.3 | 20.8 | 8.8 | 29.2 | 27.4 | 13.9 | 16.1 | 14.7 | 3.1 | 752.9 | 2356.2 | 2367.2 | 647.3 | 2235.8 | 3102.0 | 29.0 | 103.6 | 179.8 |

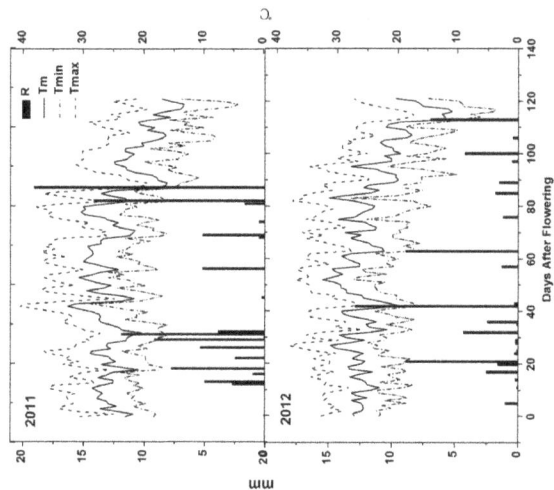

**Figure 1.** Evolution of mean, maximum and minimum daily temperature and rainfall from flowering to harvest in the study area (2011 and 2012). R, Rainfall; Tm, mean temperature; Tmax, maximum temperature; Tmin, minimum temperature.

155

**Table 2.** Physical and chemical properties of the selected soils in two vineyards at different depths. Data are mean values of three replications.

| Vineyards | Depth (cm) | Clay (%) | Slit (%) | Sand (%) | Textural Class | pH | EC (ms/cm) | CEC (cmol/kg) | Organic Matter (%) | Bulk Density (kg/m³) | Water Content (%) |
|---|---|---|---|---|---|---|---|---|---|---|---|
| Yuanyi farm | 0–20 | 8.59b | 49.40b | 41.99a | Silt loam | 7.9a | 0.18b | 16.40b | 1.40b | 1.49a | 18.82b |
| | 20–40 | 9.82b | 61.66b | 28.52a | Silt loam | 7.8a | 0.20b | 13.18b | 0.99b | 1.65a | 26.49b |
| | 40–60 | 11.37a | 54.13a | 34.50a | Silt loam | 8.1a | 0.19b | 12.82b | 0.56b | 1.41a | 26.83b |
| | 60–80 | 14.63a | 49.87b | 35.49a | Silt loam | 8.0a | 0.20b | 12.46b | 0.54b | 1.50a | 25.61b |
| | 80–100 | 9.02a | 55.72b | 35.26a | Silt loam | 7.9a | 0.18b | 11.74b | 0.41b | 1.55a | 27.97b |
| Guangdong farm | 0–20 | 15.99a | 63.97a | 20.03b | Silt loam | 7.5b | 0.39a | 23.21a | 6.38a | 0.98b | 29.55a |
| | 20–40 | 11.45a | 63.51a | 25.04b | Silt loam | 7.3b | 0.54a | 25.36a | 8.42a | 1.21b | 44.04a |
| | 40–60 | 8.30b | 55.69a | 36.01a | Silt loam | 7.4b | 0.56a | 22.14a | 6.85a | 1.29b | 38.89a |
| | 60–80 | 9.17b | 54.35a | 36.48a | Silt loam | 7.2b | 0.98a | 19.63a | 5.02a | 1.47b | 36.13a |
| | 80–100 | 9.68a | 58.12a | 32.20b | Silt loam | 7.2b | 0.66a | 18.55a | 3.69a | 1.51b | 33.74a |

Values within a column followed by different letters differ significantly in the same depth between two vineyards (t-test, $p < 0.05$).

The soil pH at Yuanyi farm ranged from 7.8 to 8.1, which classified them as slightly alkaline soils, while the pH at Guangdong ranged from 7.2 to 7.5, which defined them as very slightly alkaline soils. At pH values above 5.0 to 5.5, grapevine performance should not be seriously impeded [19], and none of the soils in Yuanyi farm or Guangdong farm showed any signs of sodicity. Soil electrical conductivity (EC) is a parameter which represents the amount of salts in soils (soil salinity). In our study, the EC of soils from Guangdong farm was significantly higher than at Yuanyi farm at each soil depth (54%–80%). According to the classes of salinity and EC in the NRCS Soil Survey Handbook, all the selected soils from the two vineyards belonged to the non-saline (0<EC<2) class. The relationship between cation-exchange capacity (CEC) and organic matter at the two vineyards showed a positive correlation (Figure 2), and both of them decreased as the soil depth deepened (Table 2). As indicated in Table 2, the soils from Yuanyi farm had significantly higher bulk density at each depth than the Guangdong farm ones, since total pore space in sands is less than is silt or clay soils. For Guangdong farm, bulk density typically increased with soil depth since the subsurface layers contained less pore spaces and higher organic matter levels. Higher organic material may result in less compact soils. From the survey of bulk density, Yuanyi farm had more compact soils than Guangdong farm. Thus, the soils from Guangdong farm contained significantly higher percentages of organic matter than those of Yuanyi farm. In previous research, the organic matter content of soils in two vineyards located in Catalonia, Spain was 0.4%–2.0% [20]. Another report pointed out that soils with 0.7%–0.9% organic matter were considered to be poor, whereas those with 1.7%–1.8% organic matter were considered to be rich soils [21]. Thus, the organic matter content of soils in Guangdong farm was excessive (3.69%–8.42%) considering the general range of a rich soil. Nevertheless, the organic matter content of soils in Yuanyi farm was within a reasonable range (0.41%–1.40%). Water content of soils from Guangdong farm was significantly higher than at Yuanyi farm at each soil depth (17.2%–39.9%). Those could be attributed to organic matter increasing the water holding ability of soils directly and indirectly. On the other hand, higher bulk density consequently reduced the water holding capacity of soils.

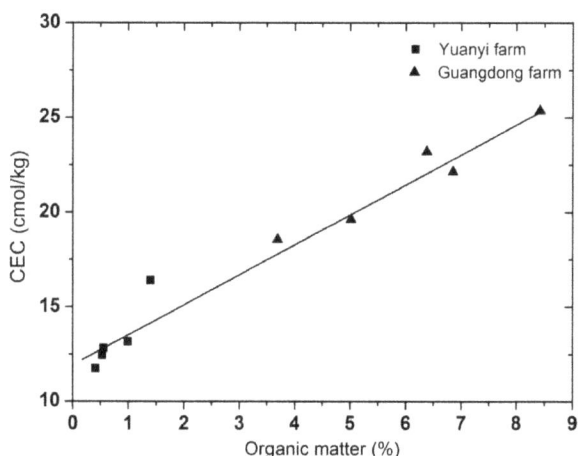

**Figure 2.** The relationship between cation-exchange capacity (CEC) and organic matter at Yuanyi farm and Guangdong farm. Data are mean values of three replications.

Principal component analysis (PCA) of soil properties from the two vineyards was conducted by using the variables including total, effective and readily available elements variables (Figure 3). PC1 explained 42.9% of total variance, and was characterized by soils of all depths from Guangdong farm on the positive side. PC1 separated the soil samples of Yuanyi farm from Guangdong farm, and was mostly explained by total N, effective Na and available Fe having positive loadings. Specifically, the concentrations of these elements from soil samples were significantly higher in Guangdong farm (Table 3). Total K and available Mn had negative loadings, and their concentrations were significantly higher in soil samples from Yuanyi farm (Table 3). PC2 explained 30.1% of the variance and separated top soils (0–20 cm) of the two vineyards from the middle and bottom soils (40–100 cm). The top soils of two vineyards had abundant available elements such as P, Mn, Zn and B (Table 3).

**Table 3.** Chemical properties of the selected soils in two vineyards at different depths. Data are mean values of three replications.

| Vineyards | Depth (cm) | Total Elements (%) | | | Effective Elements (mg/kg) | | | | | Available Elements (mg/kg) | | | | | |
|---|---|---|---|---|---|---|---|---|---|---|---|---|---|---|---|
| | | N | P | K | Na | Ca | Mg | N | P | K | Cu | Fe | Mn | Zn | B |
| Yuanyi farm | 0–20 | 0.04b | 0.08a | 1.83a | 21.07b | 3494.46b | 140.62b | 50.86b | 10.10b | 133.29b | 0.75b | 1.61b | 1.01a | 1.40a | 0.47b |
| | 20–40 | 0.07b | 0.08a | 1.85a | 33.40b | 3689.73b | 219.12b | 76.93a | 3.09b | 122.70a | 0.85a | 1.09b | 0.59a | 1.29a | 0.11b |
| | 40–60 | 0.04b | 0.08a | 1.80a | 31.33b | 3811.77a | 233.70b | 59.05b | 1.45b | 123.05a | 0.75a | 0.89b | 0.65a | 0.66a | Nd |
| | 60–80 | 0.05b | 0.07a | 1.81a | 39.48b | 3760.86b | 266.62a | 53.24b | 4.33b | 152.14a | 0.72b | 0.83b | 0.65a | 0.38a | Nd |
| | 80–100 | 0.07b | 0.07a | 1.85a | 39.77b | 3800.32a | 295.58a | 55.61b | 4.36b | 134.72a | 0.76b | 1.10b | 0.91a | 0.41a | Nd |
| Guangdong farm | 0–20 | 0.30a | 0.11a | 1.73b | 51.83a | 3789.53a | 525.98a | 61.74a | 75.42a | 175.10a | 0.85a | 19.68a | 0.58b | 1.35b | 2.51a |
| | 20–40 | 0.40a | 0.10a | 1.34b | 57.28a | 3775.11a | 490.15a | 69.34b | 18.25a | 129.57a | 0.58b | 22.65a | 0.45b | 0.82b | 1.03a |
| | 40–60 | 0.30a | 0.06a | 1.40b | 66.10a | 3776.49b | 334.47a | 91.94a | 5.68a | 100.67b | 0.61b | 29.79a | 0.41b | 0.59b | 0.21 |
| | 60–80 | 0.24a | 0.06a | 1.57b | 63.41a | 3824.79a | 265.22a | 86.13a | 8.59a | 78.92b | 0.84a | 33.92a | 0.40b | 0.35b | 0.25 |
| | 80–100 | 0.19a | 0.08a | 1.66b | 52.72a | 3769.12b | 219.11b | 90.66a | 8.93a | 40.57b | 0.85a | 33.26a | 0.39b | 0.26b | 0.75 |

Values within a column followed by different letters differ significantly in the same depth between two vineyards ($t$-test, $p < 0.05$). Nd, not detected.

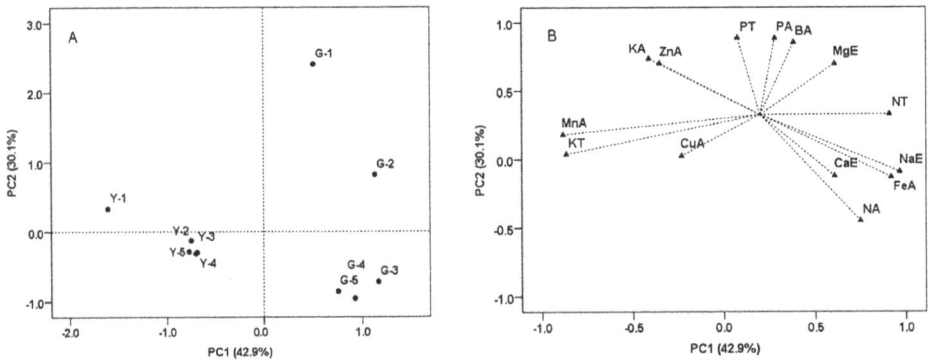

**Figure 3.** Discrimination of the soil samples from two vineyards by some kind of elements attributes and illustrated by the score (**A**) and loading (**B**) plots from principal component analysis (PCA).Y, Yuanyi farm; G, Guangdong farm; 1 = 0–20 cm; 2 = 20–40 cm; 3 = 40–60 cm; 4 = 60–80 cm; 5 = 80–100 cm; T, Total; E, Effective; A, Readily available.

## 2.3. Vine Water and Nitrogen Status

Ambient $CO_2$ contains 98.9% of $^{12}C$ isotope and 1.1% of $^{13}C$ isotope. $^{12}C$ is more easily used by the enzymes of photosynthesis in their production of hexoses. Therefore, the sugar produced by photosynthesis contains a higher rate of the $^{12}C$ isotope than ambient $CO_2$. This process is called 'isotope discrimination'. When grapevines are submitted to water deficits, isotope discrimination is reduced due to stomatal closure [22]. Therefore, the $^{12}C/^{13}C$ ratio (so-called $\delta^{13}C$) allows a very sensitive detection of grapevine water status during grape ripening [23].

With respect to vine water deficit thresholds [24], there was no water deficit in vines from Guangdong farm during the two seasons ($\delta^{13}C < -26$). However, vines grown in Yuanyi farm showed weak water deficit for its higher $\delta^{13}C$ values ($-24.5 < \delta^{13}C < -26$) corresponding to its soils with lower water-holding capacity compared with Guangdong farm (Table 4). Thus, the water content in soils affected the vine water status, which was confirmed by a significant negative correlation between $\delta^{13}C$ and water content in soils (Table S1). Furthermore, N level was significantly lower in the vines of Yuanyi farm than at Guangdong farm. Soubeyrand *et al.* [3] used the chlorophyll content of the leaves to estimate the plant nitrogen status for the positive correlation between them. The chlorophyll content of the leaves from vines grown in Guangdong farm was significantly higher than that in Yuanyi farm. In addition, there was a positive correlation between N status of the plant-leaf Chlorophyll, N status of the plant-organic matter in soils (Table S1). This confirmed that the higher nitrogen content in the soils of Guangdong farm led to an increase in nitrogen uptake and assimilation by the vines. Aside from other considerations, examination of these results showed the fact that there were differences in vine nitrogen and water status between the soils of the two vineyards.

Table 4. Differences of vine water and nutrient status between Yuanyi and Guangdong farm during two seasons. Data are mean values of three replications at least.

| Year | Vineyards | $\delta^{13}C$ (‰) | N (%) | Leaf Chlorophyll |
|------|-----------|--------------------|-------|------------------|
| 2011 | Yuanyi farm | −25.34a | 0.47b | 45.01b |
|      | Guangdong farm | −26.76b | 0.54a | 46.95a |
| 2012 | Yuanyi farm | −25.06a | 0.46b | 45.75b |
|      | Guangdong farm | −26.51b | 0.49a | 46.54a |

Values within a column followed by different letters differ significantly between two vineyards in the same year (*t*-test, $p < 0.05$).

*2.4. Some Parameters of Vine, Shoot, Cluster and Berries from Two Vineyards in Two Vintages*

Although there were no significant differences between the two vineyards regarding shoot number, yield, yield/pruning weigh and leaf area/yield in 2011 and 2012 (Table 5), vines grown in Guangdong farm had significantly longer shoots and less number of clusters per shoot than Yuanyi farm ones (Table 5). In addition, Yuanyi farm had significantly lighter and looser clusters compared with Guangdong farm (Table 5). From Table S1, there was a positive correlation between cluster compactness-organic matter in soils, cluster compactness-water content in soils. Thus, the clusters from the soils with less organic matter and lower water content would be characterized by their looser clusters, and this conclusion is verified in the discussion about anthocyanins.

Table 5. Survey of some parameters about yields from two vineyards at harvest in 2011 and 2012. Data are mean values of three replications at least.

| Years | Vineyards | Survey of Shoots | | | Survey of Clusters | | Yield (ton/hectare) | Yield/Pruning Weigh (kg/kg) | Leaf Area/Yield (m²/kg) |
| | | Average Shoot Length (cm) | Shoot Number/m | Cluster Number/Shoot | Cluster Weight (g) | Cluster Compactness (OIV rating) | | | |
|---|---|---|---|---|---|---|---|---|---|
| 2011 | Yuanyi farm | 122b a | 13.85a | 1.80a | 107.10b | 4.00b | 9.14a | 5.49a | 1.98a |
| | Guangdong farm | 127a | 14.29a | 1.64b | 111.22a | 5.67a | 8.45a | 5.32a | 2.08a |
| 2012 | Yuanyi farm | 132b | 13.55a | 1.79a | 107.97b | 3.67b | 10.48a | 5.66a | 2.03a |
| | Guangdong farm | 136a | 15.38a | 1.58b | 119.20a | 5.33a | 11.59a | 5.78a | 2.12a |

Values within a column followed by different letters differ significantly between two vineyards in the same year ($t$-test, $p < 0.05$).

Some physical and chemical characteristics of berries were measured for grape samples from the two vineyards during different developmental stages (Figure 4). During the grape development in 2011 and 2012, fresh weight per berry increased until harvest in Yuanyi and Guangdong farm (Figure 4A). During the early stages of berry development, berries grew faster at Yuanyi farm than at Guangdong farm between 4 and 11 weeks after flowering (WAF). At harvest, fresh weight per berry in Guangdong farm was higher than Yuanyi farm in the two-year study. Overall, berries had a heavier weight in 2011 than in 2012. Skin weight per berry showed an increasing trend from 4 to 11 WAF, then decreased from 13 WAF to harvest (Figure 4B).

Low nitrogen would be expected to cause a lower rate of shoot growth and chlorophyll formation [25]. The growth of the shoots is strongly affected by soil conditions and they grow longer where soils are more fertile and can hold more water [26]. In a previous report, cluster weight showed a negative correlation with the number of clusters per grapevine, because of the increasing berry numbers per cluster and weight caused by the decreasing cluster number [27]. Thus similar result patterns were seen in the two vineyards regarding shoot and cluster parameters as in previous research. Berry weight at Yuanyi farm was lighter in the soils with lower water content and weak water deficit for vines, which was consistent with a low water supply causing a reduction in berry weight [28]. In each vintage, the value of skin weight for Yuanyi farm was higher than for Guangdong farm. Previous research on deficit irrigation has shown an increase in skin weight [29], and a significant negative correlation between skin fresh weight and leaf Chlorophyll in our study (Table S1). Thus, the differences in the soil and vine parameters at Yuanyi and Guangdong farm make it easy to draw conclusions on the basis of these results.

TSS exhibited a continuous increase from 4 to 15 WAF (Figure 4C). The maximum values of TSS for Yuanyi farm and Guangdong farm were 23.0 and 23.2 °Brix in 2011, but 26.8 and 25.9 °Brix in 2012. On the other hand, the changes of TA in juice of grapes from two vineyards in 2011 and 2012 are shown in Figure 4D. Grapes from the two vineyards in 2012 had higher TA than in 2011, and for each year, Guangdong farm had more TA in grape juice compared with Yuanyi farm.

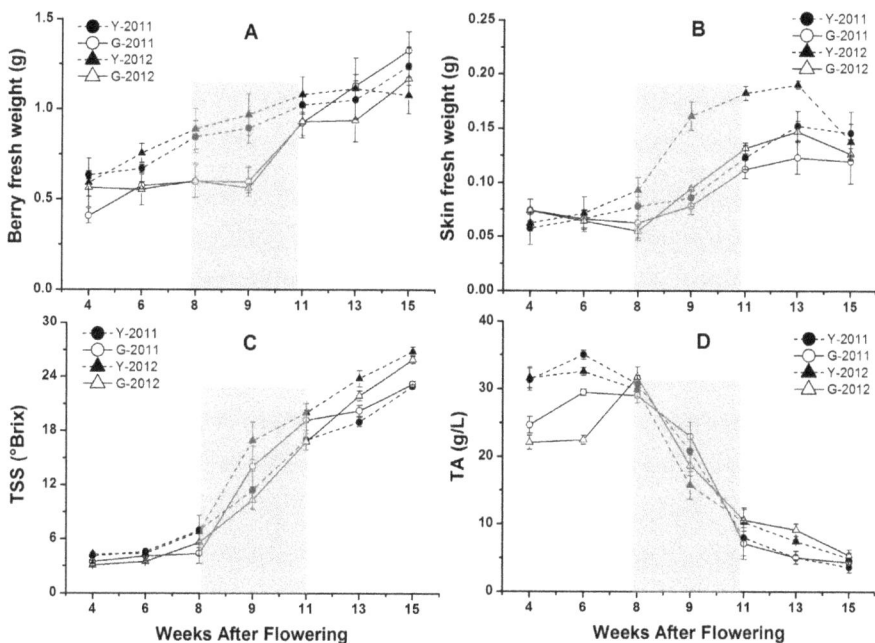

**Figure 4.** Some physical and chemical characteristics of berry changed with the grape development in two vineyards in 2011 and 2012. Data are mean values of three replications. Error bars show standard error (SE). Light grey background represents the phase of bunch turning colour (veraison) from 10% to 100% of coloured berries.

Adequate ripening may also be problematic in soils that are too wet [19]. While, in this study, TSS was equivalent in the two vineyards at harvest in 2011, the TSS of grapes from Yuanyi farm was higher than at Guangdong farm in 2012. A higher TA in the grapes from Guangdong farm was consistent with the other findings where canopies with excessive soil moisture had higher TA [30]. On the other hand, grapes with lower TSS and higher TA at the Guangdong farm were attributed to wetter soils, but this could also have been an effect of excessive organic matter.

*2.5. The Concentrations of Anthocyanins in Grape Skins from Two Vineyards in 2011 and 2012*

The anthocyanins extracted from the grape skins during crushing, pressing, and fermentation are the major components responsible for red wine color [31]. The total concentration of anthocyanins (TCA) and each kinds of anthocyanin with different modifications showed continuous increases from 8 to 15 WAF (Figure 5A). The late stage of veraison (9 to 11 WAF) was a period of rapid accumulation for anthocyanin compounds. The maximum TCA values for Yuanyi farm were 873.3 mg/kg at harvest in 2011 and 880.4 mg/kg in 2012, but those for Guangdong farm were 568.9 mg/kg in 2011 and 695.2 mg/kg in 2012 (Figure 5A). In 2011 and 2012, grapes harvested from Yuanyi farm accumulated 35% and 21% more anthocyanins, respectively, than those from Guangdong farm, if anthocyanin content was expressed as concentration in fresh berry (mg/kg of berry fresh weight). This indicated that the effect of vintage was more significant in Guangdong farm when discussing TCA during the two seasons, while the regional differences always illustrated that grapes from Yuanyi farm could accumulate much more TCA. Meanwhile, TCA in grape skins is more variable than TSS (Figure 4C) between the two vineyards at each sampling point, although the evolution of anthocyanins was parallelled by the accumulation of sugars.

In previous studies, some reported vintage variation [32], whereas others reported minimal influence of the season on anthocyanin accumulation [31]. Anthocyanin accumulation in grapes is light and temperature sensitive [32], so air temperatures might also have an influence on anthocyanin accumulation. Considering the climatic parameters such as mean temperature, the maximum and minimum temperature and growing degree days, 2011 was a cooler year with shorter sunlight duration and more rainfall compared with the same period in 2012 (Table 1), but extreme temperatures (>35 °C) were more frequent in 2011 than in 2012 (Figure 1). Although the lowest anthocyanin content was detected during the hottest year [20], this conclusion could not be verified in our study in each vineyard over two consecutive seasons. The total concentrations of anthocyanins at harvest in 2012 were higher than in 2011, especially in Guangdong farm, and their concentration increased by 22% (Figure 5).

Furthermore, the concentrations of anthocyanins expressed in mg/kg grapes were positively related to berry number and total skin surface per kilogram of grapes, resulting in the smallest berries being characterized by the highest content of anthocyanins [33]. In fact, exceedingly high temperatures (>35 °C) are particularly inhibitory to anthocyanin synthesis [32], so the increase of TCA in 2012 for each vineyard could be attributed to the decrease of berry fresh weight at harvest time (Figure 5), and less number of extreme temperature (>35 °C) days. Moreover, there was a significant positive correlation between TCA and $\delta^{13}$C, but negative correlation between TCA and N (Table S1). Thus, higher vine water status and N levels were also major causes for the low coloration of grapes in 2011.

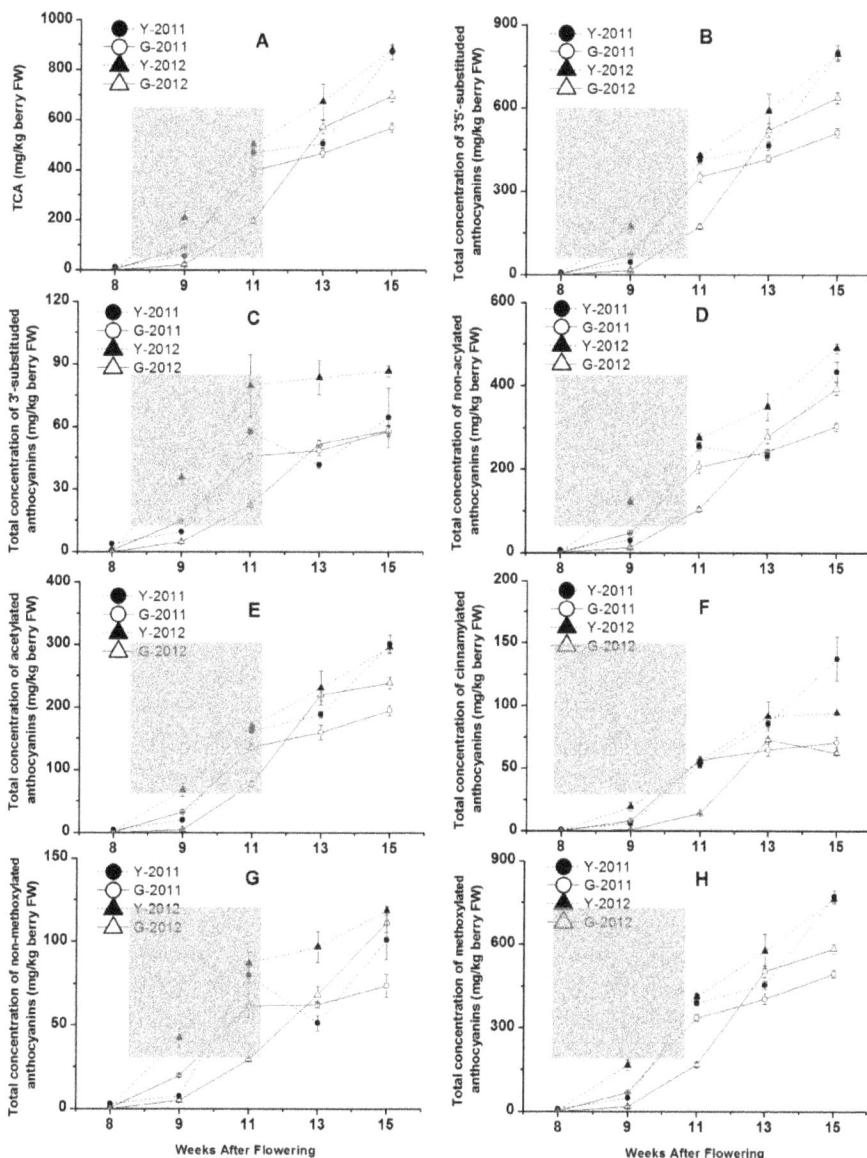

**Figure 5.** The concentration of total anthocyanins, 3'5'-substituted and 3'-substituted anthocyanins, non-acylated, acetylated and cinnamylated anthocyanins, non-methoxylated and methoxylated anthocyanins changed with the grape development in two vineyards in 2011 and 2012. Data are mean values of three replications. Error bars show standard error (SE). Light grey background represents the phase of bunch turning colour (veraison) from 10% to 100% of coloured berries.

163

In this survey, the effect of the soils on vine behaviour was mediated through varying water content and nitrogen levels of the grapevines. With regard to the effect of water status, moderate and not severe water stress or drought stress have been reported to increase anthocyanin concentrations [34]. For the different growth rate of the skins and flesh responses to the change of water conditions, the mild water deficits increase the concentrations of skin tannins and anthocyanins through increasing skin weight, relative skin weight per berry, and therefore amounts of skin-localised solutes [35]. In the present study, Yuanyi farm had higher value of berry skins than Guangdong farm at harvest in each experimental season (Figure 4B). Furthermore, there was a significant positive correlation between TCA and skin fresh weight (Table S1). Research on terroir shows that a moderate nitrogen deficiency (like a mild water deficit) has been correlated with improved grape quality [36]. For the experimental samples, not only did Yuanyi farm have a faster speed of anthocyanin accumulation during berry development, but also a higher TCA at harvest in the two years (Figure 5). The above conclusions were also associated with lower water content of soils and vine water status at Yuanyi farm (Table S1).

F3′H (flavonoid 3′-hydroxylase) and F3′5′H (flavonoid 3′5′-hydroxylase) are involved in the biosynthetic pathway of cyanidin- and delphinidin-based anthocyanins, respectively. 3′5′-Substituted anthocyanins contain delphinidin, petunidin, malvidin anthocyanins and their derivatives. A higher accumulation of delphinidin-based anthocyanins in grape skins is expected to produce more intensely purple coloured berry and dark-colored red wine [37,38]. The concentrations of 3′5′-substituded anthocyanins showed an increasing trend during the developmental phase as well as TCA for each vineyard (Figure 5B). However, the concentration of 3′5′-substituded anthocyanins at Guangdong farm was 511 mg/kg, only 64% and 80% of that at Yuanyi farm at harvest in 2011 and 2012. On the other hand, the concentrations of 3′-substituded anthocyanins increased from veraison to harvest, although they remained at low ranges compared with 3′5′-substituded anthocyanins (Figure 5C). Additionally, 2012 showed higher concentrations of 3′5′-substituted anthocyanins than 2011. It was worth noting that the concentration of 3′-substituded anthocyanins in grape skins from Yuanyi farm in 2012 was the highest among all the developmental phases. From the analysis of correlation coefficients between variables, both of 3′5′-substituded and 3′-substituded anthocyanins showed significant positive correlations with skin fresh weight or $\delta^{13}C$, but significant negative correlations with vine N status (Table S1).

Considering further modification, two types of acylated anthocyanins are produced by acylation and cinnamylation in the grape skins of 'Cabernet Sauvignon'. The concentrations of non-acylated, acetylated and cinnamylated anthocyanins accumulated and peaked at harvest stage for each vineyard (Figure 5D–F). For non-acylated and acetylated anthocyanins, the concentrations in the two vineyards in 2012 were higher than those in 2011 (Figure 5D–E), but the opposite was true for the concentrations of cinnamylated anthocyanins in each vineyard in 2011 and 2012 (Figure 5F). The concentration of each class of anthocyanins at Yuanyi farm was higher than at Guangdong farm at harvest, which could be associated with vine N status, cluster compactness and skin fresh weight (Table S1).

In addition, $O$-methylation of the 3' position of the anthocyanidins cyanidin and delphinidin leads to the formation of peonidin and petunidin. Finally, $O$-methylation of positions 3' and 5' of delphinidin, leads to the formation of petunidin and malvidin. The methylated anthocyanins included peonidin, petunidin, malvidin anthocyanins and their derivatives. Because of their phenolic B ring substitution, peonidin and malvidin are relatively stable and represent the major anthocyanin pools in mature grapes [39]. The concentrations of methylated anthocyanins were 772 and 495 mg/kg in Yuanyi and Guangdong farm at harvest time in 2011, and 762 and 584 mg/kg in 2012, respectively (Figure 5G–H). Thus, the modification of anthocyanins influences their resulting wines by the colour, and grapes from the soils with less water content and organic matter are expected to produce more dark-colored red wine in the future (Table S1).

*2.6. Principal Component Analysis (PCA) of the Grape Samples from Two Vineyards at Harvest*

Previous research suggested that the anthocyanin fingerprint of grapes was related to cultivar and weather conditions of the growing season according to PCA [40]. In the present study, PCA was used to examine the effects of soil and vintage on the composition of anthocyanins (Figure 6). Nineteen different anthocyanins detected at harvest in two vintages were used as the variables, and each result contained three replicates, although malvidin-3-$O$-(6-$O$-caffeoyl)-glucoside was not detected in grape skins from Guangdong farm (Table 6). PC1 explained 59.6% of total variance, and characterized all grape samples from Yuanyi farm on the positive side. PC1 separated all the grape samples of Yuanyi farm from Guangdong farm, and could be mostly explained by malvidin-3-$O$-glucoside and malvidin-3-$O$-(6-$O$-acetyl)-glucoside having positive loadings. It is worth noting that malvidin-3-$O$-glucoside was the most prevalent anthocyanin and malvidin-3-$O$-(6-$O$-acetyl)-glucoside was the major acylated anthocyanin in all the grape samples from the two vineyards. Some similar observations are obtained from research conducted on 'Merlot' grape clusters in Prosser [41]. While petunidin-3-$O$-(6-$O$-caffeoyl)-glucoside and cyanidin-3-$O$-(6-$O$-coumaryl)-glucoside had negative loadings, and their concentrations were higher in grape samples from Guangdong farm (Table 6). PC2 explained 22.7% of the variance and separated grape samples of 2011 from 2012. In previous research, the percentages of dephinidin-3-$O$-glucoside in warm years was lower than in a relatively cool year [40]. In the present study, the grape samples of 2012 were abundant in dephinidin-3-$O$-glucoside, cyanidin-3-$O$-glucoside, dephinidin-3-O-(6-O-acetyl)-glucoside, cyanidin-3-O-(6-O-acetyl)-glucoside, and those of 2011 had higher concentrations of petunidin-3-O-(6-O-coumaryl)-glucoside, peonidin-3-O-(*trans*-6-O-coumaryl)-glucoside, malvidin-3-O-(*cis*-6-O-coumaryl)-glucoside and malvidin-3-O-(*trans*-6-O-coumaryl)-glucoside (Table 6). Most anthocyanins in grape skins from Yuanyi farm showed higher concentrations at harvest in each year, and for each vineyard, 2011 had higher concentrations of coumarylated anthocyanins than 2012, so by using the concentrations of individual anthocyanin detected at harvest we could clearly discriminate the effects of vineyard and vintage.

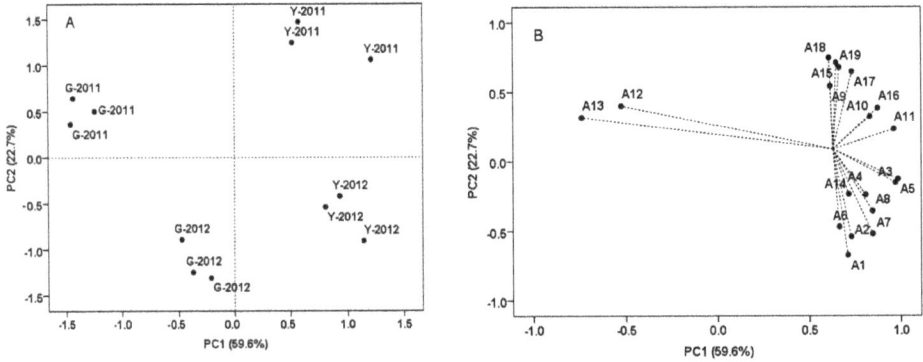

**Figure 6.** Discrimination of the grape samples from two vineyards by the concentration of individual anthocyanin detected at harvest in two vintages, and illustrated by the score (A) and loading (B) plots from principal component analysis (PCA). Each point means a biological replicate in plot A. Y, Yuanyi farm; G, Guangdong farm.

**Table 6.** The concentration of individual anthocyanin (mean ± se; n = 3) detected in grape skins of "Cabernet Sauvignon" from two vineyards at harvest in 2011 and 2012.

| Anthocyanins | 2011 | | 2012 | |
|---|---|---|---|---|
| | Yuanyi farm (mg/kg berry FW) | Guangdong farm (mg/kg berry FW) | Yuanyi farm (mg/kg berry FW) | Guangdong farm (mg/kg berry FW) |
| dephinidin-3-O-glucoside (A1) | 52.31 ± 3.7 | 37.65 ± 1.28 | 66.83 ± 0.86 | 63.20 ± 2.48 |
| cyanidin-3-O-glucoside (A2) | 6.97 ± 0.52 | 6.24 ± 0.27 | 8.49 ± 0.75 | 7.31 ± 0.21 |
| petunidin-3-O-glucoside (A3) | 42.36 ± 3.09 | 31.94 ± 1.61 | 43.40 ± 1.35 | 38.92 ± 1.29 |
| peonidin-3-O-glucoside (A4) | 28.71 ± 2.22 | 23.47 ± 1.39 | 39.28 ± 2.22 | 25.05 ± 1.44 |
| malvidin-3-O-glucoside (A5) | 302.77 ± 14.46 | 201.11 ± 6.17 | 332.49 ± 10.28 | 259.38 ± 10.72 |
| dephinidin-3-O-(6-O-acetyl)-glucoside (A6) | 24.94 ± 5.45 | 17.73 ± 5.19 | 28.22 ± 4.10 | 27.84 ± 3.43 |
| cyanidin-3-O-(6-O-acetyl)-glucoside (A7) | 5.45 ± 0.95 | 2.45 ± 0.18 | 7.91 ± 0.51 | 6.13 ± 0.65 |
| petunidin-3-O-(6-O-acetyl)-glucoside (A8) | 17.74 ± 1.76 | 12.51 ± 1.01 | 18.24 ± 0.49 | 17.95 ± 0.92 |
| dephinidin-3-O-(6-O-coumaryl)-glucoside (A9) | 7.67 ± 1.42 | 4.18 ± 0.40 | 4.98 ± 0.67 | 4.88 ± 0.47 |
| peonidin-3-O-(6-O-acetyl)-glucoside (A10) | 18.20 ± 0.86 | 13.99 ± 0.51 | 19.32 ± 0.34 | 12.58 ± 0.58 |
| malvidin-3-O-(6-O-acetyl)-glucoside (A11) | 234.36 ± 8.87 | 148.40 ± 2.63 | 221.82 ± 3.50 | 174.58 ± 5.01 |
| petunidin-3-O-(6-O-caffeoyl)-glucoside (A12) | 0.49 ± 0.15 | 1.07 ± 0.28 | 0.55 ± 0.00 | 0.31 ± 0.01 |
| cyanidin-3-O-(6-O-coumaryl)-glucoside (A13) | 2.05 ± 0.21 | 4.39 ± 0.54 | 2.00 ± 0.74 | 2.00 ± 0.76 |
| malvidin-3-O-(6-O-caffeoyl)-glucoside (A14) | 0.38 ± 0.10 | Nd | 1.21 ± 0.05 | Nd |
| petunidin-3-O-(6-O-coumaryl)-glucoside (A15) | 6.28 ± 0.43 | 3.75 ± 0.23 | 4.36 ± 0.07 | 3.60 ± 0.17 |
| peonidin-3-O-(cis-6-O-coumaryl)-glucoside (A16) | 0.65 ± 0.05 | 0.41 ± 0.01 | 0.64 ± 0.03 | 0.38 ± 0.01 |
| peonidin-3-O-(trans-6-O-coumaryl)-glucoside (A17) | 12.88 ± 1.99 | 6.85 ± 0.39 | 8.95 ± 0.32 | 4.85 ± 0.19 |
| malvidin-3-O-(cis-6-O-coumaryl)-glucoside (A18) | 4.88 ± 0.41 | 2.62 ± 0.16 | 3.65 ± 0.22 | 2.31 ± 0.09 |
| malvidin-3-O-(trans-6-O-coumaryl)-glucoside (A19) | 102.01 ± 16.81 | 47.64 ± 2.77 | 68.04 ± 0.60 | 43.95 ± 1.22 |

Each value represents the mean of three replicates and their standard. Nd, not detected.

## 3. Experimental Section

### 3.1. The Experimental Site and Plant Material

The field experiments were performed at two commercial "Cabernet Sauvignon" vineyards (Yuanyi farm and Guangdong farm) over two seasons (2011 and 2012). They are both located in Manas County (belonging to Shihezi City), the wine-producing region of Xinjiang, China. Yuanyi farm is located at 44°17′55″ North, 86°12′2″ East and at a altitude of 475 m, and Guangdong farm is located at 44°18′58″ North, 86°13′1″ East and at a altitude of 470 m. The growth area of 'Cabernet Sauvignon' was 2000 hectares, which accounted for around 80% of the wine grape cultivation in this county. The vines were planted in 2000, furrow irrigated, and grown on their own roots in a north–south row orientation. All the vines were trained to a slope trunk with a vertical shoot positioning trellis system, which we called Modified VSP (M-VSP) spaced at 2.5 m × 1.0 m (Figure S1), with a spur-pruned cordon retaining 15 nodes per linear metre. Crop load was normalized to approximately 30 bunches per plant. Grapevines in each vineyard were replicated on 45 vines, arranged in three consecutive rows.

The vines were managed according to industry standards for nutrition and pest management in the two vineyards. As in the preceding years, no nitrogen fertiliser was added in 2011 and 2012, hence vine nitrogen uptake was mainly from mineralization of organic matter. Pests and diseases were managed with application of lime sulphur, potassium dihydrogen phosphate and fertilizer containing iron and zinc in May and June, but diniconazole, mancozeb, and imidacloprid were used in July and August. In either vineyard, the management was the same, and there were no artificially induced differences between the two vineyards.

## 3.2. Meteorological Survey

All the meteorological datas of 2011 and 2012 were obtained from the local meteorological administration. Sunlight duration (h), temperature (°C) and rainfall (mm) were recorded daily. Both of the studied plots had a similar mesoclimate, since these two localities were within a distance of 5 km from each other and they had very similar topographic characteristics (altitude, slope and orientation).

## 3.3. Soil Sampling and Analysis

The chosen soils belonging to the two vineyards were collected from five different depths: 0–20 cm, 20–40 cm, 40–60 cm, 60–80 cm, 80–100 cm. The 10 soil sampling points were in a Z-shape, and 500 g of soil was collected from each depth and point, maintaining a distance of 100 cm away from the trunk. Then, all the soils from the same depth were mixed and 1 kg was retained for every sample based on "quartering". A Mastersizer 2000E laser particle size analyzer was used in the determination of physical properties of the soil samples from the two vineyards. The soil texture sorting method followed the criteria of the Soil Survey Manual [42]. The soil bulk density and water content were measured and calculated by the cutting ring method, air-dried and then heated at 105 °C till they reached constant weight. The bulk density was obtained from the ratio between the dried mass and the volume of the cylinder, and more compacted soils with less pore space will have much higher bulk density .While water content of soils was calculated by the difference between the fresh and dried mass. Organic matter, cation exchange capacity (CEC) and pH of soil samples were measured by the potassium dichromate volumetry, EDTA-ammonium salt and potentiometry methods, respectively. On the other hand, nutrient elements in soils were determinated by plasma spectrometry (Vista-AX, Varian, Palo Alto, CA, USA) after extraction and digestion.

## 3.4. Vine Water and Nutrient Status

For each vineyard, $\delta^{13}C$ measurements were carried out on three individual samples of grape at harvest using an Isotope Ratio Mass Spectrometer (Thermo Fisher Scientific Inc., Waltham, MA, USA). The $^{12}C/^{13}C$ ratio in the sample is compared to that in an international standard, the so-called PDB standard which is a rock in which this ratio is particularly stable. The results vary from −20‰ (severe water deficit stress) to −27‰. $\delta^{13}C$ are well-correlated to stem water potential [24].

Berry total nitrogen content can be used as a physiological indicator of vine nitrogen status [36]. Vine nutrient status was assessed on berries collected from the target vines at harvest time in the 2011 and 2012 vintages. The total nutrient contents were analysed using an elemental analyzer (Flash EA1112 HT, Thermo Fisher Scientific Inc., Waltham, MA, USA). Previous research has revealed that the chlorophyll content of the leaves is closely related to the N status of the plant [3]. Thus, the leaf chlorophyll content was measured non-destructively on leaves using a Minolta-502 dual wavelength chlorophyll (SPAD) meter (Minolta Co. Ltd, Osaka, Japan). In addition, thirty random measurements were made across each farm to give an average value used to indicate the vine N status.

## 3.5. Yield and Grape Composition

The berries were collected from 4 weeks after flowering till commercial harvest, and the sampling dates were scheduled at 2 week intervals. Forty five vines were selected on the basis of uniformity of shoot growth and cluster development in each vineyard, and the same plants were used at each sampling point. At veraison, the sampling points were increased at the stages of 10%, 50% and 100% coloured berries in both seasons. The sampling schedule in 2012 spanned the same developmental stages as in 2011. Harvest dates were determined by the cooperating winery. All the grape samples were harvested at technological ripeness, 23th and 24th Sep for Yuanyi farm and Guangdong farm in 2011, and 20th and 24th Sep for Yuanyi farm and Guangdong farm in 2012. For all the shoots from 10 vines, leaves were counted and the main and lateral leaves in each vineyard were measured by the portable leaf area meter (Yaxin-1242, Beijing, China) at veraison in the two seasons. On the commercial harvest day, average shoot length, shoot number per meter, cluster number per shoot were recorded for description of growth potential. Each survey was carried on 10 vines as parallel experiments. In each year, at the end of October, canes from 45 representative vines per vineyard were pruned and weighed to estimate the annual vine growth. These data was then used to calculate the yield-to-pruning weight ratio (kg/kg). Vine balance was also assessed by calculating the total leaf area-to-yield ratio in both vineyards.

Yield data were collected by hand-harvesting fruit from vines of each replicate on the commercial harvest day (18th Sep, 2011 and 20th Sep, 2012) in the two years. Bunches per vine and yield per vine were recorded, from which average bunch weights were determined. After clusters were weighed and compactness described according to OIV rating [43], 300 berries (100 per replicate) were then randomly separated from the pedicle of each vineyard using scissors, ensuring that berries were taken from all parts of the clusters. Berry samples (150, 50 per replicate) were used for determination of berry weight, then another 150 berry samples (50 per replicate) were crushed in a hand press through two layers of cheesecloth. Berry skins were obtained by carefully removing seeds and mesocarp. Berry skins were rinsed in deionised water, then were weighed after blotting excess water. Total soluble solids (TSS) was measured using a PAL-1 Digital Hand-held "Pocket" Refractometer (Atago, Tokyo, Japan). Titratable acidity (TA) was measured by titration with NaOH to the end point of pH 8.2 and expressed as tartaric acid equivalent. Each experiment was carried out in triplicate for berry and cluster samples from the two vineyards.

## 3.6. Extraction of Anthocyanins in Grape Skins

Anthocyanin analyses were done on frozen grapes after removing the stems. Three replicates of samples of grapes (80 berry per replicate) were selected for each developmental stage. While the grapes were still frozen, skins were separated from the pulp. Berry skins were frozen, crushed and then freeze-dried at $-40\,°C$. Both the wet and dry weights were recorded.

The extraction of anthocyanins in grape skins was carried out according to the previously published method of He *et al.* [44]. Grape skin powder (0.50 g) was immersed in methanol (10 mL) containing 2% formic acid. This extraction was performed with the aid of ultrasound for 10 min, and then the mixture was shaken in the dark at 25 °C for 30 min at a rate of 150 rpm. The homogenate was centrifuged at $8,000 \times g$ for 10 min and the supernatant was collected. The residues were re-extracted four times. All the supernatants were mixed, concentrated to dryness using a rotary evaporator and then redissolved in 10 mL of solvent mixed with 90% mobile phase A and 10% mobile phase B (see below). The resulting suspensions were filtered through 0.22 µm filters (Micro Pes, Membrana, Wuppertal, Germany) prior to HPLC-MS analysis. Each sample was subjected to three independent extractions from three biological repeats.

## 3.7. Chemicals and Standards

The standard, malvidin-3-*O*-glucoside was purchased from Sigma-Aldrich Co. (St. Louis, MO, USA). HPLC grade methanol, formic acid and acetonitrile were purchased from Fisher (Fairlawn, NJ, USA). Analytical grade methanol and formic acid were purchased from the Beijing Chemical Reagent Plant (Beijing, China).

## 3.8. HPLC-MS Analyses of Anthocyanins

An Agilent 1100 series LC-MSD trap VL (Agilent, Santa Clara, CA, USA), equipped with a G1379A degasser, a G1312BA QuatPump, a G1313A ALS, a G1316A column, a G1315A DAD and a Kromasil 100–5 C18 column (250 × 4.6 mm, 5 µm) was used for anthocyanins detection. A flow rate of l mL/min at ambient temperature was used. Solvent A was 2% (v:v) formic and 6% acetonitrile in water, and solvent B was acetonitrile containing 2% formic acid and 44% water. Proportions of solvent B varied as follows: 1–18 min, 10% to 25%; 18–20 min, 25%; 20–30 min, 25% to 40%; 30–35 min, 40% to 70% and 35–40 min, 70% to 100%. Injection volumes were 30 µL, and the detection wavelength was 525 nm. The column temperature was 50 °C. MS conditions were as follows: Electrospray ionisation (ESI) interface, positive ion model, 30 psi nebulizer pressure, 12 mL/min dry gas flow rate, 300 °C dry gas temperature, and scans at $m/z$ 100–1500. Anthocyanins were quantified at 525 nm as malvidin-3-$O$-glucoside using calibration curves obtained within a concentration range between 0.5 and 500 mg/L, with linear correlation coefficients greater than 0.999 in the two years. Both standards and samples were determined in triplicate.

*3.9. Statistical Analysis*

Significant differences were determined at $p < 0.05$, according to independent $t$-test. Principal component analysis (PCA) of soil elements and anthocyanins in grape skins was used to achieve a obvious discrimination of different soil depths, regions and vintages. Statistical analysis was performed by SPSS (SPSS Inc., Chicago, IL, USA) for Windows, version 20.0. All the figures were drawn using the Origin 8.0 software.

## 4. Conclusions

In the present research,,, the effects of soil and climatic conditions were studied on 'Cabernet Sauvignon' grapes in two vintages (2011 and 2012). The results showed that the soils with less water and organic matter produced looser clusters, heavier berry skins and higher TSS, which contributed to the excellent performance of the grapes. Compared with 2011, the increased anthocyanin concentrations in 2012 for each vineyard could be attributed to a lesser number of extreme temperature (>35 °C) days and rainfall, lower vine water status and N level. The higher anthocyanin concentrations in grape skins from the soils with less water and organic matter were explained by vine status differences, lighter berry weight and heavier skin weight at harvest. In particular, grapes from the soils with less water and organic matter had higher levels of 3'5'-substituted-$O$-methylated- and acylated anthocyanins, which represented a positive characteristic responsible for more stable pigments in the corresponding wine in the future. In summary, the soils with less water and organic matter produce "Cabernet Sauvignon" grapes with much better quality in some important aspects of winemaking quality, such as berry characteristics and anthocyanin profiles.

**Supplementary Materials:** Supplementary materials can be accessed at: http://www.mdpi.com/1420-3049/19/9/13683/s1.

**Acknowledgments:** This research was supported by the China Agriculture Research System for Grape Industry (CARS-30). The authors would like to thank College of Resources and Environment, Northwest A&F University for technical assistance and laboratory facilities during this study. We also thank Citic Guoan Wine Co., Ltd for its technical support in 2011 and 2012.

**Author Contributions:** Important contributions to design and also to prepare the manuscript: G.C., J.W. and Z.-W.Z. Contributed to soil and anthocyanins analysis experiments: G.C., T.-X.Y. and Y.-N.H. Analysis of the experimental data: G.C., J.W. and Z.-W.Z. Revising it critically for important intellectual content: G.C., Y.-N.H., T.-X.Y., J.W. and Z.-W.Z. All authors helped preparing the paper and approved the final version.

**Conflicts of Interest:** The authors declare no conflict of interest.

## References

1. Downey, M.O.; Dokoozlian, N.K.; Krstic, M.P. Cultural practice and environmental impacts on the flavonoid composition of grapes and wine: A review of recent research. *Am. J. Enol. Vitic.* **2006**, *57*, 257–268.
2. Kennedy, J.A. *Managing Wine Quality: Viticulture and Wine Quality*; Reynolds, A.G., Ed.; Woodhead Publishing: Cambridge, UK, 2010; pp. 73–104.
3. Soubeyrand, E.; Basteau, C.; Hilbert, G.; van Leeuwen, C.; Delrot, S.; Gomès, E. Nitrogen supply affects anthocyanin biosynthetic and regulatory genes in grapevine cv. Cabernet-Sauvignon berries. *Phytochemistry* **2014**, *103*, 38–49.
4. Baldi, A.; Romani, A.; Mulinacci, N.; Vincieri, F.; Caseta, B. HPLC/MS Application to Anthocyanins of *Vitis vinifera* L. *J. Agric. Food. Chem.* **1995**, *43*, 2104–2109.
5. Monages, M.; Nunez, V.; Bartolome, B.; Gomez-Cordoves, C. Anthocyanin-derived pigments in Graciano, Tempranillo, and Cabernet Sauvignon wines produced in Spain. *Am. J. Enol. Vitic.* **2003**, *54*, 163–169.
6. Deloire, A.; Vaudour, E.; Carey, V.; Bonnardot, V.; van Leeuwen, C. Grapevine responses to terroir, a global approach. *J. Int. Vigne. Vin.* **2005**, *39*, 149–162.
7. Van Leeuwen, C.; Friant, C.; Friant, F.; Chone, X.; Tregoat, O.; Koundouras, S.; Dubourdieu, D. Influence of climate, soil and cultivar on terroir. *Am. J. Enol. Vitic.* **2004**, *55*, 207–217.
8. Mazza, G. Anthocyanins in grape products. *Crit. Rev. Food Sci. Nutr.* **1995**, *35*, 341–371.
9. Pereira, G.E.; Gaudillere, J.P.; van Leeuwen, C.; Hilbert, G.; Maucourt, M.; Deborde, C. H$^1$-NMR metabolite fingerprints of grape berry: Comparison of vintage and soil effects in Bordeaux grapevine growing areas. *Anal. Chim. Acta* **2006**, *563*, 346–352.
10. Ferrandino, A.; Lovisolo, C. Abiotic stress effects on grapevine (*Vitis vinifera* L.): Focus on abscisic acid-mediated consequences on secondary metabolism and berry quality. *Environ. Exp. Bot.* **2014**, *103*, 138–147.
11. Cohen, S.D.; Tarara, J.M.; Kennedy, J.A. Diurnal Temperature Range Compression Hastens Berry Development and Modifies Flavonoid Partitioning in Grapes. *Am. J. Enol. Vitic.* **2012**, *63*, 112–120.
12. Bucchetti, B.; Matthews, M.A.; Falginella, L.; Peterlungera, E.; Castellarin, S.D. Effect of water deficit on Merlot grape tannins and anthocyanins across four seasons. *Sci. Hortic.* **2011**, *128*, 297–305.
13. Li, J.H.; Guan, L.; Fan, P.G.; Li, S.H.; Wu, B.H. Effect of Sunlight Exclusion at Different Phenological Stages on Anthocyanin Accumulation in Red Grape Clusters. *Am. J. Enol. Vitic.* **2013**, *64*, 349–356.
14. Castellarin, S.D.; Matthews, M.A.; di Gaspero, G.; Gambetta, G.A. Water deficits accelerate ripening and induce changes in gene expression regulating flavonoid biosynthesis in grape berries. *Planta* **2007**, *227*, 101–112.

15. Castellarin, S.D.; Pfeiffer, A.; Sivilotti, P.; Degan, M.; Peterlunger, E.; di Gaspero, G. Transcriptional regulation of anthocyanin biosynthesis in ripening fruits of grapevine under seasonal water deficit. *Plant Cell. Environ.* **2007**, *30*, 1381–1399.

16. Dokoozlian, N.K.; Kliewer, W.M. Influence of light on grape berry growth and composition varies during fruit development. *J. Am. Soc. Hortic. Sci.* **1996**, *121*, 869–874.

17. Hunter, J.J.; Ruffner, H.P.; Volschenk, C.G.; le Roux, D.J. Partial defoliation of *Vitis vinifera* L. cv. Cabernet Sauvignon/99 Richter: Effect on root growth, canopy efficiency, grape composition, and wine quality. *Am. J. Enol. Vitic.* **1995**, *46*, 306–314.

18. Bergqvist, J.; Dokoozlian, N.; Ebisuda, N. Sunlight exposure and temperature effects on berry growth and composition of Cabernet Sauvignon and Grenache in the central San Joaquin Valley of California. *Am. J. Enol. Vitic.* **2001**, *52*, 1–7.

19. Conradie, W.J. Liming and choice of rootstocks as cultural techniques for vines in acid soils. *S. Afr. J. Enol. Vitic.* **1983**, *4*, 39–44.

20. Josep, M.U.; Xavier, S.; Alicia, Z.; Rosa, M.P. Effects of soil and climatic conditions on grape ripening and wine quality of Cabernet Sauvignon. *J. Wine Res.* **2010**, *21*, 1–17.

21. De Andrés-de Prado, R.; Yuste-Rojas, M.; Sort, X.; Andrés-Lacueva, C.; Torres, M.; Lamuela-Raventós, R.M. Effect of soil type on wines produced from *Vitis vinifera* L. cv. Grenache in commercial vineyards. *J. Agric. Food. Chem.* **2007**, *55*, 779–786.

22. Farquhar, G.; Ehleringer, J.; Hubick, K. Carbon isotope discrimination and photosynthesis. *Annu. Rev. Plant Physiol. Plant Mol. Biol.* **1989**, *40*, 503–537.

23. Gaudillère, J.P.; van Leeuwen, C.; Ollat, N. Carbon isotope composition of sugars in grapevine, an integrated indicator of vineyard water status. *J. Exp. Bot.* **2002**, *53*, 757–763.

24. Van Leeuwen, C.; Trégoat, O.; Choné, X.; Bois, B.; Pernet, D.; Gaudillère, J.P. Vine water status is a key factor in grape ripening and vintage quality for red Bordeaux wine. How can it be assessed for vineyard management purposes? *J. Int. Sci. Vigne. Vin.* **2009**, *43*, 121–134.

25. King, P.D.; Smart, R.E.; Mcclellan, D.J. Within-vineyard variability in vine vegetative growth, yield, and fruit and wine composition of Cabernet Sauvignon in Hawke's Bay, New Zealand. *Aust. J. Grape Wine Res.* **2014**, *20*, 234–246.

26. Reynolds, A.G.; Senchuk, I.V.; van der Reest, C.; de Savigny, C. Use of GPS and GIS for elucidation of the basis for terroir, spatial variation in an Ontario Riesling vineyard. *Am. J. Enol. Viti.* **2007**, *8*, 145–162.

27. Reynolds, A.G.; Edwards, C.G.; Wardle, D.A.; Webster, D.R.; Dever, M. Shoot density affects "Riesling" grapevines I. Vine performance. *J. Am. Soc. Hortic. Sci.* **1994**, *119*, 874–880.

28. Peyrot des Gachons, C.P.; van Leeuwen, C.; Tominaga, T.; Soyer, J.P.; Gaudillère, J.P.; Dubourdieu, D. Influence of water and nitrogen deficit on fruit ripening and aroma potential of *Vitis vinifera* L. cv Sauvignon blanc in field conditions. *J. Sci. Food. Agric.* **2005**, *85*, 73–85.

29. Kennedy, J.A.; Matthews, M.A.; Waterhouse, A.L. The effect of maturity and vine water status on grape skin flavonoids. *Am. J. Enol. Vitic.* **2002**, *3*, 268–274.

30. Jackson, D.I.; Lombard, P.B. Environmental and management-practices affecting grape composition and wine quality-a review. *Am. J. Enol. Vitic.* **1993**, *44*, 409–430.

31. Mazza, G.; Fukumoto, L.; Delaquis, P.; Girard, B.; Ewert, B. Anthocyanins, Phenolics, and Color of Cabernet Franc, Merlot, and Pinot Noir Wines from British Columbia. *J. Agric. Food Chem.* **1999**, *47*, 4009–4017.

32. Spayd, S.E.; Tarara, J.M.; Mee, D.L.; Ferguson, J.C. Separation of sunlight and temperature effects on the composition of *Vitis Vinifera* cv. Merlot berries. *Am. J. Enol. Vitic.* **2002**, *53*, 171–182.

33. Barbagallo, M.G.; Guidoni, S.; Hunter, J.J. Berry size and qualitative characteristics of *vitis vinifera*. cv. Syrah. *S. Afr. J. Enol. Vitic.* **2011**, *32*, 129–136.

34. Ojeda, H.; Andary, C.; Kraeva, E.; Carbonneau, A; Deloire, A. Influence of pre- and postveraison water deficit on synthesis and concentration of skin phenolic compounds during berry growth of *Vitis vinifera* cv. Shiraz. *Am. J. Enol. Vitic.* **2002**, *53*, 261–267.

35. Roby, G.; Harbertson, J.F.; Adams, D.A.; Matthews, M.A. Berry size and vine water deficits as factors in winegrape composition: Anthocyanins and tannins. *Aust. J. Grape. Wine Res.* **2004**, *10*, 100–107.

36. Chone, X.; van Leeuwen, C.; Chery, P.; Ribereau-Gayon, P. Terroir influence on water status and nitrogen status of non-irrigated Cabernet Sauvignon (*Vitis vinifera*). Vegetative development, must and wine composition (example of a Medoc top estate vineyard, Saint Julien area, Bordeaux, 1997). *S. Afr. J. Enol. Vitic.* **2001**, *21*, 8–15.

37. Matsuyama, S.; Tanzawa, F.; Kobayashi, H.; Suzuki, S.; Takata, R.; Saito, H. Leaf Removal Accelerated Accumulation of Delphinidin-based Anthocyanins in "Muscat Bailey A" [*Vitis × labruscana* (Bailey) and *Vitis vinifera* (Muscat Hamburg)] Grape Skin. *J. Jpn. Soc. Hortic. Sci.* **2014**, *83*, 17–22.

38. He, J.; Giusti, M.M. Anthocyanins: Natural colorants with healthpromoting properties. *Annu. Rev. Food. Sci. Technol.* **2010**, *1*, 163–187.

39. Roggero, J.P.; Coen, S.; Ragonnet, R. High performance liquid chromatography survey on changes in pigment content in ripening grapes of Syrah. An approach to anthocyanin metabolism. *Am. J. Enol. Vitic.* **1986**, *37*, 77–83.

40. Ryan, J.M.; Revilla, E. Anthocyanin composition of Cabernet Sauvignon and Tempranillo grapes at different stages of ripening. *J. Agric. Food. Chem.* **2003**, *51*, 3372–3378.

41. Tarara, J.M.; Lee, J.; Spayd, S.E.; Scagel, C.F. Berry Temperature and Solar Radiation Alter Acylation, Proportion, and Concentration of Anthocyanin in Merlot Grapes. *Am. J. Enol. Vitic.* **2008**, *59*, 235–247.

42. Soil Survey Staff. *Soil Survey Manual*; Soil Conservation Service: Washington, DC, USA, 1993.

43. International Organisation of Vine and Wine. *OIV Descriptor List for Grape Varieties and Vitis Species*, 2nd ed.; OIV: Paris, France, 2007.

44. He, J.J.; Liu, Y.X.; Pan, Q.H.; Cui, X.Y.; Duan, C.Q. Different anthocyanin profiles of the skin and the pulp of Yan73 (Muscat Hamburg × Alicante Bouschet) grape berries. *Molecules* **2010**, *15*, 1141–1153.

**Sample Availability:** *Sample Availability*: Samples of grapes are available from the authors.

# Simple Rain-Shelter Cultivation Prolongs Accumulation Period of Anthocyanins in Wine Grape Berries

Xiao-Xi Li, Fei He, Jun Wang, Zheng Li and Qiu-Hong Pan

**Abstract:** Simple rain-shelter cultivation is normally applied during the grape growth season in continental monsoon climates aiming to reduce the occurrence of diseases caused by excessive rainfall. However, whether or not this cultivation practice affects the composition and concentration of phenolic compounds in wine grapes remains unclear. The objective of this study was to investigate the effect of rain-shelter cultivation on the accumulation of anthocyanins in wine grapes (*Vitis vinifera* L. Cabernet Sauvignon) grown in eastern China. The results showed that rain-shelter cultivation, compared with the open-field, extended the period of rapid accumulation of sugar, increased the soluble solid content in the grape berries, and delayed the senescence of the green leaves at harvest. The concentrations of most anthocyanins were significantly enhanced in the rain-shelter cultivated grapes, and their content increases were closely correlated with the accumulation of sugar. However, the compositions of anthocyanins in the berries were not altered. Correspondingly, the expressions of *VvF3'H*, *VvF3'5'H*, and *VvUFGT* were greatly up-regulated and this rising trend appeared to continue until berry maturation. These results suggested that rain-shelter cultivation might help to improve the quality of wine grape berries by prolonging the life of functional leaves and hence increasing the assimilation products.

Reprinted from *Molecules*. Cite as: Li, X.-X.; He, F.; Wang, J.; Li, Z.; Pan, Q.-H. Simple Rain-Shelter Cultivation Prolongs Accumulation Period of Anthocyanins in Wine Grape Berries. *Molecules* **2014**, *19*, 14843–14861.

## 1. Introduction

The Chinese wine industry has developed with unprecedented speed in recent years in terms of both production and consumption. Currently, it is one of the top 10 wine-producing countries in the world regarding area under vine and wine volume produced. The national appetite for wine has correspondingly more than doubled in the past two decades [1]. China's wine regions spread across the breadth of the country. On the humid and monsoonal east coast, the provinces of Shandong and Hebei and Tianjin City are responsible for a large amount of China's national production. These areas have formed the characteristic wine industry clusters. The terroir of Shandong Province avoids the harsh continental extremes of the center of

China, and instead has a maritime climate with cooler summers and warmer winters. The Shandong wine-producing regions are affected by the East Asian Monsoon, a weather system that brings cool and moist air from the Pacific Ocean to the shores of the province, causing summer rains. Fungal vine diseases caused by high rainfall are an important concern for vignerons in the late summer and early autumn.

Rain-shelter cultivation is a common kind of canopy management consisting of building a polyethylene (PE) film roof one meter from the top of the grapevine canopy. In China, rain-shelter cultivation is usually implemented in the production of table grape berries along the midstream and downstream areas of Yangtze River and in the south [2]. A large amount of viticulture practice has indicated that rain-shelter cultivation can effectively eliminate the incidence of major diseases, such as downy mildew (*Plasmoparaviticola*), powdery mildew (*Uncinulanecator*), botrytis (*Botrytiscinerea*), rip rot (*Glomerellacingulata*), and sour rot (imperfect yeasts), by keeping rainwater away from leaves and fruits. Under this circumstance, the use of pesticides could be remarkably reduced [3].

Anthocyanins are a crucial class of phenolic pigments in grapes and wine, and they contribute to the appearance (color), sensory quality (chromaticity and color tone), stability (aging potential), and potential human health benefits [4–6]. Anthocyanins are accumulated predominantly in the skins of grapes from the beginning of véraison. The biosynthesis of anthocyanins has two important branches, the 3'-substituted anthocyanin synthesis (for example, cyanidin-3-*O*-glucoside and peonidin-3-*O*-glucoside) led by flavonoid-3'-hydroxylase (F3'H), and the 3'5'-substituted anthocyanin synthesis (for example, delphindin-3-*O*-glucoside, petunidin-3-*O*-glucoside and malvidin-3-*O*-glucoside) regulated by flavonoid-3'5'-hydroxylase (F3'5'H) (Figure 1).

Some literature has reported that rain-shelter cultivation effectively delayed the maturation of grapes [7], slowed the sugar accumulation by reducing photosynthetically active radiation (PAR), significantly increased berry and cluster weight, and improved economic returns [8]. Although rain-shelter cultivation has been studied for its commercial value on table grapes and other fruits during the past several decades [8–11], few studies have focused on wine grapes. Some problems remain to be solved. For example, (i) does rain-shelter cultivation delay the maturity of wine grape berries and decrease their quality?; (ii) what is the effect on the composition and content of anthocyanins in the skins of grape berries under rain-shelter cultivation? This study aimed to investigate the effect of simple rain-shelter cultivation on the accumulation of anthocyanins in *Vitis vinifera* L. Cabernet Sauvignon berries, and to evaluate the prospects for application in the production of high-quality wine grapes in rainy regions.

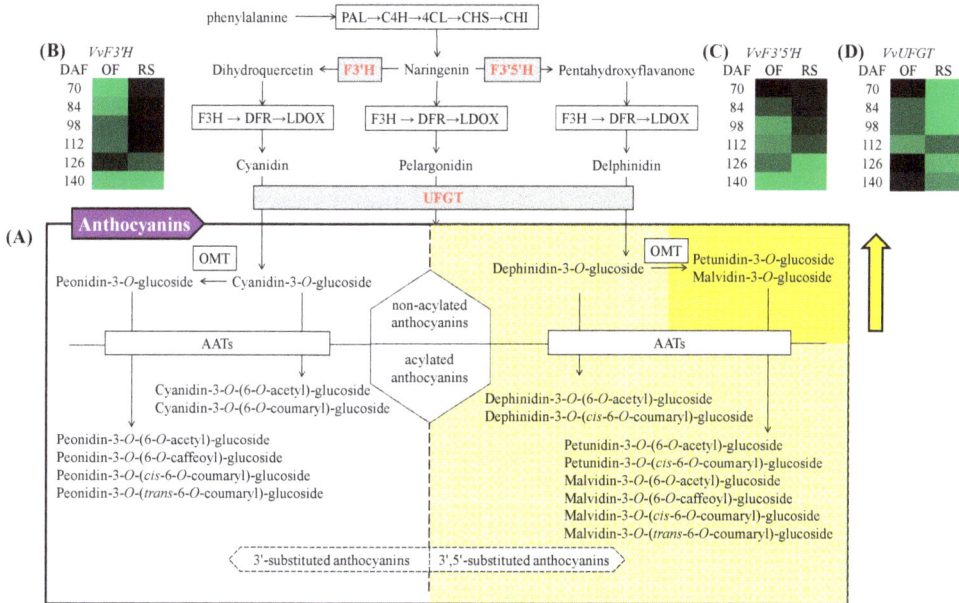

**Figure 1.** Biosynthetic pathway of anthocyanidins in grape (based on [12], only detected anthocyanins are listed in this figure, (**A**)). The relative expression amount of *VvF3'H* (**B**), *VvF3'5'H* (**C**) and *VvUFGT* (**D**) is shown using three heat maps. Green squares from dark to bright represent the gene expression levels from high to low during the mature period (DAF70 to 140) with respect to both open-field cultivation (OF) and rain-shelter cultivation (RS). The yellow square shown in the anthocyanins block indicates that the compounds from this branch have higher concentration in berry skins under rain-shelter cultivation in comparison to open-field at commercial harvest. Abbreviations: DAF, days after flowering; PAL, phenylalanine ammonia-lyase; C4H, cinnamate 4-hydroxylase; 4CL, 4-coumarate: CoA ligase; CHS, chalcone synthase; CHI, chalcone isomerase; F3'H, flavonoid 3'-hydroxylase; F3'5'H, flavonoid 3',5'-hydroxylase; F3H, flavanone 3-hydroxylase; DFR, dihydroflavonol reductase; LDOX, leucoanthocyanidin dioxygenase; UFGT, UDP-glucose: flavonoid 3-O-glucosyltransferase; OMT, o-methyltransferase; AATs, anthocyanin acyltransferases.

## 2. Results and Discussion

### 2.1. Total Soluble Solid and pH Value

Figure 2 shows the variation of total soluble solids (°Brix) and pH value in the grape berries under rain-shelter and in open-field cultivation from 2 weeks before véraison (56 DAF) to the technological harvest (126 DAF) and post-maturity (140 DAF), respectively. From 56 DAF, total soluble solids in the grape berries under the open-field cultivation rapidly accumulated by 98 DAF and then slowly increased until the end of the experiment. Correspondingly, under rain-shelter cultivation, the rapid accumulation period of soluble solids in the grape berries was extended to 126 DAF and the peak value appeared at 126 DAF. During post-maturity, total soluble solids in the berries of rain-shelter cultivation showed no significant changes. At harvest, the content of sugar in the berries under rain-shelter cultivation was significantly higher than the control, reaching 21.5 °Brix compared with 16.0 °Brix in the open-field cultivated berries. Similarly, the maximum pH values were observed at 98 DAF and 112 DAF, under the open-field cultivation and rain-shelter cultivation respectively. After the end of véraison, pH values in the berries showed no significant differences between the two cultivation modes.

The longer and greater accumulation of sugar in the berries under rain-shelter cultivation was possibly related to a delay in canopy leaf senescence. We observed that the grapevines under the rain-shelter cultivation still had more green leaves, even at harvest, in comparison with the vines in the open-field cultivation (Figure 2C,D). This indicates that the rain-shelter cultivation could effectively prolong the life of functional leaves and therefore enhance the production ability of photosynthetic assimilates of whole vines. The visible prolonging of functional leaves is inferred to be related to a lower ultraviolet radiation (UVR) and lower canopy temperature inside the rain-shelter. Simple rain-shelter cultivation improves the quality of a great number of horticultural products, such as tomato, pepper, oat, blueberry, loquat and table grape [10,13–17]. Previous research indicated that the plastic covering materials used in rain-shelter cultivation can filter out part of UVR and PAR, and help avoid the direct sunlight at noon [7,10]. In a field, temperature above the top of canopy was about 5°C higher than inside the canopy [18], and the photosynthesis in grape leaves was inhibited by UVR, specifically by UV-B radiation (UVBR) [19]. Similar inhibitory effects also had been observed in other plant tissues. For example, high irradiance on bayberry leaves resulted in photo-inhibition and photo-damage by inactivation of photosystem II reaction centers [12]. These results obtained in the previous studies all imply that high irradiation can accelerate leaf senescence. Accordingly, it is considered that rain-shelter cultivation can effectively delay the process of leaf senescence and maintain the ability of leaf photosynthesis.

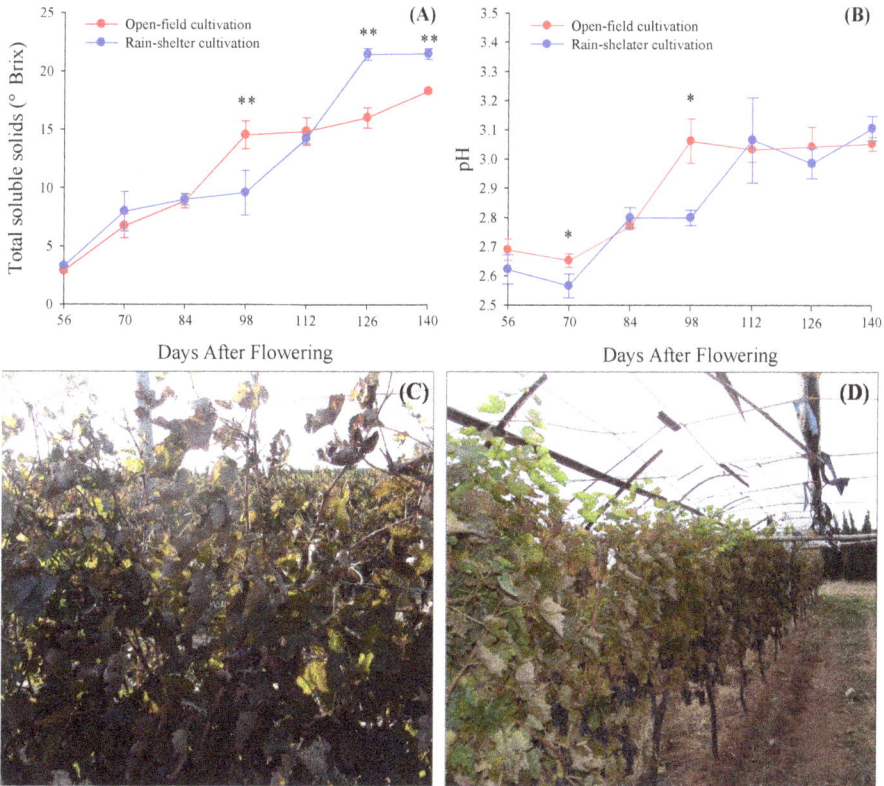

**Figure 2.** The variations of total soluble solids **(A)** and pH **(B)** of grape juice between vines grown under open-field cultivation and rain-shelter cultivation; and canopies of vines grown under open-field cultivation **(C)** and rain-shelter cultivation **(D)** at commercial harvest. Data are mean ± standard deviation. The symbol "**" or "*" above each set of columns represents that there are significant difference between rain-shelter cultivation and open-field cultivation by Duncun analysis at 0.01 level ($p < 0.01$) or 0.05 level ($p < 0.05$). $n = 6$.

## 2.2. Accumulation of Anthocyanins and Expression of VvUFGT

The rain-shelter cultivation did not alter the composition of anthocyanins in the grape berries (Table 1). The content of anthocyanins increased along with berry maturation period, and a total of 19 anthocyanins were identified in the berry skins at 140 DAF. The rain-shelter cultivation could advance the generation of new anthocyanins during the maturation period of the grape berry. Malvidin-3-O-(6-O-caffeoyl)-glucoside and peonidin-3-O-(6-O-caffeoyl)-glucoside were quantified in the rain-shelter cultivated grapes at 105 DAF and 112 DAF, respectively, whereas both of them were detected only at 140 DAF (post-maturity stage) in the open-field cultivated grapes.

Table 1. Content of individual anthocyanins in grape skins under rain-shelter cultivation and open-field cultivation.

| Anthocyanins (mg/kg DW skin) | 70 DAF | | 84 DAF | | 91 DAF | | 98 DAF | | 105 DAF | | 112 DAF | | 126 DAF | | 140 DAF | |
|---|---|---|---|---|---|---|---|---|---|---|---|---|---|---|---|---|
| | Rain-Shelter | Open-Field | Rain-Shelter | Open-Field | Rain-Shelter | Open-Field | Rain-Shelter | Open-Field | Rain-Shelter | Open-Field | Rain-Shelter | Open-Field | Rain-Shelter | Open-Field | Rain-Shelter | Open-Field |
| Delphinidin-3-$O$-glucoside | 14.5 ± 0.4a | 7.8 ± 0.9b | 77.6 ± 0.9b | 172.3 ± 5.5a | 212.4 ± 7.2b | 420.8 ± 18.6a | 94.8 ± 13.2b | 238.4 ± 7.2a | 149.4 ± 3.2b | 286.4 ± 8.0a | 191.7 ± 4.9a | 174.0 ± 4.9b | 491.5 ± 26.2a | 268.7 ± 11.6b | 403.7 ± 13.6a | 181.2 ± 7.1b |
| Petunidin-3-$O$-glucoside | 9.8 ± 0.4a | 5.3 ± 0.7b | 50.3 ± 3.5b | 95.5 ± 3.8a | 146.5 ± 11.4b | 252.4 ± 8.4a | 75.5 ± 12.4b | 166.5 ± 11.5a | 106.1 ± 9.9b | 189.4 ± 5.2a | 140.6 ± 5.0a | 109.9 ± 2.8b | 303.8 ± 17.3a | 164.5 ± 7.7b | 231.6 ± 1.6a | 122.1 ± 10.7b |
| Malvidin-3-$O$-glucoside | 34.8 ± 2.0a | 12.9 ± 0.8b | 280.9 ± 1.0b | 392.6 ± 13a | 895.3 ± 21.2b | 1291.2 ± 32.2a | 720.8 ± 103.0b | 1329.1 ± 22.6a | 882.9 ± 0.9b | 1252.6 ± 25.7a | 1366.1 ± 59.2a | 998.3 ± 1.8a | 1831 ± 23.2a | 1291.3 ± 1.6b | 1595.4 ± 1.3a | 1323.3 ± 5.6b |
| Delphinidin-3-$O$-(6-$O$-acetyl)-glucoside | 8.1 ± 0.4a | 3.8 ± 0.1b | 28.8 ± 1.1b | 51.8 ± 1.7a | 73.9 ± 0.2b | 121.8 ± 0.8a | 47.2 ± 2.4b | 85.5 ± 3.9a | 59.6 ± 0.4b | 95.6 ± 0.8a | 79.7 ± 3.4a | 72.8 ± 1.7a | 148.6 ± 0.2a | 90.5 ± 4.8b | 135.9 ± 3.9a | 82.3 ± 0.6b |
| Petunidin-3-$O$-(6-$O$-acetyl)-glucoside | 6.1 ± 0.6a | 3.4 ± 0.06b | 19.1 ± 0.2b | 35.3 ± 1.8a | 53.0 ± 4.1b | 87.0 ± 1.7a | 29.5 ± 4.2b | 61.8 ± 2.5a | 41.3 ± 3.0b | 68.5 ± 0.1a | 54.8 ± 3.7a | 39.1 ± 2.3b | 104.9 ± 0.9a | 52.0 ± 1.5b | 89.5 ± 5.3a | 41.1 ± 4.2b |
| Malvidin-3-$O$-(6-$O$-acetyl)-glucoside | 28.0 ± 3.1a | 10.9 ± 0.5b | 153.0 ± 8.2b | 196.6 ± 6.1a | 475.2 ± 2.7b | 647.8 ± 10.3a | 472.2 ± 4.6b | 782.6 ± 3.3a | 516.3 ± 13.8b | 709.5 ± 0.8a | 782.3 ± 48.2a | 582.4 ± 3.0b | 883.6 ± 1.7a | 673.8 ± 2.4b | 836.4 ± 10.9a | 700.2 ± 10.1b |
| Delphinidin-3-$O$-($cis$-6-$O$-coumaryl)-glucoside | - | - | 4.9 ± 0.1b | 9.4 ± 0.8a | 16.6 ± 1.7b | 27.3 ± 3.7a | 10.2 ± 4.8b | 20.4 ± 0.9a | 13.8 ± 0.8b | 22.8 ± 0.3a | 17.4 ± 1.8a | 14.6 ± 0.1a | 30.4 ± 0.5a | 18.5 ± 0.3b | 25.6 ± 1.6a | 16.0 ± 0.4b |
| Malvidin-3-$O$-(6-$O$-caffeoyl)-glucoside | - | - | - | - | - | - | - | - | 5.3 ± 0.5 | - | 7.4 ± 0.8 | - | 7.7 ± 0.5 | - | 11.4 ± 1.7a | 9.4 ± 0.2b |
| Petunidin-3-$O$-($cis$-6-$O$-coumaryl)-glucoside | - | - | 4.4 ± 0.3b | 7.9 ± 0.8a | 12.7 ± 0.3b | 19.4 ± 0.8a | 7.8 ± 1.2b | 15.3 ± 0.4a | 9.8 ± 0.4b | 18.0 ± 0.0a | 11.2 ± 1.4a | 12.1 ± 0.3a | 21.5 ± 0.5a | 14.0 ± 0.1b | 16.8 ± 1.4a | 11.8 ± 0.4b |
| Malvidin-3-$O$-($cis$-6-$O$-coumaryl)-glucoside | - | - | 5.9 ± 0.3b | 7.7 ± 1.1a | 17.1 ± 1.2a | 21.1 ± 1.1a | 17.9 ± 2.3b | 25.0 ± 0.5a | 18.4 ± 0.2a | 20.7 ± 0.4a | 26.2 ± 3.2a | 25.4 ± 1.0a | 17.2 ± 1.4b | 23.6 ± 1.9a | 16.7 ± 0.9b | 27.8 ± 1.4a |
| Malvidin-3-$O$-($trans$-6-$O$-coumaryl)-glucoside | 4.5 ± 0.2a | 3.0 ± 0.3b | 38.6 ± 4.3b | 52.5 ± 5.3a | 156.5 ± 1.0b | 221.2 ± 6.8a | 163.1 ± 29.1b | 274 ± 5.3a | 175.8 ± 11.3b | 250.2 ± 2.7a | 252.1 ± 18.4a | 255.3 ± 6.5a | 294.3 ± 3.3a | 267.4 ± 9.9a | 249.5 ± 2b | 330.2 ± 9.1a |
| Cyanidin-3-$O$-glucoside | 6.7 ± 0.2a | 4.9 ± 0.5a | 24.9 ± 2.2b | 64.0 ± 2.4a | 51.2 ± 2.2b | 122.2 ± 3.9a | 26.6 ± 4.1b | 47.7 ± 9.1a | 35.0 ± 4.9b | 71.7 ± 1.1a | 43.2 ± 1.5a | 35.1 ± 3.4b | 106.2 ± 2.9a | 59.9 ± 1.5b | 68.1 ± 3.6a | 33.2 ± 2.9b |
| Peonidin-3-$O$-glucoside | 13.0 ± 0.5a | 7.8 ± 0.5b | 61.0 ± 0.3b | 129.2 ± 4.6a | 148.0 ± 6.6b | 299.2 ± 6.9a | 108.9 ± 17.9b | 188.7 ± 5.3a | 140.1 ± 2.8b | 225.8 ± 4.0a | 178.5 ± 7.8a | 160.1 ± 1.2b | 316 ± 6.8a | 213.7 ± 0.8b | 212.3 ± 3.9a | 176.0 ± 5.7b |
| Cyanidin-3-$O$-(6-$O$-acetyl)-glucoside | 3.5 ± 0.2a | 2.6 ± 0.0a | 8.0 ± 0.6b | 17.8 ± 0.2a | 17.7 ± 2.6b | 33.3 ± 0.6a | 12.7 ± 4.6b | 21.6 ± 2.4a | 15.9 ± 2.7b | 24.9 ± 0.2a | 21.3 ± 1.9a | 13.5 ± 1.3b | 34.5 ± 0.3a | 19.6 ± 1.3b | 27.1 ± 3.0a | 17.0 ± 2.6b |
| Peonidin-3-$O$-(6-$O$-acetyl)-glucoside | 7.4 ± 1.0a | 4.4 ± 0.3b | 21.3 ± 0.6b | 41.6 ± 1.9a | 53.3 ± 1.5b | 97.5 ± 0.9a | 46.9 ± 0.9b | 75.1 ± 0.8a | 56.1 ± 0.6b | 83.8 ± 0.5a | 70.2 ± 4.1a | 61.5 ± 1.1b | 101.7 ± 0.1a | 72.7 ± 0.1b | 75.8 ± 2.5a | 63.9 ± 1.4b |

# Table 1. Cont.

| Anthocyanins (mg/kg DW skin) | 70 DAF | | 84 DAF | | 91 DAF | | 98 DAF | | 105 DAF | | 112 DAF | | 126 DAF | | 140 DAF | |
|---|---|---|---|---|---|---|---|---|---|---|---|---|---|---|---|---|
| | Rain-Shelter | Open-Field | Rain-Shelter | Open-Field | Rain-Shelter | Open-Field | Rain-Shelter | Open-Field | Rain-Shelter | Open-Field | Rain-Shelter | Open-Field | Rain-Shelter | Open-Field | Rain-Shelter | Open-Field |
| Peonidin-3-O-(6-O-caffeoyl)-glucoside | - | - | - | - | - | - | - | - | - | - | $5.5 \pm 0.2$ | - | $5.3 \pm 0.3$ | - | $5.8 \pm 1.1$ | Trace |
| Cyanidin-3-O-(6-O-coumaryl)-glucoside | - | - | - | $7.7 \pm 0.7$ | $7.2 \pm 0.4b$ | $15.5 \pm 0.3a$ | $4.7 \pm 0.7b$ | $8.2 \pm 0.0a$ | $6.2 \pm 0.0b$ | $12.7 \pm 1.5a$ | $7.2 \pm 0.7$ | $6.9 \pm 0.1$ | $11.5 \pm 0.4a$ | $10.8 \pm 1.3a$ | $8.4 \pm 0.8a$ | $6.4 \pm 0.2a$ |
| Peonidin-3-O-(cis-6-O-coumaryl)-glucoside | - | - | $2.5 \pm 0.1a$ | $4.0 \pm 0.4a$ | $4.4 \pm 0.3b$ | $7.0 \pm 0.4a$ | $4.4 \pm 0.2a$ | $5.9 \pm 0.1a$ | $5.1 \pm 0.2a$ | $6.0 \pm 0.6a$ | $6.3 \pm 1.2a$ | $5.5 \pm 0.3a$ | $5.7 \pm 0.9a$ | $6.1 \pm 0.5a$ | $5.3 \pm 1.1a$ | $5.5 \pm 0.2a$ |
| Peonidin-3-O-(trans-6-O-coumaryl)-glucoside | $2.7 \pm 0.2a$ | $2.3 \pm 0.1a$ | $8.9 \pm 1.0b$ | $19.4 \pm 2.4a$ | $26.7 \pm 0.2b$ | $55.2 \pm 3.0a$ | $24.0 \pm 4.0b$ | $42.5 \pm 0.5a$ | $31.2 \pm 1.9b$ | $52.0 \pm 0.0a$ | $38.9 \pm 4.1a$ | $43.6 \pm 0.1a$ | $59.2 \pm 1.1a$ | $50.6 \pm 2.6b$ | $41 \pm 0.3b$ | $48.0 \pm 3.3a$ |

Malvidin-3-O-glucoside was used as the external standard for quantification. Data are mean ± standard deviation. "-" represents "not detected". Different letters in each category represent that the concentrations of this anthocyanin had significant difference between the rain-shelter cultivation and the open-field cultivation at $p \le 0.05$ by Duncun analysis.

181

The accumulation of anthocyanins displayed different trends in the open-field and rain-shelter cultivated grapes (Figure 3A). Under open-field cultivation conditions, the total concentration of anthocyanins in the grape skins increased to the highest value at 91 DAF (3rd week after coloring), and then decreased slightly. However, the increasing trend of the total anthocyanin content in the rain-shelter cultivated grapes was extended to 126 DAF. Interestingly, the total concentration of anthocyanins in the rain-shelter cultivated grapes was always lower than that in the open-field cultivated grapes from 70 to 105 DAF, but the contrary occurred during the following ripening phases. UDP-glucose:flavonoid 3-O-glucosyltransferase (UFGT) has been demonstrated to be a critical enzyme for anthocyanin biosynthesis in grape berries, and it can transfer the glucose residues of the UDP-glucose molecule to the 3 position of anthocyanidin in the C ring to form anthocyanins [20]. The expression of *VvUFGT* was significantly up-regulated in the rain-shelter cultivated grapes during the whole ripening period, with the exception of the 112 DAF (Figure 3B). The variation of the total anthocyanin concentration paralleled well that of the soluble solids in the process of berry maturation and to the accumulative expression of *VvUFGT* under either of cultivation modes (Figure 3C,D). For the rain-shelter cultivated grapes, the correlation was more significant in terms of the correlation coefficient (the $R^2$ value was shown in Figure 3C,D).

These results showed that rain-shelter cultivation did not alter the compositional proportion of anthocyanins, but increased the concentrations of almost all the anthocyanin components detected in the ripening berries. This suggested that rain-shelter cultivation could promote the whole anthocyanin biosynthetic pathway rather than a specific branch. This promotion was possibly achieved by enhancing the supplementation of photo-assimilates from the grape leaves since the senescence of the leaves under the rain-shelter cultivation was delayed. Previous investigations have shown that rain-shelter cultivation contributes to a higher yield [8,9], but few of them were focused on fruit quality. For example, Meng *et al.* [2] investigated the effect of rain-shelter cultivation mode on accumulation of phenolic compounds and incidence of grape diseases, and found that both phenolic compounds of grape skins and incidence of grape diseases decreased under rain-shelter cultivation compared with open-field cultivation. They explained that lower solar radiation and higher air humidity in the rain-shelter could produce a larger influence on the anthocyanin accumulation than air temperatures. However, in the present study, the accumulation periods of sugar and anthocyanins were both prolonged in the grape berry under the rain-shelter cultivation, which might be attributed to the delay of senescence of grape leaves and the change of sink-source relationship in this cultivation mode.

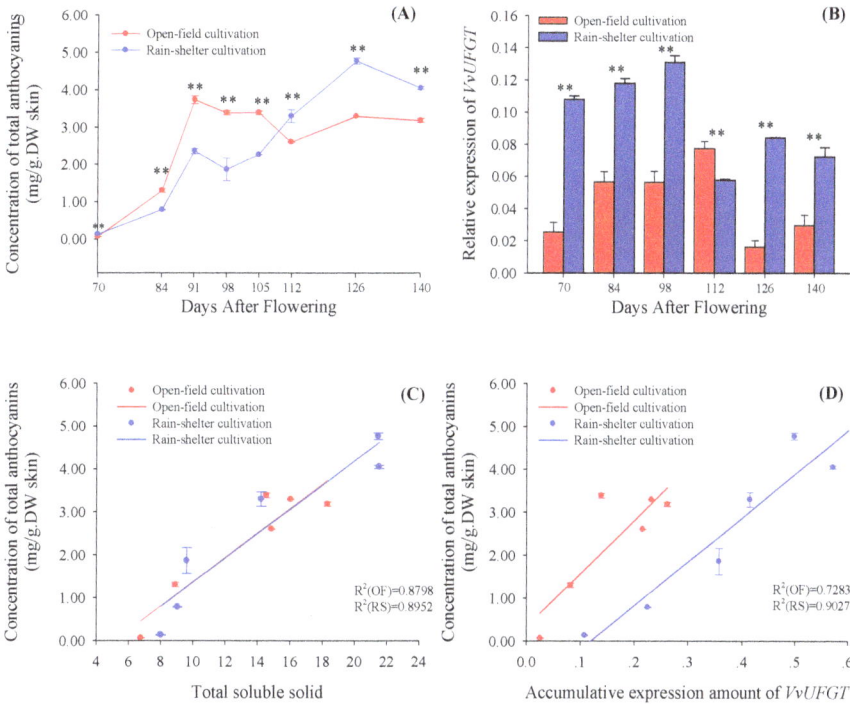

**Figure 3.** The variation of concentration of total anthocyanins (**A**) and relative expression of *VvUFGT* (**B**) in berry skins of vines grown under open-field cultivation (OF) and rain-shelter cultivation (RS). The linear correlations between evolution of total anthocyanins and total soluble solids (**C**), as well as between total anthocyanins and *VvUFGT* expression (**D**). Data are mean ± standard deviation. The symbol "**" or "*" above each set of columns represents that there are significant difference between rain-shelter cultivation and open-field cultivation by Duncun analysis at 0.01 level ($p < 0.01$) or 0.05 level ($p < 0.05$). $n = 6$.

The promotion in all kinds of anthocyanins was caused by both an up-regulation of *VvUFGT* expression and a continuous accumulation of sugar. In fact, the expression level of *VvUFGT* did not correspond to the production of anthocyanins in the early stage of berry maturation under the rain-shelter cultivation and the accumulation speed of anthocyanin in the rain-shelter treated berries during véraison was lower than that in the open-field berries. These results suggest that during this period anthocyanin synthesis was still restricted by the production of upstream metabolites (like sugar). This also demonstrated a close correlation between the supplement of photo-assimilates and the accumulation of anthocyanins.

183

## 2.3. 3′5′-Substituted and 3′-Substituted Anthocyanins

In order to help to understand the influence of rain-shelter cultivation on the different branches of the anthocyanin synthetic pathway in grape berries, we grouped the detected anthocyanins into two types according to their B-ring substitutions: 3′5′-substituted anthocyanins produced from F3′5′H branch (also called delphinidin-type) and 3′-substituted anthocyanins from F3′H branch (also called cyanidin-type). Of the 19 anthocyanins identified in this study, there were eleven 3′5′-substituted anthocyanins, accounting for greater than 65% of total anthocyanin concentration, and eight 3′-substituted ones (Table 1). Like the variation of the total anthocyanin concentration, these two types of anthocyanins under rain-shelter cultivation showed lower concentrations from 70 to 105 DAF, but higher concentrations in the following maturity stages compared to those under open-field cultivation (Figure 4A,B). Moreover, the amplitude of variation of 3′-substituted anthocyanins was similar to that of 3′5′-substituted ones in post-véraison period, indicating that these were similar effects of rain-shelter cultivation on the two branch pathways of anthocyanin biosynthesis.

Generally, 3′5′-substituted and 3′-substituted anthocyanins are the downstream products of flavonoid metabolism, in which flavonoid 3′-hydroxylase (F3′H) and flavonoid 3′5′-hydroxylase (F3′5′H) lead to two branch pathways (Figure 1). The transcriptional expressions of *VvF3′H* and *VvF3′5′H* in the berry skins showed different trends under these two cultivation modes during the ripening period (Figure 4C,D). The relative expression increase of *VvF3′H* was observed in the berries under the rain-shelter cultivation. However, the open-field cultivated grape berries showed a decreasing trend on the expression of *VvF3′H* from véraison through ripening, and then was recovered at the post-maturity period. The relative expression of *VvF3′5′H* was enhanced during the ripening stages in the berries under both the open-field and rain-shelter cultivations.

It should be observed that the evolution of 3′-substituted and 3′5′-substituted anthocyanins along with berry maturation were roughly accompanied with the accumulative expression pattern of *VvF3′H* and *VvF3′5′H* under both of the cultivations, respectively (Figure 4E,F). Comparably, the relationship between the accumulation of anthocyanins and the expression of the genes tended to be closer in the berries under the rain-shelter cultivation than the open-field cultivation. These observations suggested that rain-shelter cultivation might promote the synthetic branches of both 3′5′-substitued and 3′-substitued anthocyanins in grape berries. The results indicated that the rain-shelter cultivation can promote the two branches of anthocyanin biosynthesis to a similar extent.

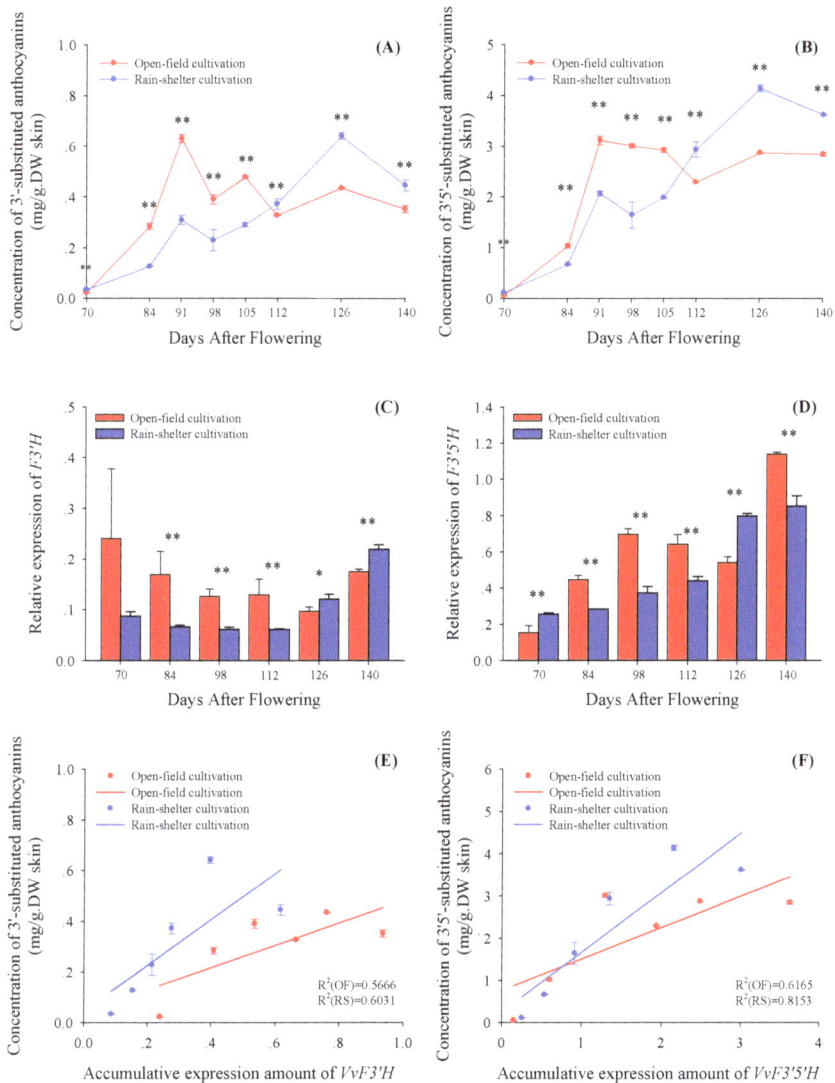

**Figure 4.** The variation of concentration of 3'-substituted anthocyanins (**A**), 3'5'-substituted anthocyanins (**B**), relative expression of *VvF3'H* (**C**) and *VvF3'5'H* (**D**) in berry skins of vines grown under open-field cultivation (OF) and rain-shelter cultivation (RS) in the process of maturing. And the correlation between content of total 3'-substituted anthocyanins and accumulative expression amount of *VvF3'H* (**E**), as well as between 3'5'-substituted anthocyanins and *VvF3'5'H* expression (**F**). Data are mean ± standard deviation. The symbol "**\*\***" or "**\***" above each set of columns represents that there are significant differences between rain-shelter cultivation and open-field cultivation by Duncun analysis at 0.01 level ($p < 0.01$) or 0.05 level ($p < 0.05$). $n = 6$.

*2.4. Acylating and Methylating Modification of 3'5'-Substituted and 3'-Substituted Anthocyanins*

Regarding the acylation on glycosyl residues of anthocyanins, the compounds can be divided into two kinds: acylated and non-acylated anthocyanins. Compared with homologous non-acylated anthocyanins, the acylated ones have a better stability in water [21]. Under these two cultivation modes, the concentration of acylated anthocyanins accounted for about 40%, most of which were acetylated. The proportion presented a decline initially, followed by an increase. Besides the acylated delphindin-type anthocyanins (3'5'-substituted), the other three kinds of anthocyanins had similar accumulation patterns (Figure 5A–D). Under the open-field cultivation, the highest concentration appeared at 91 DAF, followed by a big reducing concentration by 126 DAF. However, under rain-shelter cultivation the accumulation increased slowly and reached the highest at 126 DAF. The four kinds of anthocyanins (acylated and non-acylated cyanidin-type anthocyanins and delphindin-type anthocyanins, respectively) under the rain-shelter cultivation had lower levels than those under the open-field during 70–105 DAF, but showed significantly higher levels during the following ripening time.

Non-acylated delphindin-type anthocyanins showed the highest concentration and the greatest change, which played the primary role in leading to a higher concentration of total anthocyanins under the rain-shelter cultivation. Combined with the results in Table 1, it was found that the main compounds causing the great change of 3'5'-substitued anthocyanins under rain-shelter cultivation were malvidin-3-*O*-(6-*O*-acetyl)-glucoside and malvidin-3-*O*-glucoside.

The methylation of the B-ring on anthocyanin can bring down the chemical activity of the hydroxyl groups [22], and reduce the activity of hydroxyl groups of flavonoid substances [23]. The 3'-methylation and 3',5'-methylation of the B-ring can also increase the wavelength of anthocyanin maximum absorption. Figure 5E–H show the trends of the effect of cultivation modes on the accumulation of four kinds of anthocyanins (methylated/non-methylated of cyanidin-type/delphindin-type). These trends were similar with that of the impact on the total concentration of anthocyanins. That is, the rain-shelter cultivation reduced the accumulation of these four kinds of anthocyanins during 70–105 DAF, but enhanced the accumulation at the following ripening stages.

Acylation, especially acetylation, can significantly increase the stability of anthocyanins. The absorbing wavelength of acylated anthocyanins is blue shifted relative to the corresponding non-acylated ones, owing to the formation of fold structure inside aromatic nucleus molecules [21]. Methylated anthocyanins also provide wine with redder colorations [24]. In this study, the berries under the rain-shelter cultivation had higher concentration of acylated anthocyanins at the technological harvest and post-maturity dates. This indicated that this cultivation mode in the experimental region could help to improve the quality of grape berry and wine. In addition, the previous studies suggested that grapes under rain-shelter cultivation were suggested to be late-harvested because of the delay of ripeness and sugar accumulation [11,25]. In regions with grape growing season such as Penglai, late harvest under rain-shelter cultivation is not required.

## 2.5. Climate Characters

The meteorological data of 2010 in Penglai region is shown in Figure 6. During véraison (70–98 DAF), the average relative humidity was above 80% (Figure 6B), while the high temperature appeared at the same period (Figure 6D). There were four heavy rain episodes during véraison, at the end of véraison, at 127 DAF, and at the post-maturity stage. Rainfall was accompanied by an increase of air humidity, a decrease of diurnal temperature, and a lack of sunshine. The daily average temperature declined gradually, from above 25 °C to around 10 °C. Taken together, there were two brief periods during time series from véraison through ripening, DAF 91 to 98 and 112 to 126, which were similar with "drier with cool-warm climate", which contributed to the promotion of anthocyanin concentration, with higher temperature and sunshine hours, and lower rainfall and relative humidity [26].

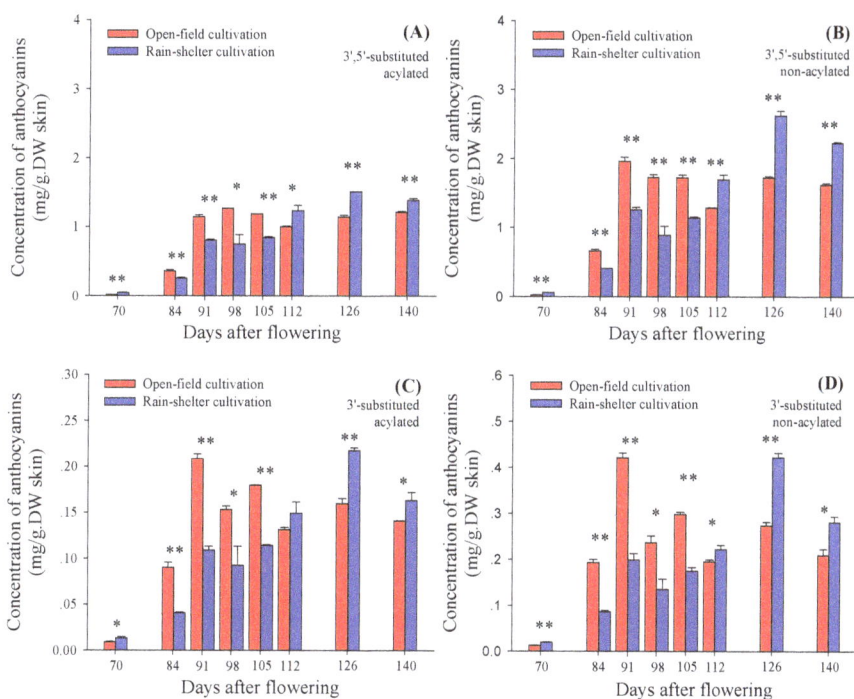

**Figure 5.** *Cont.*

187

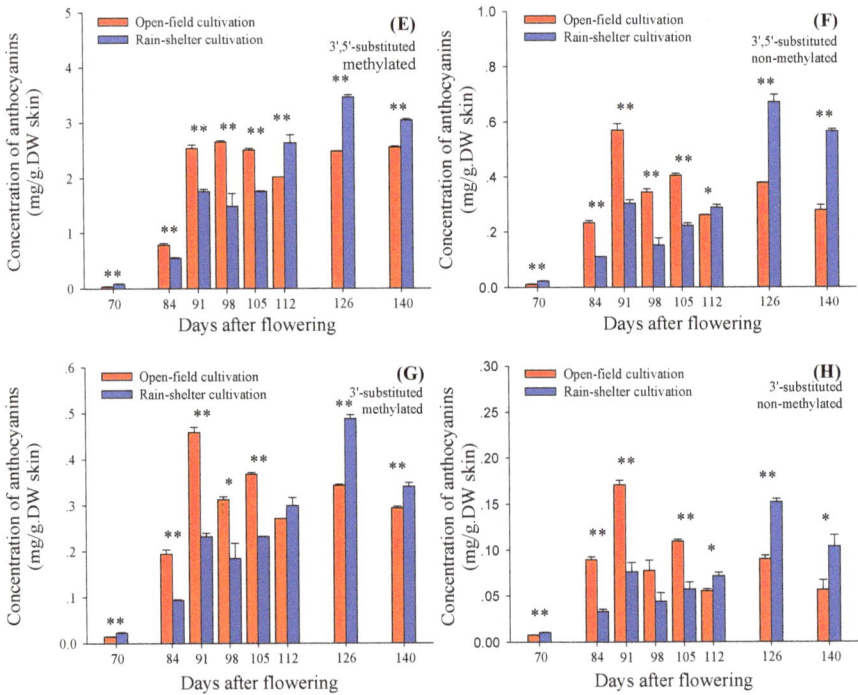

**Figure 5.** The variation in the concentrations of different kinds of anthocyanins in berry skins of vines grown under open-field cultivation and rain-shelter cultivation during berry maturation. These compounds are divided into acylated (**A**) and non-acylated (**B**) anthocyanins from F3'5'H branch (3',5'-substituted) as well as acylated (**C**) and non-acylated (**D**) anthocyanin from F3'H branch (3'-substituted). Another classification is made according to methylation in 3'- or 3',5'-position(s) of (**E**) and non-methylation (**F**) anthocyanins from F3'5'H branch, as well as methylation in 3'-position of (**G**) and non-methylation (**H**) anthocyanins from F3'H branch, respectively. Data are mean ± standard deviation. The symbol "**" or "*" above each set of columns represents that there are significant differences between rain-shelter cultivation and open-field cultivation by Duncun analysis at 0.01 level ($p < 0.01$) or 0.05 level ($p < 0.05$). $n = 6$.

188

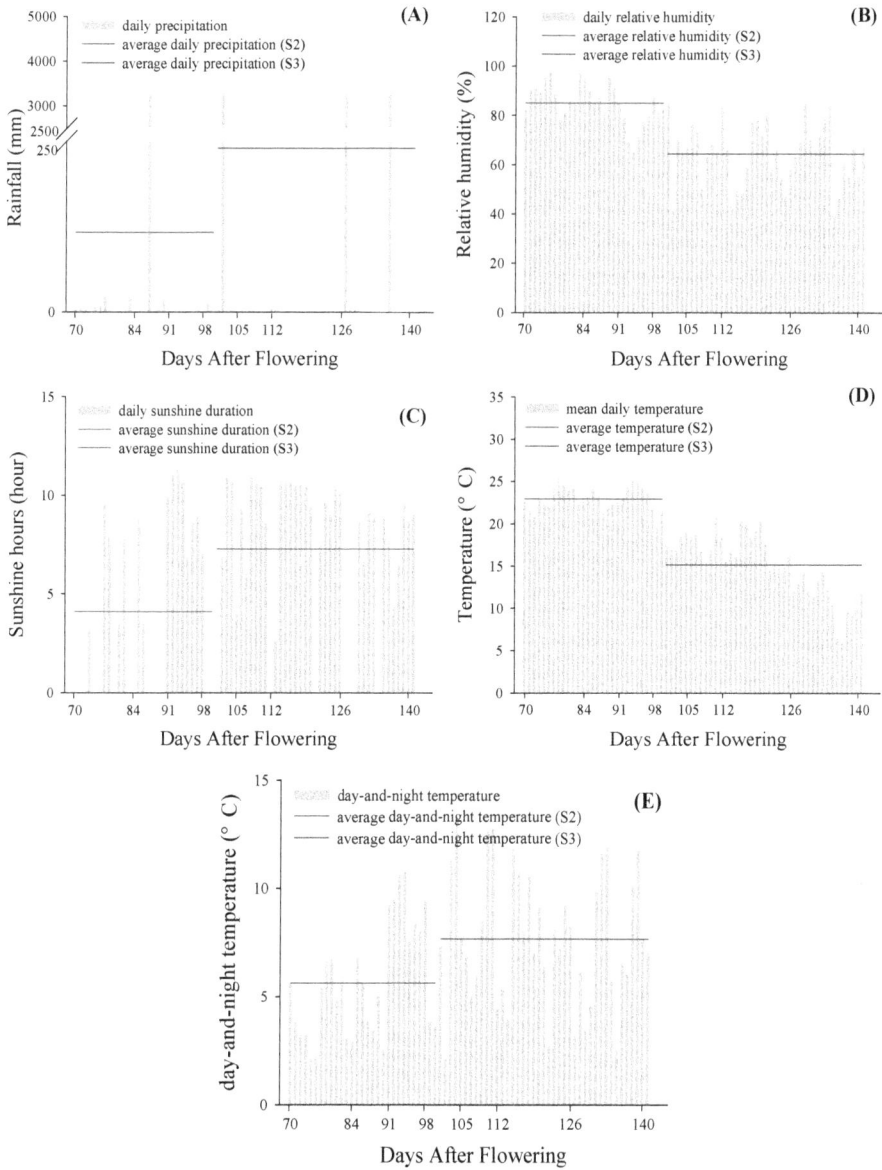

**Figure 6.** Rainfall (**A**), relative humidity (**B**), sunshine hours (**C**), mean daily temperature (**D**) and day-and-night temperature (**E**) during veraison (stage 2) and maturation period (stage 3) in 2010, Penglai region, Shandong Province, China.

189

## 3. Experimental Section

### 3.1. Field Treatment and Sample Collection

The field work was conducted during the grape growth season of 2010 in a ten-year-old commercial vineyard of own-rooted *Vitis vinifera* L. Cabernet Sauvignon in Penglai ($37°48'N$; $120°45'E$), located in the east coast of Shandong Province, China. The experimental plot comprised fifteen 30-meter-length rows (40 vines per row) oriented north-south. The inter-rows had 2.5 m width. Vines were double-trunked, trained to a bilateral cordon at 0.8 m aboveground, spur-pruned annually, and covered with soil to assure overwinter protection. Shoots were trained upwards and each vine carried *ca.* 20 clusters with uniform management. Experimental rows were selected from these 12 center rows of the plot, and divided into six groups to create three rain-shelter treatment replicates and three open-field control replicates (two rows per replicate). All the experimental units were treated using similar production management practices.

The rain-shelter roof with a colorless polyethylene (PE) film was set up at 14th day after flowering (DAF, 28 June). For each row, rain shelter frame, with a height of 1.85 m from the ground, a width of 1.1 m, and an arc length of 1.25 m, was designed as a jackroof to eliminate the rain and provide the natural ventilation. The rows without the rain-shelter treatment were used as the control (also called open-field cultivation) in the same vineyard. At least one control row was left between the rain-shelter treated rows.

Sampling was carried out on the indicated dates. In this region, "Cabernet sauvignon" grape berries are harvested generally at about 120 DAF. In the present vintage, the grape berries started and completed coloring, and reached the technological harvest at 70 DAF, 98 DAF, and 126 DAF, respectively. To investigate whether the rain-shelter cultivation delayed the maturity of the grape berry, we harvested the grape berries by 140 DAF. Sampling time was fixed at 10:00–11:00 in the morning. The samples were placed in foam boxes and transported to the laboratory within two hours. Any physically injured, abnormal, or infected berries were removed. Each time, about 300 berries were collected from at least 100 clusters. Of these, 50 fresh berries were used for the analyses of soluble solid content and pH, whereas the rest of the samples were frozen in liquid nitrogen and stored at $-80$ °C. The berries were peeled and the skins were ground in liquid nitrogen to the fine powder, and then lyophilized for 24 h at $-50$ °C using an LGJ-10 vacuum freeze-dryer (Vibra-Schultheis, Offenbach am Main, Germany) prior to the analysis of anthocyanins.

## 3.2. Determination of Physicochemical Parameters

The soluble solids and pH value in the grape berries were determined according to a published method with minor modifications [27]. After squeezing the berries, the resultant juice was filtered through clean cheesecloth and decanted into a clean centrifuge tube. The filtrate was used for determination of the soluble solids and pH value. The percent soluble solids (Brix) was measured using a digital handheld pocket Brix refractometer (PAL-2, ATAGO, Tokyo, Japan), whereas pH value was analyzed using a pH meter (FE20, Mettler Toledo, Greifensee, Switzerland).

## 3.3. Extraction and Determination of Anthocyanins from Berry Skins

Anthocyanins in the grape skins were extracted according to the method described by Liang *et al.* [28]. Briefly, 0.5 g of the dry skin powder was extracted in methanol (10 mL) containing 2% formic acid. The extract was ultrasonicated for 10 min, followed by shaking in the dark for 30 min at 25 °C. Afterwards, the resultant mixture was centrifuged at 8000× $g$ for 10 min and the supernatant was collected. The residues were re-extracted four more times using the same procedure. Finally, all the supernatants were pooled into a distilling flask. Methanol was removed using a rotary evaporator (SY-2000, Shanghai Yarong Biochemistry Factory, Shanghai, China) at 28 °C, and the residues were re-dissolved in 10 mL of 10% ethanol solution (pH 3.7). The extracts yielded were filtered through 0.45 μm filters (cellulose acetate and nitrocellulose, CAN) and directly used for HPLC analysis. The sample from a replicate was carried out in two independent extractions. As a result, each data point in the Tables and Figures represented the average of six values consisting of three biological replicates multiplied by two extraction replicates.

The analysis of anthocyanins was performed according to a previously published method [28] using an Agilent 1100 series LC-MSD trap VL instrument (Agilent Technologies, Santa Clara, CA, USA), equipped with a diode array detector (DAD) and a reverse-phase column (Kromasil C18, 250 × 4.6, 5 μm). The injection volume was 30 μL with a flow rate of 1.0 mL/min. The mobile phase was comprised of solvent A (water/acetonitrile/formic acid, 92:6:2, v/v) and solvent B (water/acetonitrile/formic acid, 44:54:2, v/v/v). The gradient elution was applied as follows: 10% B for 1 min, from 10% to 25% B for 17 min, isocratic 25% B for 2 min, from 25% to 40% B for 10 min, from 40% to 70% B for 5 min, from 70% to 100% B for 5 min. The column temperature was set at 50 °C and the detection wavelength on DAD was 525 nm. MS conditions were described as follows: Electro-spray ionization (ESI) interface, positive ion mode; Nebulizer pressure, 35 psi; dry gas flow rate, 10 mL/min; dry gas temperature, 350 °C, and mass scan mode, all mass scan from $m/z$ 100–1000.

Five individual anthocyanins, including dephinidin-3-*O*-glucoside, petunidin-3-*O*-glucoside, malvidin-3-*O*-glucoside, cyanidin-3-*O*-glucoside, and peonidin-3-*O*-glucoside, were identified by comparing mass spectra and order of retention time with the commercially available standards. The identification of the remaining anthocyanins was achieved by analyzing the deprotonated ion and product ion of these compounds. The *cis* and *trans* isomers of the coumaroylates for peonidin-3-O-glucoside and malvidin-3-O-glucoside were distinguished by their elution time and concentrations. The *cis* isomer was generally eluted earlier on a reverse phase HPLC column and was present in a lower level compared to the *trans* isomer according to previously reported data [29–35].

### 3.4. Analysis of Transcript Level by Real-Time PCR

The total RNA of the grape berry skins was isolated using Universal Plant Total RNA Extraction Kit (Cat. # RP3301, BioTeke Co., Beijing, China), and digested with DNaseI (Code 2212, TaKaRa, Tokyo, Japan) to remove genomic DNA. The quality of RNA was verified by the existence of intact ribosomal bands following agarose gel electrophoresis and the absorbance ratios (A260/A280) of 1.8–2.0. The reverse transcription procedure followed a published method [36]. For cDNA synthesis, 2 μL of total RNA was reverse-transcribed in a 25 μL reaction mixture using M-MLV Reverse Transcriptase (M1707, Promega, Madison, WI, USA) and Oligo d(T)$_{18}$ Primen (Code 3806, TaKaRa) with the manufacturer's instruction. The synthesized cDNA was quantified and all the tested DNA samples were adjusted to the same concentration.

Relative expression of genes was measured by real-time PCR using SYBR® PremixEX Taq$^{TM}$ II (Code DRR081A, TaKaRa) on a 7300 Real Time PCR System (Applied Biosystems, Foster, CA, USA). PCR reaction mixture (20 μL) was comprised of 10 μL 2 × SYBR®PremixEx Taq$^{TM}$, 8.5 μL ddH$_2$O (TIANGEN, Beijing, China), 0.4 μL cDNA, 0.4 μL 50 × ROX Reference Dye II, and 0.7 μL primers (forward and reverse primers mixture, 10 μmol/L). The template cDNA was denatured at 95 °C for 30 s followed by 40 cycles of amplification at 95 °C for 10 s, 60 °C for 31 s, and a melt cycle from 60 °C to 95 °C. The sequences of the primers used for real-time PCR were referred from the previous studies [36–38] (Table 2). The specificity of the primers was verified by an agarose gel electrophoresis with one specific ribosomal band and dissociation curve with one specific peak. Quantification was normalized to *VvUbiquitin1* fragments amplified in the same conditions according to Bogs *et al.* [39]. There was identical amplification efficiency between target genes and internal reference gene. Each grape sample was conducted from three independent RNA extraction replicates to produce three RNA samples. Each RNA sample was performed from two technological replicates in analysis of real-time PCR.

Table 2. The primers of genes used in the Real-Time PCR assays.

| Genes | Genbank Accession | Primer Sequences (5'–3') | Size of PCR Product (bp) |
|---|---|---|---|
| *VvF3'H* | AJ880357 | F: CCAAGTTTTCGGGAAGTAAATG<br>R: TACCCCTTGAGAATCATCGTTT | 171 |
| *VvF3'5'H* | AJ880356 | F: GCATGGATGCAGTTAAGTAGAAAA<br>R: ATATGGCTTGGTGGTAGAATGAAACGA | 113 |
| *VvUFGT* | AF000372 | F: GGGATGGTAATGGCTGTGG<br>R: ACATGGGTGGAGAGTGAGTT | 253 |
| *VvUbiquitin* | BN000705 | F: GTGGTATTATTGAGCCATCCTT<br>R: AACCTCCAATCCAGTCATCTAC | 182 |

*3.5. Statistical Analysis*

Statistical analysis was performed by SPSS 11.5 software (Chicago, IL, USA). Duncun analysis was used to assess statistically significant differences in the content of various anthocyanins and the transcription levels of genes between the treated groups and the control. Sigma Plot 10.0 (Systat Software Inc., Richmond, CA, USA) was used to draft the graph. Each data point, expressed as milligram equivalent of the respective standard per kilogram of dried grape skin, was the average of three replications, $n = 6$.

## 4. Conclusions

Integrating the data from various analyses shown above, we found pattern changes between grape berry anthocyanins produced by rain-shelter cultivation and open-field cultivation. The life of green functional leaves was visibly prolonged under rain-shelter cultivation conditions. The rain-shelter cultivation did not alter the compositional proportion of anthocyanins, but increased the concentrations of almost all the anthocyanin components detected in the ripening berries. Correspondingly, the expression of *VvF3'H*, *VvF3'5'H* and *VvUFGT* was greatly up-regulated and this rising trend was kept until berry maturation. The accumulation period of both sugar and anthocyanins was prolonged by rain-shelter cultivation, which was strongly consistent with the delay of leaf senescence. In regions with grape growing season such as Penglai, the rain-shelter cultivation could increase wine quality of grape berries to a certain extent.

**Acknowledgments:** The authors want to thank Shangeli-La Winery, Shandong Province, China for their kind help with grape sampling. Thanks also go to the National Natural Science Foundation of China (Grant No. 30971980) and the China Agriculture Research System for Grape Industry (CARS-30) for their funding support.

Author Contributions: Q.-H.P., X.-X.L. and J.W. conceived and designed research; X.-X.L. and F.H. performed research and analyzed the data; X.-X.L. wrote the paper and Z.L. edited the manuscript. All authors read and approved the final manuscript.

Conflicts of Interest: The authors declare no conflict of interest.

## References

1. Staff, W.-S. *Chinese wine*. Available online: http://www.wine-searcher.com/regions-china (accessed on 3 March 2013).
2. Meng, J.-F.; Ning, P.-F.; Xu, T.-F.; Zhang, Z.-W. Effect of rain-shelter cultivation of *Vitis vinifera* cv. Cabernet gernischet on the phenolic profile of berry skins and the incidence of grape diseases. *Molecules* **2012**, *18*, 381–397.
3. Chavarria, G.; dos Santos, H.P.; Sônego, O.R.; Marodin, G.A.B.; Bergamaschi, H.; Cardoso, L.S. Incidence of diseases and needs of control in overhead covered grapes. *Rev. Bras. Frutic.* **2007**, *29*, 477–482.
4. Cortell, J.M.; Halbleib, M.; Gallagher, A.V.; Righetti, T.L.; Kennedy, J.A. Influence of vine vigor on grape (*Vitis vinifera* L. Cv. Pinot Noir) anthocyanins. 1. Anthocyanin concentration and composition in fruit. *J. Agric. Food Chem.* **2007**, *55*, 6575–6584.
5. Tarara, J.M.; Lee, J.; Spayd, S.E.; Scagel, C.F. Berry temperature and solar radiation alter acylation, proportion, and concentration of anthocyanin in merlot grapes. *Am. J. Enol. Viticult.* **2008**, *59*, 235–247.
6. Cheynier, V. Polyphenols in foods are more complex than often thought. *Am. J. Clin. Nutr.* **2005**, *81*, 223S–229S.
7. Berli, F.J.; Fanzone, M.; Piccoli, P.; Bottini, R. Solar UV-B and ABA are involved in phenol metabolism of *Vitis vinifera* L. Increasing biosynthesis of berry skin polyphenols. *J. Agric. Food Chem.* **2011**, *59*, 4874–4884.
8. Tangolar, S.G.; Tangolar, S.; Blllr, H.; Ozdemir, G.; Sabir, A.; Cevlk, B. The effects of different irrigation levels on yield and quality of some early grape cultivars grown in greenhouse. *Asian J. Plant Sci.* **2007**, *6*, 643–647.
9. Fanizza, G.R.L. The effect of vineyard overhead plastic sheet covering on some morphological and physiological characteristics in the table grape cv. Regina dei Vigneti *Vitis vinifera* L. *Agric. Mediterr.* **1991**, *121*, 239–243.
10. Júnior, M.J.P.; Hernandes, J.L.; de Souza Rolim, G.; Blain, G.C. Microclimate and yield of "niagara rosada" grapevine grown in vertical upright trellis and "y" shaped under permeable plastic cover overhead. *Rev. Bras. Frutic.* **2011**, *33*, 511–518.
11. Chavarria, G.; dos Santos, H.P.; Zanus, M.C.; Marodin, G.A.B.; Zorzan, C. Plastic cover use and its influences on physical-chemical characteristics in must and wine. *Rev. Bras. Frutic.* **2011**, *33*, 809–815.
12. Guo, Y.-P.; Guo, D.-P.; Zhou, H.-F.; Hu, M.-J.; Shen, Y.-G. Photoinhibition and xanthophyll cycle activity in bayberry (*Myrica rubra*) leaves induced by high irradiance. *Photosynthetica* **2006**, *44*, 439–446.
13. Sasaki, H.; Yano, T.; Yamasaki, A. Reduction of high temperature inhibition in tomato fruit set by plant growth regulators. *JARQ-Jpn. Agr. Res. Q.* **2005**, *39*, 135–138.

14. Yoon, J.; Ji, J.; Lim, S.; Lee, K.; Kim, H.; Jeong, H.; Lee, J. Changes in selected components and antioxidant and antiproliferative activity of peppers depending on cultivation. *J. Korean Soc. Food Sci. Nutr.* **2010**, *39*, 731–736.

15. Larsson, S.; Górny, A. Grain yield and drought resistance indices of oat cultivars in field rain shelter and laboratory experiments. *J. Agron. Crop Sci.* **1988**, *161*, 277–286.

16. Kim, J.; Jo, J.; Kim, H.; Sub-Station, N.; Ryou, M.; Kim, J.; Hwang, H.; Hwang, Y. Growth and fruit characteristics of blueberry "northland" cultivar as influenced by open field and rain shelter house cultivation. *J. Biol. Environ. Control* **2011**, *20*, 387–393.

17. Polat, A.A.; Durgac, C.; Caliskan, O. Effect of protected cultivation on the precocity, yield and fruit quality in loquat. *Sci. Hortic.* **2005**, *104*, 189–198.

18. Heraud, P.; Beardall, J. Changes in chlorophyll fluorescence during exposure of dunaliella tertiolecta to UV radiation indicate a dynamic interaction between damage and repair processes. *Photosynth. Res.* **2000**, *63*, 123–134.

19. Pfündel, E.E. Action of UV and visible radiation on chlorophyll fluorescence from dark-adapted grape leaves (*Vitis vinifera* L.). *Photosynth. Res.* **2003**, *75*, 29–39.

20. Boss, P.K.; Davies, C.; Robinson, S.P. Anthocyanin composition and anthocyanin pathway gene expression in grapevine sports differing in berry skin colour. *Aust. J. Grape Wine Res.* **1996**, *2*, 163–170.

21. Bridle, P.; Timberlake, C. Anthocyanins as natural food colours-selected aspects. *Food Chem.* **1997**, *58*, 103–109.

22. Kim, B.; Lee, H.; Park, Y.; Lim, Y.; Ahn, J.-H. Characterization of an *O*-methyltransferase from soybean. *Plant Physiol. Biochem.* **2006**, *44*, 236–241.

23. Suelves, M.; Puigdomènech, P. Specific mrna accumulation of a gene coding for an *O*-methyltransferase in almond (*Prunus amygdalus*, batsch) flower tissues. *Plant Sci.* **1998**, *134*, 79–88.

24. Bakker, J.; Timberlake, C.F. Isolation, identification, and characterization of new color-stable anthocyanins occurring in some red wines. *J. Agric. Food Chem.* **1997**, *45*, 35–43.

25. Chavarria, G.; dos Santos, H.P.; Zanus, M.C.; Marodin, G.A.B.; Chalaça, M.Z.; Zorzan, C. Grapevine maturation of moscato giallo under plastic cover. *Rev. Bras. Frutic.* **2010**, *32*, 151–160.

26. Li, Z.; Pan, Q.; Jin, Z.; Mu, L.; Duan, C. Comparison on phenolic compounds in *Vitis vinifera* cv. Cabernet Sauvignon wines from five wine-growing regions in china. *Food Chem.* **2011**, *125*, 77–83.

27. Spayd, S.E.; Tarara, J.M.; Mee, D.L.; Ferguson, J. Separation of sunlight and temperature effects on the composition of *Vitis vinifera* cv. Merlot berries. *Am. J. Enol. Viticult.* **2002**, *53*, 171–182.

28. Liang, N.-N.; Pan, Q.-H.; He, F.; Wang, J.; Reeves, M.J.; Duan, C. Phenolic profiles of *Vitis davidii* and *Vitis quinquangularis* species native to China. *J. Agric. Food Chem.* **2013**, *61*, 6016–6027.

29.  Revilla, E.; García-Beneytez, E.; Cabello, F. Anthocyanin fingerprint of clones of tempranillo grapes and wines made with them. *Aust. J. Grape Wine Res.* **2009**, *15*, 70–78.

30.  Wang, H.; Race, E.J.; Shrikhande, A.J. Anthocyanin transformation in Cabernet Sauvignon wine during aging. *J. Agric. Food Chem.* **2003**, *51*, 7989–7994.

31.  Villiers, A.D.; Vanhoenacker, G.; Majek, P.; Sandra, P. Determination of anthocyanins in wine by direct injection liquid chromatography-diode array detection-mass spectrometry and classification of wines using discriminant analysis. *J. Chromatogr. A* **2004**, *1054*, 195–204.

32.  Han, F.-L.; Zhang, W.-N.; Pan, Q.-H.; Zheng, C.-R.; Chen, H.-Y.; Duan, C.-Q. Principal component regression analysis of the relation between cielab color and monomeric anthocyanins in young Cabernet Sauvignon wines. *Molecules* **2008**, *13*, 2859–2870.

33.  García-Beneytez, E.; Cabello, F.; Revilla, E. Analysis of grape and wine anthocyanins by HPLC-MS. *J. Agric. Food Chem.* **2003**, *51*, 5622–5629.

34.  Downey, M.O.; Rochfort, S. Simultaneous separation by reversed-phase high-performance liquid chromatography and mass spectral identification of anthocyanins and flavonols in Shiraz grape skin. *J. Chromatogr. A* **1201**, 43–47.

35.  Núñez, V.; Monagas, M.; Gomez-Cordoves, M.; Bartolomé, B. *Vitis vinifera* L. cv. Graciano grapes characterized by its anthocyanin profile. *Postharvest Biol. Technol.* **2004**, *31*, 69–79.

36.  Zhang, Z.-Z.; Che, X.-N.; Pan, Q.-H.; Li, X.-X.; Duan, C.-Q. Transcriptional activation of flavan-3-ols biosynthesis in grape berries by UV irradiation depending on developmental stage. *Plant Sci.* **2013**, *208*, 64–74.

37.  Bogs, J.; Ebadi, A.; McDavid, D.; Robinson, S.P. Identification of the flavonoid hydroxylases from grapevine and their regulation during fruit development. *Plant Physiol.* **2006**, *140*, 279–291.

38.  Deluc, L.; Bogs, J.; Walker, A.R.; Ferrier, T.; Decendit, A.; Merillon, J.-M.; Robinson, S.P.; Barrieu, F. The transcription factor VvMYB5b contributes to the regulation of anthocyanin and proanthocyanidin biosynthesis in developing grape berries. *Plant Physiol.* **2008**, *147*, 2041–2053.

39.  Bogs, J.; Downey, M.O.; Harvey, J.S.; Ashton, A.R.; Tanner, G.J.; Robinson, S.P. Proanthocyanidin synthesis and expression of genes encoding leucoanthocyanidin reductase and anthocyanidin reductase in developing grape berries and grapevine leaves. *Plant Physiol.* **2005**, *139*, 652–663.

**Sample Availability:** *Sample Availability*: Not available.

# Stability of Anthocyanins from Red Grape Skins under Pressurized Liquid Extraction and Ultrasound-Assisted Extraction Conditions

Ali Liazid, Gerardo F. Barbero, Latifa Azaroual, Miguel Palma and
Carmelo G. Barroso

**Abstract:** The stability of anthocyanins from grape skins after applying different extraction techniques has been determined. The following compounds, previously extracted from real samples, were assessed: delphinidin 3-glucoside, cyanidin 3-glucoside, petunidin 3-glucoside, peonidin 3-glucoside, malvidin 3-glucoside, peonidin 3-acetylglucoside, malvidin 3-acetylglucoside, malvidin 3-caffeoylglucoside, petunidin 3-*p*-coumaroylglucoside and malvidin 3-*p*-coumaroylglucoside (*trans*). The techniques used were ultrasound-assisted extraction and pressurized liquid extraction. In ultrasound-assisted extraction, temperatures up to 75 °C can be applied without degradation of the aforementioned compounds. In pressurized liquid extraction the anthocyanins were found to be stable up to 100 °C. The relative stabilities of both the glycosidic and acylated forms were evaluated. Acylated derivatives were more stable than non-acylated forms. The differences between the two groups of compounds became more marked on working at higher temperatures and on using extraction techniques with higher levels of oxygen in the extraction media.

Reprinted from *Molecules*. Cite as: Liazid, A.; Barbero, G.F.; Azaroual, L.; Palma, M.; Barroso, C.G. Stability of Anthocyanins from Red Grape Skins under Pressurized Liquid Extraction and Ultrasound-Assisted Extraction Conditions. *Molecules* **2014**, *19*, 21034–21043.

## 1. Introduction

The anthocyanins are a group of compounds that belong to the flavonoid family and these are of great interest in the food industry, mainly due to their colouring properties [1,2]. However, interest in these compounds has increased in recent years due to their antioxidant, anti-inflammatory, antiviral, antibacterial and even anticarcinogenic properties [3–5].

Anthocyanins are found in many plants but red grapes are of particular interest due to the presence of high concentrations of a wide variety of these compounds. The stability of anthocyanins is dependent on various factors, such as the structure, whether or not they are bound to sugars, the pH, light, the presence of ions and enzymes and, most importantly, temperature [6]. The levels of these compounds obtained from plants are often influenced by the conditions used for their extraction.

Various techniques have been proposed for the extraction of anthocyanins and these involve the use of methanol, ethanol, acetone, water, or mixtures of these solvents [7]. The addition of small quantities of hydrochloric or formic acid has been used as a way to improve the extraction outcome [8]. The techniques used for extraction from semi-solid or viscous samples (such as grape must) have mainly involved extraction by maceration with solvents [9], followed by liquid-liquid extraction or the use of solid phase extraction (SPE) as a step for cleaning and/or concentration of the extract. In any case, such techniques require long extraction times and have limited efficiency.

For the reasons outlined above, new and more efficient techniques are currently being used that enable a reduction in both the extraction time and the consumption of organic solvents. Pressurized Liquid Extraction (PLE) and Ultrasound-Assisted Extraction (UAE) are two techniques that can meet these requirements. In these cases, the first step should be to test the stability of the compounds to be extracted at different extraction temperatures used in these techniques in order to determine a reliable working range.

UAE has been used in the extraction of organic compounds from soil, plant tissues and packing materials [10]. Ultrasound has a mechanical effect that enables greater penetration of the solvents into the matrix and increases the surface contact between the solid and the liquid. In addition, the occurrence of cavitation leads to cell breakage and this can increase the speed of extraction. An application of UAE for the extraction of anthocyanins from red raspberries was found in a review of the literature [11]; however, studies on the stability of these compounds during the extraction process were not found.

PLE is a technique that is used to prepare samples when pressure and temperature are the main extraction variables in the design of fast methods for extraction from solid or semi-solid materials [12]. Pressure is used to increase the contact between the liquid and the sample and also to maintain the solvent in a liquid state when working at temperatures above its boiling point [13]. At these temperatures, breakage of the analyte-matrix bonds is facilitated. Additionally, the temperature can have a significant effect on the properties of the solvent and may lead to a change in its dielectric constant, thereby affecting the selectivity of the extraction. Applications of PLE for the determination of anthocyanins in grapes were not found in our review of the literature.

The aim of the study described here was to determine the effects of various possible extraction temperatures on the stability of anthocyanins from grape skins. UAE and PLE working conditions were therefore assessed using anthocyanins previously extracted from grape skins. The results were compared with those obtained previously on using microwave-assisted extraction (MAE).

## 2. Results and Discussion

For each extraction technique studied, a range of different temperatures was evaluated according to the working range available. It must be noted that extraction processes were not run because a solid matrix was not used, rather the investigation was carried out on liquid samples, *i.e.*, a standardized extract. The standardized extract was prepared using grape skins as the solid material and methanol as the solvent under very mild extraction conditions. The solid material contained all types of anthocyanins previously determined in grape skins. UAE and PLE conditions were applied to the standardized extracts at different working temperatures. The experiments were run in order to assess the stability of both glucosylated anthocyanins and their derivatives in the sample under given extraction conditions. Therefore, full recovery would be expected if degradation processes did not occur under the working conditions. On the other hand, recoveries of less than 100% with reference to the starting values in the extract would indicate degradation of anthocyanins during the experiments. All experiments were evaluated statistically in order to establish reliable results.

Temperatures above 75 °C were not studied in the ultrasound-assisted extraction as this would lead to a low level of reproducibility, principally due to the evaporation of water. As a consequence, it was considered that higher temperatures would not be applicable in practice, even if the anthocyanins were stable under such conditions. As far as PLE is concerned, temperatures were studied up to the point where clear degradation of the anthocyanins was found.

For each technique, the standardized extract from the grape skins was used as a reference at the same dilution as developed for each extraction technique. Two references were processed, one at the beginning and one at the end of each day on which the extractions were carried out with each technique. The average of the chromatographic areas of each of the peaks was obtained from the chromatogram and a value of 100 was assigned to the mean area of each.

One of the most likely degradation processes to occur under the extraction conditions is the breakage of the bonds between the anthocyanins and the different acids to which they are bound. As a consequence, the effects of the extractions on the anthocyanins under investigation were divided into two groups for evaluation: (i) glucosylated compounds and (ii) acylated derivatives of these compounds.

### 2.1. Ultrasound-Assisted Extraction

Extraction temperatures between 0 °C and 75 °C were evaluated. The presence of methanol as the solvent for the standardized extract prevented freezing of the sample during the extraction process at 0 °C. Temperatures above 75 °C were also assayed but, due to the high variability in the results and the need for the addition of further solvent during the experiment, these higher temperatures were not taken into account when analyzing the results. The average recoveries of the compounds quantified in the chromatograms are shown in Figure 1.

No statistically significant differences were found between the results obtained at the different temperatures studied. This finding implies that both the glucosylated compounds and their derivatives are stable under the conditions used in the ultrasound extraction up to 75 °C. Similar results were obtained for other phenolics [14], but results on the stability of anthocyanins under UAE working conditions have not been reported previously.

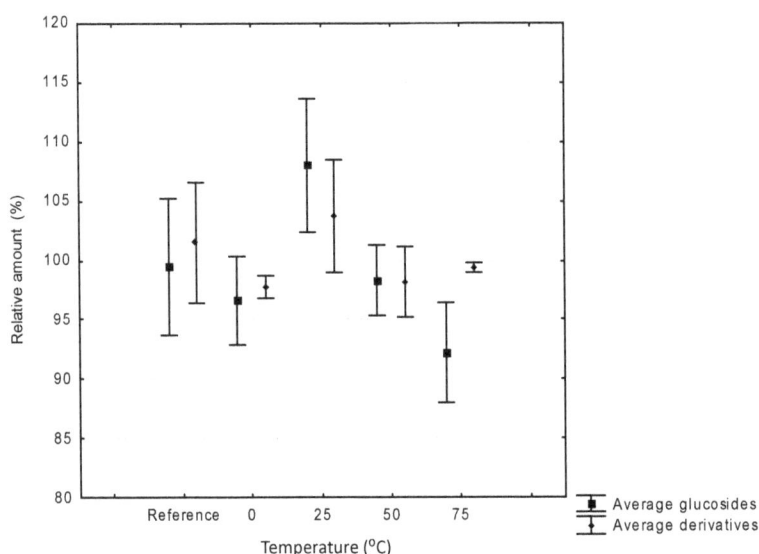

**Figure 1.** Average recoveries (mean ± standard deviation) of anthocyanins at different working temperatures (0, 25, 50 and 75 °C) and the reference in the ultrasound-assisted extraction system.

## 2.2. Pressurized Liquid Extraction

The PLE technique was used at temperatures from 75 to 125 °C. Lower temperatures were not tested because the aim of the work was to find methods that could, where appropriate, produce faster extractions than the more commonly used methods, meaning that it was essential that high extraction temperatures were used. The results, which are grouped into compound families, are shown in Figure 2. It can be seen from Figure 2 that both the glucosylated anthocyanins and their acylated derivatives are stable up to 100 °C. This means that extractions can be carried out up to this temperature without having an adverse effect on the stability of the anthocyanins. It must also be noted that significant differences were not found between the glucosides and their acylated derivatives in terms of their behaviour under these extraction conditions.

On working at temperatures around 100 °C dramatic degradation was observed for both glucosides and their acylated derivatives. Degradation levels between 40% and 50% were reached. The degradation observed at 125 °C is probably the result of oxidation reactions promoted by the very high temperature. It should be noted that conversion from acyl derivatives to the glucosyl forms was not observed. This result indicates that PLE would produce unreliable results if the system was operated above 100 °C for the extraction of anthocyanins.

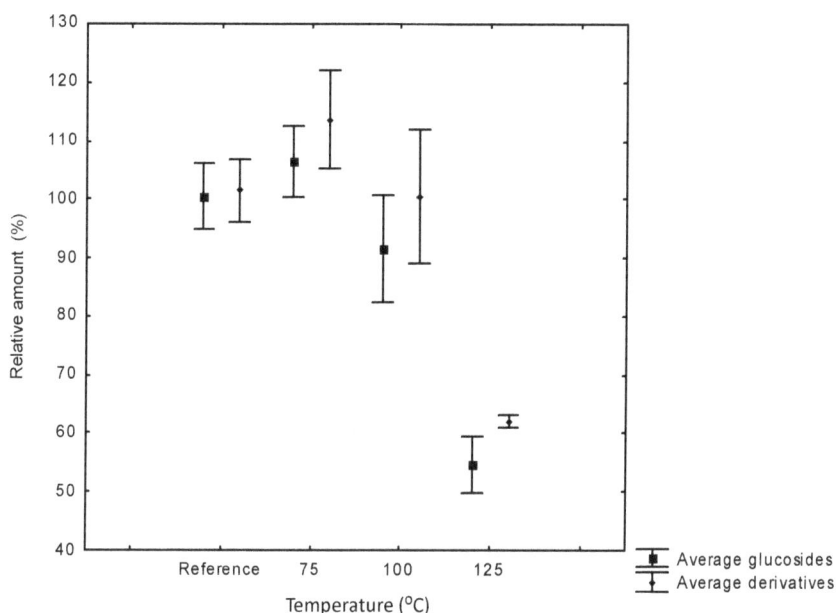

**Figure 2.** Average recoveries (mean ± standard deviation) of anthocyanins at different working temperatures (75, 100 and 125 °C) and reference sample in the pressurized liquid extraction system.

## 2.3. Effects on Individual Anthocyanins

In order to develop extraction methods for specific anthocyanins present in the grape extract, it would be of interest to determine the sensitivity to degradation of compounds on an individual basis, rather than globally, and for this reason an attempt was made to determine the differences between the different compounds analyzed. In this respect, significant differences have been described in the literature for related compounds.

The influence of the specific functional groups in the chemical structure on the stability during extractions can also be obtained. This information would be of interest when estimating the stability of other related compounds.

In order to achieve this goal, an analysis was carried out on the results obtained in two different systems, PLE and MAE, carried out at the same temperature and where partial degradation of the anthocyanins occurred, namely 125 °C. The MAE system was selected because the extraction under these conditions can be run at the same temperature as PLE but with a higher level of oxygen in the medium. This study would enable the effects of high oxygen levels to be evaluated. The results obtained are shown in Figure 3 and all of these compounds gave values that are significantly lower than that of the reference.

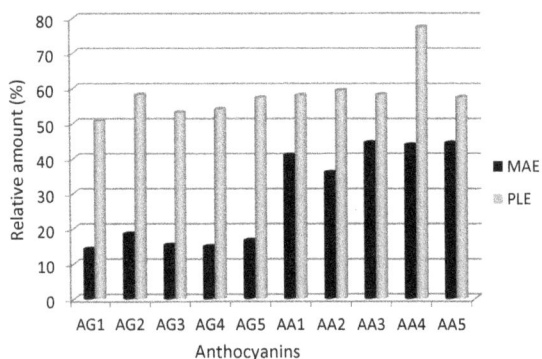

**Figure 3.** Recoveries of anthocyanins at 125 °C using microwave-assisted extraction and pressurized liquid extraction conditions. AG1–AG5: glycosylated anthocyanins, AA1–AA5 acylated anthocyanin derivatives. See Table 1 for a full identification of anthocyanins.

**Table 1.** Amounts of anthocyanins in the standardized extract.

| Anthocyanin | mg/L * |
|---|---|
| Delphinidin 3-glucoside (AG1) | 26.8 |
| Cyanidin 3-glucoside (AG2) | 11.2 |
| Petunidin 3-glucoside (AG3) | 36.1 |
| Peonidin 3-glucoside (AG4) | 130.0 |
| Malvidin 3-glucoside (AG5) | 381.2 |
| Peonidin 3-acetylglucoside (AA1) | 14.5 |
| Malvidin 3-acetylglucoside (AA2) | 46.4 |
| Malvidin 3-caffeoylglucoside (AA3) | 7.9 |
| Petunidin 3- p-coumaroylglucoside (AA4) | 7.1 |
| Malvidin 3- p-coumaroylglucoside (trans) (AA5) | 61.6 |

*: Expressed as malvidin-3-glucoside.

Comparison of the five glycosylated compounds (AG1–AG5) shows that, regardless of their chemical structure, *i.e.*, the ring substituents, their sensitivity to degradation is practically the same both under the high oxidation (MAE at 125 °C) and the low oxidation (PLE 125 °C) conditions. On the other hand, acylated derivatives of the glucosides were generally found to be more stable than the glucosides themselves when the more oxidative conditions were used (MAE 125 °C), whereas the degradation percentages were similar when the only factor favourable to degradation was temperature. In any case, as in the case of glucosides, differences were not observed between the different susceptibilities to degradation for the five acyl derivatives studied individually, thus showing that this behaviour is also independent of chemical substitution in the aromatic ring. The nature of the chemical substituents in the aromatic rings does not influence the stability of anthocyanins during the extraction process, *i.e.*, the maximum extraction temperatures found in this study could be applied to other anthocyanins present in different samples even in cases where different functional groups are present in the aromatic rings.

## 3. Experimental Section

### 3.1. Samples

The grape skins were obtained from red grapes of the Tintilla de Rota variety grown in Jerez (Spain). Levels for anthocyanins in the grape skins (mg Kg$^{-1}$ FW) were as it follows: delphinidin 3-glucoside: 1.26; cyanidin 3-glucoside: 0.87; petunidin 3-glucoside: 2.31; peonidin 3-glucoside: 4.33; malvidin 3-glucoside: 16.99; peonidin 3-acetylglucoside: 1.21; malvidin 3-acetylglucoside: 5.52; malvidin 3-caffeoylglucoside: 0.66; petunidin 3-*p*-coumaroylglucoside: 1.96; malvidin 3-*p*-coumaroylglucoside (*trans*): 12.46. The skins were separated from the seeds and pomace, and then milled in a coffee grinder for 2 min, in bursts of 15 s in order to avoid sample heating. The sample was stored at –20 °C prior to extraction. A standardized extract with a known anthocyanin concentration was used in order to evaluate the stability accurately. This extract was prepared because standards for most of the anthocyanins were not available and, as a consequence, they had to be obtained from real samples in order to assess their stabilities. This extract was obtained by solid-liquid extraction of the ground grape skin in an ultrasonic bath. In order to obtain a sufficient amount of extract, approximately 100 g of ground grape skin was steeped in 250 mL of 100% methanol for 30 min at 25 °C. Re-extractions of the solid sample were carried out. Extractions obtained using this protocol provided approximately 1 L of grape skin extract. The extract was concentrated using a nitrogen stream at 40 °C. It was not dried but used in liquid form. The resulting concentrated extract was stored at –20 °C until it was used in the stability studies. Concentrations of anthocyanins in the concentrated extract are shown in Table 1.

Values are expressed as malvidin-3-glucoside equivalents because of most standards were not commercially available. Identification for those compounds were previously developed by the authors by HPLC-MS [15].

### 3.2. Chemicals and Solvents

All of the reagents used were of analytical grade: methanol and formic acid were obtained from Merck (Darmstadt, Germany). HPLC grade water was obtained from a Milli-Q system (Millipore, Bedford, MA, USA). All samples were filtered through a 0.45 µm nylon syringe filter (Millipore) before chromatographic analysis. All extractions were performed in triplicate.

### 3.3. Ultrasound-Assisted Extraction

UAE extraction conditions were applied in a water bath at 400 W (J.P. Selecta, Barcelona, Spain). The extraction protocol was carried out on 1.5 g of the standardized grape skin extract in approximately 9 mL of water for 20 min. The experiments were performed at constant temperature by means of a temperature controller coupled to the ultrasonic bath. Four temperatures were assayed: 0, 25, 50 and 75 °C. After each extraction, the volume of extract was made up to 10 mL with water.

### 3.4. Pressurized Liquid Extraction

Extraction conditions were applied in a Dionex ASE 200 extractor (Dionex, Sunnyvale, CA, USA). The standardized grape skin extract (8 g) was mixed with sea sand (Panreac, Barcelona, Spain) and placed in an 11 mL stainless steel extraction cell. A cellulose filter (Dionex, Sunnyvale, CA, USA) was placed at the bottom of the extraction cell. Nitrogen was used to purge and dry the extraction chambers during the extractions.

The extraction chamber was filled with water, pressurized to 100 atm, and heated to temperatures ranging from 75 to 150 °C for 20 min. The extracts were topped up to 50 mL with water, and these samples were then analyzed by RP-HPLC.

### 3.5. Ultra-Performance Liquid Chromatography (UPLC)

UPLC analyses were performed on a Waters Acquity Ultra Performance Liquid Chromatographic system (Waters, Milford, MA, USA) equipped with a PDA detector, an autosampler and a column oven to control the temperature of the analytical column (35 °C). Data were collected and processed by Empower chromatographic software (Waters). Chromatographic separation was achieved in an Acquity UPLC BEH C18 column (100 mm × 2.1 mm, 1.7 µm, Milford, MA, USA) equipped with an in-line 0.2 µm Acquity filter. Mobile phase A was 5% formic acid in water and mobile phase B was methanol. The gradient was as follows: 0 min, 5% B; 11 min,

29.6% B; 12 min, 30% B; 12.5 min, 100% B. A flow of 0.7 mL· min$^{-1}$ was used. A typical chromatogram of the diluted standardized extract is shown in Figure 4.

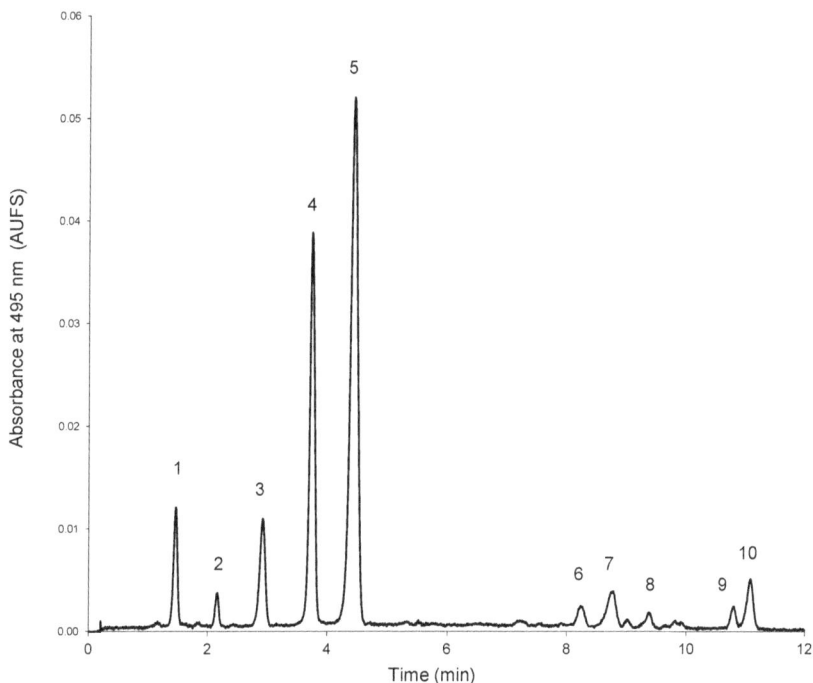

**Figure 4.** Typical chromatogram of the standardized extract. **1**: Delphinidin 3-glucoside; **2**: cyanidin 3-glucoside; **3**: petunidin 3-glucoside; **4**: peonidin 3-glucoside; **5**: malvidin 3-glucoside; **6**: peonidin 3-acetylglucoside; **7**: malvidin 3-acetylglucoside; **8**: malvidin 3-caffeoylglucoside; **9**: petunidin 3-p-coumaroylglucoside; **10**: malvidin 3-p-coumaroyl-glucoside (*trans*).

## 4. Conclusions

The stabilities of anthocyanins under the working conditions used for UAE and PLE were specifically studied for the first time. In view of the results, all assayed anthocyanins showed full stability under the UAE conditions, including working temperatures from 0 °C to 75 °C. It is therefore possible to employ extraction methods based on UAE up to 75 °C without degradation of the components.

The high temperatures usually applied in PLE methods led to significant degradation of anthocyanins. Specifically, the use of temperatures above 100 °C led to 50% degradation for most anthocyanins. It is therefore recommended that PLE for anthocyanins should only be employed up to 100 °C.

The glucosyl anthocyanins proved to be more susceptible to degradation than the acyl derivatives, with the difference in susceptibility to degradation being greater on working at higher temperature and in the presence of greater levels of oxygen. The results show that, depending on the specific compounds present in the samples (glucosides or acyl derivatives), PLE or UAE can be selected as the extraction method and the stability of anthocyanins would be guaranteed during the extraction process.

The stabilities of individual anthocyanins were also assessed and significant differences were not found with respect to the susceptibility to degradation for the studied compounds. Therefore, a relationship between the type of substituent present in either glycosylated or acyl derivatives and the susceptibility to degradation was not established. This means that information about stability of compounds beyond the scope of this study cannot be extrapolated from the results obtained on the assayed compounds.

**Acknowledgments:** This study was carried out within the research project AGR6874 financed by the Junta de Andalucía.

**Author Contributions:** AL, MP and CGB designed research; AL and GFB performed research; AL, LA and MP analyzed the data; MP and CGB wrote the paper. All authors read and approved the final manuscript.

**Conflicts of Interest:** The authors declare no conflict of interest.

## References

1. De Carvalho Alves, A.P.; Correa, A.D.; Marques Pinheiro, A.C.; de Oliveira, F.C. Flour and anthocyanin extracts of jaboticaba skins used as a natural dye in yogurt. *Int. J. Food Sci. Technol.* **2013**, *48*, 2007–2013.
2. Pina, F.; Melo, M.J.; Laia, C.A.T.; Parola, A.J.; Lima, J.C. Chemistry and applications of flavylium compounds: A handful of colours. *Chem. Soc. Rev.* **2012**, *41*, 869–908.
3. Kruger, M.J.; Davies, N.; Myburgh, K.H.; Lecour, S. Proanthocyanidins, anthocyanins and cardiovascular diseases. *Food Res. Int.* **2014**, *59*, 41–52.
4. Martin Bueno, J.; Saez-Plaza, P.; Ramos-Escudero, F.; Maria Jimenez, A.; Fett, R.; Asuero, A.G. Analysis and Antioxidant Capacity of Anthocyanin Pigments. Part II: Chemical Structure, Color, and Intake of Anthocyanins. *Crit. Rev. Anal. Chem.* **2012**, *42*, 126–151.
5. Pojer, E.; Mattivi, F.; Johnson, D.; Stockley, C.S. The Case for Anthocyanin Consumption to Promote Human Health: A Review. *Compr. Rev. Food Sci. Food Saf.* **2013**, *12*, 483–508.
6. Sui, X.; Dong, X.; Zhou, W. Combined effect of pH and high temperature on the stability and antioxidant capacity of two anthocyanins in aqueous solution. *Food Chem.* **2014**, *163*, 163–70.
7. Welch, C.R.; Wu, Q.; Simon, J.E. Recent advances in anthocyanin analysis and characterization. *Curr. Anal. Chem.* **2008**, *4*, 75–101.
8. Turker, N.; Erdogdu, F. Effects of pH and temperature of extraction medium on effective diffusion coefficient of anthocynanin pigments of black carrot (*Daucus carota* var. L.). *J. Food Eng.* **2006**, *76*, 579–583.

9.  Karvela, E.; Makris, D.P.; Kalogeropoulos, N.; Karathanos, V.T. Deployment of response surface methodology to optimise recovery of grape (*Vitis vinifera*) stem polyphenols. *Talanta* **2009**, *79*, 1311–1321.
10. Barbero, G.F.; Liazid, A.; Palma, M.; Barroso, C.G. Ultrasound-assisted extraction of capsaicinoids from peppers. *Talanta* **2008**, *75*, 1332–1337.
11. Chen, F.; Sun, Y.; Zhao, G.; Liao, X.; Hu, X.; Wu, J.; Wang, Z. Optimization of ultrasound-assisted extraction of anthocyanins in red raspberries and identification of anthocyanins in extract using high-performance liquid chromatography-mass spectrometry. *Ultrason. Sonochem.* **2007**, *14*, 767–778.
12. Mendiola, J.A.; Herrero, M.; Cifuentes, A.; Ibañez, E. Use of compressed fluids for sample preparation: Food applications. *J. Chromatogr. A* **2007**, *1152*, 234–246.
13. Carabias-Martinez, R.; Rodriguez-Gonzalo, E.; Revilla-Ruiz, P.; Hernandez-Mendez, J. Pressurized liquid extraction in the analysis of food and biological samples. *J. Chromatogr. A* **2005**, *1089*, 1–17.
14. Palma, M.; Piñeiro, Z.; Barroso, C.G. Stability of phenolic compounds during extraction with superheated solvents. *J. Chromatogr. A* **2001**, *921*, 169–174.
15. Guerrero, R.F.; Liazid, A.; Palma, M.; Puertas, B.; González-Barrio, R.; Gil-Izquierdo, A.; Barroso, C.G.; Cantos-Villar, E. Phenolic characterisation of red grapes autochthonous to Andalusia. *Food Chem.* **2009**, *112*, 949–955.

**Sample Availability:** *Sample Availability*: Not available.

# A New Solid Phase Extraction for the Determination of Anthocyanins in Grapes

Marta Ferreiro-González, Ceferino Carrera, Ana Ruiz-Rodríguez, Gerardo F. Barbero, Jesús Ayuso, Miguel Palma and Carmelo G. Barroso

**Abstract:** A method for the concentration and cleaning of red grape extracts prior to the determination of anthocyanins by UPLC-DAD has been developed. This method is of special interest in the determination of phenolic maturity as it allows the analysis of the anthocyanins present in grapes. Several different SPE cartridges were assessed, including both C-18- and vinylbenzene-based cartridges. C-18-based cartridges presented a very low retention for the glucosylated anthocyanidins while vinylbenzene-based cartridges showed excellent retention for these compounds. The optimized method involves the initial conditioning of the cartridge using 10 mL of methanol and 10 mL of water, followed by loading of up to 100 mL of red grape extract. Ten mL of water was used in the washing step and anthocyanins were subsequently eluted using 1.5 mL of acidified methanol at pH 2. This method simplifies the determination of individual anthocyanins as, on the one hand, it cleans the sample of interference and, on the other hand, it increases the concentration to up to 25:1.5. The developed method has been validated with a range of different grapes and it has also been tested as a means of determining the different anthocyanins in grapes with different levels of maturity.

Reprinted from *Molecules*. Cite as: Ferreiro-González, M.; Carrera, C.; Ruiz-Rodríguez, A.; Barbero, G.F.; Ayuso, J.; Palma, M.; Barroso, C.G. A New Solid Phase Extraction for the Determination of Anthocyanins in Grapes. *Molecules* **2014**, *19*, 21398–21410.

## 1. Introduction

Anthocyanins are natural pigments that belong to the flavonoids group and those compounds that are responsible for the red color of many fruits, flowers and food, especially in red grapes and wines, are the most abundant phenolic compounds in the skin of red grapes [1,2].

The main anthocyanins present in grapes are the monoglucosides of five anthocyanins, which are called delphinidin, cyanidin, petunidin, peonidin and malvidin. Caffeoyl, coumaroyl and acetyl derivatives of glucosidic forms are also found in grapes [3] (Figure 1).

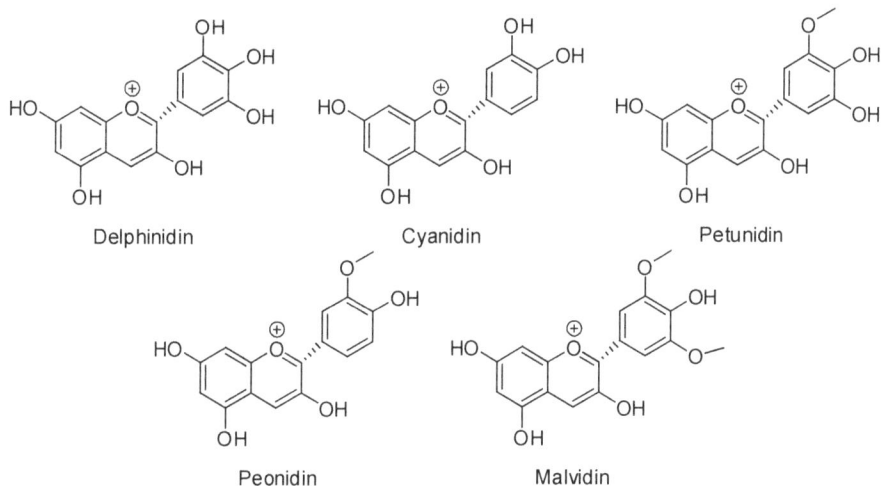

**Figure 1.** Chemical structures of anthocyanins.

The quantity and the profile of polyphenolic compounds in grapes, particularly in the anthocyanins in red grapes, depend on the grape variety. However, these parameters are highly influenced by climatic factors and viticulture practices and techniques [4,5].

The phenolic compounds present in the grape play a fundamental role in the sensory properties of wine [6] and are also related to various health benefits associated with the consumption of wine [7]. The total contents of phenolic compounds and the ratio between the different types of polyphenols, including anthocyanins, in the red grape varieties are strongly related to the quality of resulting wines. Therefore, the determination of phenolic levels provides very interesting information in setting the best harvest date. The control of the phenolic maturity of the berries is one of the most critical stages in the elaboration of red wines.

In a previous study, a new ultrasound-assisted extraction technique was developed. Ultrasound-assisted extraction provides an alternative to the classical maceration for the extraction of polyphenols, anthocyanins and tannins. This method can be applied to grapes during the ripening process and allows the quantitative and reproducible extraction of the phenolic compounds (total phenolic, total anthocyanins and condensed tannins) present in grapes. The method only requires a short time (6 min) and ethanol/water is employed as the extraction solvent [8]. Extracts should also be suitable for the determination of individual phenolic compounds, although the low levels found for individual phenolics would make a prior concentration step necessary. A solid phase extraction (SPE) method allows both cleaning of the sample by removing sugars and also concentration of the anthocyanins prior to determination by chromatographic techniques.

Solid phase extraction is one of the most common and least expensive purification techniques in terms of the preparation of samples for analysis of both major and minor food components [9–11]. This technique allows the development of rapid and automated methods. In addition to the food industry, this technique is used in the pharmaceutical industry [12] and environmental research [13].

The SPE technique has previously been applied for the extraction of anthocyanins. He and Giusti [14] used this technique to obtain fractions of high-purity in anthocyanins from different fresh fruits and vegetables (blueberry, raspberry, strawberry and red radishes), as well as commercial extracts enriched in anthocyanins. For biological samples, this technique is even more useful as it allows the removal of matrix components such as proteins, carbohydrates or lipids as well as the concentration of the analyte. Martí *et al.* [15] used the SPE technique to pre-concentrate anthocyanins in the plasma of rats fed with grape pulp extract.

In this paper, a method is proposed in which solid phase extraction is used to concentrate grape extracts obtained by ultrasound-assisted extraction for subsequent analysis by UPLC with UV-Vis detection. Different solid phases (C-18- and polymer-based phases) were assessed and working conditions were optimized to guarantee full recovery of phenolics from the extracts.

## 2. Results and Discussion

### 2.1. Comparison of SPE Cartridges

The first step in the development of the SPE method was the selection of the most appropriate SPE cartridge. The SPE protocol described in Section 2.4 was used. The results for the relative anthocyanin concentrations in the sample loading step, washing step and elution step obtained with all of the assayed SPE cartridges are shown in Table 1. These values are quoted relative to the amount of each anthocyanin in the original extract (100%).

The presence of anthocyanins in the resulting liquid residues from the loading sample (sample residues) and washing (washing residues) steps is indicative of inadequate retention by the cartridge. As can be seen, there are significant differences in the retention of anthocyanins within the assayed SPE cartridges. Indeed, some cartridges retained all anthocyanins whereas others had notable anthocyanin losses during the sample loading step.

**Table 1.** Relative anthocyanin concentrations (% ± RSD) for sample and wash residues and recoveries obtained with the assayed SPE cartridges ($n = 2$).

| Solid Phase | Steps | Relative Anthocyanins Concentration (% ± RSD) | | | | | | | | |
|---|---|---|---|---|---|---|---|---|---|---|
| | | D3G | Pt3G | Pd3G | M3G | PtAG | MAG | MCafG | PtCG | M3tCG |
| DSC-18 | Sample residue | 23.6 ± 10.9 | 19.1 ± 7.8 | - | 10.3 ± 4.1 | - | - | - | - | - |
| | Wash residue | - | - | - | - | - | - | - | - | - |
| | Recovery | 7.3 ± 0.2 | 6.5 ± 0.1 | 39.9 ± 22.0 | 16.2 ± 3.0 | 49.0 ± 6.3 | 19.2 ± 1.4 | 71.0 ± 4.0 | 59.1 ± 11.6 | 31.5 ± 1.2 |
| VC-18 | Sample residue | 34.7 ± 8.8 | 43.6 ± 1.2 | - | 23.48 ± 0.1 | - | - | - | - | - |
| | Wash residue | - | - | - | - | - | - | - | - | - |
| | Recovery | 3.6 ± 5.1 | 3.7 ± 5.2 | - | 10.7 ± 0.5 | 70.7 ± 28.1 | 46.1 ± 0.8 | 85.9 ± 7.1 | 69.9 ± 13.5 | 33.1 ± 1.8 |
| VEN | Sample residue | 29.6 ± 2.7 | 30.6 ± 11.5 | - | 22.2 ± 0.7 | - | - | - | - | - |
| | Wash residue | - | - | - | - | - | - | - | - | - |
| | Recovery | - | - | - | 15.2 ± 1.1 | 54.8 ± 9.9 | 38.0 ± 0.4 | 67.9 ± 0.4 | 58.1 ± 3.4 | 34.4 ± 1.2 |
| Strata X | Sample residue | - | - | - | - | - | - | - | - | - |
| | Wash residue | - | - | - | - | - | - | - | - | - |
| | Recovery | 44.7 ± 7.4 | 55.9 ± 1.8 | 88.7 ± 0.2 | 44.0 ± 1.1 | 24.5 ± 3.5 | 27.6 ± 0.2 | 47.7 ± 2.1 | 55.1 ± 5.0 | 23.6 ± 3.3 |
| EN | Sample residue | - | - | - | - | - | - | - | - | - |
| | Wash residue | - | - | - | - | - | - | - | - | - |
| | Recovery | 19.2 ± 0.3 | 23.3 ± 1.6 | - | 33.2 ± 0.6 | 47.7 ± 0.8 | 35.6 ± 0.4 | 62.44 ± 1.8 | 52.5 ± 1.7 | 18.6 ± 0.0 |

Because of their polarities, the solid phases used should produce better results for the less polar anthocyanins, therefore special attention must be paid to the most polar components during sample loading. The most polar compounds, *i.e.*, glucosylated anthocyanins, could be not retained by the solid phases, at least partially. If not retained, low recoveries would be obtained. The C-18-based cartridges showed losses of the glucosylated anthocyanins during sample loading step and these losses reached more than 40% on using VC-18 and almost 20% on using DSC-18 in the case of Pt3G. M3G (10%–23%) and D3G (23%–34%) were also lost during sample loading. Therefore, the C-18-based cartridges were ruled out for use in the subsequent method optimization. Of the vinylbenzene based cartridges, VEN cartridges also showed losses of the glucosylated anthocyanins during the sample loading step (22% for M3G to 30% for D3G and Pt3G) and, consequently, the use of VEN cartridges was also ruled out. In contrast, Strata X and EN cartridges showed excellent performance, retaining approximately 100% of the anthocyanins without observable losses either during the sample loading or the washing steps. It means, Strata X and EN cartridges are the only cartridges that are able to retain the total amount of anthocyanins from the sample. Therefore, both of these cartridges were used for the optimization in the next step, *i.e.*, the determination of the breakthrough volume of the cartridges.

The breakthrough volume study was carried out by increasing the amount of sample (from 10 mL of the extract to 100 mL) to an extent where losses can be observed, a point that indicates saturation of the sorbent bed. A high breakthrough volume will guarantee that losses will not occur during sample loading and washing steps.

Both of the selected cartridges (Strata X and LiChrolut EN) showed full retention of the anthocyanins with a sample loading up to 100 mL. Although both cartridges showed similar results, the Strata X cartridge was selected for optimization of the extraction parameters as it has previously shown excellent reproducibility [16].

## 2.2. Sample Loading Flow Rate

A long analysis time would be expected because high volumes of samples must be used to obtain high concentration ratios. Therefore, in order to reduce the duration of the method, different high loading flow rates were assessed by increasing the loading flow from 0.5–15 mL·min$^{-1}$, 20 mL·min$^{-1}$ and 25 mL·min$^{-1}$. Recoveries of anthocyanins from the sample at the different rates are shown in Table 2. Values are relative to the total amount of each anthocyanin recovered at a flow of 0.5 mL·min$^{-1}$.

Regardless of the sample loading flow used, the recoveries were greater than 95% except for M3tCG with a sample loading flow of 25 mL·min$^{-1}$, for which the recovery decreased to 85%. A flow rate of 25 mL·min$^{-1}$ was selected for the optimization as this flow rate allowed the total loading time to be reduced from 200 min to 4 min *versus* the flow rate previously used (0.5 mL·min$^{-1}$). The relatively low recovery for M3tCG is not a concern since it was expected that this could be increased later after optimizing other extraction variables. Loading flows above 25 mL· min$^{-1}$ resulted in saturation of the cartridge and subsequent overpressure, which stopped the SPE system.

Table 2. Effect of sample loading flow on the recovery of anthocyanins ($n$ = 3).

| Sample Loading Flow (mL· min$^{-1}$) | Relative Anthocyanins Concentration (% ± RSD) | | | | |
|---|---|---|---|---|---|
| | M3G | PtAG | MAG | PtCG | M3tCG |
| 15 | 102.6 ± 11.3 | 114.6 ± 14.5 | 118.9 ± 3.8 | 104.3 ± 5.1 | 116.4 ± 3.1 |
| 20 | 109.3 ± 8.7 | 113.2 ± 0.5 | 108.9 ± 4.9 | 108.6 ± 4.4 | 95.2 ± 5.6 |
| 25 | 96.3 ± 12.4 | 97.8 ± 3.4 | 98.7 ± 1.8 | 97.0 ± 1.6 | 85.6 ± 2.8 |

## 2.3. Amount of Eluting Solvent

An important aspect to be considered when developing any method is to minimize the quantities of solvents required, thereby obtaining a higher concentration ratio and also a higher signal in the subsequent determination of the anthocyanins. However, a reduction in the amount of elution solvent may result in an incomplete or inadequate recovery of the anthocyanins from the cartridge. For this reason, it is necessary to optimize the amount of elution solvent.

Elutions of anthocyanins with 1 mL and 1.5 mL of MeOH at pH = 2 were compared. The recoveries obtained with 2 mL of acidified methanol (MeOH pH = 2) were used as reference values. The results obtained on using different elution volumes are shown in Table 3. Recoveries obtained with 1.5 mL of MeOH at pH = 2 were approximately 100% for almost all anthocyanins. However, when 1 mL of MeOH was used, the recoveries of esterified anthocyanins dramatically decreased and this mainly concerned the recovery of the PtCG (37 ± 64.1) and M3tCG (31.6 ± 70.3). These anthocyanins have a low polarity. It must also be noted that repeatability dramatically decreased for the extraction using 1 mL of MeOH as the eluting solvent. In some cases, RSD values of around 70% were found for some components in the samples.

Anthocyanins present an acid-base equilibrium and, as a result, an adjustment in pH values in the final elution step could also be of interest to increase selectivity and recovery. A decrease in the pH of the elution solvent to below pH = 2 would allow an easier removal of anthocyanins retained in the sorbent, thus allowing a reduction in the elution solvent volume. With this aim in mind, the pH of the eluting solvent (MeOH) was decreased to pH = 1.5 and pH = 1. However, despite the decrease in pH (Table 3) of the eluting solvent differences were not observed in the recovery of anthocyanins in the pH range tested and it was not possible to reduce the elution solvent volume. Therefore, the elution solvent volume selected was 1.5 mL of MeOH at pH = 2.

**Table 3.** Effect of eluting solvent volume and pH on the recovery of anthocyanins ($n = 3$).

| Elution Solvent | Elution Volume | Relative Anthocyanins Concentration (% ± RSD) | | | | |
|---|---|---|---|---|---|---|
| | | M3G | PtAG | MAG | PtCG | M3tCG |
| MeOH $_{pH = 2}$ | 1 mL | 111.3 ± 6.5 | 96.1 ± 2.5 | 70.4 ± 25.8 | 37.1 ± 64.1 | 31.6 ± 70.3 |
| | 1.5 mL | 106.8 ± 10.8 | 97.0 ± 1.8 | 103.0 ± 0.8 | 118.2 ± 10.8 | 105.6 ± 7.4 |
| MeOH $_{pH = 1.5}$ | 1mL | 90.8 ± 9.0 | 83.9 ± 7.0 | 48.4 ± 22.0 | 24.7 ± 27.5 | 21.3 ± 18.4 |
| | 1.5 mL | 92.8 ± 1.8 | 102.2 ± 3.1 | 104.8 ± 6.1 | 77.1 ± 4.2 | 87.8 ± 6.3 |
| MeOH $_{pH = 1}$ | 1 mL | 100.7 ± 3.8 | 100.2 ± 38.0 | 52.3 ± 77.2 | 21.0 ± 30.9 | 19.9 ± 10.8 |
| | 1.5 mL | 81.7 ± 29.1 | 97.6 ± 19.8 | 89.5 ± 44.5 | 82.4 ± 20.3 | 72.2 ± 48.1 |

The final result of the extraction method can be seen in Figure 2, which shows the resulting chromatogram of the sample before (original extract) and after application of the method. As can be seen, the signal is almost 25 times higher for the compounds found in the chromatogram after using the optimized method.

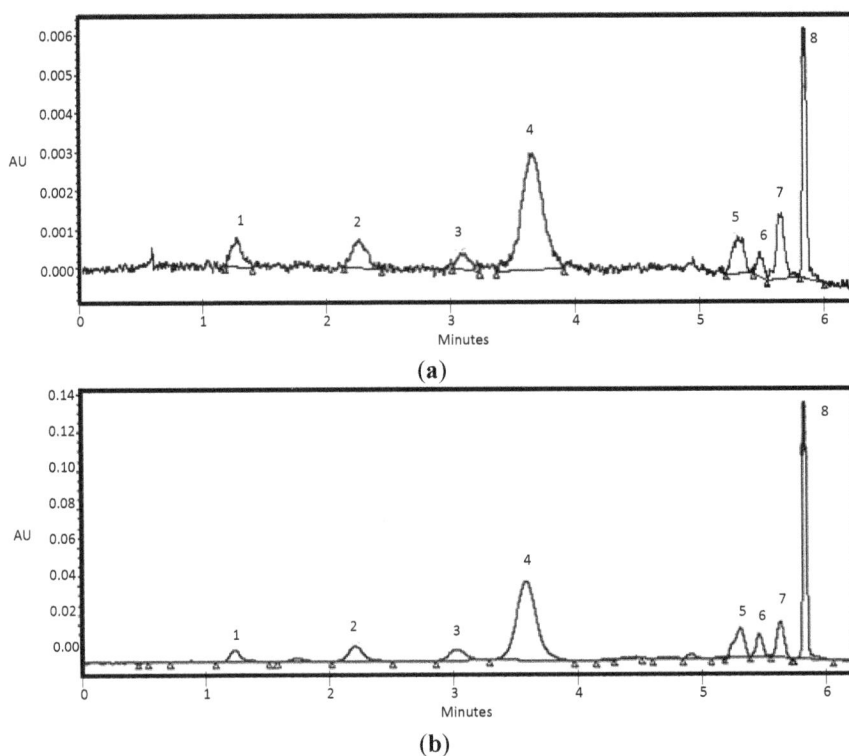

(a)

(b)

**Figure 2.** Chromatograms obtained before (a) and after (b) applying the optimized SPE method. 1: D3g, 2: Pt3G, 3: Pd3G, 4: M3G, 5: MAG, 6: MCafG, 7: PtCG and 8: M3tCG.

## 2.4. Repeatability and Intermediate Precision

The repeatability and intermediate precision were determined by running the developed SPE method for 15 extractions on three different days with the same sample: nine extractions the first day and three extractions on the two following days. Intra-day and inter-day residual standard deviation (RSD) was calculated for different types of anthocyanins.

The RSD found for repeatability ranged from 3.9% for glycosylated anthocyanins and 6.7% for acyl anthocyanins and cinnamyl derivatives. Regarding reproducibility, the results for RSD ranged from 9.4% for the glucosylated anthocyanins to 9.6% for acyl anthocyanins and cinnamyl derivatives.

## 2.5. Application to Real Samples

It has been reported previously that different red grape varieties contain different ratios of individual anthocyanins and also different chemical forms of the same anthocyanin, *i.e.*, glycosyl, cinnamyl or acyl forms [17].

The suitability of the method developed in this work was evaluated on real samples by using four different grape varieties: Petit Verdot (PV), Cabernet Sauvignon (CS), Syrah (SY) and Tintilla de Rota (TR). Different anthocyanin levels were obtained for the different grape varieties as shown in Figure 3. Tintilla de Rota grapes showed the highest values for the main glucosyl derivatives, *i.e.*, M3G and Pd3G, while Petit Verdot showed the lowest values for these forms. Syrah and Cabernet Sauvignon showed intermediate values for the main glucosyl forms but Syrah showed the highest level for Pt3G. The highest levels of anthocyanins were found for acetyl and coumaroyl derivatives in most varieties, with only Tintilla de Rota showing a higher level for the glucosyl derivative of maldivin than for acetyl/coumaroil forms. For Syrah, Cabernet Sauvignon and Petit Verdot, both MAG and M3tCG were present in the highest levels of the anthocyanins found in the samples. Levels for Syrah samples were particularly high: 21 and 23 mg of MAG and M3tCG, respectively, per 100 g of samples.

With the aim of evaluating the applicability of the method to monitor the evolution of anthocyanins during the ripening process and to determine the effects of different cultivar practices, the Tempranillo grape variety was obtained from different cultivar practices [cluster thinning (CT) and vines intercropped with cover crops (CC)] on different dates (August 25, September 13 and September 26), *i.e.*, 30 days and 13 days before harvest and also on the harvest day.

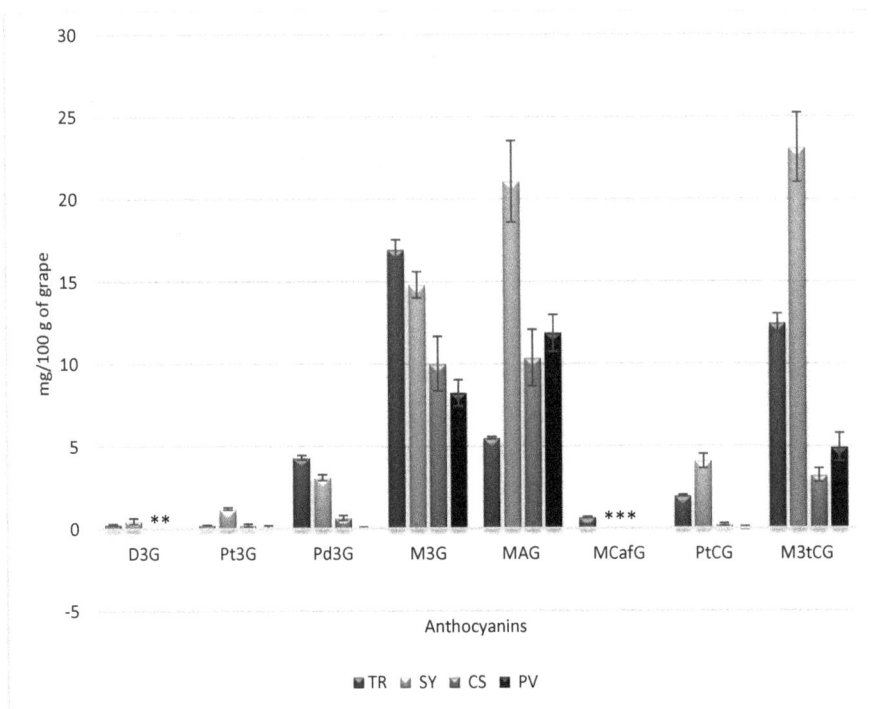

**Figure 3.** Anthocyanin levels obtained for the different grape varieties (TR: Tintilla de Rota, Sy: Syrah, CS: Cabernet Sauvignon, PV: Petit Verdot). **: There are two non detected compounds; ***: There are three non detected compounds.

Cultivar practices affect the final levels of anthocyanins in grapes [18]. In particular, cluster thinning at early ripening stages is used to increase the total amount of anthocyanins in grapes for Syrah [18] and Tempranillo varieties [19]. This technique works because fewer grapes are produced and harvested and, as a consequence, the anthocyanins are present in higher concentrations in the grapes [18,20]. On the other hand, it has been also described the effects of cover crops practices on total anthocyanins of Cabernet Sauvignon grapes [21,22], however no information about the effects of cover crops practices on individual anthocyanin levels has been found in the revised literature. These two cultivar practices should also modify the relative levels of anthocyanins. The resulting values for anthocyanins are shown in Figure 4. Information from three different sampling zones (1, 2 and 3) in the vineyard is also presented in Figure 4. It can be seen that both parameters clearly affect the final levels for anthocyanins. For minor anthocyanins, non-significant differences were found, however for the two main anthocyanins specific significant differences were found. Grapes cultivated using the cluster thinning cultivar practice show higher values for M3G than those cultivated using cover crops when cultivated in vineyard zones 1 or 2, although lower values are obtained for cultivation in vineyard zone 3. Regarding the other major anthocyanin, *i.e.*, M3tCG, it was found that cover crops led to higher values by between 20% and 25% in zone 1 and 3 than cluster thinning, whereas in zone 2 differences were not obtained. Therefore, winemakers would be able to manage some additional information about grape composition. It must be noted that even grapes from the same cultivar practices could show different anthocyanin values when cultivated in different areas of the same vineyard.

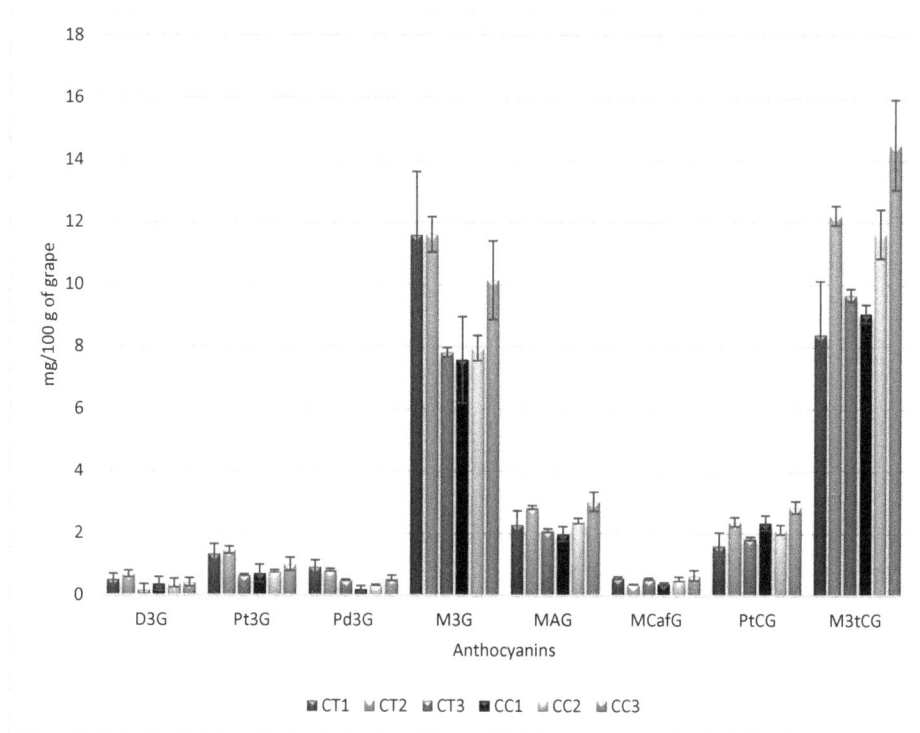

**Figure 4.** Evolution of the anthocyanin levels during phenolic maturation using the same grape variety (Tempranillo) but from different cultivar practices: CT: cluster thinning, CC: cover crops, in three different vineyard locations.

## 3. Experimental Section

### 3.1. Chemicals and Solvents

Methanol (Merck, Darmstadt, Germany) and ethanol (Panreac, Barcelona, Spain) were HPLC grade. Ultra pure water was obtained from a Milli-Q water purification system from Millipore (Bedford, MA, USA).

## 3.2. Grape Samples

The red grape var. Tempranillo was employed to develop the extraction method. The grape samples were obtained from local vineyards. The full berries (skin, pulp and seeds) were triturated with a conventional beater until a homogeneous sample was obtained for analysis. The resulting triturated sample was stored in a freezer at $-20\,°C$ prior to analysis.

## 3.3. Extraction Procedure

The extraction of anthocyanins originating from red grapes was performed by using ultrasound and following the procedure previously described by Carrera *et al.* [8].

In previous studies, it was demonstrated that this technique is sensitive to the ethanol concentration [23] and the extracts obtained were therefore diluted so that the ethanol concentration in each sample tested was below 15%.

## 3.4. Solid Phase Extraction

The development of the SPE method was performed on a Zymark Rapid Trace (Caliper, Hopkinton, MA, USA) automated system. Selection of the appropriate cartridge was based on the evaluation and comparison of five different cartridges from several suppliers and with different stationary phases. The main characteristics of the cartridges used are given in Table 4.

**Table 4.** Characteristics of evaluated solid phase extraction (SPE) cartridges.

| Commercial Brand | Abbreviation | Solid Phase | Amount of Solid Phase (mg) | Supplier |
|---|---|---|---|---|
| Discovery DSC-18 | DSC-18 | Octadecyl silica | 500 | Supelco |
| Bond Elut C-18 | VC-18 | Octadecyl silica | 500 | Varian |
| Bond Elut ENV | VEN | Styrene-divinylbenzene | 200 | Varian |
| Strata X | Strata X | Modified divinylbenzene | 200 | Phenomenex |
| LiChrolut EN | EN | Ethyl-vinyl-benzene styrene-divinylbenzene | 200 | Merck |

The protocol used to evaluate all SPE cartridges was as follows: the cartridge was conditioned with 10 mL of methanol and 10 mL of water (10 mL· min$^{-1}$) and the extract (10 mL) was loaded onto the cartridges (1 mL· min$^{-1}$). The cartridge was washed with 10 mL of water (10 mL· min$^{-1}$) and eluted with 2 mL (10 mL· min$^{-1}$) of methanol (pH = 2). Samples and wash residues were collected and analyzed to evaluate losses during these steps.

## 3.5. Ultra-Performance Liquid Chromatography (UPLC)

The determination of anthocyanins was carried out by ultra-performance liquid chromatography (UPLC) on a Waters system (Waters, Milford, MA, USA). An

ACQUITY UPLC C-18 column (2.1 mm internal diameter, 100 mm length and 1.7 microns particle size) was used. The temperature of the column was kept constant at 50 °C. The mobile phases were acidified water (5% formic acid) (solvent A) and methanol (solvent B) and a flow-rate of 0.5 mL· min$^{-1}$ was used. The gradient used for the separation was as follows: 0 min 15% B, 3.30 min 20% B, 3.86 min 30% B, 5.05 min 40% B, 5.35 min 55% B, 5.64 min 60% B, 5.94 min 95% B.

The anthocyanins identified and quantified were as follows: delphinidin-3-glucoside (D3G), petunidin-3-glucoside (Pt3G), peonidin-3-glucoside (Pd3G), malvidin-3-glucoside (M3G), malvidin-3-acetylglucoside (MAG), malvidin-3-caffeoyl glucoside (MCafG), petunidin-3-coumaroyl glucoside (PtCG) and malvidin-3-trans-coumaroyl glucoside (M3tCG).

The quantification of each anthocyanin was carried out by integrating the area of the peaks at 500 nm with a linear response between 0.5 and 27 mg· L$^{-1}$ (7 points) and a correlation coefficient (R2) of 0.997. Malvidin chloride (Sigma-Aldrich, St. Louis, MO, USA) was the standard used for the calibration curve.

The limits of detection and quantification were established by measuring the area at lower concentration for D3G after running the extraction six times. The LOD values and LOQ values were 1.21 and 4.05 mg· L$^{-1}$, respectively, for extracts prior to solid phase extraction and 0.19 and 0.64 mg· L$^{-1}$, respectively, for extracts after solid phase extraction.

## 4. Conclusions

A large variation was found in the retention of anthocyanins within the assayed SPE cartridges. During the sample loading step, the C-18 based cartridges showed noticeable losses for the glucosylated anthocyanins, which are the most polar. The best retention of anthocyanins was achieved with the vinylbenzene-based cartridges.

The optimized extraction method is fast (less than 10 min) and reproducible, with high anthocyanin recoveries achieved from grape extracts. The method developed in this study also concentrates the extract from the grape by up to 16.6 times (25:1.5), thus allowing measurement of anthocyanins at low concentrations and providing cleaner extracts that are less detrimental to the chromatographic column than original samples.

The developed method can be applied for the individual determination of anthocyanin compounds in grapes during ripening. The method takes only 25 min (UAE + SPE + UPLC) to complete and requires very little solvent. Winemakers could receive a more detailed information about grape composition; therefore, if needed, they could manage grapes using different winemaking conditions depending on their specific composition. Therefore, the information obtained from the SPE method could be helpful for wine production.

**Acknowledgments:** This study was carried out within the research project AGR6874 financed by the Junta de Andalucía.

**Author Contributions:** MF, MP and CGB designed research; MF, CC, AR and GFB performed research; MF, JA and MP analyzed the data; MF, MP and CGB wrote the paper. All authors read and approved the final manuscript.

**Conflicts of Interest:** The authors declare no conflict of interest.

## References

1.    Yoo, M.-A.; Kim, J.-S.; Chung, H.-K.; Park, W.-J.; Kang, M.-H. The antioxidant activity of various cultivars of grape skin extract. *Food Sci. Biotechnol.* **2007**, *16*, 884–888.

2.    He, F.; Mu, L.; Yan, G.-L.; Liang, N.-N.; Pan, Q.-H.; Wang, J.; Reeves, M.J.; Duan, C.-Q. Biosynthesis of anthocyanins and their regulation in colored grapes. *Molecules* **2010**, *15*, 9057–9091.

3.    Guerrero, R.F.; Liazid, A.; Palma, M.; Puertas, B.; Gonzalez-Barrio, R.; Gil-Izquierdo, A.; Garcia-Barroso, C.; Cantos-Villar, E. Phenolic characterisation of red grapes autochthonous to Andalusia. *Food Chem.* **2008**, *112*, 949–955.

4.    Downey, M.O.; Dokoozlian, N.K.; Krstic, M.P. Cultural practice and environmental impacts on the flavonoid composition of grapes and wines: A review of recent research. *Am. J. Enol. Vitic.* **2006**, *57*, 257–268.

5.    Segade, S.R.; Vazquez, E.S.; Orriols, I.; Giacosa, S.; Rolle, L. Possible use of texture characteristics of winegrapes as markers for zoning and their relationship with anthocyanin extractability index. *Int. J. Food Sci. Technol.* **2011**, *46*, 386–394.

6.    Gawel, R. Red wine astringency: A review. *Aust. J. Grape Wine Res.* **1998**, *4*, 74–95.

7.    Santos-Buelga, C.; Scalbert, A. Proanthocyanidins and tannin-like compounds—Nature, occurrence, dietary intake and effects on nutrition and health. *J. Sci. Food Agric.* **2000**, *80*, 1094–1117.

8.    Carrera, C.; Ruiz-Rodriguez, A.; Palma, M.; Barroso, C.G. Ultrasound assisted extraction of phenolic compounds from grapes. *Anal. Chim. Acta* **2012**, *732*, 100–104.

9.    Puoci, F.; Curcio, M.; Cirillo, G.; Iemma, F.; Spizzirri, U.G.; Picci, N. Molecularly imprinted solid-phase extraction for cholesterol determination in cheese products. *Food Chem.* **2007**, *106*, 836–842.

10.   Grigoriadou, D.; Androulaki, A.; Psomiadou, E.; Tsimidou, M.Z. Solid phase extraction in the analysis of squalene and tocopherols in olive oil. *Food Chem.* **2007**, *105*, 675–680.

11.   Zhu, Y.; Chiba, K. Determination of cadmium in food samples by ID-ICP-MS with solid phase extraction for eliminating spectral-interferences. *Talanta* **2012**, *90*, 57–62.

12.   Tian, M.; Li, S.; Row, K. Molecularly imprinted polymer for solid-phase extraction of ecteinascidin 743 from sea squirt. *Chin. J. Chem.* **2012**, *30*, 43–46.

13.   Erdogan, H.; Yalcinkaya, O.; Tuerker, A.R. Determination of inorganic arsenic species by hydride generation atomic absorption spectrometry in water samples after preconcentration/separation on nano $ZrO_2/B_2O_3$ by solid phase extraction. *Desalination* **2011**, *280*, 391–396.

14. He, J.; Giusti, M.M. High-purity isolation of anthocyanins mixtures from fruits and vegetables—A novel solid-phase extraction method using mixed mode cation-exchange chromatography. *J. Chromatogr. A* **2011**, *1218*, 7914–7922.

15. Martí, M.-P.; Pantaleon, A.; Rozek, A.; Soler, A.; Valls, J.; Macia, A.; Romero, M.-P.; Motilva, M.-J. Rapid analysis of procyanidins and anthocyanins in plasma by microelution SPE and ultra-HPLC. *J. Sep. Sci.* **2010**, *33*, 2841–2853.

16. Rostagno, M.A.; Palma, M.; Barroso, C.G. Solid-phase extraction of soy isoflavones. *J. Chromatogr. A* **2005**, *1076*, 110–117.

17. Romero-Cascales, I.; Ortega-Regules, A.; Lopez-Roca, J.M.; Fernandez-Fernandez, J.I.; Gomez-Plaza, E. Differences in anthocyanin extractability from grapes to wines according to variety. *Am. J. Enol. Vitic.* **2005**, *56*, 212–219.

18. Peña-Neira, A.; Caceres, A.; Pastenes, C. Low molecular weight phenolic and anthocyanin composition of grape skins from cv.syrah (*Vitis vinifera* L.) in the maipo valley (Chile): Effect of clusters thinning and vineyard yield. *Food Sci. Technol. Int.* **2007**, *13*, 153–158.

19. Diago, M.P.; Vilanova, M.; Blanco, J.A.; Tardaguila, J. Effects of mechanical thinning on fruit and wine composition and sensory attributes of Grenache and Tempranillo varieties (*Vitis vinifera* L.). *Aust. J. Grape Wine Res.* **2010**, *16*, 314–326.

20. Zalamena, J.; Cassol, P.C.; Brunetto, G.; Panisson, J.; Marcon, J.L.; Schlennper, C. Productivity and composition of grapes and wine of vines intercropped with cover crops. *Pesqui. Agropecu. Bras.* **2013**, *48*, 182–189.

21. Wheeler, S.J.; Black, A.S.; Pickering, G.J. Vineyard floor management improves wine quality in highly vigorous Vitis vinifera "Cabernet Sauvignon" in New Zealand. *N. Z. J. Crop Hortic. Sci.* **2005**, *33*, 317–328.

22. Lopes, C.M.; Monteiro, A.; Machado, J.P.; Fernandes, N.; Araujo, A. Cover cropping in a sloping non-irrigated vineyard: II—Effects on vegetative growth, yield, berry and wine quality of "Cabernet Sauvignon" grapevines. *Cienc. Tec. Vitivinic.* **2008**, *23*, 37–43.

23. Jeffery, D.W.; Mercurio, M.D.; Herderich, M.J.; Hayasaka, Y.; Smith, P.A. Rapid isolation of red wine polymeric polyphenols by solid-phase extraction. *J. Agric. Food Chem.* **2008**, *56*, 2571–2580.

**Sample Availability:** *Sample Availability*: Not available.

# Section 3:
# Anthocyanin Biosynthesis and Regulation

# Support for a Photoprotective Function of Winter Leaf Reddening in Nitrogen-Deficient Individuals of *Lonicera japonica*

Kaylyn L. Carpenter, Timothy S. Keidel, Melissa C. Pihl and Nicole M. Hughes

**Abstract:** Plants growing in high-light environments during winter often exhibit leaf reddening due to synthesis of anthocyanin pigments, which are thought to alleviate photooxidative stress associated with low-temperature photoinhibition through light attenuation and/or antioxidant activity. Seasonal high-light stress can be further exacerbated by a limited photosynthetic capacity, such as nitrogen-deficiency. In the present study, we test the following hypotheses using three populations of the semi-evergreen vine *Lonicera japonica*: (1) nitrogen deficiency corresponds with reduced photosynthetic capacity; (2) individuals with reduced photosynthetic capacity synthesize anthocyanin pigments in leaves during winter; and (3) anthocyanin pigments help alleviate high-light stress by attenuating green light. All populations featured co-occurring winter-green and winter-red leafed individuals on fully-exposed (high-light), south-facing slopes in the Piedmont of North Carolina, USA. Consistent with our hypotheses, red leaves consistently exhibited significantly lower foliar nitrogen than green leaves, as well as lower total chlorophyll, quantum yield efficiency, carboxylation efficiency, and photosynthesis at saturating irradiance ($A_{sat}$). Light-response curves measured using ambient sunlight *versus* red-blue LED (*i.e.*, lacking green wavelengths) demonstrated significantly reduced quantum yield efficiency and a higher light compensation point under sunlight relative to red-blue LED in red leaves, but not in green leaves, consistent with a (green) light-attenuating function of anthocyanin pigments. These results are consistent with the hypothesis that intraspecific anthocyanin synthesis corresponds with nitrogen deficiency and reduced photosynthetic capacity within populations, and support a light-attenuating function of anthocyanin pigments.

Reprinted from *Molecules*. Cite as: Carpenter, K.L.; Keidel, T.S.; Pihl, M.C.; Hughes, N.M. Support for a Photoprotective Function of Winter Leaf Reddening in Nitrogen-Deficient Individuals of *Lonicera japonica*. *Molecules* **2014**, *19*, 17810–17828.

## 1. Introduction

Anthocyanins are vacuolar, flavonoid pigments synthesized via the shikimic acid pathway that impart red to purplish colors in plant tissues [1]. Of special interest to plant physiologists is the synthesis of anthocyanin pigments in photosynthetic

225

tissues during periods of high-light stress, which may be defined generally as seasons, ontogenetic stages, and/or environmental conditions corresponding with an imbalance of light capture relative to energy processing (for reviews see [1–3]). For example, anthocyanin synthesis has been observed under high light in combination with: cold temperatures [1,4,5], drought stress [6–8], leaf development [9–12], and senescence [13,14]. However, the functional significance of leaf reddening remains a matter of debate (discussed in further detail below; for reviews see [5,14,15]). Furthermore, why some individuals or species synthesize red pigments, while others do not, is also not yet fully understood [5,16–18].

There are currently two functional explanations for anthocyanin synthesis in leaves—photoprotection and ecological defense. According to the photoprotection hypothesis, anthocyanins protect photosynthetic tissues vulnerable to high-light stress through antioxidant activity, and/or by intercepting green quanta, thereby alleviating excess chlorophyll excitation pressure in underlying cells [1–3]. According to the ecological defense hypothesis, anthocyanins function to reduce damage by potential herbivores or pathogens by either: (a) reducing visibility to herbivores lacking a red photoreceptor (*i.e.*, camouflage); (b) signaling low leaf quality (e.g., high investment in chemical defenses, low nitrogen content) [10,19–21]; (c) undermining herbivorous insect camouflage [22]; and/or (d) inhibiting fungal growth [23,24]. Because plant-insect interactions are generally less frequent during the winter, we focus here on the putative photoprotective function of anthocyanin pigments, as this function seems most directly relevant to the high-light, cold temperature conditions in which winter-leaf reddening frequently occurs [5].

During winter, high-light in combination with cold temperatures results in excess energy capture by chlorophylls relative to (reduced) energetic demands of the Calvin cycle [25]. The resulting photooxidative damage and associated photoinhibition of photosynthesis further reduce carbon gain, and plants have evolved photoprotective strategies to alleviate this imbalance accordingly. Such strategies include: increases in xanthophyll-cycle pigments, increased conversion of violaxanthin to zeaxanthin, selective degradation and/or sustained-phosphorylation of D1/D2 protein and whole PSII cores, increased antioxidant pools, vertical leaf orientation, and/or synthesis of photoprotective anthocyanin pigments [5,26–33]. As would be expected, relative engagement of photoprotection has been shown to be inversely correlated with energy processing capacity [34,35]. Hence, anthocyanin synthesis might be expected to occur in individuals or species with diminished capacity for photosynthesis and/or energy dissipation.

Recent studies on intraspecific populations featuring co-occurring red and green individuals have demonstrated that red-leafed individuals tend to exhibit symptoms of photosynthetic inferiority relative to co-occurring green-leafed individuals, including lower leaf nitrogen, lower photosynthetic capacity, and greater photoinhibition of photosynthesis [36–42]. Because foliar nitrogen levels are directly correlated with molecular and enzymatic pools involved in photosynthesis, including Rubisco, chlorophyll, and chlorophyll binding protein [43–45], nitrogen deficiency not only reduces a plant's capacity for light capture and processing [46], but also increases its need for photoprotection [34,35]. The photosynthetic-inferiority hypothesis posits that individuals suffering from physiological limitations to energy processing, such as nitrogen deficiency, should synthesize anthocyanins as a means of alleviating this photosynthetic imbalance [5,39,42,47]. To date, this idea has only been tested in a few species, and evidence linking nitrogen deficiency, photosynthetic capacity, anthocyanin production, and photoprotection all within an individual study system are sparse in the literature.

The objective of this study was to test the photosynthetic-inferiority hypothesis for leaf reddening using co-occurring red and green populations of Japanese honeysuckle, *Lonicera japonica* Thunb. *Lonicera japonica* is a non-native, semi-evergreen vine that is invasive to the USA, that often synthesizes anthocyanins in sun-exposed leaves during winter under high-light conditions [16]. We utilize three separate, high-light field sites in the Piedmont of North Carolina featuring co-occurring winter-red (anthocyanic) and winter-green (acyanic) populations of *L. japonica* to test the hypothesis that red-leafed individuals correspond with lower leaf nitrogen content and associated photosynthetic deficiencies (e.g., lower carboxylation efficiency, reduced chlorophyll content, reduced capacity for photosynthesis) relative to co-occurring green individuals. We further test whether light attenuation by anthocyanin results in physiologically significant reductions in green light absorption in red-leafed individuals, which would support a photoprotective function for leaf reddening.

## 2. Results and Discussion

### 2.1. Leaf Nitrogen

Winter-red individuals at RR, WE and I40 had significantly (23% on average) lower leaf N content than winter green-leafed individuals ($p < 0.01$ at WE, $p < 0.001$ at RR and I40 and $p < 0.0001$ when combined; Figure 1A). When individual sites were compared, winter leaf N content was highest at the RR site, with mean N content of 2.8% and 2.1% for green and red leaves respectively. At WE, green leaves had a mean N content of 2.1% during winter, and red leaves, 1.74%; I40 green leaves had a mean N content of 2.24%, and red leaves, 1.59% during winter. When summer (all green) leaves were compared from I40, leaves on the winter-green side of the embankment continued to exhibit higher average N content relative to leaves on the winter-red side, though this difference was only marginally significant ($p = 0.12$). Leaves from the winter-green side of the embankment showed no significant difference in foliar

N between winter and the following summer ($p = 0.96$), while the leaves from the winter-red side red exhibited significant increases in percent nitrogen between winter and the following summer ($p < 0.001$; Figure 1B).

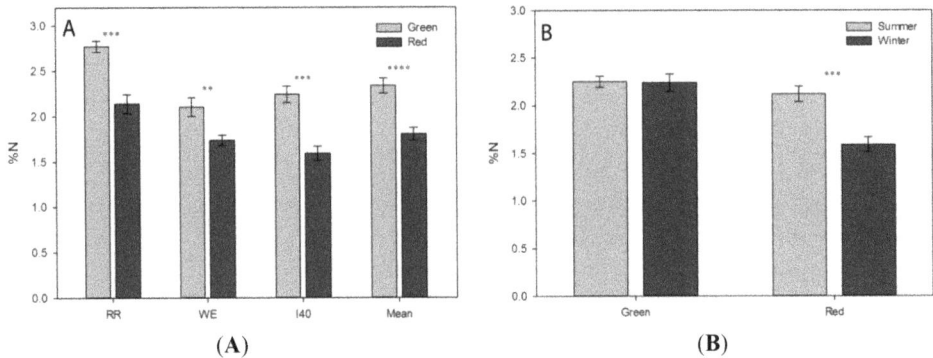

(A) (B)

**Figure 1.** Mean nitrogen content in red *versus* green *L. japonica* leaves. (**A**) Winter mean percent nitrogen content ($\pm$SE) at RR ($n_{green} = 5$, $n_{red} = 5$), WE ($n_{green} = 7$, $n_{red} = 6$), I40 ($n_{green} = 6$, $n_{red} = 6$), and combined means from all sites. Significant differences between red and green leaves denoted by asterisks (* $p < 0.05$; ** $p < 0.01$, *** $p < 0.001$, and **** $p < 0.0001$); (**B**) Mean percent leaf nitrogen ($\pm$SE) from leaf tissues collected at I40 during summer ($n_{winter-green} = 6$, $n_{winter-red} = 6$) and winter ($n_{green} = 6$, $n_{red} = 6$).

## 2.2. Chlorophyll Content

Red leaves exhibited consistently lower (40% on average) total chl during winter compared to co-occurring green leaves ($p < 0.01$ for WE, $p < 0.001$ at RR and I40, combined sites $p < 0.0001$ see Figure 2A). Trends in chl *a/b* ratios were less consistent between sites (Figure 2A). Chl *a/b* was significantly lower in red leaves relative to green during winter at RR ($p < 0.05$), but there were no statistically significant differences at either WE or I40. Analysis of combined winter data from all sites showed no significant difference in chl *a/b* between winter-red and winter-green leaves.

Leaves collected during summer from the winter-green side of the I40 embankment contained significantly higher total chl per unit leaf area and lower chl *a/b* than leaves on the winter-red side ($p < 0.05$ for both), consistent with trends observed at this site the previous winter (Figure 2B). However, differences in total chl were much smaller in magnitude than the values obtained during winter. When comparing summer *versus* winter total chl at I40 (Figure 2B), leaves on the winter-green side of the embankment exhibited significantly (18%) higher total chl content during winter relative to the summer ($p < 0.05$). However, the opposite was observed in leaves on the red-leafed side of the embankment, where leaves exhibited significant increases (42%) in chlorophyll content during summer relative to the previous winter ($p < 0.01$, Figure 2B). Chlorophyll $a/b$ ratios were slightly lower in the winter-red leaves during winter than summer ($p < 0.1$), but winter-green leaves showed no notable differences.

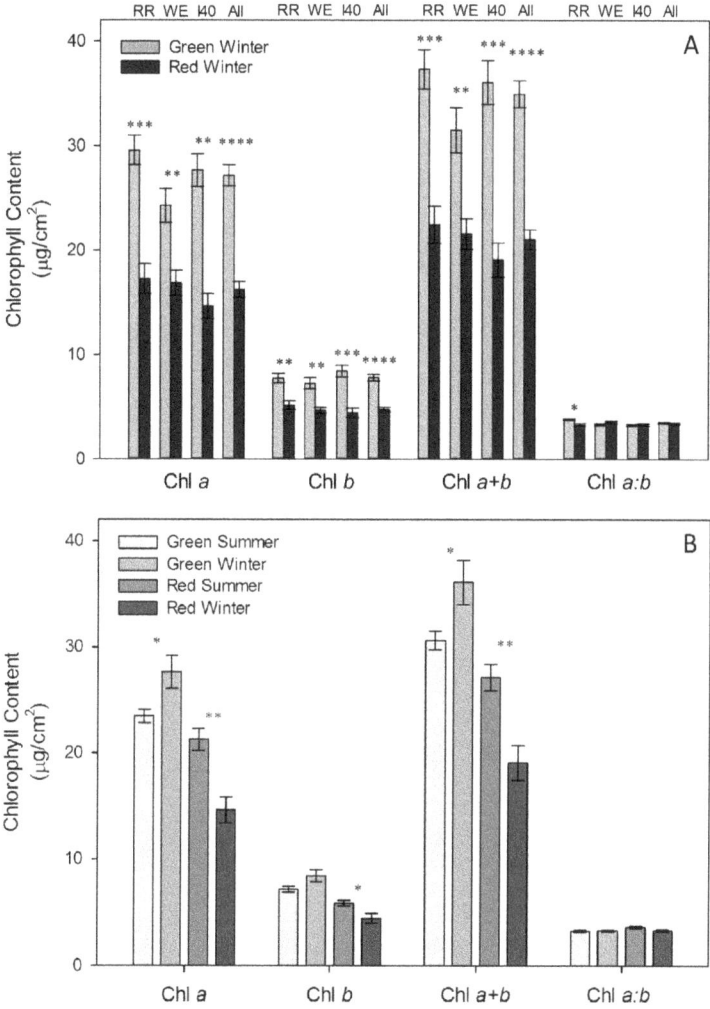

Figure 2. (A) Mean chlorophyll content per unit leaf area during winter for green ($n = 5$) compared to red ($n = 5$) leaves $\pm$ SE at each site for RR, WE, I40, and combined sites (see labels at top of figure); (B) Mean summer chlorophyll content for winter-red ($n = 6$) and winter-green ($n = 6$) portions of I40 site compared with winter chlorophyll content ($n = 5$ for both sides) $\pm$ SE. Significant differences between red *versus* green leaves (A); and summer *versus* winter leaves (B) denoted by asterisks (* $p < 0.05$; ** $p < 0.01$, *** $p < 0.001$, and **** $p < 0.0001$).

## 2.3. Photosynthetic Gas Exchange

### 2.3.1. Diurnal Measurements

Diurnal photosynthetic gas exchange measurements at both fields sites (Figure 3) showed significantly reduced photosynthesis in red leaves relative to green under saturating red/blue LED irradiance throughout the day, with the only exception being the early morning measurement at the RR site (Figure 3A, RR: $p < 0.001$ at 1200, $p < 0.05$ at 1600, Figure 3B WE: $p < 0.001$ at 1000 and 1500, $p < 0.05$ at 1200). In general, photosynthesis tended to decrease in all plants during the day, corresponding with declines in leaf stomatal conductance to water vapor (g). On average, green leaves tended to have higher g than red-leaves at both sites, however, this difference was only significant in two measurements made at the RR site ($p < 0.05$ for midday and 1600, Figure 3C).

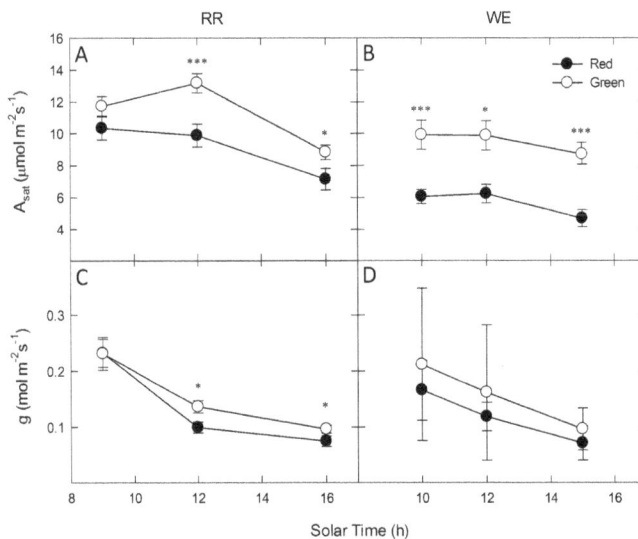

**Figure 3.** Diurnal measurements of photosynthetic gas exchange for co-occurring red leaves (closed circles) and green leaves (open circles) under saturating red/blue LED irradiance. Panels (**A**) and (**B**) show photosynthesis at saturating irradiance ($A_{sat}$) and (**C**) and (**D**) show stomatal conductance (g) at RR (left column) and WE (right column). Points represent means of 3–15 individuals of each color $\pm$ SE. Significant differences between red and green leaves at each time point denoted by asterisks (* $p < 0.05$; ** $p < 0.01$, *** $p < 0.001$).

### 2.3.2. Light-Response Curves

Light-response curves derived using red/blue LED *versus* ambient sunlight allowed for comparison of photosynthetic parameters with and without interference by the (green-light absorbing) anthocyanic layer (Figure 4, Table 1). Red leaves showed significantly reduced QYE (30% lower on average) under ambient sunlight relative to red/blue LED measurements at both field sites (WE $p < 0.05$, RR $p < 0.001$, combined $p < 0.0001$; Figure 4A,C). Additionally, significantly (180%) more PAR was required to reach LCP under ambient sunlight than under LED (WE and RR $p < 0.05$, combined $p < 0.001$). For green leaves, significant (albeit less dramatic) differences in QYE and LCP were observed under LED *versus* sunlight at RR ($p < 0.05$ for both; Figure 4D), but no significant differences were observed at WE (Figure 4B). Upon analyzing combined site data for winter-green leaves, no difference was found in the QYE between sunlight and LED light sources, though significantly more (68%) light was required to reach LCP under ambient sunlight compared to LED ($p < 0.05$). Both red and green leaves at RR had significantly greater photosynthesis at saturating irradiance ($A_{sat}$) under LED relative to ambient sunlight ($p < 0.01$ for both), while no differences were found at WE for either red or green leaves (Table 1). When data from both field sites were combined, $A_{sat}$ did not significantly differ when LED or sunlight was used as a saturating light source in red leaves, though green leaves had significantly (14%) higher $A_{sat}$ under LED light ($p < 0.05$). Dark respiration measurements made following red/blue LED *versus* sunlight light response curves did not significantly differ in green leaves at either site, or in red leaves at RR; however, in red leaves at WE, DR was significantly lower (*i.e.*, greater respiration) following measurements made with ambient sunlight relative to measurements made with the red/blue LED ($p < 0.05$).

Statistical analyses were also used to compare photosynthetic parameters for red *versus* green leaves within each field site, under the same type of light (rather than between types of light). No significant differences were observed in dark respiration (DR) between red and green leaves at either field site, or when site values were combined ($p > 0.4$ for all; Table 1). PAR intensities required to reach LCP also did not differ for red *versus* green leaves at either site under red/blue LED or ambient sunlight ($p > 0.4$ for both; Table 1). However, when site data were combined, LCP was 60% higher in red leaves, but only under ambient sunlight ($p < 0.05$). Green leaves exhibited significantly higher QYE and $A_{sat}$ relative to red leaves at both sites under red/blue LED (QYE: WE $p < 0.01$, RR $p < 0.05$; $A_{sat}$: $p < 0.01$ at both sites; Table 1). When data from both field sites were combined, QYE values were an average of 32% higher in green *versus* red leaves, and $A_{sat}$ was 40% higher in green *versus* red under red/blue LED ($p < 0.001$ for both, Table 1). Under sunlight, mean QYE and $A_{sat}$ were also significantly higher in green leaves compared to red at both sites (77% and 33% higher on average, respectively; QYE: WE $p < 0.01$, RR and combined sites $p < 0.0001$; $A_{sat}$: WE $p < 0.05$, RR $p < 0.001$, combined sites $p < 0.0001$, Table 1).

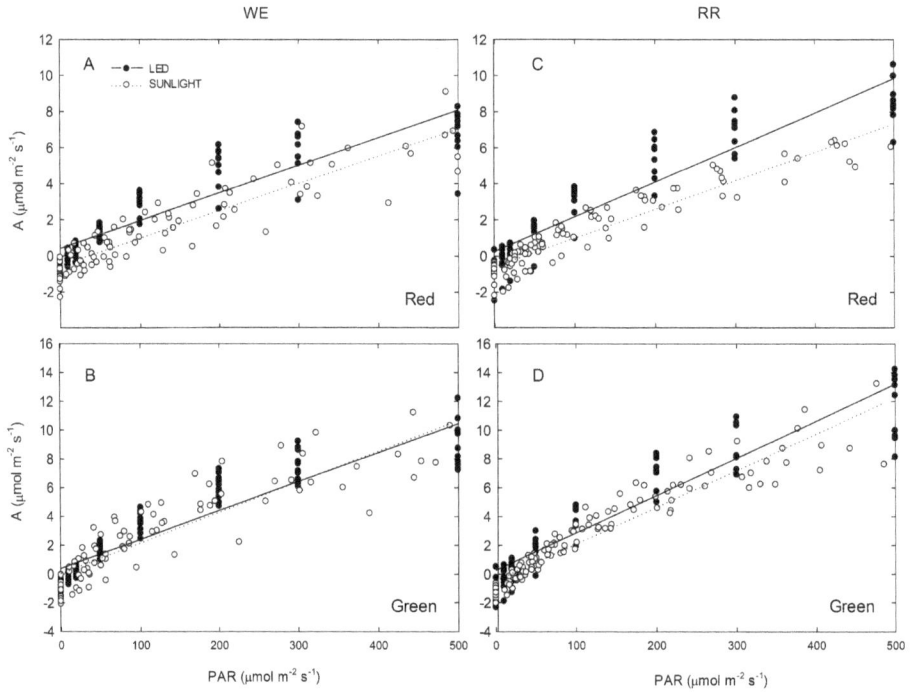

**Figure 4.** Linear (light-dependent) portion of light response curves measured under red/blue LED (closed symbols, solid line) *versus* ambient sunlight (open symbols, dashed line) at WE (left column) and RR (right column). (**A**) Red leaves at WE, $n_{LED} = 10$ $n_{Sun} = 10$; (**B**) Green leaves at WE, $n_{LED} = 12$ $n_{Sun} = 9$; (**C**) Red leaves at RR, $n_{LED} = 9$ $n_{Sun} = 10$; (**D**) Green leaves at RR, $n_{LED} = 9$ $n_{Sun} = 13$. All measurements made during winters of 2011–2013.

### 2.3.3. A/C$_i$ Curves

A *versus* C$_i$ curves measured during winter showed significantly greater carboxylation efficiency (CE) in green *versus* red leaves at both RR and WE sites ($p < 0.05$ for both sites individually, $p < 0.01$ when sites were combined; Figure 5 and Table 2). On average, CE in winter-green leaves was 30% higher than in winter-red leaves. Winter-green leaves also had higher maximum photosynthesis ($A_{max}$) under saturating irradiance and $CO_2$ than winter-red leaves at both study sites (Table 2); these differences were significant at WE ($p < 0.05$) and marginally significant at RR ($p = 0.08$). When data from both sites were combined, winter-green leaves had significantly (30%) higher $A_{max}$ than winter-red leaves ($p < 0.01$). No differences in calculated stomatal limitation (I) or $CO_2$ compensation point were observed between red and green leaves at either site ($p_{WE} = 0.18$, $p_{RR} = 0.6$, $p_{COMB} = 0.36$; $p_{WE} = 0.71$, $p_{RR} = 0.72$, $p_{COMB} = 0.61$ respectively).

**Table 1.** Data derived from light response curves for red *versus* green leaves measured during winters of 2011–2013. Data are means derived during winter at West End (WE) and Railroad (RR) field sites, using either red/blue LED as a light source, or ambient sunlight. Data include: dark respiration rate (DR), light compensation point (LCP), quantum yield efficiency (QYE), and photosynthesis under saturating irradiance ($A_{sat}$). Asterisks denote statistical significance between means ($\pm$SE) of red and green leaves at each site (* $p < 0.05$, ** $p < 0.01$, *** $p < 0.001$, **** $p < 0.0001$).

| | $n$ | DR ($\mu mol \cdot m^{-2} \cdot s^{-1}$) | LCP ($\mu mol \cdot m^{-2} \cdot s^{-1}$) | QYE | $A_{sat}$ ($\mu mol \cdot m^{-2} \cdot s^{-1}$) |
|---|---|---|---|---|---|
| **Red/Blue LED** | | | | | |
| Red WE | 10 | $-0.603 \pm 0.27$ | $14.7 \pm 8.9$ | $0.0350 \pm 0.0056$ ** | $7.88 \pm 1.7$ ** |
| Green WE | 12 | $-0.870 \pm 0.39$ | $15.6 \pm 8.0$ | $0.0442 \pm 0.0076$ | $11.1 \pm 2.4$ |
| Red RR | 9 | $-0.697 \pm 0.76$ | $18.0 \pm 20$ | $0.0362 \pm 0.0080$ * | $11.4 \pm 2.1$ ** |
| Green RR | 9 | $-1.03 \pm 0.83$ | $18.9 \pm 15$ | $0.0504 \pm 0.013$ | $16.4 \pm 3.7$ |
| Red AVG | 19 | $-0.648 \pm 0.55$ | $16.3 \pm 15$ | $0.0356 \pm 0.0067$ *** | $9.53 \pm 2.6$ **** |
| Green AVG | 21 | $-0.939 \pm 0.61$ | $17.0 \pm 11$ | $0.0469 \pm 0.010$ | $13.4 \pm 4.0$ |
| **Ambient Sunlight** | | | | | |
| Red WE | 10 | $-1.12 \pm 0.71$ | $49.6 \pm 39$ | $0.0255 \pm 0.0094$ ** | $8.74 \pm 2.4$ * |
| Green WE | 9 | $-1.20 \pm 0.82$ | $26.7 \pm 23$ | $0.0502 \pm 0.023$ | $11.0 \pm 1.7$ |
| Red RR | 10 | $-1.04 \pm 0.52$ | $41.9 \pm 25$ | $0.0241 \pm 0.0045$ **** | $8.96 \pm 1.4$ *** |
| Green RR | 13 | $-1.28 \pm 0.36$ | $29.8 \pm 9.0$ | $0.0398 \pm 0.0049$ | $12.4 \pm 2.7$ |
| Red AVG | 20 | $-1.08 \pm 0.60$ | $45.8 \pm 32$ * | $0.0248 \pm 0.0072$ **** | $8.85 \pm 1.9$ **** |
| Green AVG | 22 | $-1.24 \pm 0.57$ | $28.6 \pm 16$ | $0.0440 \pm 0.015$ | $11.8 \pm 2.4$ |

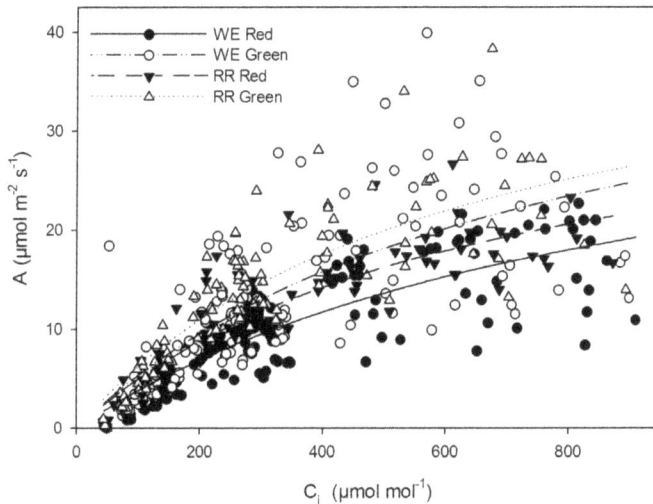

**Figure 5.** Photosynthesis (A) *versus* internal $CO_2$ concentrations ($C_i$) for red (solid symbols) *versus* green (open symbols) leaves at WE (circles) and RR (triangles). WE $n_{red} = 16$, $n_{green} = 15$; RR $n_{red} = 12$, $n_{green} = 11$.

**Table 2.** Mean values obtained from $A/C_i$ curves at individual and combined sites. Data include: carboxylation efficiency (CE), relative stomatal limitation (I), carbon dioxide compensation point ($CO_2$ CP), and $A_{max}$ as maximum photosynthesis as measured at 1000 $\mu$mol$\cdot$mol$^{-1}$ [$CO_2$] and saturating irradiance ($\mu$mol$\cdot$m$^{-2}$ s$^{-1}$). Asterisks denote statistical significance between means ($\pm$SE) of red and green leaves (* $p < 0.1$, ** $p < 0.05$, *** $p < 0.01$).

|  | n | CE | I | $CO_2$ CP | $A_{max}$ |
|---|---|---|---|---|---|
| Red WE | 16 | 0.0475 $\pm$ 0.011 ** | 0.170 $\pm$ 0.071 | 49.0 $\pm$ 13 | 17.7 $\pm$ 4.4 ** |
| Green WE | 15 | 0.0649 $\pm$ 0.027 | 0.185 $\pm$ 0.15 | 47.4 $\pm$ 11 | 23.8 $\pm$ 8.3 |
| Red RR | 12 | 0.0613 $\pm$ 0.014 ** | 0.183 $\pm$ 0.079 | 41.1 $\pm$ 10 | 19.4 $\pm$ 3.0 * |
| Green RR | 11 | 0.0757 $\pm$ 0.018 | 0.167 $\pm$ 0.059 | 39.8 $\pm$ 7.2 | 24.0 $\pm$ 6.6 |
| Red AVG | 28 | 0.0534 $\pm$ 0.014 *** | 0.171 $\pm$ 0.076 | 45.6 $\pm$ 12 | 18.4 $\pm$ 3.9 *** |
| Green AVG | 26 | 0.0694 $\pm$ 0.024 | 0.195 $\pm$ 0.11 | 44.2 $\pm$ 10 | 23.9 $\pm$ 7.5 |

## 2.4. Discussion

This study provides strong support for the photosynthetic-inferiority hypothesis for intraspecific leaf reddening during winter. According to this hypothesis, individuals with a reduced photosynthetic capacity (e.g., nitrogen-deficient individuals) synthesize anthocyanin pigments under high-light, cold-temperature (winter) conditions as a means of balancing energy capture with reduced demand. Specifically, we demonstrate that winter-red leaves exhibit significantly lower leaf N, reduced chlorophyll content, and lower photosynthetic capacity compared to co-occurring winter-green leaves, and also, that anthocyanin pigments attenuate a physiologically-significant portion of photosynthetically active radiation (PAR).

Red leaves of *L. japonica* contained significantly less (23% on average) leaf nitrogen than green leaves during winter at all three field sites (Figure 1A). These differences persisted during summer at the I40 site, which was the only field site where summer N measurements were made (Figure 1B). We suspect that reddening only manifested during winter due to the additional photoinhibitory stress imparted by cold temperatures [5,33]. These findings are consistent with previous studies reporting an inverse correlation between N content and anthocyanin synthesis in leaves within species [37–42]. Although determining the ultimate cause for the difference in nitrogen content between individuals examined was beyond the scope of this study, two anecdotal observations suggest that above-ground damage/defoliation may be responsible for reductions in leaf N, and consequent leaf reddening, in winter-red *L. japonica*. First, at RR, the portion of the slope featuring the highest density of winter-red individuals appeared to have been sprayed with an unknown, broad-spectrum herbicide during the summer following our measurements, resulting in complete necrosis of all above-ground plant matter; however, the area where winter-green individuals were present in higher frequency remained verdant (presumably not being sprayed). If this same spatial pattern of herbicide application also occurred at some point prior to our experiment, it could explain the reduction in N among the red-leafed individuals, as translocation of foliar nitrogen from leaves would not have been possible prior to their abrupt senescence. Similarly, a colleague anecdotally reported that pruning of above-ground *L. japonica* on one side of a walkway resulted in subsequent winter-reddening of new growth on the pruned side, but not the un-pruned side [48]. Regardless of the ultimate cause, significant reductions in foliar nitrogen were observed in winter-red individuals of *L. japonica* at all three field sites, and we believe that this is a proximate cause for the photosynthetically-inferior characteristics of winter-red individuals described below.

Ribulose-1,5-bisphosphate carboxylase/oxygenase (Rubisco) accounts for roughly 50% of photosynthetic N [43,49], resulting in a strong correlation between leaf N content and photosynthetic capacity [43,46,50]. Nitrogen deficiency is also known to correlate with a decrease in proteins involved in synthesis of chlorophyll, and chlorophyll *a/b* binding protein [44,51,52], which would limit the photon-capturing capacity of the photosystems and further reduce photosynthetic capacity. Consistent with these symptoms of N limitation, we demonstrate that winter-red *L. japonica* leaves exhibited significantly reduced photosynthetic capacity, carboxylation efficiency, quantum yield efficiency, and chlorophyll content relative to green-leafed individuals during winter. Specifically, $A/C_i$ curves illustrate that red-leafed individuals exhibited significant (23% on average) reductions in maximum photosynthesis ($A_{max}$) and carboxylation efficiency (CE) (23%) relative to green-leafed individuals (Figure 5, Table 2). Stomatal limitation did not significantly differ between red and green-leafed individuals in $A/C_i$ curves, suggesting limitations to photosynthesis were biochemical. Similarly, diurnal measurements of photosynthetic gas exchange in the field during winter showed that red-leafed individuals generally exhibited significantly lower $A_{sat}$ relative to green-leafed individuals, despite similar values of g (Figure 3C,D). Winter-red individuals also had lower total chlorophyll content per unit leaf area on average relative to green-leafed individuals both during summer (11% lower) and winter (40% lower) (Figure 2), as well as significantly reduced QYE and $A_{sat}$ per unit leaf area (Table 1). These results corroborate previous reports demonstrating a photosynthetic inferiority (both in terms of reduced capacity for energy capture and processing) in winter-red individuals relative to green [40].

The significant reduction in photosynthetic capacity in N-deficient individuals provides a physiological basis for anthocyanin synthesis. As previously described, engagement of photoprotection has been shown to be inversely correlated with energy processing capacity [34,35]. Hence, an increase in anthocyanin content (which imparts photoprotective light-attenuating and antioxidant functions) would seem a suitable response for N-limited individuals [39]. Indeed, it has previously been demonstrated that N deficiency corresponds with up-regulation of expression of genes involved in synthesis of anthocyanin [44,53,54]. In another study, transcription levels of genes involved in the anthocyanin pathway were increased 7.6 to 49.2 fold under N-limiting conditions [55].

In order to assess whether anthocyanins attenuate a physiologically-significant amount of sunlight, light-response curves were derived using red/blue LED and compared to curves derived using ambient sunlight (Figure 4). This allowed for comparison of photosynthetic parameters with and without potential interference by the (green-light absorbing) anthocyanic layer. Consistent with our hypothesis, under red/blue LED, red-leaves exhibited a significant increase in quantum yield efficiency (QYE), and a significant reduced light compensation point (LCP) relative to measurements derived using ambient sunlight (Figure 4A and C). In green-leafed individuals, there was no significant difference in QYE or LCP under LED relative to ambient sunlight at WE, though significant (albeit substantially less dramatic) differences were observed at RR (Figure 4B and D respectively). When data from both field sites were combined, red leaves had a significant, 30% mean reduction in QYE under ambient sunlight compared to LED, while no significant difference was observed in green leaves. Similarly, red leaves exhibited a 180% higher LCP on average under sunlight compared to LED, while green leaves only presented a 68% increase, representing a three-fold difference between the two groups; both of these differences were statistically different.

## 3. Experimental Section

### 3.1. Plant Material and Field Sites

*Lonicera japonica* Thunb. (Japanese honeysuckle) is an invasive vine found in the majority of the continental United States [56]. Three south-facing slopes in the Piedmont of North Carolina, USA featuring vines with both winter-red and winter-green leaves were utilized during this study. The Railroad site (RR) (36°10'21.24" N, −80°26'31.83" W) consisted of a fully-exposed, south-facing embankment located approximately 10 m from a roadside, situated along a railroad track. Red and green individuals co-occurred irregularly throughout the field site, although a distinct, uniformly green population occurred in one section of the embankment lying adjacent to a land-bridge overhanging a stream. All individuals were fully exposed to sun for >6 h per day during winter. The West End site (WE) (36°09'30.87" N, −80°26'27.16" W) consisted of two slopes, one southeast-facing, the other southwest-facing, in a residential area. The site was exposed to full sunlight during most of the day during winter, although presence of some evergreen trees resulted in brief, punctuated (1–2 h) shade intervals throughout the day. Red and green-leafed vines at this site were heavily intertwined, resulting in no clear definition between red and green populations. A third site along Interstate 40 Westbound (I40)

(36°06'18.52" N, −80°27'62.84" W) was added in January 2013 for additional nitrogen and chlorophyll measurements (site pictured in Figure 6A,B). Field measurements at this site were limited due to the close (<5 m) proximity to high-speed vehicles, hence, no gas exchange measurements were made at this site. Red and green populations at I40 were distinctly separated, with green-leafed individuals being located primarily on the west side of the slope, and red-leafed individuals on the east side (Figure 6A).

**Figure 6.** Photographs of *Lonicera japonica*. (**A**) I40 site during winter; green (**left**) and red (**right**) "color sides" are visibly distinct; (**B**) I40 site during summer; all *L. japonica* individuals presenting with green leaves; (**C**) Co-occurring red and green individuals of *L. japonica* as found *in situ* at RR; (**D**) Cross section of red *L. japonica* leaf, with anthocyanin pigments in the uppermost palisade layer.

In all sites, red and green leaves were similar in size, featured similar leaf orientations, and occurred in seemingly identical environments with respects to azimuth, sunlight exposure, and precipitation. Visible leaf reddening in *L. japonica* leaves began during mid-December, and remained until new leaves developed the following spring. All leaves used in this study were fully-developed, and were either distinctly red or green (e.g., Figure 6C); leaves with intermediate concentrations of anthocyanin were not used in this study.

To view anatomical distribution of anthocyanin pigments, sample red *Lonicera japonica* leaves were hand-sectioned and mounted on a Zeiss Axioplan upright microscope (Carl Zeiss Inc., Thornwood, NY, USA). Sections were viewed under bright-field microscopy, and images captured using a Hamamatsu C5810 three-chip cooled color CCD camera (Hamamatsu Photonics; Hamamatsu City, Japan).

### 3.2. Leaf Nitrogen

Red and green leaves were harvested from WE in February 2011, RR in January 2012, and I40 in January 2013. Since I40 showed dramatic spatial separation of red and green populations, measurements could be made on winter-red *versus* winter-green sides of the slope during the summer as well (August 2013) when both "color sides" were green. Five to seven shoots (4–6 leaves per shoot) of each phenotype were randomly sampled from all sites on each measurement date. Leaf tissues were stored briefly in wet paper towels, then (within 2–3 h) homogenized in liquid nitrogen using a mortar and pestle, and oven-dried at 60 °C. Percent leaf nitrogen was quantified using a CHN 2400 Elemental Analyzer (Perkin Elmer Corporation, Norwalk, CT, USA). A NIST (National Institute of Standards and Technology) standard was also run every 22 samples to ensure accuracy of measurements. Normality was assessed using the Shapiro-Wilk test using JMP (3.2.2) statistical package (SAS Institute Inc., Cary, NC, USA) with normality defined as $p > 0.05$. Means were compared within individual sites using a one-tailed Student's $t$ test in Microsoft Excel (14.3.5) (Microsoft Corporation, Redmond, WA, USA). Data sets from the three sites were also combined and analyzed using a randomized complete block design ANOVA using Statistix (9.0) (Analytical Software, Tallahassee, FL, USA).

### 3.3. Chlorophyll Content

Fresh leaf tissues were collected from RR, WE, and I40 sites on the morning of 10 January 2013 and immediately transported to the laboratory in a wet paper towel within a plastic bag. Five replicates of each colored leaf from each site were obtained (30 samples total). Additionally, on 27 August 2013 (when all plants were green), six leaves from each "color side" of I40 were collected and analyzed. For all assays, three 0.317 cm$^2$ hole-punched tissue sections were excised from each leaf, and immediately extracted in 3 mL $N,N$ dimethylformamide in the dark at room temperature for 24 h. The absorbance of the supernatant was then determined spectrophotometrically (Ocean Optics, USB4000-UV-VIS with USB-ISS-UV/VIS attachment, Dunedin, FL, USA). Chlorophyll $a$ and $b$ were estimated on a per unit leaf area basis using equations from Porra [57]. Data were tested for normality via Shapiro-Wilk test using JMP

(3.2.2). Means within individual sites were compared using a two-tailed Student's *t* test in Microsoft Excel (14.3.5), and randomized complete block design ANOVAs were run using Statistix (9.0) for combined data sets.

## 3.4. Photosynthetic Gas Exchange

Diurnal photosynthetic gas exchange measurements for red and green-leafed individuals were made using a Li-Cor 6400XT (Li-Cor, Lincoln, NE, USA) with Li-6400-02B red/blue LED chamber with PAR (photosynthetically active radiation) set to 1500 $\mu mol \cdot m^{-2} \cdot s^{-1}$ during one warm (low temp > 0 °C), mostly sunny winter day at RR (13 January 2012) and WE (6 February 2011). Measurements were collected at WE at 1000, 1200, 1500, and at RR at 0900, 1200, and 1600. During all measurement intervals, 3–15 individuals of each leaf color were sampled, with measurements alternating randomly between leaf color.

Photosynthetic light-response curves (LRC) were derived on warm (low temp > 0 °C), sunny days during the winters of 2011–2013 using a Li-Cor 6400XT. Curves were derived separately using either ambient sunlight (Li-6400 standard clear-top chamber plus neutral-density shade films) or the red/blue LED light source. Curves were derived both with sunlight and red/blue LED in order to compare the light response of photosynthesis with and without the putative light-attenuating effects of anthocyanins, which absorb strongly in the green wavelengths. Separate leaves were chosen at random for each light-response curve. Within individual days, measurements alternated between red and green leaves. The light-source used in light-response measurements was randomized between days. Measurements were taken at ambient temperature and humidity, between 0700 and 1330. For LRC utilizing the LED light source, each curve was initiated at 2000 $\mu mol \cdot m^{-2} \cdot s^{-1}$ and was subsequently decreased incrementally in a total of 12 stepwise reductions until 0 $\mu mol \cdot m^{-2} \cdot s^{-1}$ was reached. The same protocol was used for LRC utilizing ambient light and neutral density shade screens, though maximum PAR values ranged from 1031 to 1799 $\mu mol \cdot m^{-2} \cdot s^{-1}$. For each curve, the following parameters were determined: dark respiration rate (DR), determined as $CO_2$ flux at 0 $\mu mol \cdot m^{-2} \cdot s^{-1}$, light compensation point (LCP), the level of PAR corresponding with 0 $\mu mol \cdot m^{-2} \cdot s^{-1}$ net $CO_2$ uptake, quantum yield efficiency (QYE), estimated as the slope of the linear, light-limited portion of the curve between 0 and 200 $\mu mol \cdot m^{-2} \cdot s^{-1}$ PAR, and photosynthesis at saturating irradiance ($A_{sat}$).

Photosynthetic $CO_2$ response ($A/C_i$) curves were derived using Li-Cor 6400XT, equipped with Li-6400-02B red/blue LED light source set at saturating irradiance (1500 $\mu mol \cdot m^{-2} \cdot s^{-1}$). Measurement protocol (with regards to randomization and replication) was similar to that described above for LRC. Measurements were made between 0700 and 1330 under ambient temperature and humidity conditions. Measurements began near ambient [$CO_2$] levels (400 $\mu mol \cdot mol^{-1}$), then decreased incrementally in five stages to 50 $\mu mol \cdot mol^{-1}$, returned to 400 $\mu mol \cdot mol^{-1}$, and then increased in three increments until 1000 $\mu mol \cdot mol^{-1}$ was reached. Measurements were taken after allowing approximately 60–90 s following each $CO_2$ adjustment to allow leaves to acclimate. For each curve, carboxylation efficiency (CE), relative stomatal limitation (I), and $CO_2$ compensation point ($CO_2$ CP) were determined as described by Ku and Edwards [58], Farquhar *et al.* [59], and von Caemmerer and Farquhar [60].

All photosynthetic gas exchange data were tested for normality as previously described. Red *versus* green diurnal measurements were compared at each time point within each site individually using a two-tailed Student's *t* test, using Microsoft Excel (14.3.5). Parameters derived from light response curves were analyzed by comparing red *versus* green-leaf values under each light source, and LED *versus* sunlight values according to leaf color. Within sites, mean DR, LCP, QYE, and $A_{sat}$ for were compared using a two-tailed Student's *t* test, and a randomized complete block design ANOVA via Statistix (9.0) was used to analyze combined-site data. For A/$C_i$ curves, mean CE, I, $CO_2$ CP and $A_{max}$ values for red *versus* green leaves were compared within each site individually using a two-tailed Student's *t* test, and combined using a randomized complete block design ANOVA via Statistix (9.0).

## 4. Conclusions

Results presented here support the photosynthetic-inferiority hypotheses for intraspecific winter-leaf reddening, which posits that winter-red leaves suffer from reduced nitrogen, photosynthetic capacity, and/or chlorophyll content. Furthermore, we demonstrate that anthocyanin pigments attenuate a physiologically-significant amount of green-light.

The association between winter-reddening and photosynthetic deficiency could potentially serve as a useful diagnostic for identifying photosynthetically-inferior individuals within a population, which could have valuable applications to crop management (e.g., see [61,62]). We are hesitant, however, to extend these results to explain interspecific differences in leaf reddening at the community level. Previous studies comparing co-occurring winter-red and winter-green species have yet to demonstrate significant differences in photosynthetic capacity or leaf N [16,18,63]. It is likely that confounding differences in anatomy and/or physiology between different species make such comparisons difficult, though our results combined with those of previous studies certainly encourage further investigation of this hypothesis at the community level.

**Acknowledgments:** The authors would like to thank Elizabeth McCorquodale of the Chemistry and Physics department at High Point University for assistance with spectrophotometry. K. Carpenter was supported in her work on this project by the National

Science Foundation under Grant Number 1122064, awarded to N. Hughes and High Point University. Publication fees were paid for by High Point University.

**Author Contributions:** Nicole M. Hughes conceived and designed the experiments; Melissa C. Pihl, Kaylyn L. Carpenter, Nicole M. Hughes and Timothy S. Keidel performed the experiments; Kaylyn L. Carpenter analyzed the data; Kaylyn L. Carpenter and Nicole M. Hughes wrote the paper.

**Conflicts of Interest:** The authors declare no conflicts of interest and that the funding sponsors had no role in the design of the study; in the collection, analyses, or interpretation of data; in the writing of the manuscript, and in the decision to publish the results.

# References

1. Chalker-Scott, L. Environmental significance of anthocyanins in plant stress responses. *Photochem. Photobiol.* **1999**, *70*, 1–9.
2. Close, D.C.; Beadle, C.L. The ecophysiology of foliar anthocyanin. *Bot. Rev.* **2003**, *69*, 149–161.
3. Gould, K.S. Nature's Swiss Army Knife: The Diverse Protective Roles of Anthocyanins in Leaves. *J. Biomed. Biotechnol.* **2004**, *5*, 314–320.
4. Ruelland, E.; Vaultier, M.-N.; Zachowski, A.; Hurry, V. Cold signaling and cold adaptation in plants. *Adv. Bot. Res.* **2009**, *49*, 35–150.
5. Hughes, N.M. Winter leaf reddening in angiosperm "evergreen" species. *New Phytol.* **2011**, *190*, 573–581.
6. Spyropoulos, C.G.; Mavrommatis, M. Effect of Water Stress on Pigment Formation in *Quercus* Species. *J. Exp. Bot.* **1978**, *29*, 473–477.
7. Chalker-Scott, L. Do anthocyanins function as osmoregulators in leaf tissues? *Adv. Bot. Res.* **2002**, *37*, 104–129.
8. Sperdouli, I.; Moustakas, M. Interaction of proline, sugars, and anthocyanins during photosynthetic acclimation of *Arabidopsis thaliana* to drought stress. *J. Plant Physiol.* **2012**, *169*, 577–585.
9. Manetas, Y.; Drinia, A.; Petropoulou, Y. High contents of anthocyanins in young leaves are correlated with low pools of xanthophyll cycle components and low risk of photoinhibition. *Photosynthetica* **2002**, *40*, 349–354.
10. Karageorgou, P.; Manetas, Y. The importance of being red when young: Anthocyanins and the protection of young leaves of *Quercus coccifera* from insect herbivory and excess light. *Tree Physiol.* **2006**, *26*, 613–621.
11. Liakopoulos, G.; Nikolopoulos, D.; Klouvatou, A.; Vekkos, K.-A.; Manetas, Y.; Karabourniotis, G. The photoprotective role of epidermal anthocyanins and surface pubescence in young leaves of grapevine (*Vitis vinifera*). *Ann. Bot.* **2006**, *98*, 257–265.
12. Hughes, N.M.; Morley, C.B.; Smith, W.K. Coordination of anthocyanin decline and photosynthetic maturation in developing leaves of three deciduous tree species. *New Phytol.* **2007**, *175*, 675–685.
13. Kozlowski, T.T.; Pallardy, S.G. *Physiology of Woody Plants*; Academic Press: San Diego, CA, USA, 1997; pp. 159–172.

14. Archetti, M.; Doring, T.F.; Hagen, S.B.; Hughes, N.M.; Leather, S.R.; Lee, D.W.; Lev-Yadun, S.; Manetas, Y.; Ougham, H.J.; Schaberg, P.G.; *et al.* Adaptive explanations for autumn colours- an interdisciplinary approach. *Trends Ecol. Evol.* **2009**, *24*, 166–173.

15. Manetas, Y. Why some leaves are anthocyanic and why most anthocyanic leaves are red? *Flora* **2006**, *201*, 163–177.

16. Hughes, N.M.; Smith, W.K. Seasonal photosynthesis and anthocyanin production in ten broadleaf evergreen species. *Funct. Plant Biol.* **2007**, *34*, 1072–1079.

17. Hughes, N.M.; Reinhardt, K.; Gierardi, A.; Feild, T.S.; Smith, W.K. Association between winter anthocyanin production and drought stress in angiosperm evergreen species. *J. Exp. Bot.* **2010**, *61*, 1699–1709.

18. Hughes, N.M.; Burkey, K.O.; Cavender-Bares, J.; Smith, W.K. Seasonal xanthophyll cycle and antioxidant properties of red (anthocyanic) and green (acyanic) angiosperm evergreen species. *J. Exp. Bot.* **2012**, *63*, 1895–1905.

19. Hamilton, W.D.; Brown, S.P. Autumn tree colours as a handicap signal. *Proc. R. Soc. Lond. B* **2001**, *268*, 1489–1493.

20. Archetti, M.; Brown, S.P. The coevolution theory of autumn colours. *Proc. R. Soc. Lond. B* **2004**, *271*, 1219–1223.

21. Archetti, M.; Leather, S.R. A test of the coevolution theory of autumn colours: Colour preference of *Rhopalosiphum padi* on *Prunus padus. Oikos* **2005**, *110*, 339–343.

22. Lev-Yadun, S.; Dafni, A.; Flaishman, M.A.; Inbar, M.; Izhaki, I.; Katzir, G.; Ne'eman, G. Plant coloration undermines herbivorous insect camouflage. *Bioessays* **2004**, *26*, 1126–1130.

23. Coley, P.D.; Aide, T.M. Red coloration of tropical young leaves: A possible antifungal defence? *J. Trop. Ecol.* **1989**, *5*, 293–300.

24. Schaefer, H.M.; Rentzsch, M.; Breuer, M. Anthocyanins reduce fungal growth in fruits. *Nat. Prod. Commun.* **2008**, *3*, 1267–1272.

25. Baker, N.R. Chilling stress and photosynthesis. In *Causes of Photooxidative Stress and Amelioration of Defense Systems in Plants*; Foyer, C.H., Mullineaux, P.M., Eds.; CRC Press: Boca Raton, FL, USA, 1994; pp. 105–126.

26. Bao, Y.; Nilsen, E.T. The ecophysiological significance of leaf movements in *Rhodoendron maximum. Ecology* **1988**, *69*, 1578–1587.

27. Adams, W.W., III; Demmig-Adams, B.; Verhoeven, A.S.; Barker, D.H. "Photoinhibition" during winter stress: Involvement of sustained xanthophyll cycle-dependent energy dissipation. *Aust. J. Plant Physiol.* **1994**, *22*, 261–276.

28. Grace, S.C.; Logan, B.A. Acclimation of foliar antioxidant systems to growth irradiance in three broad-leaved evergreen species. *Plant Physiol.* **1996**, *112*, 1631–1640.

29. Verhoeven, A.S.; Adams, W.W., III; Demmig-Adams, B. Close relationship between the state of the xanthophyll cycle pigments and photosystem II efficiency during recovery from winter stress. *Physiol. Plant.* **1996**, *96*, 567–576.

30. Logan, B.A.; Grace, S.C.; Adams, W.W., III; Demmig-Adams, B. Seasonal differences in xanthophyll cycle characteristics and antioxidants in *Mahonia repens* growing in different light environments. *Oecologia* **1998**, *116*, 9–17.

31. Adams, W.W., III; Demmig-Adams, B; Rosenstiel, T.V.; Ebbert, V.; Brightwell, A.K.; Barker, D.H.; Carter, C.R. Photosynthesis, xanthophylls, and D1 phosphorylation under winter stress. In PS2001, Proceedings of the 12th International Congress on Photosynthesis, Brisbane, Queensland, Australia, 18–23 August 2001; CSIRO Publishing: Melbourne, Australia, 2001.

32. Ebbert, V.; Demmig-Adams, B.; Adams, W.W., III; Mueh, K.E.; Staehelin, L.A. Correlation between persistent forms of zeaxanthin-dependent energy dissipation and thylakoid protein phosphorylation. *Photosynth. Res.* **2001**, *67*, 63–78.

33. Hughes, N.M.; Neufeld, H.S.; Burkey, K.O. Functional role of anthocyanins in high-light winter leaves of the evergreen herb, *Galax urceolata*. *New Phytol.* **2005**, *168*, 575–587.

34. Verhoeven, A.S.; Adams, W.W., III; Demmig-Adams, B. The xanthophyll cycle and acclimation of *Pinus ponderosa* and *Malva neglecta* to winter stress. *Oecologia* **1999**, *118*, 277–287.

35. Close, D.C.; Beadle, C.L.; Hovenden, M.J. Interactive effects of nitrogen and irradiance on sustained xanthophyll cycle engagement in *Eucalyptus nitens* leaves during winter. *Oecologia* **2003**, *134*, 32–36.

36. Woodall, G.S.; Dodd, I.C.; Stewart, G.R. Contrasting leaf development within the genus Syzygium. *J. Exp. Bot.* **1998**, *49*, 79–87.

37. Schaberg, P.G.; van Den Berg, A.K.; Murakami, P.F.; Shane, J.B.; Donnelly, J.R. Factors influencing red expression in autumn foliage of sugar maple trees. *Tree Physiol.* **2003**, *23*, 325–333.

38. Hormaetxe, K.; Hernández, A.; Becerril, J.M.; García-Plazaola, J.I. Role of red carotenoids in photoprotection during winter acclimation in *Buxus sempervirens* leaves. *Plant Biol.* **2004**, *6*, 325–332.

39. Kytridis, V.-P.; Karageorgou, P.; Levizou, E.; Manetas, Y. Intra-species variation in transient accumulation of leaf anthocyanins in *Cistus creticus* during winter: Evidence that anthocyanins may compensate for an inherent photosynthetic and photoprotective inferiority of the red-leaf phenotype. *J. Plant Physiol.* **2008**, *165*, 952–959.

40. Zeliou, K.; Manetas, Y.; Petropoulou, Y. Transient winter leaf reddening in *Cistus creticus* characterizes weak (stress-sensitive) individuals, yet anthocyanins cannot alleviate the adverse effects on photosynthesis. *J. Exp. Bot.* **2009**, *60*, 3031–3042.

41. Nikiforou, C.; Zeliou, K.; Kytridis, V.-P.; Kyzeridou, A.; Manetas, Y. Are red leaf phenotypes more or less fit? The case of winter leaf reddening in *Cistus creticus*. *Environ. Exp. Bot.* **2010**, *67*, 509–514.

42. Nikiforou, C.; Nikolopoulos, D.; Manetas, Y. The winter-red-leaf syndrome in *Pistacia lentiscus*: Evidence that the anthocyanic phenotype suffers from nitrogen deficiency, low carboxylation efficiency and high risk of photoinhibition. *J. Plant Physiol.* **2011**, *168*, 2184–2187.

43. Evans, J.R.; Seemann, J.R. The allocation of protein nitrogen in the photosynthetic apparatus: Costs, consequences and control. In *Photosynthesis*; Briggs, W.R., Ed.; Alan R. Liss: New York, NY, USA, 1989; pp. 183–205.

44. Martin, T.; Oswald, O.; Graham, I.A. Arabidopsis seedling growth, storage lipid mobilization, and photosynthetic gene expression are regulated by carbon:nitrogen availability. *Plant Physiol.* **2002**, *128*, 472–481.

45. Hikosaka, K. Nitrogen partitioning in the photosynthetic apparatus of *Plantago asiatica* leaves grown under different temperature and light conditions: Similarities and differences between temperature and light acclimation. *Plant Cell Physiol.* **2005**, *46*, 1283–1290.

46. Hikosaka, K. Interspecific difference in the photosynthesis-nitrogen relationship: Patterns, physiological causes, and ecological importance. *J. Plant Res.* **2004**, *117*, 481–494.

47. Nikiforou, C.; Manetas, Y. Strength of winter leaf redness as an indicator of stress vulnerable individuals in *Pistacia lentiscus*. *Flora* **2010**, *205*, 424–427.

48. Campbell, J.W. *Personal Communication*; High Point University: High Point, NC, USA, 2013.

49. Spreitzer, R.J.; Salvuccia, M.E. RUBISCO: Structure, regulatory interactions, and possibilities for a better enzyme. *Annu. Rev. Plant Biol.* **2002**, *53*, 449–475.

50. Field, C.B.; Mooney, H.A. The photosynthesis-nitrogen relationship in wild plants. In *The Economy of Plant form and Function*; Givnish, T.J., Ed.; Cambridge University Press: Cambridge, UK, 1986; pp. 25–55.

51. Zhao, D.; Reddy, K.R.; Kakani, V.G.; Reddy, V.R. Nitrogen deficiency effects on plant growth, leaf photosynthesis and hyperspectral reflectance properties of sorghum. *Eur. J. Agron.* **2005**, *22*, 391–403.

52. Peng, M.; Hannam, C.; Gu, H.; Bi, Y.-M.; Rothsteain, S.J. A mutation in *NLA*, which encodes a RING-type ubiquitin ligase, disrupts the adaptability of *Arabidopsis* to nitrogen limitation. *Plant J.* **2007**, *50*, 320–337.

53. Lea, U.S.; Slimestad, R.; Smedvig, P.; Lillo, C. Nitrogen deficiency enhances expression of specific MYB and bHLH transcription factors and accumulation of end products in the flavonoid pathway. *Planta* **2007**, *225*, 1245–1253.

54. Scheible, W.-F.; Morcuende, R.; Czechowski, T.; Fritz, C.; Osuna, D.; Palacios-Rojas, N.; Schindelasch, D.; Thimm, O.; Udvardi, M.K.; Stitt, M. Genome-wide reprogramming of primary and secondary metabolism, protein synthesis, cellular growth processes, and the regulatory infrastructure of Arabidopsis in response to nitrogen. *Plant Physiol.* **2004**, *136*, 2483–2499.

55. Peng, M.; Bi, Y.-M.; Zhu, T.; Rothstein, S.J. Genome-wide analysis of *Arabidopsis* responsive transcriptome to nitrogen limitation and its regulation by the ubiquitin ligase gene *NLA*. *Plant Mol. Biol.* **2007**, *65*, 775–797.

56. USDA NRCS National Plant Data Team. *Lonicera japonica* Thunb. Japanese honeysuckle, the PLANTS Database. Available online: http://plants.usda.gov/core/profile?symbol= LOJA (accessed on 13 March 2013).

57. Porra, R.J. The chequered history of the development and use of simultaneous equations for the accurate determination of chlorophylls *a and b*. *Photosyn. Res.* **2002**, *73*, 149–156.

58. Ku, S.-B.; Edwards, G. Oxygen inhibition of photosynthesis, II. Kinetic characteristics as affected by temperature. *Plant Physiol.* **1977**, *59*, 991–999.

59. Farquhar, G.D.; von Caemmerer, S.; Berry, J.A. A biochemical model of photosynthetic $CO_2$ assimilation in leaves of $C_3$ species. *Planta* **1980**, *149*, 78–90.

60. Von Caemmerer, S.; Farquhar, G.D. Some relationships between the biochemistry of photosynthesis and the gas exchange of leaves. *Planta* **1981**, *153*, 376–387.

61. Lawanson, A.O.; Akindele, B.B.; Fasalojo, P.B.; Akpe, B.L. Time-course of anthocyanin formation during deficiencies of nitrogen, phosphorus and potassium in seedlings of Zea mays Linn. var. E.S. 1. *Z. Pflanzenphysiol.* **1972**, *66*, 251–253.

62. Nittler, L.W.; Kenny, T.J. Effect of ammonium to nitrate ratio on growth and anthocyanin development of perennial ryegrass cultivars. *Agron. J.* **1976**, *68*, 680–682.

63. Oberbauer, S.F.; Starr, G. The role of anthocyanins for photosynthesis of Alaskan arctic evergreens during snowmelt. *Adv. Bot. Res.* **2002**, *37*, 129–145.

**Sample Availability:** *Sample Availability:* Samples of the compounds are not available from the authors.

# The Role of Acyl-Glucose in Anthocyanin Modifications

Nobuhiro Sasaki, Yuzo Nishizaki, Yoshihiro Ozeki and Taira Miyahara

**Abstract:** Higher plants can produce a wide variety of anthocyanin molecules through modification of the six common anthocyanin aglycons that they present. Thus, hydrophilic anthocyanin molecules can be formed and stabilized by glycosylation and acylation. Two types of glycosyltransferase (GT) and acyltransferase (AT) have been identified, namely cytoplasmic GT and AT and vacuolar GT and AT. Cytoplasmic GT and AT utilize UDP-sugar and acyl-CoA as donor molecules, respectively, whereas both vacuolar GT and AT use acyl-glucoses as donor molecules. In carnation plants, vacuolar GT uses aromatic acyl-glucoses as the glucose donor *in vivo*; independently, vacuolar AT uses malylglucose, an aliphatic acyl-glucose, as the acyl-donor. In delphinium and *Arabidopsis*, p-hydroxybenzoylglucose and sinapoylglucose are used *in vivo* as bi-functional donor molecules by vacuolar GT and AT, respectively. The evolution of these enzymes has allowed delphinium and *Arabidopsis* to utilize unique donor molecules for production of highly modified anthocyanins.

Reprinted from *Molecules*. Cite as: Sasaki, N.; Nishizaki, Y.; Ozeki, Y.; Miyahara, T. The Role of Acyl-Glucose in Anthocyanin Modifications. *Molecules* **2014**, *19*, 18747–18766.

## 1. Introduction

Angiosperms display a wide range of flower colors and many species are exploited for horticultural purposes because of this characteristic. Flower colors commonly depend on three major plant pigments, namely, flavonoids/anthocyanins, betalains and carotenoids [1]. Of these, anthocyanins are responsible for the widest array of color varieties. The fundamental color of anthocyanins depends on the aglycon type. The six common anthocyanin aglycons are produced in angiosperms that have different pattern of the hydroxylation and methylation on B-ring [1]. Although angiosperms have only six anthocyanin aglycons (the basic skeletal form of anthocyanins), their flowers show considerable variability in color. Anthocyanins are subjected to diverse types of modifications, such as the position within the molecule of the modifying moiety, the type of attached sugar or organic acid, or a combination of these variable factors; through this diversity of possible modifications, a large range of differently colored anthocyanins can be generated. Furthermore, these modifications play important roles in the chemical stabilization of anthocyanin molecules in the vacuolar sap and in determining the color of the flowers in response to variations in vacuolar sap pH or to interactions with metal ions or aromatic organic acids in the vacuole [2]. The understanding of the reaction mechanisms to generate those complex anthocyanin molecules would help us to know how the plants acquire them in their evolutional process and to clarify the other metabolite biosynthetic pathways. The DNA manipulation technique based on that information would be useful to create a flower crops presenting the new color.

Some plant species can produce complicated molecular anthocyanin structures, termed polyacylated anthocyanins, by attachment of multiple sugar and organic acid moieties (see Figure 1). As each step of anthocyanin modification is mediated by specific enzymes, the diversity of anthocyanin molecular structures is believed to have resulted from divergent evolution of those enzymes. In recent years, most of the enzymes, and the genes encoding them, for each reaction step to generate anthocyanin aglycons have been identified, and, currently, attempts to change and modulate flower colors using the information on gene sequences have been initiated [3].

Glycosylation of anthocyanidin and anthocyanin is largely catalyzed by family 1 glycosyltransferases (UGTs) that utilize UDP-sugars as the sugar donor and that are active in the cytosol [4]. To date, UGTs have been identified in many plant species [4]. A second modification system, acylation, is commonly catalyzed in the cytosol by anthocyanin acyltransferase (AT) that utilizes acyl-CoA as the acyl-donor. This type of acyltransferase is classified as a BAHD family protein, which characteristically displays benzoylalcohol acetyl transferase, anthocyanin-O-hydroxycinnamoyltransferase, anthoranilate-N-hydroxybenzoyl/benzoyltransferase and deacetyl-vindoline acetyltransferase activities [5]. The final modified structures of some anthocyanins, such as the polyacylated anthocyanin gentiodelphin of Japanese gentian (Figure 1), are accomplished through the activities of UGTs and an AT. All of the reaction steps necessary to synthesize gentiodelphin have now been identified [6]. In gentian flower petals, the synthesized anthocyanin aglycon is glucosylated at the 3-position and the 5-position. Two pathways have been identified: acylation of the glucose moiety at the 5-position in advance of 3'-glucosylation; and, 3'-glucosylation before the acylation of the glucose moiety at the 5-position [7]. Interestingly, acylation of both glucoses at the 5- and 3'-positions is catalyzed by the same enzyme [8–10]. Hence, gentiodelphin formation is achieved through modification of the aglycon by UGTs and a BAHD type AT (BAHD-AT).

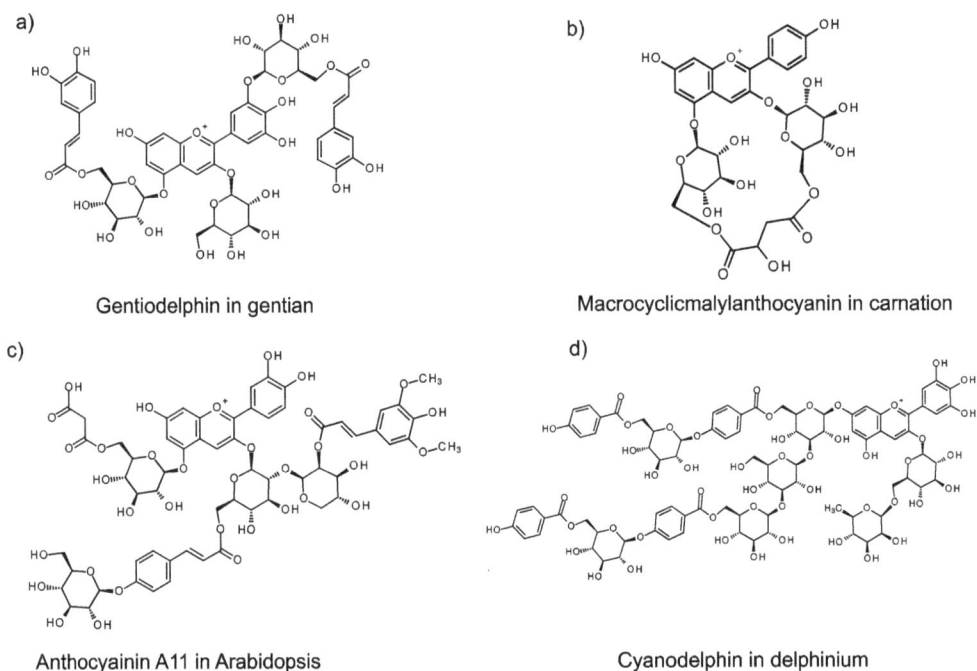

a) Gentiodelphin in gentian

b) Macrocyclicmalylanthocyanin in carnation

c) Anthocyainin A11 in Arabidopsis

d) Cyanodelphin in delphinium

**Figure 1.** Representative anthocyanins showing modification by sugar and acyl groups.

Anthocyanin acylation has also been reported to be mediated by an enzyme that is clearly distinct from BAHD-AT. Carrot cells in suspension cultures produce the anthocyanin cyanidin 3-*O*-(6"-*O*-(6'''-*O*-sianpoyl-glucosyl)-2"-*O*-xylosyl)-galactoside (Cya 3-(Xyl-sinapoyl-Glc-Gal). Crude enzyme extracts prepared from these cells can catalyze the transfer of the sinapoyl group from sinapoylglucose to the 6-position of the glucose attached to the anthocyanin (Figure 2) [11,12]. Likewise, in *Chenopodium rubrum* cells in suspension cultures and in *Lampranthus sociorum* petals, *p*-coumaroylglucose and feruloylglucose are utilized as acyl-donors for betacyanins, another major class of plant pigment [13]. Identification of the genes for these acyl-glucose-dependent ATs was only achieved much later. In 2000, acyl-glucose-dependent ATs involved in diacyl-glucose biosynthesis in the wild tomato (*Lycopersicon pennellii*) and sinapoylmalate biosynthesis in *Arabidopsis thaliana* were characterized [14,15]. Analyses of their amino acid sequences revealed that these acyl-glucose dependent acyltransferases belong to the serine carboxypeptidase-like (SCPL) protein family. More interestingly, immunological and cell biological analyses showed the transition of this type of AT (termed here SCPL-AT) into the vacuole where they are expected to act to accumulate acyl-glucose(s) [16,17].

Initially, it was believed that the glycosylation reactions for production of anthocyanins were only catalyzed by UGTs, which can glycosylate a wide variety of plant secondary metabolites. However, another type of anthocyanin glucosyltransferase (GT) has been discovered [18]. Surprisingly, this new type of GT belongs to a separate protein family from UGTs, namely glycoside hydrolase family 1 (GH1), which characteristically hydrolyzes glycosides. This GH1 type glucosyltransferase (GH1-GT) transfers a glucosyl group to an anthocyanin using an acyl-glucose as the glucosyl donor. Glucosylation reactions mediated by a GH1-GT have now been identified in several species [17,19–21].

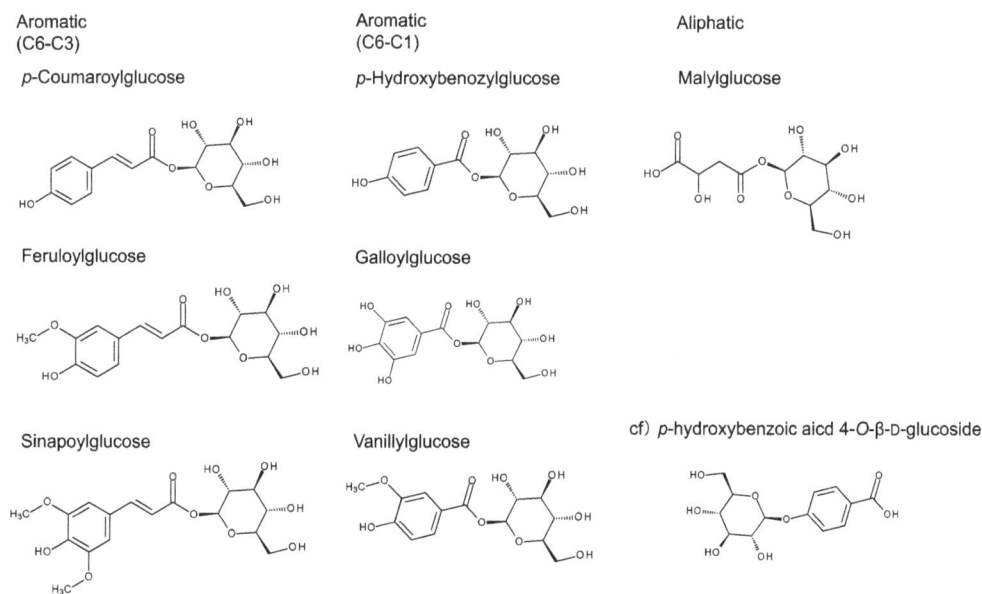

Figure 2. Representative chemical structures of acyl-glucoses.

Furthermore, the acyl-glucose dependent anthocyanin glucosyltransferase (AAGT) protein contains a signal peptide at its *N*-terminal end for translocation into the vacuole [17,18]. As a consequence of recent investigations, four types of enzyme have now been identified as mediating anthocyanin modification: two glycosyltransferases (GTs) and two acyltransferases (ATs), with one of each type functioning in the cytosol and the other in the vacuole [3] (Figure 3). Thus, the variable anthocyanin molecules in higher plants are created by the combined activities of these enzymes. Notably, both vacuolar enzymes, *i.e.*, GH1-GT and SCPL-AT, make use of acyl-glucoses as the donor molecule [3].

Cytoplasmic type | Vacuolar type

Glycosylations

| UDP-sugar dependent glycosyltransferase (UGT) | Acyl-glucose dependent glucosyltransferase (GH1-GT) |
|---|---|
| Glucose, galactose, rhamnose, etc. | Glucose |
| Acyl-CoA dependent acyltransferase (BAHD-AT) | Acyl-glucose dependent acyltransferase (SCPL-AT) |
| Aliphatic and aromatic organic acid | Aliphatic and aromatic organic acid |

Acylations

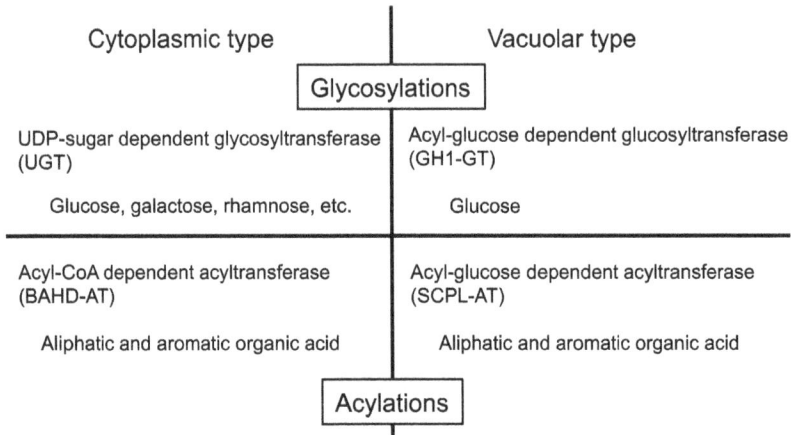

Figure 3. Classification of enzymes involved in anthocyanin modification.

As is outlined above, recent studies have clarified the contribution of acyl-glucoses to anthocyanin generation in higher plants. In the remainder of this review, we focus on the roles of acyl-glucoses as glucosyl and acyl donors in the anthocyanin biosynthetic pathways of carnation, delphinium and *Arabidopsis* in which the involvement of acyl-glucoses in both acylation and glucosylation of anthocyanin were shown.

## 2. UGTs Involved in Acyl-Glucose Generation

As described above, acyl-glucose contributes to the diversity of secondary metabolites by acting as an energy-rich donor molecule in both transacylation and transglucosylation reactions. It is known that the biosynthesis of glucose esters is performed by UGTs (Figure 4). Acyl-glucoses mainly accumulate in the vacuole, although the glucosyltransferase reactions catalyzed by UGTs take place in the cytosol [22]. To date, a number of UGTs have been reported as being involved in the formation of acyl-glucose. In *Arabidopsis*, many of the UGTs necessary for synthesis of the different types of acyl-glucose have been identified. Lim *et al.* performed biochemical analyses of recombinant proteins *in vitro* and identified UGTs that exhibit GT activity to form an ester bond [23]. That study listed UGT84A1, UGT84A2 and UGT84A3 as showing significant activity to form glucose ester conjugates with hydroxycinnamic acids (HCAs), such as cinnamic acid, *p*-coumaric acid, caffeic acid, ferulic acid and sinapic acid. Although UGT84A1, UGT84A2 and UGT84A3 functionally overlap in their acceptor preferences *in vitro*, subsequent investigations have shown that UGT84A1 is involved in *p*-hydroxybenzoylglucose (pHBG) synthesis rather than HCA-glucose synthesis [24]. Analyses using overexpressing or functionally-deficient mutant lines show that UGT84A2 supplies sinapoylglucose, which acts as a bi-functional donor for anthocyanin modification [25], and that UGT84A3 seems to be involved in the synthesis of *p*-coumaroylglucose, which is associated with cell wall structure [26].

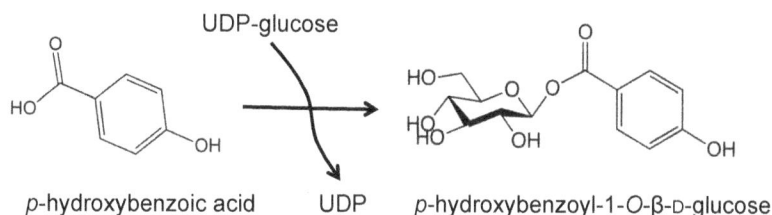

**Figure 4.** Representative UDP-glucose dependent glucosyltransferase (UGT) reaction for acyl-glucose synthesis.

Phylogenetic analysis of the amino acids sequences of UGTs may provide some insights into their metabolic functions. An example of such an analysis of UGTs responsible for glucose ester and *O*-glucoside synthesis in higher plants is presented in Figure 5. This analysis shows that UGTs capable of glucose ester formation largely form a cluster (marked with an asterisk in Figure 5), although there are exceptions.

For example, UGT75B1 is located in a cluster composed of anthocyanin 5-*O*-glucosyltransferases; Lim *et al.* showed that UGT75B1 acts on the carboxyl group of hydroxybenzoic acids (HBAs), such as benzoic acid, 3-hydroxybenzoic acid, *p*-hydroxybenzoic acid (pHBA) and 3,4-dihydroxybenzoic acid [24]. UGT75B1 also catalyzes glucose ester formation in *p*-aminobenzoic acid (pABA), which contributes to the storage of pABA in the vacuole in the glucosylated form [27]. Although UGT74F1 can form a benzoate glucose ester *in vitro* [24], the *in vivo* role of this enzyme is suggested to be 2-*O*-glucosylation of salicylic acid [28]. Both VlRSgt of *Vitis labrusca* and CuLGT of *Citrus unshiu* have *O*-glucosyltransferase activity, although they are positioned in the ester-forming GT clade in the phylogenetic tree (Figure 5). It has been reported that VlRSgt also shows significant activity in the formation of glucose ester conjugates with both HCAs and HBAs [29,30]. Many UGTs seem to exhibit broad substrate specificities *in vitro* and to show enzymatic activity for substrates other than those in their *in vivo* metabolic pathway(s) [31–33]. Sometimes, such discrepancies lead to confusion in determining the *in vivo* roles of the enzymes. Thus, conclusions on the function of a UGT need to be based not only on *in vitro* enzymatic activity but also on the results from genetic, physiological and molecular biological analyses.

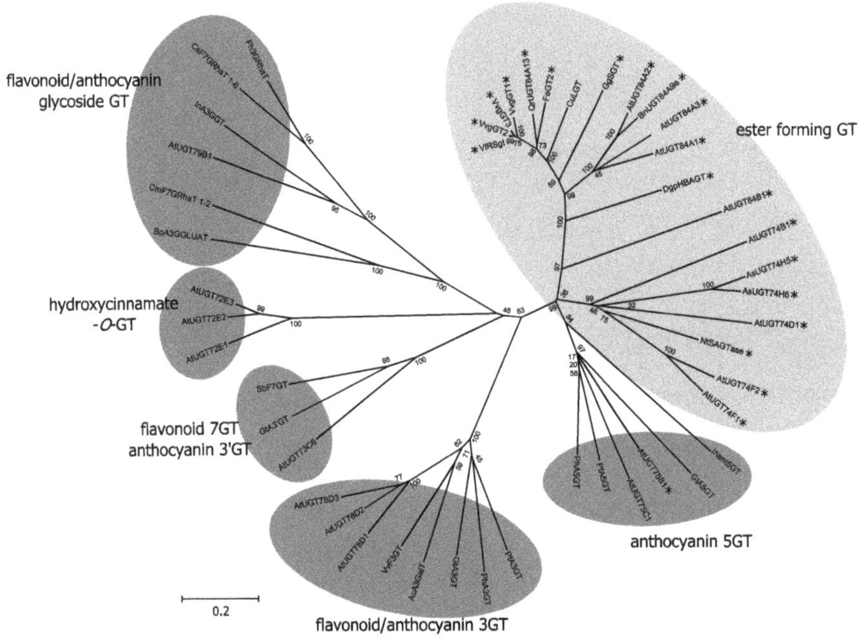

The accession numbers of the amino acid sequences are as follows: Ac3AGalT (AB103471), AtUGT72E1 (AL049862), AtUGT72E2 (AB018119), AtUGT72E3 (AF077407), AtUGT73C6 (NM_129234), AtUGT74B1 (BT001160), AtUGT74D1 (NP_180734), AtUGT74F1 (NP_973682), AtUGT74F2 (BT010327), AtUGT75B1 (NP_563742), AtUGT75C1 (AK226538), AtUGT78D1 (NM_102790), AtUGT78D2 (NM_121711), AtUGT78D3 (NM_121709), AtUGT79B1 (BT033073), AtUGT84A1 (BT015796), AtUGT84A2 (Q9LVF0), AtUGT84A3 (NP_193284), AtUGT84B1 (NM_127890), AsUGT74H5 (EU496509), AsUGT74H6 (EU496520), BnUGT84A9a (AF287143), BpA3GGLUAT (AB190262), CuLGT (Q9MB73), CmF7GRhaT 1-2 (AY048882), CsF7GRhaT 1-6 (DQ119035), DgpHBAGT (AB889521), FaGT2 (AY663785), GgSGT (AB362221), GtA3GT (D85186), GtA3'GT (AB076697), GtA5GT (AB363839), Ihant5GT (AB113664), In3GGT (AB192314), NtSAGTase (AF190634), PfA3GT (AB002818), PfA5GT (AB013596), PhA3GT (AB027454), PhA5GT (AB027455), Ph3GRhaT (X71059), QrUGT84A13 (KF527849), SbF7GT (AB031274), VlRSgt (ABH03018), VvF3GT (AF000371), VvgGT1 (JN164679), VvgGT2 (JN164680) and VvgGT3 (JN164681). Ac, *Aralia cordata*; As, *Avena strigosa*; At, *Arabidopsis thaliana*; Bn, *Brassica napus*; Bp, *Bellis perennis*; Cm, *Citrus maxima*; Cs, *Citrus sinensis*; Cu, *Citrus unshiu*; Dg, *Delphinium grandiflorum*; Fa, *Fragaria* × *ananassa*; Gg, *Gomphrena globosa*; Gt, *Gentiana triflora*; Ih, *Iris* × *hollandica*; In, *Ipomoea nil*; Nt, *Nicotiana tabacum*; Pf, *Perilla frutescens*; Ph, *Petunia* × *hybrida*; Qr, *Quercus robur*; Sb, *Scutellaria baicalensis*; Vl, *Vitis labrusca*; Vv, *Vitis vinifera*. Bar = 0.2 amino acid substitutions/site.

**Figure 5.** Phylogenetic tree analysis of UGTs mediating acyl-glucose production. The tree was constructed from multiple amino acid sequences using the Neighbor-Joining algorithm. The clade of the UGTs involved in acyl-glucose biosynthesis is shown with a gray circle and the others with dark gray circles. UGTs that exhibit acyl-glucose synthesis activity *in vitro* are marked with asterisks.

pHBG, the key compound for cyanodelphin and violdelphin biosynthesis in delphiniums, is supplied by a UGT, DgpHBAGT, that exhibits broad substrate specificity *in vitro* [34]. In DgpHBAGT depletion cultivars, delphinidin 3-O-rutinoside, *i.e.*, 7-unmodified anthocyanin, is accumulated as the major anthocyanin because lack of pHBG prevents enzymes such as DgAA7GT generating anthocyanin 7-polyacylation [34]. Interestingly, these cultivars accumulate *p*-hydroxybenzoic acid 4-O-β-D-glucoside instead of pHBG (Figure 1). This may be the result of other UGT(s) showing broad substrate specificities and acting to glucosylate excess pHBA to detoxify and transport it into the vacuole. In wild type cultivars, the pHBA synthesized in the cytosol is efficiently glucosylated by DgpHBAGT for transport into the vacuole. Such flexibility in the metabolic roles of UGTs *in vivo* might be one mechanism for producing the diversity of secondary metabolites in plants. In addition to aromatic acyl-glucose, aliphatic acyl-glucose also acts as an active donor. Malylglucose is used as the donor of a malyl moiety in anthocyanin modification in carnations (described above). In addition to anthocyanin biosynthesis, isobutyrylglucose is known to be an energy-rich acyl-donor for glucose polyester generation by LpSCPL in the wild tomato (*L. pennellii*) [15,35]. Despite the importance of aliphatic acyl-glucoses, relatively little is known of the UGTs that are responsible for aliphatic acyl-glucose synthesis. FaGT2, which was first shown to be involved in cinnamoyl- and *p*-coumaroyl-glucose synthesis during the ripening of strawberry fruits, also exhibits the ability to glucosylate sorbic acid at very high rates and to glucosylate HCAs and xenobiotic compounds [36,37]. Future research on GTs involved in aliphatic acyl-glucose will undoubtedly provide new insights into the *in vivo* roles of aliphatic acylated compounds and their biosynthetic pathways in higher plants.

## 3. Acyl-Glucoses as Potential Intermediates for Secondary Metabolite Biosynthesis

Plant secondary metabolites commonly exist in their glycoside form in which aglycons bind to sugar group(s) through glycosidic (ether) linkages in the vacuole [38]. Some metabolites accumulate in the C-glycoside form [39], whereas others form compounds in which organic acid and glucose are bound to each other by ester bonding. Typical acyl-glucoses (acyl-1-O-β-D-glucose) are illustrated in Figure 2 (these substances are generally termed acyl-glucoses rather than glucosides). The major category of acyl-glucoses is that conjugating to HCAs and HBAs. For example, caffeoylglucose and coumaroylglucose accumulation occur in the fresh flowers of *Camellia reticulate* [40], while large amounts of sinapoylglucose and feruloylglucose respectively accumulate in carrot and silver beachtop cells in suspension cultures [11,12]. Likewise, accumulation of HCA-glucoses and HBA-glucoses has been reported in carnation petals [18]. All these acyl-glucoses are thought to be stored in the vacuole [41]. Plant species of the order Brassicaceae accumulate sinapoylglucose; additionally, acylated sinapoylglucose has also been detected in *Brassica napus* seeds [42]. Polyacylated acyl-glucose has been found in the leaves of *Rhus typhina* and dimeric *p*-coumaroylglucose has been extracted from *Petrorhagia velutina* [43,44]. Some studies have also reported other acyl-glucoses that are not derived from HCAs or HBAs. In the carnation, in addition to various aromatic acyl-glucoses, aliphatic acyl-glucose has also been found [45]. Aliphatic acyl-glucose also occurs in the glandular trichomes of the wild tomato (*Lycopersicon pennellii*), which secrete 2,3,4-tri-isobutyrylglucose to aid insect resistance [35].

It is clear from the above description that acyl-glucose molecules are widespread in the plant kingdom. These acyl-glucoses have a variety of functions, such as antifeedants, phytoanticipins and phytoalexins, signaling molecules, and UV protectants [46]. Thus, acyl-glucoses are physiologically important compounds in plants as they are the final, functional products of secondary metabolites. A possible role for acyl-glucoses has been proposed from the estimation of Gibb's free energy changes that occur in acyl-glucose hydrolysis: acyl-glucoses may function as high-energy intermediates in the production of acylated secondary metabolites [47]. As described above, acyl-glucose dependent acyltransferase activities have been detected in several plants such as the carrot, silver beachtop, wild tomato, *C. rubrum*, and *L. sociorum*. One well-studied metabolic pathway involving acyl-glucose dependent acyltransferase is the biosynthesis of sinapoyl-derivatives in brassicaceous species such as *Arabidopsis*. Sinapoylmalate is a major sinapate ester in *Arabidopsis* that accumulates in the vacuoles of cells in the sub-epidermal tissue to provide shielding against the deleterious effects of UV-B irradiation [26]. Formation of the sinapoylmalate is mediated by sinapoylglucose:malate sinapoyltransferase (an SCPL-AT family protein). Sinapoylcholine, a major phenylpropanoid that accumulates in seeds and is utilized during germination, is generated by a similar enzyme, SCPL19, that transfers a sinapoyl group to choline from acyl-glucose [48]. SCPL19 can also transfer a benzoyl group to choline from benzoylglucose [49]. Arabidopsis seeds contain a range of related compounds, such as benzoylated glucosinolate and sinapoylated glucosinolate. A knockout mutant analysis showed that *SCPL17* was a candidate gene for benzoylated glucosinolate production. Sinapoylglucose dependent sinapoyltransferase can also synthesize the more complicated sinapoyl ester, 1,2-disinapoylglucose [46].

Recent investigations in our laboratory have identified another role for acyl-glucoses in anthocyanin modification. We found that several plant species glucosylate some anthocyanins in the vacuole with an enzyme that utilizes acyl-glucose as the glucosyl-donor [17–21]. That is, acyl-glucoses can be used not only as acyl-donors but also as glucosyl-donors. Our studies showed that pHBG can be used as a bi-functional donor for both acyl- and glucosyl-moieties for anthocyanin modification in delphinium [17]. The properties of the enzymes for this modification are described below.

## 4. Carnations

Carnations are the one of the most popular and big-market flower crops worldwide. Flower colors in carnations result from modification of the anthocyanin molecule, macrocyclicmalylanthocyanin (Figure 1b) [50–53]. The anthocyanin can be modified by two glucose molecules at the 3- and 5-positions and by formation of a bridge by a single molecule of malate between these moieties; this modified anthocyanin compound is thought to be limited to the family Caryophyllaceae [51].

Aliphatic acyl-group transfer reactions are generally catalyzed by a BAHD-AT enzyme. For example, bonding of a malonyl residue is mediated by an anthocyanin malonyltransferase, a type of enzyme that is now well characterized in several plant species [54]. The modification of anthocyanin with succinic acid, an aliphatic organic acid similar to malic acid (they differ in the hydroxyl group at the C-2 position), has been identified in the cornflower (*Centaurea cyanus*) [55–57]. The enzymatic activity of succinyl-CoA:anthocyanin 3-glucoside succinyltransferase has been characterized [58]; however, to date, the gene encoding this enzyme has not been identified. Although the information on the activity of this enzyme seems to support the view that the malyl transfer reaction is also catalyzed by an acyl-CoA dependent AT in carnations, a study in 2008 showed that the reaction was actually mediated by a distinctly different type of AT [45]. In a screen for the compound that acts as the malyl-donor for malyltransferase in carnation petals (Figure 6), Abe *et al.* identified malylglucose (not malyl-CoA) as the malyl-donor molecule in the anthocyanin malyl transfer reaction.

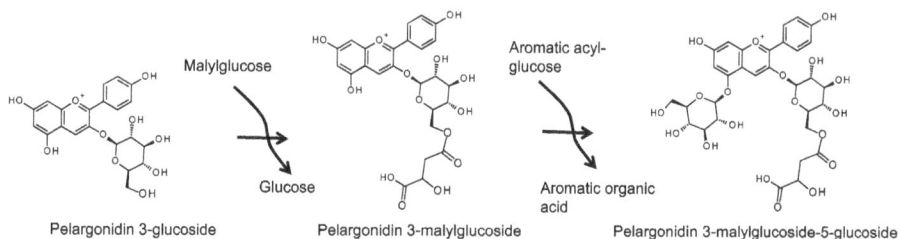

**Figure 6.** Anthocyanin malylation and glucosylation catalyzed by acyl-glucose dependent transferases in carnation.

A candidate gene for anthocyanin malyltransferase (AMalT) was identified in variegated carnation flowers using a transposon-tagging method [59]. AMalT is a member of the SCPL-AT family, which includes proteins that are known to act in the vacuole [16]. Interestingly, crude protein extracts prepared from carnation petals show AMalT activity toward anthocyanidin 3-glucosides, although only trace activity occurs when anthocyanin 3,5-diglucosides are used as the acyl acceptor. This suggests that 5-glucosylation may occur in the vacuole; this reaction is expected to be the next step after malylation catalyzed by a vacuolar enzyme.

The cDNA for the anthocyanin 3-glucosyltransferase that mediates the first modification step in macrocyclicmalylanthocyanin has been isolated from carnation petals [60]. Additionally, 18 UGT homologous cDNAs have also been isolated; however, the enzyme that mediates the 5-glucosylation step has not been identified, although anthocyanin 5-glucosyltransferases of other plant species have been identified [54]. Recently, Matsuba et al. reported that anthocyanin 5-glucosylation is performed by an enzyme that is distinctly different to UGT [18]. They showed that anthocyanin 5-glucosyltransferase activity could be detected in a crude enzyme extract prepared from carnation petals combined with an alcohol extract containing low molecular substances expected to include glucose donor substance(s). Vanillylglucose, a type of acyl-glucose, was identified as the glucosyl-donor molecule for the anthocyanin 5-glucosylation reaction in carnation petals (Figure 6). Subsequently, an enzyme showing acyl-glucose dependent anthocyanin 5GT (DcAA5GT) activity was purified and the amino acid sequence of its N-terminus was determined. The observation that this sequence lacks a first methionine implies that the protein contains a signal peptide at its N-terminus. Analysis of the deduced amino acid sequence corresponding to the full-length cDNA suggests that the enzyme belongs to the glycoside hydrolase family 1 (GH1); this family of proteins are known to mediate glycoside hydrolysis and to have a putative signal peptide at their N-terminal ends. In a transient expression experiment in onion epidermal cells, AA5GT fused to green fluorescence protein was found to localize to the vacuole [18].

The final steps of anthocyanin modification in carnations are likely to require the activities of both AMalT and DcAA5GT in the vacuole. AMalT is able to utilize anthocyanin 3-glusoside as the acceptor molecule but is less efficient at utilizing anthocyanin 3,5-diglucosides [45]. By contrast, DcAA5GT can recognize both anthocyanin 3-glucoside and 3-malylglucoside [18], suggesting that glucosylation of the 5-position occurs after malylation of the glucose moiety at the 3-position. However, it is still unclear why the AMalT reaction using anthocyanin 3-glucoside occurs prior to AA5GT reaction in vivo. Although both AMalT and DcAA5GT recognize anthocyanin 3-glucosides, the modification steps to generate macrocyclicmalylanthocyanin as the final product are catalyzed in the sequence of malylation and then glucosylation at the 5-position. If glucosylation of anthocyanin 3-glucoside occurs (an activity that has been observed in vitro) prior to malylation in vivo, then both macrocyclicmalylanthocyanin and anthocyanin 3,5-diglucosides would be expected to be synthesized and accumulated in petals. Petals of wild type carnations accumulate macrocyclicmalylanthocyanin but few 3,5-diglucosides could be detected; however, anthocyanin 3,5-diglucosides are synthesized and accumulated in AMalT defective mutants [59,61]. The order of this sequential reaction might result from the intrinsic properties of the enzymes or from the formation of weakly associated enzyme complex in the vacuole [62]. Future analyses of the proteins will undoubtedly provide further insights into this phenomenon.

## 5. Delphiniums

Delphiniums are popular ornamental plants because of their characteristic blue flowers. Several mechanisms have been proposed to explain how anthocyanin molecules produce the bluish coloration of the sepals [2]. Delphinium sepals contain two major 7-polyacylated anthocyanins termed violdelphin and cyanodelphin [63,64]. The formation of face-to-face intramolecular stacking structures between anthocyanin chromophores and aromatic groups attached to the anthocyanins through their sugar groups can generate blue pigment complexes [2,64,65]. Violdelphin consists of a structure that contains two glucose moieties and pHBA groups at the 7-position (Figure 7). It is known that species such campanula (*Campanula medium*) and cineraria (*Senecio cruentus*), which have violet-blue flowers, accumulate delphinidin-based anthocyanins decorated with multiple aromatic acyl groups at the 7-position [2]. However, characterization of anthocyanin 7-glucosylation, the expected first step of modification at the 7-position, was not achieved until 2010. At that time, the anthocyanin 5-glucosylation mechanism was first clarified in carnations; subsequently, anthocyanin 7-glucosyltransferase was identified in delphinium by means of a homology based cloning method [18]. Anthocyanin 7-glucosyltransferase in delphinium (DgAA7GT) is a GH1-GT, as is DcAA5GT. AA7GTs have also recently been identified in agapanthus and campanula [20,21]. Agapanthus is a monocot, while delphinium and campanula belong to different dicot orders. The facts that AA7GTs are present in such evolutionarily diverse plant species and that anthocyanin 7-glucosylation mediated by UGTs has not been reported [54], supports the conclusion that anthocyanin 7-glucosylation is commonly mediated by GH1-GTs.

In delphinium, the modification steps after 7-glucosylation have been clarified in the biosynthesis of violdelphin [delphinidin 3-O-rutinoside-7-O-(6-O-(4-O-(6-O-(p-hydroxybenzoyl)-glucosyl)-oxybenzoyl)-glucoside)]. Analyses using a crude protein extract prepared from delphinium sepals showed that these modification steps are carried out in a step-by-step sequence in the order p-hydroxybenzoylation, a second glucosylation, and then a second p-hydroxybenzoylation [17]. All of the enzymatic reactions to generate violdelphin after the 7-glucosylation reaction are mediated by acyl-glucose dependent enzymes (SCPL-ATs and GH1-GT): p-hydroxybenzoylation is catalyzed by AA7G-AT; glucosylation is catalyzed by AA7BG-GT; and the second p-hydroxybenzoylation is catalyzed by AA7GBG-AT. The deduced amino acid sequences of these SCPL-ATs and GH1-GT contain a putative signal peptide for localization in the vacuole at their N-termini. This attribute of these enzymes was expected since it is reasonable to predict that the reactions that follow 7-glucosylation, which is mediated by a vacuolar type GH1-GT, should also occur in the vacuole. For the biosynthesis of violdelphin, all of the biosynthetic reactions are expected to occur sequentially and be dependent on pHBG (Figure 7). pHBG behaves as a bi-functional donor providing the glucosyl moiety for AA7GT and AA7BG-GT, and the acyl moiety for AA7G-AT and AA7GBG-AT; this bi-functionality has led to pHBG being referred to as a "Zwitter donor" in analogy to the concept of a Zwitterion [17].

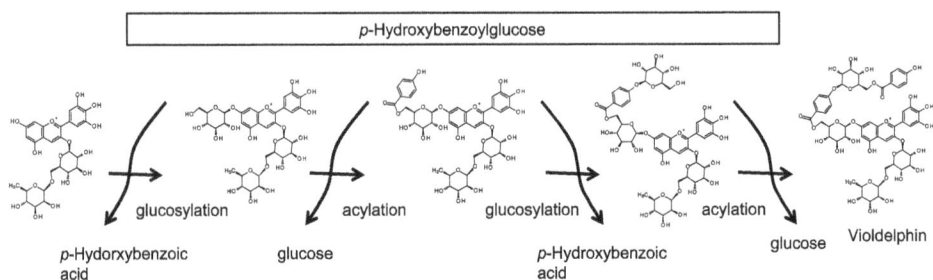

**Figure 7.** Concatenated units in violdelphin synthesized in a step-by-step reaction utilizing acyl-glucose as both glucosyl and acyl donor in delphinium.

Bluish delphinium sepals accumulate a remarkable amount of pHBG, but do not accumulate other aromatic acyl-glucoses, such as vanillyl-, isovanillyl-, caffeoyl-, p-coumaroyl-, sinapoyl- or feruloyl-glucose to a detectable level. Recombinant DgAA7GT and DgAA7BG-GTs (DgAA7BG-GT1 and DgAA7BG-GT2) proteins in *Escherichia coli* showed a donor preference for pHBG over other aromatic acyl-glucoses [17]. However, both AA7G-AT and AA7GBG-AT in crude extracts of delphinium sepals can use other HCA-glucose species as well as pHBG [17]. Despite this ability, there is no evidence that anthocyanins in delphiniums are modified using HCG molecules other than pHBG. The moieties attached to anthocyanins in delphiniums appear to be determined by the acyl-glucose species that accumulates in the vacuole rather than by the substrate preference of the AT. Thus, carrot cells and beach silvertop cells in suspension cultures accumulate Cya 3-(Xyl-sinapoyl-Glc-Gal) and Cya 3-(Xyl-feruloyl-Glc-Gal), respectively. Both anthocyanins are produced by acylation mediated by an acyl-glucose-dependent AT and, in both species, the ATs have a preference for feruloylglucose over sinapoyl-, caffeoyl- or p-coumaroyl-glucose *in vitro*, although they can acylate Cya 3-(Xyl-Glc-Gal) using different HCGs. Carrot cells mainly synthesize sinapoylglucose *in vivo*, while beach silvertop cells synthesize feruloylglucose; these accumulated acyl-glucoses are used as acyl-donors to modify Cya 3-(Xyl-Glc-Gal) resulting in the determination of the major acylated anthocyanin molecule *in vivo* [12]. Furthermore, recombinant delphinium AA7BG-GTs do not recognize artificially synthesized delphinidin 3-*O*-rutinoside-7-*O*-(6-*O*-(*p*-coumaroyl)-glucoside) as an acceptor substance *in vitro* [17]. The evolution of the ability of GT and AT to use pHBG as a common donor substance for glucosylation and acylation may underlie the generation of complicated anthocyanin species like violdelphin and cyanodelphin in delphiniums.

## 6. *Arabidopsis*

The leaves and stems of *Arabidopsis* accumulate anthocyanins that have been modified with multiple sugar and organic acid moieties when they are cultivated under stressful conditions. The exact function(s) of anthocyanin induced by biotic and abiotic stress have still remained controversial, while the contribution to protect against several stresses such as light, UV, and free radicals are proposed [66]. The major anthocyanin structure is cyanidin 3-*O*-(2"-*O*-(2'''-sinapoyl)-xylosyl)-6"-*O*-(*p*-*O*-(glucosyl)-*p*-coumaroyl)-glucoside) 5-*O*-(6''''-*O*-(malonyl) glucoside), generally termed A11 (Figure 1c) [67]; the ten precursor structural anthocyanins of A11 have also been determined [68]. Three UGTs, one AAGT, two BAHD-ATs and one SCPL-AT are required for A11 construction. The genes encoding these anthocyanin modification enzymes have also been identified [19].

Anthocyanin modification in *Arabidopsis* commences with the glucosylation of the 3-position of the anthocyanin aglycon, cyanidin. UGT78D2 (At5g17050) catalyzes 3-glucosylation of both flavonol and anthocyanidin in the cytosol [69,70]. After this glucosylation step, two possible reactions may occur. Two anthocyanin ATs (At3AT1: At1g03940, At3AT2: At1g03495) that mediate *p*-coumaroylation of the C6 position of the glucosyl group on anthocyanin have been identified. These At3ATs are BAHD-AT type enzymes that use acyl-CoA as the acyl-donor and have a preference for cyanidin 3-glucoside over cyanidin 3,5-diglucoside [71]. Xylosylation at the 2"-position of cyanidin 3-glucoside is catalyzed by UGT79B1 (At5g54060). UGT79B1 also recognizes cyanidin 3-glucoside as an acceptor substrate but not cyanidin 3,5-diglucoside [25]. In an *Arabidopsis* mutant deficient for the *UGT79B1* gene, anthocyanin A11 lacking the sinapoylxylose moiety accumulated. This suggests the possibility of two reaction routes: xylosylation occurs first; or, *p*-coumaroylation occurs in advance of xylosylation. However, the modification steps at the 5-position are believed to occur after these reactions because At3AT1, At3AT2 and UGT79B1 are all able to use cyanidin 3-glucoside as an acceptor substance but not cyanidin 3,5-diglucoside [25,71]. The 5-position of anthocyanidin is glucosylated by UGT75C1 (At4g14090) and then malonylated by malonyl-CoA: anthocyanidin 5-*O*-glucoside-6"-*O*-malonyltransferase (At5MAT: At3g29590) [68,72]. A *UGT75C1* knockout mutant was found to accumulate deglucosylated anthocyanins at the 5-position but its 3-position was modified similarly to A11 [72]. An *At5MAT* knockout mutant was found to accumulate de-malonylated A11. These results suggest that the modification at the 5-position occurs irrespective of the 3-position modification. Since these modification steps are mediated by enzymes that are active in the cytoplasm and that use UDP-sugars as UGTs or acyl-CoA for BAHD-ATs (Figure 3), then construction of cyanidin 3-(2"-*O*-(xylosyl)-6"-*O*-(*p*-coumaroyl)-glucoside) 5-(6"'-*O*-(malonyl) glucoside), termed A5, is also likely to be carried out in the cytoplasm (Figure 8). A5 is then subjected to further modification reactions by vacuolar type enzymes.

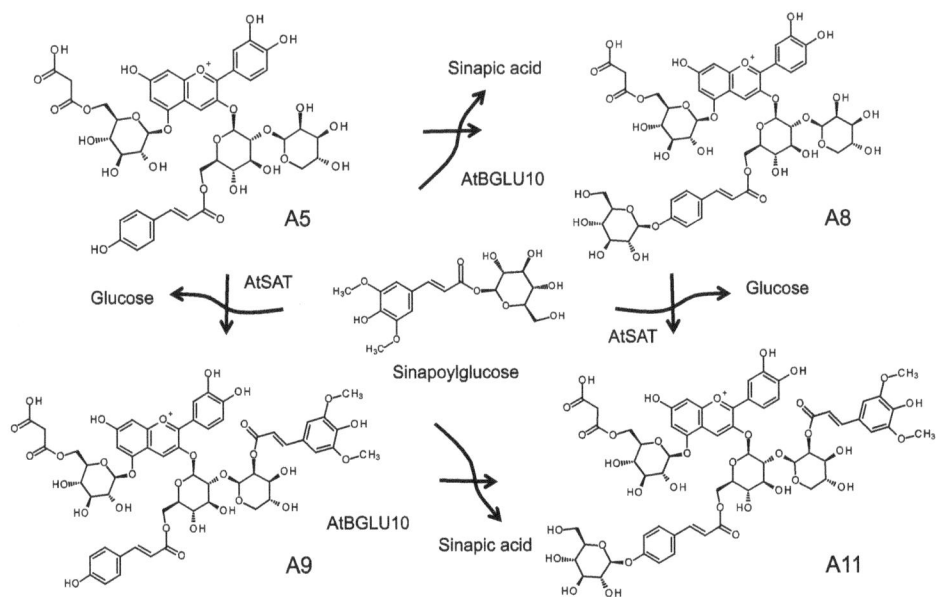

**Figure 8.** Sinapoylation and the glucosylation of the *p*-coumaroyl moiety on the anthocyanin A11 is catalyzed by enzymes using sinapoylglucose as a Zwitter donor in *Arabidopsis*.

Sinapoylation of the xylosyl group is catalyzed by sinapoylglucose:anthocyanin sinapoyltransferase (AtSAT: At2g23000), an SCPL-AT type of protein. Analysis of an *AtSAT* knockout mutant showed the accumulation of the de-sinapoylated anthocyanins A5 and A8 (glucosylated A5) (Figure 8) [46]. Until recently, the glucosylation mechanism of the hydroxyl group on *p*-coumaroyl residue was unclear. However, *AtBGLU10* (At4g27830) has now been identified as the gene encoding the enzyme for this glucosylation [19]. AtBGLU10 had been identified as a β-glucosidase with an unknown function. Acyl-glucose-dependent GT activity toward A9 has been found in a crude protein extract prepared from wild type *Arabidopsis* leaves that had been induced to synthesize anthocyanin under high light intensity conditions. In this reaction, sinapoylglucose was also used as the glucosyl donor for glucosylation at the *p*-coumaroyl group in the *in vitro* enzyme reaction. No activity is detected in protein extracts prepared from similarly treated leaves of *AtBGLU10* knockout mutant plants that mainly accumulate A9 (sinapoylated at the xylosyl group of A5; Figure 8) [19]. On the basis of these results, we conclude that AtBGLU10 mediates glucosylation of the *p*-coumaroyl group of A9. In addition, the fact that *AtSAT* knockout mutant line accumulates both A5 and A8 anthocyanins indicates that AtBGLU10 can recognize both A5 and A9 *in vivo* [46]. Hence, analysis of anthocyanin structures in *Arabidopsis* knockout mutants indicates that vacuolar type enzymes might have a broad acceptor recognition that leads to the formation of a metabolic grid, unlike the situation in carnations in which there is a strict sequential order of acylation followed by glucosylation at the 5-position (Figure 8) [25]. Evidence from reverse genetic studies and biochemical analyses indicates these anthocyanin modification enzymes may be able to complete A11 through multiple pathways, e.g., it is possible that 5-modification occurs after the 3-modification. However, considering the sub-cellular localization of the enzymes involved in the processes, it seems reasonable to suggest that A5 is first generated in the cytoplasm and that sinapoylation and glucosylation subsequently occur in the vacuole after anthocyanin transportation. Both vacuolar type anthocyanin modification enzymes utilize acyl-glucose as the donor molecule. The exclusive accumulation of sinapoylglucose in *Arabidopsis* suggests that this compound is a unique donor molecule unlike pHBG in delphinium which is a Zwitter donor [17,19,26,46].

## 7. Perspectives and Conclusions

This review has mainly focused on the roles of acyl-glucoses as substrates for anthocyanin biosynthesis. Although the vacuole was initially regarded simply as a storage organ that accumulated a wide range of substances, recent studies show that it is also the location of diverse enzymatic reactions. The modification reactions mediated by vacuolar enzymes are not limited to anthocyanins [13–16,35,41–43]. Vacuolar enzymes are also involved in acylation reactions that utilize acyl-glucoses for the biosynthesis of other compounds [73]. With regard to acyl-glucose-dependent GTs, the acceptor molecules are not limited to anthocyanins. A GH1 protein that has the ability to catalyze the attachment of a glucosyl group from an acyl-glucose to phenylpropanoids, flavonoids, and phytohormones has been identified, suggesting that GH1-GTs might be required for glucosylation of compounds other than anthocyanins [74]. However, GH1-GTs that transfer sugar moieties other than glucose and that utilize other acylated sugars, such as acyl-xylose and acyl-galactose, have not yet been identified. The question remains to be answered whether acyl-sugars (other than acyl-glucose), for example, acyl-galactose and acyl-xylose, can be used by other GH1-related GTs as sugar donors and what roles these accumulated acyl-sugars play in the metabolism of higher plants. Another important question requiring further attention is the evolution of GH1-GTs in plants. Possibly, clarification of the mechanism by which glucoside hydrolase activity is changed to glucosyl transfer activity will lead to a more complete understanding of the origin of these enzymes.

Recent studies indicate the dynamic function of the vacuole as an active site of synthesis of plant metabolites. In future work, it will be important to verify that these enzymatic reactions truly occur in the vacuole as opposed to other organelles such as the pre-vacuole [38]. The question of how the reactions are mediated efficiently in the correct order also needs to be clarified. For example, cyanodelphin in delphinium requires nine synthetic steps mediated by vacuolar proteins [17]. Possibly, the enzymes involved in this biosynthetic process in the vacuole are associated to enable sequential and efficient metabolic reactions, as has been reported for other pathways in the cytoplasm. Further biochemical and cell biological analysis of vacuolar enzymes will contribute to our understanding of many aspects of enzymatic reactions in this organelle.

**Acknowledgments:** This work was supported by Grant-in-Aid for Scientific Research (B) (23370016) to Y.O.

**Author Contributions:** N.S. wrote the article with the aid of Y.N. and T.M. Y.O. aided administrative, obtained funding and aided critical review of the manuscript.

**Conflicts of Interest:** The authors declare no conflict of interest.

## References

1. Tanaka, Y.; Sasaki, N.; Ohmiya, A. Biosynthesis of plant pigments: Anthocyanins, betalains and carotenoids. *Plant J.* **2008**, *54*, 733–749.
2. Yoshida, K.; Mori, M.; Kondo, T. Blue flower color development by anthocyanins: From chemical structure to cell physiology. *Nat. Prod. Rep.* **2009**, *26*, 884–915.
3. Sasaki, N.; Nakayama, T. Achievements and perspectives in biochemistry concerning anthocyanin modification for blue flower coloration. *Plant Cell Physiol.* **2014**, in press.
4. Yonekura-Sakakibara, K.; Hanada, K. An evolutionary view of functional diversity in family 1 glycosyltransferases. *Plant J.* **2011**, *66*, 182–193.
5. D'Auria, J.C. Acyltransferases in plants: A good time to be BAHD. *Curr. Opin. Plant Biol.* **2006**, *9*, 331–340.
6. Nakatsuka, T.; Saito, M.; Yamada, E.; Nishihara, M. Production of picotee-type flowers in Japanese gentian by CRES-T. *Plant Biotechnol.* **2011**, *28*, 173–180.

7. Fukuchi-Mizutani, M.; Okuhara, H.; Fukui, Y.; Nakao, M.; Katsumoto, Y.; Yonekura-Sakakibara, K.; Kusumi, T.; Hase, T.; Tanaka, Y. Biochemical and molecular characterization of a novel UDP-glucose:anthocyanin 3'-O-glucosyltransferase, a key enzyme for blue anthocyanin biosynthesis, from gentian. *Plant Physiol.* **2003**, *132*, 1652–1663.

8. Fujiwara, H.; Tanaka, Y.; Fukui, Y.; Nakao, M.; Ashikari, T.; Kusumi, T. Anthocyanin 5-aromatic acyltransferase from *Gentiana triflora*. Purification, characterization and its role in anthocyanin biosynthesis. *Eur. J. Biochem.* **1997**, *249*, 45–51.

9. Fujiwara, H.; Tanaka, Y.; Yonekura-Sakakibara, K.; Fukuchi-Mizutani, M.; Nakao, M.; Fukui, Y.; Yamaguchi, M.; Ashikari, T.; Kusumi, T. cDNA cloning, gene expression and subcellular localization of anthocyanin 5-aromatic acyltransferase from *Gentiana triflora*. *Plant J.* **1998**, *16*, 421–431.

10. Nakatsuka, T.; Mishiba, K.; Kubota, A.; Abe, Y.; Yamamura, S.; Nakamura, N.; Tanaka, Y.; Nishihara, M. Genetic engineering of novel flower colour by suppression of anthocyanin modification genes in gentian. *J. Plant Physiol.* **2010**, *167*, 231–237.

11. Gläßgen, W.E.; Seitz, H.U. Acylation of anthocyanins with hydroxycinnamic acid via 1-O-acylglucosides by protein preparations from cell cultures of *Daucus carota* L. *Planta* **1992**, *186*, 582–585.

12. Matsuba, Y.; Okuda, Y.; Abe, Y.; Kitamura, Y.; Terasaka, K.; Mizukami, H.; Kamakura, K.; Kawahara, N.; Goda, Y.; Sasaki, N.; *et al.* Enzymatic preparation of 1-O-hydroxycinnamoyl-β-D-glucose and their application to the study of 1-O-hydroxycynnamoyl-β-D-glucose dependent acyltransferase in anthocyanin-producing cultured cells of *Daucus carota* and *Glehnia littorails*. *Plant Biotechnol.* **2008**, *25*, 369–375.

13. Bokern, M.; Strack, D. Synthesis of hydroxycinnamic acid esters of betacyanins via 1-O-acylglucosides of hydroxycinnamic acids by protein preparations from cell suspension cultures of *Chenopodium rubrum* and petals of *Lampranthus sociorum*. *Planta* **1988**, *174*, 101–105.

14. Lehfeldt, C.; Shirley, A.M.; Meyer, K.; Ruegger, M.O.; Cusumano, J.C.; Viitanen, P.V.; Strack, D.; Chapple, C. Cloning of the *SNG1* gene of *Arabidopsis* reveals a role for a serine carboxypeptidase-like protein as an acyltransferase in secondary metabolism. *Plant Cell* **2000**, *12*, 1295–1306.

15. Li, A.X.; Steffens, J.C. An acyltransferase catalyzing the formation of diacylglucose is a serine carboxypeptidase-like protein. *Proc. Natl. Acad. Sci. USA* **2000**, *97*, 6902–6907.

16. Hause, B.; Meyer, K.; Viitanen, P.V.; Chapple, C.; Strack, D. Immunolocalization of 1-O-sinapoylglucose: Malate sinapoyltransferase in *Arabidopsis thaliana*. *Planta* **2002**, *215*, 26–32.

17. Nishizaki, Y.; Yasunaga, M.; Okamoto, E.; Okamoto, M.; Hirose, Y.; Yamaguchi, M.; Ozeki, Y.; Sasaki, N. p-Hydroxybenzoyl-glucose is a Zwitter donor for the biosynthesis of 7-polyacylated anthocyanin in delphinium. *Plant Cell* **2013**, *25*, 4150–4165.

18. Matsuba, Y.; Sasaki, N.; Tera, M.; Okamura, M.; Abe, Y.; Okamoto, E.; Nakamura, H.; Funabashi, H.; Takatsu, M.; Saito, M.; *et al.* A novel glucosylation reaction on anthocyanins catalyzed by acyl-glucose-dependent glucosyltransferase in the petals of carnation and delphinium. *Plant Cell* **2010**, *22*, 3374–3389.

19. Miyahara, T.; Sakiyama, R.; Ozeki, Y.; Sasaki, N. Acyl-glucose-dependent glucosyltransferase catalyzes the final step of anthocyanin formation in *Arabidopsis*. *J. Plant Physiol.* **2013**, *170*, 619–624.

20. Miyahara, T.; Takahashi, M.; Ozeki, Y.; Sasaki, N. Isolation of an acyl-glucose-dependent anthocyanin 7-*O*-glucosyltransferase from the monocot *Agapanthus africanus*. *J. Plant Physiol.* **2012**, *169*, 1321–1326.

21. Miyahara, T.; Tani, T.; Takahashi, M.; Nishizaki, N.; Ozeki, Y.; Sasaki, N. Isolation of anthocyanin 7-*O*-glucosyltransferase from Canterbury bells (*Campanula medium*). *Plant Biotechnol* **2014**, in press.

22. Vogt, T.; Jones, P. Glycosyltransferases in plant natural product synthesis: characterization of a supergene family. *Trends Plant Sci.* **2000**, *5*, 380–386.

23. Lim, E.K.; Li, Y.; Parr, A.; Jackson, R.; Ashford, D.A.; Bowles, D.J. Identification of glucosyltransferase genes involved in sinapate metabolism and lignin synthesis in *Arabidopsis*. *J. Biol. Chem.* **2001**, *276*, 4344–4349.

24. Lim, E.K.; Doucet, C.J.; Li, Y.; Elias, L.; Worrall, D.; Spencer, S.P.; Ross, J.; Bowles, D.J. The activity of *Arabidopsis* glycosyltransferases toward salicylic acid, 4-hydroxybenzoic acid, and other benzoates. *J. Biol. Chem.* **2002**, *277*, 586–592.

25. Yonekura-Sakakibara, K.; Fukushima, A.; Nakabayashi, R.; Hanada, K.; Matsuda, F.; Sugawara, S.; Inoue, E.; Kuromori, T.; Ito, T.; Shinozaki, K.; *et al.* Two glycosyltransferases involved in anthocyanin modification delineated by transcriptome independent component analysis in *Arabidopsis thaliana*. *Plant J.* **2012**, *69*, 154–167.

26. Meißner, D.; Albert, A.; Bottcher, C.; Strack, D.; Milkowski, C. The role of UDP-glucose: Hydroxycinnamate glucosyltransferases in phenylpropanoid metabolism and the response to UV-B radiation in *Arabidopsis thaliana*. *Planta* **2008**, *228*, 663–674.

27. Eudes, A.; Bozzo, G.G.; Waller, J.C.; Naponelli, V.; Lim, E.K.; Bowles, D.J.; Gregory, J.F., III; Hanson, A.D. Metabolism of the folate precursor *p*-aminobenzoate in plants: Glucose ester formation and vacuolar storage. *J. Biol. Chem.* **2008**, *283*, 15451–15459.

28. Dean, J.V.; Delaney, S.P. Metabolism of salicylic acid in wild-type, *ugt74f1* and *ugt74f2* glucosyltransferase mutants of *Arabidopsis thaliana*. *Physiol. Plantarum* **2008**, *132*, 417–425.

29. Hall, D.; de Luca, V. Mesocarp localization of a bi-functional resveratrol/hydroxycinnamic acid glucosyltransferase of Concord grape (*Vitis labrusca*). *Plant J.* **2007**, *49*, 579–591.

30. Kita, M.; Hirata, Y.; Moriguchi, T.; Endo-Inagaki, T.; Matsumoto, R.; Hasegawa, S.; Suhayda, C.G.; Omura, M. Molecular cloning and characterization of a novel gene encoding limonoid UDP-glucosyltransferase in *Citrus*. *FEBS Lett.* **2000**, *469*, 173–178.

31. Ford, C.M.; Boss, P.K.; Hoj, P.B. Cloning and characterization of *Vitis vinifera* UDP-glucose: Flavonoid 3-*O*-glucosyltransferase, a homologue of the enzyme encoded by the maize *Bronze-1* locus that may primarily serve to glucosylate anthocyanidins *in vivo*. *J. Biol. Chem.* **1998**, *273*, 9224–9233.

32. Lee, H.I.; Raskin, I. Purification, cloning, and expression of a pathogen inducible UDP-glucose: Salicylic acid glucosyltransferase from tobacco. *J. Biol. Chem.* **1999**, *274*, 36637–36642.

33. Taguchi, G.; Imura, H.; Maeda, Y.; Kodaira, R.; Hayashida, N.; Shimosaka, M.; Okazaki, M. Purification and characterization of UDP-glucose:hydroxycoumarin 7-O-glucosyltransferase, with broad substrate specificity from tobacco cultured cells. *Plant Sci.* **2000**, *157*, 105–112.

34. Nishizaki, Y.; Sasaki, N.; Yasunaga, M.; Miyahara, T.; Okamoto, E.; Okamoto, M.; Hirose, Y.; Ozeki, Y. Identification of the glucosyltransferase gene that supplies the *p*-hydroxybenzoyl-glucose for 7-polyacylation of anthocyanin in delphinium. *J. Exp. Bot.* **2014**, *65*, 2495–2506.

35. Li, A.X.; Eannetta, N.; Ghangas, G.S.; Steffens, J.C. Glucose polyester biosynthesis. Purification and characterization of a glucose acyltransferase. *Plant Physiol.* **1999**, *121*, 453–460.

36. Lunkenbein, S.; Bellido, M.; Aharoni, A.; Salentijn, E.M.; Kaldenhoff, R.; Coiner, H.A.; Munoz-Blanco, J.; Schwab, W. Cinnamate metabolism in ripening fruit. Characterization of a UDP-glucose: Cinnamate glucosyltransferase from strawberry. *Plant Physiol.* **2006**, *140*, 1047–1058.

37. Landmann, C.; Fink, B.; Schwab, W. FaGT2: A multifunctional enzyme from strawberry (*Fragaria* x *ananassa*) fruits involved in the metabolism of natural and xenobiotic compounds. *Planta* **2007**, *226*, 417–428.

38. Ozeki, Y.; Matsuba, Y.; Abe, Y.; Umemoto, N.; Sasaki, N. Pigment biosynthsis I. Anthoycanins. In *Plant Metabolism and Biotechnology*, 1st ed.; Ashihara, H., Crozeir, A., Komamine, A., Eds.; Wiley: West Sussex, UK, 2011; pp. 321–342.

39. Veitch, N.C.; Grayer, R.J. Flavonoids and their glycosides, including anthocyanins. *Nat. Prod. Rep.* **2008**, *25*, 555–611.

40. Teng, X.F.; Yang, J.Y.; Yang, C.R.; Zhang, Y.J. New flavonol glycosides from the fresh flowers of *Camellia reticulate*. *Helve. Chim. Acta* **2008**, *91*, 1305–1312.

41. Sharma, V.; Strack, D. Vacuolar localization of 1-sinapolglucose: L-Malate sinapoyltransferase in protoplasts from cotyledons of *Raphanus sativus*. *Planta* **1985**, *163*, 563–568.

42. Baumert, A.; Milkowski, C.; Schmidt, J.; Nimtz, M.; Wray, V.; Strack, D. Formation of a complex pattern of sinapate esters in *Brassica napus* seeds, catalyzed by enzymes of a serine carboxypeptidase-like acyltransferase family? *Phytochemistry* **2005**, *66*, 1334–1345.

43. Frohlich, B.; Niemetz, R.; Gross, G.G. Gallotannin biosynthesis: Two new galloyltransferases from *Rhus typhina* leaves preferentially acylating hexa- and heptagalloylglucoses. *Planta* **2002**, *216*, 168–172.

44. D'Abrosca, B.; Scognamiglio, M.; Tsafantakis, N.; Fiorentino, A.; Monaco, P. Phytotoxic chlorophyll derivatives from *Petrorhagia velutina* (Guss) Ball et Heyw leaves. *Nat. Prod. Commun.* **2010**, *5*, 99–102.

45. Abe, Y.; Tera, M.; Sasaki, N.; Okamura, M.; Umemoto, N.; Momose, M.; Kawahara, N.; Kamakura, H.; Goda, Y.; Nagasawa, K.; *et al*. Detection of 1-*O*-malylglucose: Pelargonidin 3-*O*-glucose-6"-*O*-malyltransferase activity in carnation (*Dianthus caryophyllus*). *Biochem. Biophys. Res. Commun.* **2008**, *373*, 473–477.

46. Fraser, C.M.; Thompson, M.G.; Shirley, A.M.; Ralph, J.; Schoenherr, J.A.; Sinlapadech, T.; Hall, M.C.; Chapple, C. Related *Arabidopsis* serine carboxypeptidase-like sinapoylglucose acyltransferases display distinct but overlapping substrate specificities. *Plant Physiol.* **2007**, *144*, 1986–1999.

47. Mock, H.P.; Strack, D. Energetics of the uridine 5-diphosphoglucose: Hydroxycinnamic acid acyltransferase reaction. *Phytochemistry* **1993**, *32*, 515–519.

48. Shirley, A.M.; McMichael, C.M.; Chapple, C. The *sng2* mutant of *Arabidopsis* is defective in the gene encoding the serine carboxypeptidase-like protein sinapoylglucose:choline sinapoyltransferase. *Plant J.* **2001**, *28*, 83–94.

49. Lee, S.; Kaminaga, Y.; Cooper, B.; Pichersky, E.; Dudareva, N.; Chapple, C. Benzoylation and sinapoylation of glucosinolate R-groups in *Arabidopsis*. *Plant J.* **2012**, *72*, 411–422.

50. Bloor, S.J. A macrocyclic anthocyanin from red mauve carnation flowers. *Phytochemistry* **1997**, *49*, 225–228.

51. Nakayama, M.; Koshioka, M.; Yoshida, H.; Kan, Y.; Fukui, Y.; Koike, A.; Yamaguchi, M. Cyclic malyl anthocyanins in *Dianthus caryophyllus*. *Phytochemistry* **2000**, *55*, 937–939.

52. Terahara, N.; Yamaguchi, M. $^{1}$H-NMR spectral analysis of the malylated anthocyanins from *Dianthus*. *Phytochemistry* **1986**, *25*, 2906–2907.

53. Terahara, N.; Yamaguchi, M.; Takeda, K.; Harborne, J.B.; Self, R. Anthocyanins acylated with malic acid in *Dianthus caryophyllus* and *D. deltoides*. *Phytochemistry* **1986**, *25*, 1715–1717.

54. Yonekura-Sakakibara, K.; Nakayama, T.; Yamazaki, M.; Saito, K. Modification and stabilization of anthocyanins. In *Anthoycanins Biosynthesis, Functions, and Applications*, 1st ed.; Gould, K., Davies, K., Winefield, C., Eds.; Springer Science+Business: New York, NY, USA, 2009; pp. 169–190.

55. Takeda, K.; Kumegawa, C.; Harborne, J.B.; Self, R. Pelargonidin 3-(6"-succinyl glucoside)-5-glucoside from pink *Centaurea cyanus* flowers. *Phytochemistry* **1988**, *27*, 1228–1229.

56. Takeda, K.; Osakabe, A.; Saito, S.; Furuyama, D.; Tomita, A.; Kojima, Y.; Yamadera, M.; Sakuta, M. Components of protocyanin, a blue pigment from the blue flowers of *Centaurea cyanus*. *Phytochemistry* **2005**, *66*, 1607–1613.

57. Tamura, T.; Kondo, T.; Kato, Y.; Goto, T. Structures of a succinyl anthocyanin and a malonyl flavone, two constituents of the complex blue pigment of cornflower *Centaurea cyanus*. *Tetrahedron Lett.* **1983**, *24*, 5749–5752.

58. Yamaguchi, M.; Maki, T.; Ohishi, T.; Ino, I. Succinyl-coenzyme A: Anthocyanidin 3-glucoside succinyltransferase in flowers of *Centaurea cyanus*. *Phytochemistry* **1995**, *39*, 311–313.

59. Umemoto, N.; Abe, Y.; Cano, E.A.; Okamura, M.; Sasaki, N.; Yoshida, S.; Ozeki, Y. Carnation serine carboxypeptodase-like acyltransferase is important for anthocyanin malyltransferase activity and formation of anthocyanic vacuolar inclusions. In Proceedings of the 5th International Workshop on Anthocyanins, Nagoya, Japan, 16 September 2009; p. 115.

60. Ogata, J.; Itoh, Y.; Ishida, M.; Yoshida, H.; Ozeki, Y. Cloning and heterologous expression of cDNA encoding flavonoid glucosyltrasnferase from *Dianthus caryophyllus*. *Plant Biotechnol.* **2004**, *21*, 367–375.

61. Okamura, M.; Nakayama, M.; Umemoto, N.; Cano, E.A.; Hase, Y.; Nishizaki, Y.; Sasaki, N.; Ozeki, Y. Crossbreeding of a metallic color carnation and diversification of the peculiar coloration by ion-beam irradiation. *Euphytica* **2013**, *191*, 45–56.

62. Winkel, B.S. Metabolic channeling in plants. *Annu. Rev. Plant Biol.* **2004**, *55*, 85–107.

63. Hashimoto, F.; Tanaka, M.; Maeda, H.; Fukuda, S.; Shimizu, K.; Sakata, Y. Changes in flower coloration and sepal anthocyanins of cyanic delphinium cultivars during flowering. *Biosci. Biotechnol. Biochem.* **2002**, *66*, 1652–1659.

64. Honda, T.; Saito, N. Recent progress in the chemistry of polyacylated anthocyanins as flower color pigments. *Heterocycles* **2002**, *56*, 633–692.

65. Goto, T.; Kondo, T. Structure and molecular stacking of anthocyanins-flower color varietion. *Angew. Chem. Int. Ed. Engl.* **1991**, *30*, 17–33.

66. Haiter, J.H.B.; Goud, K.S. Anthocyanin function in vesitative organs. In *Anthoycanins Biosynthesis, Functions, and Applications*, 1st ed.; Gould, K., Davies, K., Winefield, C., Eds.; Springer Science+Business: New York, NY, USA, 2009; pp. 1–12.

67. Bloor, S.J.; Abrahams, S. The structure of the major anthocyanin in *Arabidopsis thaliana*. *Phytochemistry* **2002**, *59*, 343–346.

68. Tohge, T.; Nishiyama, Y.; Hirai, M.Y.; Yano, M.; Nakajima, J.; Awazuhara, M.; Inoue, E.; Takahashi, H.; Goodenowe, D.B.; Kitayama, M.; *et al.* Functional genomics by integrated analysis of metabolome and transcriptome of *Arabidopsis* plants over-expressing an MYB transcription factor. *Plant J.* **2005**, *42*, 218–235.

69. Kubo, H.; Nawa, N.; Lupsea, S.A. *Anthocyaninless1* gene of *Arabidopsis thaliana* encodes a UDP-glucose:flavonoid-3-*O*-glucosyltransferase. *J. Plant Res.* **2007**, *120*, 445–449.

70. Lee, Y.; Yoon, H.R.; Paik, Y.S.; Liu, J.R.; Chung, W.; Choi, G. Reciprocal regulation of *Arabidopsis* UGT78D2 and BANYULS is critical for regulation of the metabolic flux of anthocyanidins to condensed tannins in developing seed coats. *J. Plant Biol.* **2005**, *48*, 356–370.

71. Luo, J.; Nishiyama, Y.; Fuell, C.; Taguchi, G.; Elliott, K.; Hill, L.; Tanaka, Y.; Kitayama, M.; Yamazaki, M.; Bailey, P.; *et al.* Convergent evolution in the BAHD family of acyl transferases: Identification and characterization of anthocyanin acyl transferases from *Arabidopsis thaliana*. *Plant J.* **2007**, *50*, 678–695.

72. D'Auria, J.C.; Reichelt, M.; Luck, K.; Svatos, A.; Gershenzon, J. Identification and characterization of the BAHD acyltransferase malonyl CoA: Anthocyanidin 5-*O*-glucoside-6"-*O*-malonyltransferase (At5MAT) in *Arabidopsis thaliana*. *FEBS Lett.* **2007**, *581*, 872–878.

73.  Liu, Y.; Gao, L.; Liu, L.; Yang, Q.; Lu, Z.; Nie, Z.; Wang, Y.; Xia, T. Purification and characterization of a novel galloyltransferase involved in catechin galloylation in the tea plant (*Camellia sinensis*). *J. Biol. Chem.* **2012**, *287*, 44406–44417.

74.  Luang, S.; Cho, J.I.; Mahong, B.; Opassiri, R.; Akiyama, T.; Phasai, K.; Komvongsa, J.; Sasaki, N.; Hua, Y.L.; Matsuba, Y.; *et al.* Rice Os9BGlu31 is a transglucosidase with the capacity to equilibrate phenylpropanoid, flavonoid, and phytohormone glycoconjugates. *J. Biol. Chem.* **2013**, *288*, 10111–10123.

**Sample Availability:** *Sample Availability*: Samples of the acyl-glucoses are available from the authors.

# The Regulation of Anthocyanin Synthesis in the Wheat Pericarp

Olesya Y. Shoeva, Elena I. Gordeeva and Elena K. Khlestkina

**Abstract:** Bread wheat producing grain in which the pericarp is purple is considered to be a useful source of dietary anthocyanins. The trait is under the control of the *Pp-1* homoealleles (mapping to each of the group 7 chromosomes) and *Pp3* (on chromosome 2A). Here, *TaMyc1* was identified as a likely candidate for *Pp3*. The gene encodes a MYC-like transcription factor. In genotypes carrying the dominant *Pp3* allele, *TaMyc1* was strongly transcribed in the pericarp and, although at a lower level, also in the coleoptile, culm and leaf. The gene was located to chromosome 2A. Three further copies were identified, one mapping to the same chromosome arm as *TaMyc1* and the other two mapping to the two other group 2 chromosomes; however none of these extra copies were transcribed in the pericarp. Analysis of the effect of the presence of combinations of *Pp3* and *Pp-1* genotype on the transcription behavior of *TaMyc1* showed that the dominant allele *Pp-D1* suppressed the transcription of *TaMyc1*.

Reprinted from *Molecules*. Cite as: Shoeva, O.Y.; Gordeeva, E.I.; Khlestkina, E.K. The Regulation of Anthocyanin Synthesis in the Wheat Pericarp. *Molecules* **2014**, *19*, 20266–20279.

## 1. Introduction

The anthocyanins represent a class of secondary metabolites synthesized by most higher plants. They are responsible for the pigmentation of flowers and fruits, and function as attractors for the vectors of pollen and seeds. Their presence in vegetative tissue is associated with the response to biotic and abiotic stress [1,2], enabled by their ability to neutralize free radicals, chelate heavy metal ions, and aid in osmoregulation and photoprotection [1–7]. In addition, their inclusion in the human diet is beneficial in numerous ways [8–13]. The main source of dietary anthocyanins is berries and fruits, but in recent years, cereals are also being considered as additional sources of these compounds [14–19].

In bread wheat (*Triticum aestivum* L., $2n = 6x = 42$, BBAADD) grain the anthocyanins reside either in the pericarp or in aleurone layer; the grain of some accessions has a purple or blue appearance as a result of the anthocyanin content of one or other of these tissues [20]. The genetic basis of purple grain pigmentation resides in the action of the homoeallelic *Pp-1* genes and *Pp3* [21–25]. The former map to the short arms of the homeologous group 7 chromosomes [21–25], and the latter to chromosome arm 2AL [21–23]. Comparative mapping has shown that the

269

*Pp-1* genes are orthologs of both maize *C1* and rice *OsC1*, which encode MYB-like transcription factors (TFs) responsible for the activation of structural genes encoding various enzymes participating in anthocyanin synthesis [26–28] (Supplementary Table S1). Similarly, *Pp3* has been shown to be orthologous to both *Pb/Ra* in rice [29,30] and *Lc/R* in maize [31], which encode MYC-like TFs underlying the regulation of anthocyanin synthesis (Supplementary Table S1). Regulatory role of the *Pp* genes has been confirmed by functional analysis of the anthocyanin synthesis structural genes in wheat near-isogenic lines (NILs) differing by the allelic state of the *Pp-1* and *Pp3* genes (both genes were in dominant or recessive state) [32]. Here, the nucleotide sequence of *Pp3* has been determined, and a functional characterization of the gene has been described.

## 2. Results

### 2.1. Identification and Chromosome Location of Wheat Myc-Like Sequences

The BLAST search based on the maize *Lc* and rice *Ra* sequences identified a matching sequence on *T. urartu* BAC clone 404H6 (GenBank accession number EF081030, Supplementary Figure S1), and this sequence allowed the design of a wheat primer pair targeting the *Myc*-like sequences (Supplementary Table S2, Figure S1 and Figure S2). When gDNA from the NIL "i:S29*Pp-A1Pp-D1Pp3*$^P$" (Table 1) was amplified using this primer pair, four distinct sequences were generated (Supplementary Figure S3). The pair-wise level of homology between the four sequences varied from 86.7% to 95.8% (Supplementary Table S3 and Figure S3). Three distinct sequences were amplified from *T. durum* gDNA in the same way. The eight sequences (four from *T. aestivum*, three from *T. durum* and one from *T. urartu*) formed three clusters: one grouped *TaMyc1* and *TdMyc1*, the second *TaMyc2*, *TdMyc2* and the *T. urartu* sequence, and the third *TaMyc3*, *TdMyc3* and *TaMyc4* (Figure 1). The sequence information was used to design a series of copy-specific primer pairs (Supplementary Table S2), which when applied to the aneuploid stocks of cv. "Chinese Spring", allowed *TaMyc1* and *TaMyc2* to be assigned to chromosome arm 2AL, *TaMyc3* to chromosome arm 2BL and *TaMyc4* to chromosome arm 2DL (Figure 2, Supplementary Figure S4).

**Table 1.** Genetic stocks used to characterize the transcription of the $Myc$-like genes in the wheat grain pericarp (controlled by $Pp3$ and $Pp-1$), the culm ($Pc$), the leaf ($Plb$), and the coleoptile ($Rc$). D: dominant allele, R: recessive allele, NIL: near-isogenic line. *, **: genotypes, in which the pericarp is, respectively, dark and light purple.

| Name | Alternative Name | Description | Pp -A1 | Pp -B1 | Pp -D1 | Plb -A1 | Plb -B1 | Plb -D1 | Pc -A1 | Pc -B1 | Pc -D1 | Rc -A1 | Rc -B1 | Rc -D1 | References |
|---|---|---|---|---|---|---|---|---|---|---|---|---|---|---|---|
| ▲ i:S29 Pp-A1pp-D1pp3 | "Saratovskaya 29" ("S29") | Russian spring wheat | D | R | R | D | R | R | D | R | R | D | R | R | [25,33] |
| ▲ i:S29 Pp-A1Pp-D1pp3^PF* | i:S29 Pp1Pp2^PF | wheat NIL developed on "S29", donor—"Purple Feed" | D | R | D | D | R | D | D | R | D | D | R | D | [21,24,25] |
| ▲ i:S29 Pp-A1Pp-D1Pp3^P* | i:S29 Pp1Pp3^P | wheat NIL developed on "S29", donor—"Purple" | D | R | D | D | R | D | D | R | D | D | R | D | [21,24,25] |
| ▲ i:S29 Pp-A1pp-D1Pp3^PF** | no | wheat NIL developed on "S29", donor—"Purple Feed" | D | R | R | D | R | D | D | R | D | D | R | R | [25] |
| ▲ i:S29 Pp-A1Pp-D1pp3^PF | no | wheat NIL developed on "S29", donor—"Purple Feed" | D | R | D | D | R | D | D | R | D | D | R | D | [25] |
| ▲ i:S29 Pp-A1Pp-D1Pp3^P** | no | wheat NIL developed on "S29", donor—"Purple" | D | R | D | D | R | D | D | R | D | D | R | R | [25] |
| ▲ i:S29 Pp-A1Pp-D1Pp3^P | no | wheat NIL developed on "S29", donor—"Purple" | D | R | D | D | R | D | D | R | D | D | R | D | [25] |
| ▼ i:S29 mp-A1pp-D1pp3 | line 140;"S29" ("YP" 4D*7A) | wheat NIL developed on "S29", donor—"Yanetzkis Probat" | R | R | D | R | R | D | R | R | D | R | R | R | [25,33] |
| "Novosibirskaya 67" ("N67") | no | Russian spring wheat | R | R | D | R | R | D | R | R | D | R | R | D | [24,34] |
| "Purple"* | no | Australian spring wheat "k-46990" | R | R | D | R | R | D | R | R | D | R | R | D | [24] |
| "Purple Feed"* | no | Canadian spring wheat "k-49426" | R | R | D | R | R | D | R | R | D | R | R | D | [24] |

▲ names for NILs obtained on "Saratovskaya 29" with dominant alleles $Pp-D1$ and/or $Pp3$ inherited from cultivars "Purple" (P) or "Purple Feed" (PF); in these lines dominant allele of $Pp-A1$ is from "Saratovskaya 29"; ▼ name for NIL obtained on "Saratovskaya 29" with its own recessive alleles $pp-D1$ and recessive $pp3$ and recessive $pp-A1$ inherited from "Yanetzkis Probat".

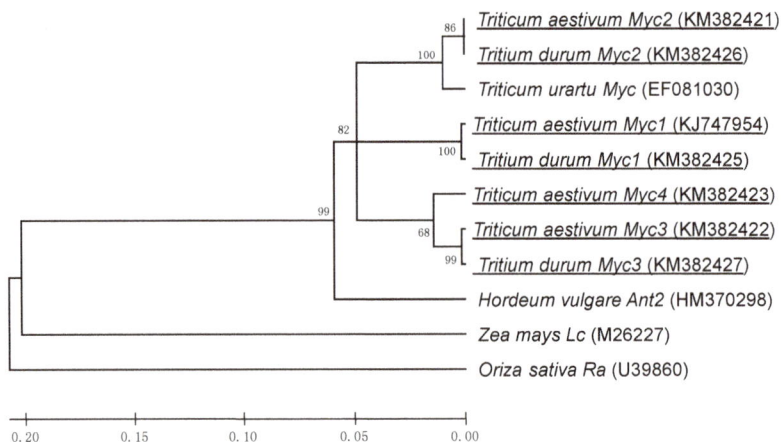

**Figure 1.** *Myc*-like sequence phylogeny. The sequences shown underlined were isolated in the current study, while the remainders were downloaded from GenBank.

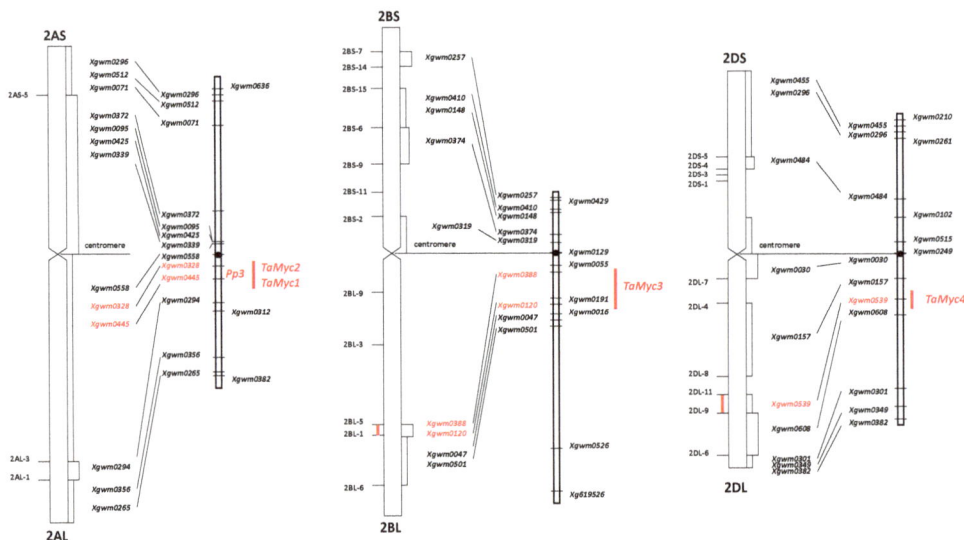

**Figure 2.** Chromosome location of wheat *Myc* genes.

## 2.2. Functional Activity of the Wheat Myc Gene Copies

The transcript abundance of each *Myc* gene in the grain pericarp was derived by RT-PCR. Only *TaMyc1* was strongly represented in the pericarp transcriptome of genotypes harboring dominant alleles at *Pp-1* and *Pp3*. The transcription profile of *TaMyc1* was similar to that of the anthocyanin synthesis pathway genes encoding for flavanone 3-hydroxylase (F3H), dihydroflavonol-4-reductase (DFR) and anthocyanidin synthase (ANS) (Figure 3).

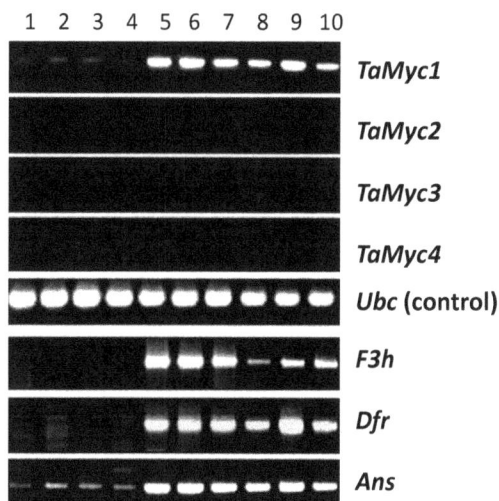

**Figure 3.** Transcription of the *Myc* gene copies in the pericarp of cv. "Novosibirskaya 67" (**1**); cv. "Saratovskaya 29" (**2–4**); "i:S29*Pp-A1Pp-D1Pp3*$^P$" NIL (**5–6**); "i:S29*Pp-A1Pp-D1Pp3*$^{PF}$" NIL (**7,8**); cv. "Purple" (**9**) and cv. "Purple Feed" (**10**).

The quantitative RT-PCR analysis showed that *TaMyc1* transcript was more abundant in both NILs "i:S29*Pp-A1Pp-D1Pp3*$^P$", "i:S29*Pp-A1Pp-D1Pp3*$^{PF}$" than in the parental cultivar "Saratovskaya 29" in all tissues investigated, while the level present in the pericarp was two-three orders of magnitude higher than elsewhere in the plant (Figure 4).

### 2.3. The Full-Length Sequence of TaMyc1

Sets of overlapping amplicons were generated to obtain the full sequence of the *TaMyc1* copy present in the NIL "i:S29*Pp-A1Pp-D1Pp3*$^P$" (Supplementary Figure S2). The 5381 nt sequence was shown via a comparison of the gDNA and cDNA sequences to be split into nine exons (Figure 5a). The first intron lay in the 5' untranslated region as was determined by 5'RACE method. The length of the open reading frame was 1707 nt, and the predicted product was a 568 residue protein (Figure 5b) harboring a conserved basic helix-loop-helix (bHLH) domain encoded by exons 7 and 8 (Figure 5c).

The bHLH domain consisted of 56 residues, split into a 13 residue segment dominated by basic amino acids and a longer segment predicted to form two amphipathic α helices separated by a 6 residue loop. The basic region contained the conserved residues $H^5$-$E^9$-$R^{13}$, thought to be critical for DNA binding [35]. The highly conserved hydrophobic residues in helix 1 and 2 are believed to be necessary for achieving the dimerization of a pair of bHLH proteins [35]. An alignment of MYC-like proteins participating in anthocyanin synthesis revealed that they all (including that encoded by the *TaMyc1* gene) shared, in addition to their bHLH domain, a highly conserved run of 200 residues at their N terminal end (Supplementary Figure S5). This segment has been implicated as being important for the proteins' interaction with R2R3-MYB TFs [36]. The structure of *TaMyc1*, the position of its bHLH domain and the presence of certain other conserved regions are all consistent with its involvement in the regulation of anthocyanin synthesis [35,37].

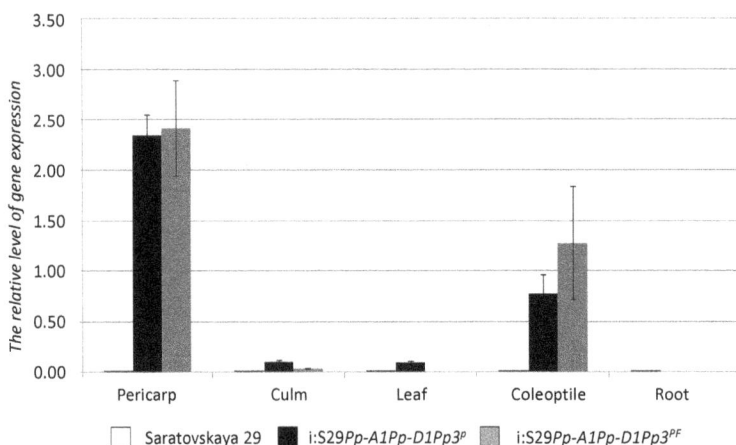

**Figure 4.** Transcription of *TaMyc1* in various parts of the wheat plant. Statistical analysis of transcript abundances given in Supplementary Table S4.

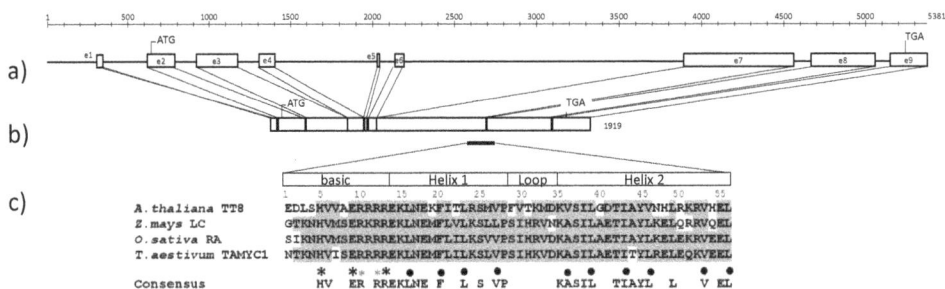

**Figure 5.** (a) Gene structure of *TaMyc1*; (b) mRNA identified in the pericarp of the NIL "i:S29Pp-A1Pp-D1Pp3$^P$"; (c) The conserved bHLH domain. The translation start site (ATG) and stop codon (TGA). Black asterisks: amino acid contacts with nucleotide bases, small gray asterisks: amino acid contacts with DNA backbone, dots: non-polar residues important for protein–protein interactions.

## 2.4. TaMyc1 Transcription as Affected by the Combination of Pp Alleles Present

Genotypes carrying dominant *Pp3* (lines 1, 2, Figure 6) were associated with the most abundant *TaMyc1* transcript, consistent with the notion that *TaMyc1* is synonymous with *Pp3*. In genotypes harboring the dominant allele at *Pp-D1*, the abundance of *TaMyc1* transcript was significantly lower than in those carrying the recessive allele (lines 1, 2, Figure 6). In lines with recessive allele at *Pp3*, dominant *Pp-D1* also reduced the abundance of *TaMyc1* transcript (lines 3, 4, Figure 6). The lowest level of *TaMyc1* transcript was observed in the line bearing recessive alleles at *Pp3* and both *Pp-A1* and *Pp-D1* (line 5, Figure 6). Described pattern of expression of the *TaMyc1* gene was also observed for the complete lines set generated on cv. "Purple Feed" as a donor of the *Pp* genes (Supplementary Table S6). These data suggested that the presence of the dominant allele at *Pp-D1* had an incomplete suppressive effect on the level of *TaMyc1* transcription.

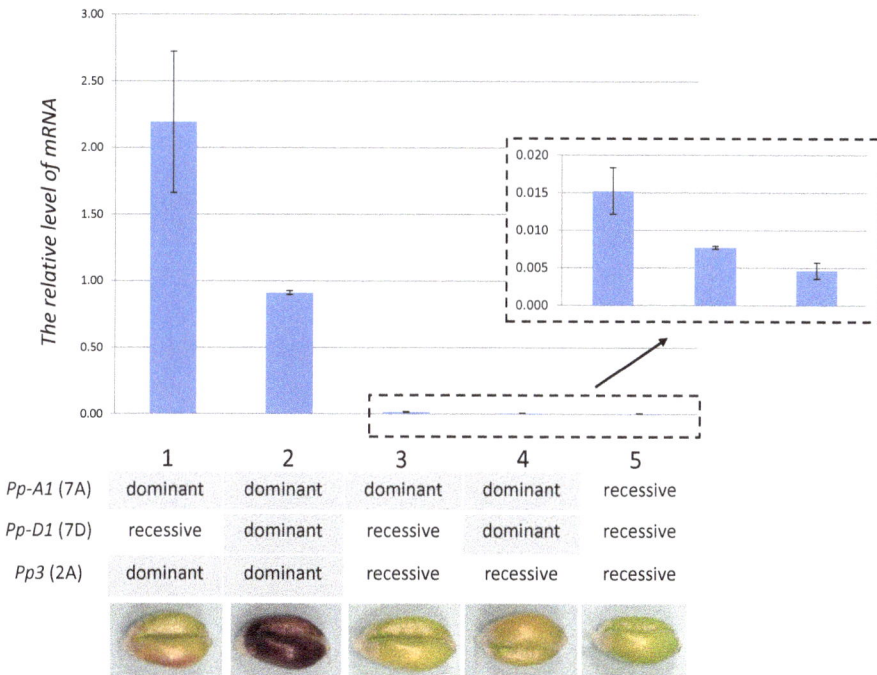

**Figure 6.** *TaMyc1* transcription in the pericarp of NILs carrying various combinations of *Pp* alleles. 1: "i:S29*Pp-A1pp-D1Pp3*P" NIL, 2: "i:S29*Pp-A1Pp-D1Pp3*P" NIL, 3: cv. "Saratovskaya 29" ("i:S29*Pp-A1pp-D1pp3*"), 4: "i:S29*Pp-A1Pp-D1pp3*P" NIL, 5: "i:S29*pp-A1pp-D1pp3*" NIL. Statistical analysis of transcript abundances given in Supplementary Table S5. The phenotypes of lines set generated on "Purple" as a donor of *Pp* genes are shown.

## 3. Discussion

The chromosomal location (Figure 2) and transcription profile (Figures 2 and 2) of *TaMyc1* were all consistent with the notion that it is *Pp3*, one of the two complementary *Pp* genes required for the synthesis of anthocyanins in the pericarp. This gene has high level of structural similarity with the others genes (Figure 5, Supplementary Figure S5), that have been shown to encode bHLH transcriptional factors, participating in anthocyanin synthesis regulation, such as maize *Lc* and *B*, rice *Ra* and *Pb*, *Arabidopsis TT8*, barley *Ant2* [29–31,38–40]. Partial sequences of the other three *Myc* copies identified all shared a high level of sequence similarity with *TaMyc1* (Figure 1, Supplementary Figure S3 and Table S3), but none of them was transcribed in the pericarp (Figure 3). Plant genomes typically harbor large numbers of bHLH-domain containing TFs, thought to have evolved via multiple duplication events followed by functional specialization [29,41,42].

The *TaMyc1* sequence is more similar to that of *T. durum Myc1* than it was to that of the A genome donor gene *TuMyc*. Assuming that *TaMyc1* and *TdMyc1* are orthologous, the implication is that *TaMyc2* and *TdMyc2* are their respective paralogs; the latter are directly related by descent to *TuMyc*. The purple grain trait has not been noted to date in any A genome diploid species [43]. As a result, it is likely that *TdMyc1* arose later than *TdMyc2* via a segmental duplication event post the formation of the BA tetraploid, and that *TaMyc1* and *TaMyc2* were transmitted from the BA tetraploid to the BAD hexaploid.

Although the functions of *TaMyc2* through *TaMyc4* have not been identified, it is possible that one or more of them do participate in anthocyanin synthesis, perhaps outside the pericarp, or in response to an external stimulus. In a number of plant species, anthocyanin synthesis is regulated by TF complexes [44,45]. The transgenic activation of the maize anthocyanin synthesis structural gene *Bz1* requires the presence of both *C1* (an R2R3-MYB TF) and *B* (a bHLH TF) [36]. Similarly, in *Petunia hybrida*, the TFs AN2 and PhJAF13 co-regulate a number of anthocyanin synthesis genes [37]. In wheat, anthocyanin synthesis in the culm, leaf and coleoptile is under the control of genes thought to be MYB family TFs [27]. *TaMyc1* was up-regulated in the coleoptile (Figure 4), which implies that it may interact with the *Myb* gene *Rc*.

Anthocyanin biosynthesis in grain pericarp is controlled by two complementary *Pp* genes, which encode for R2R3-MYB and bHLH TFs. The pericarps of genotypes harboring the dominant allele at both *Pp-1* (*R2R3-Myb*) and *Pp3* (*Myc/bHLH*) genes are pigmented from light to dark purple [25]. The presence of *Pp-A1*, inherited from cv. "Saratovskaya 29", in combination with *Pp3*, inherited from cvs "Purple" or "Purple Feed", ensures that the pericarp of each of the NILs "i:S29*Pp-A1pp-D1Pp3*$^{\text{P}}$" and "i:S29*Pp-A1pp-D1Pp3*$^{\text{PF}}$" is light purple in color, whereas the color of both "i:S29$Pp$-$A1Pp$-$D1Pp3^{P}$" and "i:S29$Pp$-$A1Pp$-$D1Pp3^{PF}$" is dark purple due to the presence of $Pp$-$D1$ [25].

The transcription behavior of $TaMyc1$ varied according to the plant's $Pp$ gene content (Figure 6). The highest abundance was noted in the combination $Pp3$ + $Pp$-$A1$, although this feature was unrelated to the intensity of pericarp pigmentation, since the relevant NIL produced a light purple rather than a dark purple pericarp. The most intense pigmentation was seen in the combination $Pp3$ + $Pp$-$A1$ + $Pp$-$D1$, although in this case, the abundance of $TaMyc1$ transcript was only about half of that observed in the $Pp3$ + $Pp$-$A1$ combination (Figure 6, Supplementary Table S5 and Supplementary Table S6). A similar interaction was noted where the recessive allele of $Pp3$ was present; in this case the overall levels of $TaMyc1$ transcript were much reduced (Figure 6, Supplementary Table S5 and Supplementary Table S6). The conclusion is that there is an influence exerted by the $Pp$-$D1$ genes on $TaMyc1$ expression, but the nature of the underlying mechanism is obscure. A possible model might involve negative feedback, in which the presence of an active R2R3-MYB/bHLH/WD40 (MBW) complex represses the transcription of $TaMyc1$ and leads to optimal proportion of partners in functional MBW complex.

Negative and positive feedback regulation of anthocyanin synthesis has also been reported in *Arabidopsis thaliana* [45]. Expression of the *TT8* gene has been shown to be positively regulated by MBW complex including the WD40 TTG1, the MYBs TT2/PAP1 and the bHLHs TT8 itself or GL3/EGL3 [46]. In addition to this positive feedback regulation two negative regulators of anthocyanin synthesis were identified (MYBL2 and CPC), both of which encode single MYB repeat proteins [47,48]. Although both MYBL2 and CPC inhibit anthocyanin accumulation by repressing the biosynthesis genes [47,49], direct suppression of the *Myb* and *bHLH* regulatory genes expression has been also reported for MYBL2 [47].

## 4. Experimental Section

### 4.1. Plant Materials

*Myc*-like sequences were identified and isolated from the near-isogenic line (NIL) "i:S29$Pp$-$A1Pp$-$D1Pp3^{P}$" (Table 1) and from *T. durum* accession TRI15744; the latter was obtained from the IPK genebank in Gatersleben (Germany). The chromosomal and intra-chromosomal locations of the wheat sequences obtained were assigned using nulli-tetrasomic, ditelosomic, and deletion lines of cv. "Chinese Spring" [50–52]. The other genetic stocks used to profile the transcription of the *Myc* genes are listed in Table 1.

## 4.2. Gene Identification, Isolation and Sequence Analysis

The maize *Lc* (GenBank accession M26227) and rice *Ra* (U39860) sequences were used as a query to identify a *Myc*-like sequence present on a *T. urartu* bacterial artificial chromosome (BAC) clone. A pair of PCR primers (pair 1: sequences given in Supplementary Table S2) was designed to amplify a segment of this gene, and was then used to recover its *T. aestivum* homologs via a PCR based on DNA extracted from fresh leaves following [53]. These and all subsequent primers were designed using OLIGO software [54]. Amplification of gDNA templates from the NIL "i:S29*Pp-A1Pp-D1Pp3*[P]" was performed in 20 μL PCRs each containing 1 U *Taq* DNA polymerase (Medigen, Novosibirsk, Russia), 1× PCR buffer (Medigen), 1.5 or 1.8 mM MgCl$_2$ (Supplementary Table S2), 0.2 mM dNTP and 0.25 μM of each primer. Amplification by different primer pairs was performed in distinct PCR conditions and amplification regimes (Supplementary Table S2). The amplified fragments were purified from a 1% agarose gel, using a DNA Clean kit (Cytokine, St. Petersburg, Russia), then cloned using a PCR Cloning kit (Qiagen, Venlo, The Netherlands). Ten clones were sequenced in both directions to exclude any PCR and/or sequencing errors. The full length *TaMyc1* sequence present in the NIL "i:S29*Pp-A1Pp-D1Pp3*[P]" was re-constructed from a series of overlapping amplicons covering the relevant stretch of genomic DNA, using primer sequences designed from the sequences of contigs 249890, 467773, 1475001 and 1821237 (http://www.cerealsdb.uk.net) [55].

A Mint RACE primer set (Evrogen, Moscow, Russia) was used to obtain the ends of *TaMyc1* transcripts present in the grain pericarp. Two rounds of 5' and 3' end amplification were conducted (primers listed in Supplementary Table S2). The resulting amplicons were cloned using a Qiagen PCR Cloning kit; a total of respectively, 35 and 12 clones obtained from the 5'- and 3'-RACE were sequenced in both directions. DNA sequencing was performed by SB RAS Genomics (Novosibirsk, Russia, http://sequest.niboch.nsc.ru). Multiple sequence alignments were carried out using Multalin v5.4.1 software [56], and the subsequent phylogenetic analysis using MEGA v5.1 software [57], based on the Neighbor-Joining algorithm and 1000 bootstrap replicates. Gene structure was determined using the FGENESH+ program [58] and confirmed by sequencing cDNAs obtained from the grain pericarp (Supplementary Table S2).

## 4.3. Chromosomal Assignment of Wheat Myc Sequences

Amplification of gDNA templates from cv. "Chinese Spring" and its aneuploid derivates was performed in 20 μL PCRs each containing 1 U *Taq* DNA polymerase (Medigen, Novosibirsk, Russia), 1× PCR buffer (Medigen), 1.5 or 1.8 mM MgCl$_2$ (Supplementary Table S2), 0.2 mM dNTP and 0.25 μM of each primer. The amplification was initiated by a denaturing step (94 °C/2 min), followed by 13 cycles of 94 °C/15 s,

65 °C/30 s (decreasing by 0.7 °C/cycle), 72 °C/45 s, 24 cycles of 94 °C/15 s, 56 °C/30 s, 72 °C/45 s and completed with a final extension step of 72 °C/5 min.

## 4.4. Transcription Analysis

A ZR Plant RNA MiniPrep™ kit (Zymo Research, Irvine, CA, USA) followed by DNAse treatment was employed to extract RNA from the grain pericarp, leaf, culm, coleoptile, and root of genotypes described in Table 1. Plants and seedlings for RNA extractions were grown, respectively, in greenhouse (ICG Greenhouse Core Facilities, Novosibirsk, Russia) or in climatic chamber "Rubarth Apparate" (RUMED GmbH, Laatzen, Germany) under 12 h of light per day at 20–25 °C. Pericarp samples for RNA extraction were peeled by scalpel from immature grains within 55th–75th day after sowing. RNA from leaf and culm were extracted within 70th–75th day after sowing. RNA samples from roots and coleoptiles were obtained on the fifth day after caryopsis germination. Single-stranded cDNA was synthesized in a 20 μL reaction from a template consisting of 0.7 μg total RNA using a (dT)15 primer and a Fermentas RevertAid™ first strand cDNA synthesis kit (Fisher Scientific, Loughborough, UK). Subsequent RT-PCRs were primed either with *Myc* copy-specific primers (Supplementary Table S2) or with the primers amplifying a segment of the genes *F3h*, *Dfr* and *Ans* [32]. A fragment of the wheat *Ubc* sequence (X56601) was used as the internal reference [59]. The PCR conditions and amplification regime were as above (Section 4.3), and the amplicons obtained were electrophoresed through 2% agarose gels. Quantitative RT-PCRs (qPCRs) were based on a SYBR Green I kit (Syntol, Moscow, Russia). Pre-determined quantities of cloned cDNA were used to generate a standard curve. Three biological replicates for each sample were run as three technical replicates. Differences in transcript abundance between lines were tested by applying the Mann-Whitney $U$-test [60], adopting a significance threshold of $p \leqslant 0.05$.

**Supplementary Materials:** Supplementary materials can be accessed at:http://www.mdpi.com/1420-3049/19/12/20266/s1.

**Acknowledgments:** We thank Galina Generalova for technical assistance, Robert Koebner (http://www.smartenglish.co.uk) for linguistic advice and valuable comments during the preparation of this manuscript, Valentina Arbuzova for near-isogenic lines, Andreas Börner for durum wheat seeds, and Marion Röder for DNA of wheat aneuploid stocks. This study was partially supported by RFBR (grant No 14-04-31637), grant from the President of the Russian Federation (MD-2615.2013.4), RAS (MCB Programme grant) and the State Budget Programme (Project No VI.53.1.5.).

**Author Contributions:** Olesya Y. Shoeva and Elena I. Gordeeva performed the molecular genetic studies and analyzed the data, Olesya Y. Shoeva wrote the paper. Elena K. Khlestkina designed and coordinated the study, contributed to interpretation of data and to revising the manuscript critically. All authors read and approved the final manuscript.

**Conflicts of Interest:** The authors declare no conflict of interest.

# References

1.  Chalker-Scott, L. Environmental significance of anthocyanins in plant stress responses. *Photochem. Photobiol.* **1999**, *70*, 1–9.
2.  Hatier, J.H.B.; Gould, K.S. Anthocyanin function in vegetative organs. In *Anthocyanins: Biosynthesis, Functions, and Applications*; Gould, K., Davies, K., Winefield, C., Eds.; Springer Science+Business Media: New York, NY, USA, 2009; pp. 1–19.
3.  Wang, H.; Cao, G.; Prior, R.L. Oxygen radical absorbing capacity of anthocyanins. *J. Agric. Food Chem.* **1997**, *45*, 304–309.
4.  Hale, K.L.; McGrath, S.P.; Lombi, E.; Stack, S.M.; Terry, N.; Pickering, I.J.; George, G.N.; Pilon-Smits, E.A. Molybdenum sequestration in *Brassica* species. A role for anthocyanins? *Plant Physiol.* **2001**, *126*, 1391–1402.
5.  Hale, K.L.; Tufan, H.A.; Pickering, I.J.; George, G.N.; Terry, N.; Pilon, M.; Pilon-Smits, E.A.H. Anthocyanins facilitate tungsten accumulation in *Brassica. Physiol. Plant.* **2002**, *116*, 351–358.
6.  Manetas, Y. Why some leaves are anthocyanic and why most anthocyanic leaves are red? *Flora* **2006**, *201*, 163–177.
7.  Khlestkina, E.K. The adaptive role of flavonoids: Emphasis on cereals. *Cereal Res. Commun.* **2013**, *41*, 185–198.
8.  Dell'Agli, M.; Busciala, A.; Bosisio, E. Vascular effects of wine polyphenols. *Cardiovasc. Res.* **2004**, *63*, 593–602.
9.  Brown, J.E.; Kelly, M.F. Inhibition of lipid peroxidation by anthocyanins, anthocyanidins and their phenolic degradation products. *Eur. J. Lipid Sci. Technol.* **2007**, *109*, 66–71.
10. Wang, H.; Nair, M.G.; Strasburg, G.M.; Chang, Y.C.; Booren, A.M.; Gray, J.I.; DeWitt, D.L. Antioxidant and antiinflammatory activities of anthocyanins and their aglycon, cyanidin, from tart cherries. *J. Nat. Prod.* **1999**, *62*, 294–296.
11. Hui, C.; Bin, Y.; Xiaoping, Y.; Long, Y.; Chunye, C.; Mantian, M.; Wenhua, L. Anticancer activities of an anthocyanin-rich extract from black rice against breast cancer cells *in vitro* and *in vivo. Nutr. Cancer* **2010**, *62*, 1128–1136.
12. Tsuda, T.; Horio, F.; Uchida, K.; Aoki, H.; Osawa, T. Dietary cyanidin 3-*O*-β-D-glucoside-rich purple corn color prevents obesity and ameliorates hyperglycemia. *J. Nutr.* **2003**, *133*, 2125–2130.
13. Howard, B.V.; Kritchevsky, D. Phytochemicals and cardiovascular disease: A statement for healthcare professionals from the American heart association. *Circulation* **1997**, *95*, 2591–2593.
14. Abdel-Aal, E.S.M.; Young, J.C.; Rabalski, I. Anthocyanin composition in black, blue, pink, purple, and red cereal grains. *J. Agric. Food Chem.* **2006**, *54*, 4696–4704.
15. Dykes, L.; Rooney, L.W. Phenolic compounds in cereal grains and their health benefits. *Cereal Foods World* **2006**, 105–111.
16. Zofajova, A.; Psenakova, I.; Havrlentova, M.; Piliarova, M. Accumulation of total anthocyanins in wheat grain. *Agriculture (Poľnohospodárstvo)* **2012**, *58*, 50–56.

17. Ficco, D.B.M.; de simone, V.; Nigro, V.F.; Finocchiaro, F.; Papa, R.; de vita, P. Genetic variability in anthocyanin composition and nutritional properties of blue, purple and red bread (*Triticum aestivum* L.) and durum (*Triticum turgidum* L. spp. *turgidum* var. *durum*) wheats. J. Agric. Food Chem. **2014**. [CrossRef]

18. Li, W; Pickard, M.D.; Trust, B. Effect of thermal processing on antioxidant properties of purple wheat bran. *Food Chem.* **2008**, *104*, 1080–1086.

19. Hirawan, R.; Diehl-Jones, W.; Trust, B. Comparative evaluation of the antioxidant potential of infant cereals produced from purple wheat and red rice grains and LC-MS analysis of their anthocyanins. *J. Agric. Food Chem.* **2011**, *59*, 12330–12341.

20. Zeven, A.C. Wheats with purple and blue grains: A review. *Euphytica* **1991**, *56*, 243–258.

21. Arbuzova, V.S.; Maystrenko, O.I.; Popova, O.M. Development of near-isogenic lines of the common wheat cultivar "Saratovskaya 29". *Cereal Res. Commun.* **1998**, *26*, 39–46.

22. Dobrovolskaya, O.B.; Arbuzova, V.S.; Lohwasser, U.; Röder, M.S.; Börner, A. Microsatellite mapping of complementary genes for purple grain colour in bread wheat (*Triticum aestivum* L.). *Euphytica* **2006**, *150*, 355–364.

23. Khlestkina, E.K.; Röder, M.S.; Börner, A. Mapping genes controlling anthocyanin pigmentation on the glume and pericarp in tetraploid wheat (*Triticum durum* L.). *Euphytica* **2010**, *171*, 65–69.

24. Tereshchenko, O.Y.; Gordeeva, E.I.; Arbuzova, V.S.; Börner, A.; Khlestkina, E.K. The D genome carries a gene determining purple grain colour in wheat. *Cereal Res. Commun.* **2012**, *40*, 334–341.

25. Gordeeva, E.I.; Shoeva, O.Y.; Khlestkina, E.K. Marker-assisted development of bread wheat near-isogenic lines carrying various combinations of *Pp* (purple pericarp) alleles. *Euphytica*, 2015. in press.

26. Saitoh, K.; Onishi, K.; Mikami, I.; Thidar, K.; Sano, Y. Allelic diversification at the *C1* (*OsC1*) locus of wild and cultivated rice: Nucleotide changes associated with phenotypes. *Genetics* **2004**, *168*, 997–1007.

27. Khlestkina, E.K. Genes determining coloration of different organs in wheat. *Russ. J. Genet. Appl. Res.* **2013**, *3*, 54–65.

28. Li, W.L.; Faris, J.D.; Chittoor, J.M.; Leach, J.E.; Hulbert, S.H.; Liu, D.J.; Chen, P.D.; Gill, B.S. Genomic mapping of defense response genes in wheat. *Theor. Appl. Genet.* **1999**, *98*, 226–233.

29. Hu, J.; Anderson, B.; Wessler, R. Isolation and characterization of rice *R* genes: Evidence for distinct evolutionary paths in rice and maize. *Genetics* **1996**, *142*, 1021–1031.

30. Wang, C.; Shu, Q. Fine mapping and candidate gene analysis of purple pericarp gene *Pb* in rice (*Oryza sativa* L.). *Chin. Sci. Bull.* **2007**, *52*, 3097–3104.

31. Ludwig, S.R.; Habera, L.F.; Dellaporta, S.L.; Wessler, S.R. *Lc*, a member of the maize *R* gene family responsible for tissue-specific anthocyanin production, encodes a protein similar to transcription activators and contains the *myc*-homology region. *Proc. Natl. Acad. Sci. USA* **1989**, *86*, 7092–7096.

32. Tereshchenko, O.Y.; Arbuzova, V.S.; Khlestkina, E.K. Allelic state of the genes conferring purple pigmentation in different wheat organs predetermines transcriptional activity of the anthocyanin biosynthesis structural genes. *J. Cereal Sci.* **2013**, *57*, 10–13.

33. Khlestkina, E.K.; Röder, M.S.; Pshenichnikova, T.A.; Börner, A. Functional diversity at the *Rc* (red coleoptile) gene in bread wheat. *Mol. Breed.* **2010**, *25*, 125–132.

34. Khlestkina, E.K.; Pshenichnikova, T.A.; Röder, M.S.; Börner, A. Clustering anthocyanin pigmentation genes in wheat group 7 chromosomes. *Cereal Res. Commun.* **2009**, *37*, 391–398.

35. Heim, M.A.; Jakoby, M.; Werber, M.; Martin, C.; Weisshaar, B.; Bailey, P.C. The basic Helix-Loop-Helix transcription factor family in plants: A genome-wide study of protein structure and functional diversity. *Mol. Biol. Evol.* **2003**, *20*, 735–747.

36. Goff, S.A.; Cone, K.C.; Chandler, V.L. Functional analysis of the transcriptional activator encoded by the maize *B* gene: Evidence for a direct functional of regulatory proteins. *Genes Dev.* **1992**, *6*, 864–875.

37. Quattrocchio, F.; Wing, J.F.; van der Woude, K.; Mol, J.N.M.; Koes, R. Analysis of bHLH and MYB domain proteins: Species specific regulatory differences are caused by divergent evolution of target anthocyanin genes. *Plant J.* **1998**, *13*, 475–488.

38. Cockram, J.; White, J.; Zuluaga, D.L.; Smith, D.; Comadran, J.; Macaulay, M.; Luo, Z.; Kearsey, M.J.; Werner, P.; Harrap, D.; *et al.* Genome-wide association mapping to candidate polymorphism resolution in the unsequenced barley genome. *Proc. Natl. Acad. Sci. USA* **2010**, *107*, 21611–21616.

39. Chandler, V.L.; Radicella, J.P.; Robbins, T.P.; Chen, J.; Turks, D. Two regulatory genes of the maize anthocyanin pathway are homologous: Isolation of *B* utilizing *R* genomic sequences. *Plant Cell* **1989**, *1*, 1175–1183.

40. Nesi, N.; Debeaujon, I.; Jond, C.; Pelletier, G.; Caboche, M.; Lepiniec, L. The *TT8* gene encodes a basic helix-loop-helix domain protein required for expression of *DFR* and *BAN* genes in *Arabidopsis* siliques. *Plant Cell* **2000**, *12*, 1863–1878.

41. Ramsay, N.A.; Glover, B.J. MYB-bHLH-WD40 protein complex and the evolution of cellular diversity. *Trends Plant Sci.* **2005**, *10*, 63–70.

42. Feller, A.; Machemer, K.; Braun, E.L.; Grotewold, E. Evolutionary and comparative analysis of MYB and bHLH plant transcription factors. *Plant J.* **2011**, *66*, 94–116.

43. Tereshchenko, O.Y.; Pshenichnikova, T.A.; Salina, E.A.; Khlestkina, E.K. Development and molecular characterization of a novel wheat genotype having purple grain colour. *Cereal Res. Commun.* **2012**, *40*, 210–214.

44. Mol, J.; Grotewold, E.; Koes, R. How genes paint flowers and seeds. *Trends Plant Sci.* **1998**, *3*, 212–217.

45. Petroni, K.; Tonelli, C. Recent advances on the regulation of anthocyanin synthesis in reproductive organs. *Plant Sci.* **2011**, *181*, 219–229.

46. Baudry, A.; Caboche, M.; Lepiniec, L. TT8 controls its own expression in a feedback regulation involving TTG1 and homologous MYB and bHLH factors, allowing a strong and cell-specific accumulation of flavonoids in *Arabidopsis thaliana*. *Plant J.* **2006**, *46*, 768–779.

47. Dubos, C.; le Gourrierec, J.; Baudry, A.; Lanet, E.; Debeaujon, I.; Routaboul, J.-M.; Alboresi, A.; Weisshaar, B.; Lepiniec, L. MYBL2 is a new regulator of flavonoid biosynthesis in *Arabidopsis thaliana*. *Plant J.* **2008**, *55*, 940–953.

48. Schellmann, S.; Schnittger, A.; Kirik, V.; Wada, T.; Okada, K.; Beermann, A.; Thumfahrt, J.; Jürgens, G.; Hülskamp, M. TRIPTYCHON and CAPRICE mediate lateral inhibition during trichome and root hair patterning in *Arabidopsis*. *EMBO J.* **2002**, *21*, 5036–5046.

49. Zhu, H.F.; Fitzsimmons, K.; Khandelwal, A.; Kranz, R.G. CPC, a single-repeat R3 MYB, is a negative regulator of anthocyanin biosynthesis in *Arabidopsis*. *Mol. Plant* **2009**, *2*, 790–802.

50. Sears, E.R. Nullisomic analysis in common wheat. *Am. Nat.* **1953**, *87*, 245–252.

51. Sears, E.R. Isochromosomes and telocentrics in *Triticum vulgare*. *Genetics* **1946**, *31*, 229–230.

52. Endo, T.R.; Gill, B.S. The deletion stocks of common wheat. *J. Hered.* **1996**, *87*, 295–307.

53. Plaschke, J.; Ganal, M.W.; Roder, M.S. Detection of genetic diversity in closely related bread wheat using microsatellite markers. *Theor. Appl. Genet.* **1995**, *91*, 1001–1007.

54. Offerman, J.D.; Rychlik, W. Oligo primer analysis software. In *Introduction to Bioinformatics: A Theoretical and Practical Approach*; Krawetz, S.A., Womble, D.D., Eds.; Humana Press: New York, NY, USA, 2003; pp. 345–361.

55. Wilkinson, P.A.; Winfield, M.O.; Barker, G.L.A.; Allen, A.M.; Burridge, A.; Coghill, J.A.; Burridge, A.; Edwards, K.J. CerealsDB 2.0: An integrated resource for plant breeders and scientists. *BMC Bioinform.* **2012**, *13*, 219.

56. Corpet, F. Multiple sequence alignment with hierarchical clustering. *Nucl. Acids Res.* **1988**, *16*, 10881–10890.

57. Tamura, K.; Peterson, D.; Peterson, N.; Stecher, G.; Nei, M.; Kumar, S. MEGA5: Molecular evolutionary genetics analysis using maximum likelihood, evolutionary distance, and maximum parsimony methods. *Mol. Biol. Evol.* **2011**, *28*, 2731–2739.

58. Solovyev, V.; Kosarev, P.; Seledsov, I.; Vorobyev, D. Automatic annotation of eukaryotic genes, pseudogenes and promoters. *Genome Biol.* **2006**, *7* (Suppl. 1), 10:1–10:12.

59. Himi, E.; Nisar, A.; Noda, K. Colour genes (*R* and *Rc*) for grain and coleoptile upregulate flavonoid biosynthesis genes in wheat. *Genome* **2005**, *48*, 747–754.

60. Mann, H.B.; Whitney, D.R. On a test of whether one of two random variables is stochastically larger than the other. *Ann. Math. Stat.* **1947**, *18*, 50–60.

**Sample Availability:** *Sample Availability*: Samples of DNA of near-isogenic lines are available from the authors.

# Distinctive Anthocyanin Accumulation Responses to Temperature and Natural UV Radiation of Two Field-Grown *Vitis vinifera* L. Cultivars

Ana Fernandes de Oliveira, Luca Mercenaro, Alessandra Del Caro, Luca Pretti and Giovanni Nieddu

**Abstract:** The responses of two red grape varieties, Bovale Grande (syn. Carignan) and Cannonau (syn. Grenache), to temperature and natural UV radiation were studied in a three-years field experiment conducted in Sardinia (Italy), under Mediterranean climate conditions. Vines were covered with plastic films with different transmittances to UV radiation and compared to uncovered controls. Light intensity and spectral composition at the fruit zone were monitored and berry skin temperature was recorded from veraison. Total skin anthocyanin content (TSA) and composition indicated positive but inconsistent effects of natural UV light. Elevated temperatures induced alterations to a greater extent, decreasing TSA and increasing the degree of derivatives acylation. In Cannonau total soluble solids increases were not followed by increasing TSA as in Bovale Grande, due to both lower phenolic potential and higher sensitivity to permanence of high temperatures. Multi linear regression analysis tested the effects of different ranges of temperature as source of variation on anthocyanin accumulation patterns. To estimate the thermal time for anthocyanin accumulation, the use of normal heat hours model had benefit from the addition of predictor variables that take into account the permanence of high (>35 °C) and low (<15 °C and <17 °C) temperatures during ripening.

Reprinted from *Molecules*. Cite as: de Oliveira, A.F.; Mercenaro, L.; Del Caro, A.; Pretti, L.; Nieddu, G. Distinctive Anthocyanin Accumulation Responses to Temperature and Natural UV Radiation of Two Field-Grown *Vitis vinifera* L. Cultivars. *Molecules* **2015**, *20*, 2061–2080.

## 1. Introduction

There has been increasing international recognition of the interactions and feedback between climate change and surface UV radiation [1,2], but the understanding of such interactions and of the induced ecosystem changes are limited since they act over medium-long time scales [3,4]. These environmental issues have led to a growing interest of the scientific community in studying plant acclimation to both light and thermal effects of solar radiation, under natural or modified climate conditions [5–11].

In viticulture, particular interest has recently been given to the effects of high temperature and UV radiation on secondary metabolite accumulation, namely of flavonoids (anthocyanins and pro-anthocyanidins), amino acids and aroma compound precursors, for their fundamental importance to berry color formation and stability, and to wine flavor and astringency [10,12–19]. Together with genetics, plants nutritional and sanitary status, canopy architecture and density assume a major role on modeling plant adaptation to climate conditions [20–24]. The composition and concentration of anthocyanins and pro-anthocyanidins vary with the cultivar, cultural practices and microclimate conditions during berry development and ripening [11,13,25]. Though light interception at the cluster zone have a positive effect on berry skin anthocyanin accumulation [26], high sunlight exposure may cause a reduction on anthocyanin concentration, due to high temperatures exposure [24,27].

Several authors have demonstrated that a reduction on anthocyanin accumulation by high temperature (>30 °C) can result both from reduced synthesis and increased degradation of previously accumulated contents [12,14,28]. Furthermore, increased temperature after veraison may alter berry composition by reducing the anthocyanin: sugar ratio of ripe berries [29], probably due to a delay on anthocyanin accumulation, which shifts the onset of the linear phase in which anthocyanin and sugar increase in parallel [30]. Recent research works focusing on anthocyanin biosynthesis dependence on temperature and light [31] have demonstrated that light increases anthocyanin content in berry skin regardless of temperature and act synergistically with low temperatures (15 °C) on the expression of flavonoid biosynthetic-related genes.

As far as anthocyanin partitioning is concerned, light and temperature effects of solar radiation seem to influence differently anthocyanin accumulation and composition [31,32]. In Azuma *et al.* [31] experiment, low temperature and light affected anthocyanin composition of Pione (V. xLabruscana), increasing peonidin and malvidin derivatives while malvidin contents decreased in the absence of light and the two derivatives were reduced under high temperature conditions (>35 °C). High temperatures seem to alter anthocyanin composition also towards a higher acylation proportion of all derivatives [12,13,17]. On the other hand, berry sunlight interception seems to have a positive effect on dihydroxylated anthocyanin synthesis and a decreasing effect on trihydroxylated derivatives, as compared to complete shadow [33]. Inconsistent results have been observed regarding the effects of UV radiation on anthocyanin synthesis [12]. Nevertheless, recent works have observed a reduction of flavonols under UV-B filtering treatment [34,35] and an increase in berry skin anthocyanin content in response to UV-B radiation [36]. It seems that the accumulation of monosubstituted flavonols is increased upon UV-B light treatments [36,37]. Carbonell-Bejerano *et al.* [37] report enhanced petunidin acetylglucoside and delphinidin coumaroylglucoside levels in Tempranillo berry skin. Moreover, working with the same variety, Martinez-Lüscher *et al.* [36] have observed an increase in trisubtituted and methylated anthocyanins under UV-B light treatment and also increases the acetylation level, both with acetic and *p*-coumaric acids. In addition, distinctive varietal responses to light may be observed in flavonols and anthocyanin accumulation and composition in berry skin [32,33,38]. The aim of this study was to analyze the effects of natural UV light radiation and of the permanence of high and low temperatures on berry skin anthocyanin contents and composition of two grapevine varieties traditionally cultivated in western Sardinia (Italy), Bovale Grande (syn. Carignan) and Cannonau (syn. Grenache), with distinct phenolic potential [32,39,40]. Both varieties were subjected to reduced UV light (*i.e.*, visible and visible + UV-A transmittance) under natural field-growing conditions. Along the three seasons of this experiment we analyzed ambient light intensity and spectral composition see [32], temperature and other variables derived from these, which are known to be basic weather variables affecting berry development and composition. We monitored berry skin temperature, from veraison until harvest, as well as berry skin total anthocyanin contents, and we used hierarchical linear regression to test the modeling effects of different ranges temperature as source of variation in anthocyanin accumulation patterns.

## 2. Results and Discussion

### 2.1. Experimental Season Thermal Conditions

In Table 1 the monthly average temperatures recorded in the study area during seasons 2009, 2010 and 2011, and the variation from the long term average, are reported. The first season was characterized by high UV intensities during the period from June to September [32,41] and by a much greater prevalence of high air temperatures (>30 °C), as compared both to the long term 30 year average and seasons 2010 and 2011 [32,41–43]. In 2009, the maximum temperature ($T_{max}$) during berry growth and development reached a monthly average value of 27.5 °C in June which was about 3 °C higher than the 30 year average and the seasons 2010 and 2011. Again, in July $T_{max}$ averaged 31.1 °C, nearly +3 °C than 2011 and long term average but about +1.5 °C higher compared to 2010.

The months of August and September 2009 continued recording temperatures higher than the 30 year average for the same period, while the seasons 2010 and 2011 were much less hot and the maximum temperature values remained close to those of long term data, or even lower than the average values (−0.7 and −0.3 °C during ripening 2010). Also mean ($T_{med}$) and minimum ($T_{min}$) temperatures registered much higher values than the average in 2009 (about +2.4 and +1.2 °C for $T_{med}$ and +1.5 and 0.4 °C for $T_{min}$ during ripening months). Conversely in 2010, a lower $T_{min}$ was registered in August and September (−0.4 °C and −0.6 °C, respectively) and in 2011,

$T_{min}$ in August remained 0.6 °C below the average value, while in September it was only slightly higher than the average (about +0.3 °C).

**Table 1.** Monthly temperature conditions during the 2009-11 growth seasons (from June to September) and long-term monthly 30-year average (1971 to 2000) in Capo Frasca, Italy [42,43]. Average values (x) and variation (Δ) between the study periods and the 30 year average.

| Variable | Period | June | | July | | August | | September | |
|---|---|---|---|---|---|---|---|---|---|
| | | x | Δ | x | Δ | x | Δ | x | Δ |
| $T_{max}$ (°C) | 2009 | 27.5 | 3.0 | 31.1 | 3.3 | 30.3 | 1.5 | 26.9 | 0.9 |
| | 2010 | 24.6 | 0.1 | 29.3 | 1.5 | 28.1 | −0.7 | 25.7 | −0.3 |
| | 2011 | 24.9 | 0.4 | 27.6 | −0.2 | 29.7 | 0.9 | 26.8 | 0.8 |
| | 30 year | 24.5 | | 27.8 | | 28.8 | | 26.0 | |
| $T_{med}$ (°C) | 2009 | 24.5 | 3.6 | 27.5 | 3.5 | 27.3 | 2.4 | 23.5 | 1.2 |
| | 2010 | 22.1 | 1.2 | 26.4 | 2.4 | 25.3 | 0.4 | 22.6 | 0.3 |
| | 2011 | 22.2 | 1.3 | 24.4 | 0.4 | 25.7 | 0.8 | 23.7 | 1.4 |
| | 30 year | 20.9 | | 24.0 | | 24.9 | | 22.3 | |
| $T_{min}$ (°C) | 2009 | 19.9 | 2.6 | 22.5 | 2.4 | 22.6 | 1.5 | 19.1 | 0.4 |
| | 2010 | 17.9 | 0.6 | 22.3 | 2.2 | 20.7 | −0.4 | 18.1 | −0.6 |
| | 2011 | 18.2 | 0.9 | 20.5 | 0.4 | 20.5 | −0.6 | 19.0 | 0.3 |
| | 30 year | 17.3 | | 20.1 | | 21.1 | | 18.7 | |

Notes: $T_{max}$, average maximum temperature; $T_{med}$, mean temperature; $T_{min}$, average minimum temperature.

## 2.2. Light Microclimate into the Fruit Zone

Under the plastic films, photosynthetically active radiation (PAR) intensity was attenuated to about 18% of the external ambient values. At the fruit zone, PAR attenuation at solar noon ranged on average from 90% to 98% of the ambient PAR values in the inner canopy layers, up to a minimum of about 51% and 67% in the external canopy layers of the Control treatments (Table 2). In the UV-screening treatments, UV-A and UV-B radiation intercepted by the clusters was reduced to about 10% and 30% of that measured in Control berries directly exposed to natural sunlight and no significant differences between canopy sides were observed at solar noon (Table 2).

The light regimes induced significant differences in cluster light microclimate. PAR and UV radiation intercepted at cluster zone were significantly attenuated under the screening films, especially at midday due to the smaller solar angle and the shading effect of the canopy above. In the Control clusters UV-A and UV-B transmittances were significantly higher than that measured under the screening films, particularly during mid-morning (at 11.00 h) and beginning of the afternoon (at 15.00 h) [32]. The UV-A intensity reached similar values at the fruit zone in Vis + UV-A and Vis treatments but the UV-B transmittance and the ratio UV-B/UV-A were statistically lower in Vis all through the day. An extended characterization of cluster light microclimate during this experiment can be found in [32]. UV filtering films did not alter significantly the permanence of elevated temperatures (>35 °C) on berry skin, except for 2010, when Control berries were exposed to high temperatures for a longer period as compared to Vis + UV-A.

**Table 2.** Photosynthetically active radiation (PAR) attenuation, UV-A and UV-B radiation intensity in East and West canopy sides, measured at solar noon in a clear sky day of veraison 2010, in Bovale Grande and Cannonau fruit zone. Mean values ($n$ = 12) $\pm$ SE.

| | | PAR Attenuation (% Reference PAR) | | UV-A (W·m$^{-2}$) | | UV-B (mW·m$^{-2}$) | |
|---|---|---|---|---|---|---|---|
| | Distance from the Canopy Centre (cm) | 0–10 | 10–20 | East | West | East | West |
| Bovale Grande | Control | 90 ± 7.4 | 51 ± 9.6 | 3.0 ± 0.70 | 2.7 ± 0.55 | 143.7 ± 59.2 | 150.0 ± 56.5 |
| | Vis + UV-A | 92 ± 2.1 | 84 ± 2.4 | 0.8 ± 0.04 | 0.4 ± 0.04 | 13.1 ± 4.6 | 1.0 ± 0.1 |
| | Vis | 94 ± 3.5 | 81 ± 4.1 | 1.0 ± 0.10 | 0.3 ± 0.11 | 7.2 ± 2.3 | 3.0 ± 2.3 |
| Cannonau | Control | 90 ± 6.0 | 67 ± 12.3 | 1.5 ± 0.52 | 1.9 ± 0.61 | 130.0 ± 44.9 | 102.3 ± 53.8 |
| | Vis + UV-A | 98 ± 0.7 | 88 ± 2.1 | 0.6 ± 0.18 | 0.8 ± 0.13 | 5.2 ± 2.1 | 12.2 ± 8.5 |
| | Vis | 96 ± 1.2 | 88 ± 2.4 | 0.6 ± 0.17 | 0.7 ± 0.11 | 1.7 ± 0.8 | 4.5 ± 1.8 |

## 2.3. Berry Skin Temperature

Figure 1 reports the permanence of defined ranges of berry skin temperature ($T_b$) during ripening. For each season, we calculated the 10th and 90th percentile of $T_b$ in order to determine specific low and high temperatures (<15 °C, <17 °C and >35 °C) that, due to their frequency, could have affected berry skin metabolism.

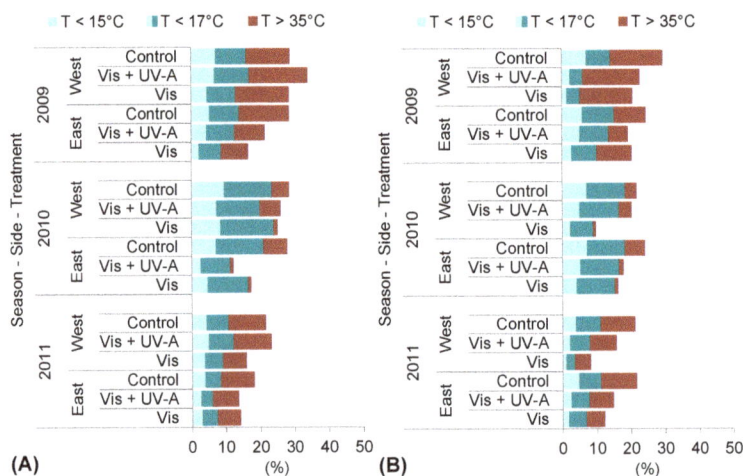

**Figure 1.** Berry skin exposure to defined ranges of temperature (<15 °C, <17 °C and >35 °C) in east and west canopy sides, during ripening in (**A**) Bovale Grande and (**B**) Cannonau.

Berries in the west side of the canopy were much more exposed to high temperatures than those in the east side. Furthermore, during ripening 2009, the prevalence of elevated temperatures (>35 °C) was significantly higher in all treatments and in both canopy sides as compared to the two following seasons. No variety effect was observed, and in Bovale Grande the permanence of such temperatures ranged from nearly 8% of the duration of ripening in East Vis and Vis + UV-A berries to 14% and 16% in Control and in West Vis and Vis + UV-A berries. In Cannonau, the results were similar, although the differences between east and west sides were higher in Control berries. In 2010, East Vis and Vis + UV-A berries were exposed to more than 35 °C for less than 1% of the entire ripening period in both varieties, while Control and West Vis + UV-A berry skin had more than 35 °C for about 6% of the time. In 2011, all treatments were subjected to more than 35 °C for a similar amount of time, the greatest difference being recorded in Cannonau between Vis and Control berries (from 5% to 10% of the time, respectively). The smallest prevalence of low temperatures (<15 °C) was observed in 2009, specially in Cannonau Vis berry skin (for about less than 2% of the duration of ripening). In that year, Control and Vis + UV-A berry skin reached less than 15 °C with higher frequency, lasting 5% and 7% of the time below this threshold in both varieties. In 2010, the percentage of time for which Control and Vis + UV-A berry skin remained with less than 15 °C was about 7% in Cannonau and 9% in Bovale Grande, and in 2011 it decreased to 4% and to 5% respectively.

Overall, the season 2011 showed the lowest permanence of $T_b$ inferior to 17 °C and also the higher permanence of milder temperatures (ranging from 17 °C to 35 °C). This result is in accordance to the previously described air temperature conditions, since among the three seasons, the third was in fact the one in which air temperatures remained closer to the 30 year average.

## 2.4. Berry Skin Anthocyanins

The effect of light regime in berry skin anthocyanin content (TSA) at harvest was inconsistent and only statistically significant in Bovale Grande during the season 2010 and in Cannonau during 2009, when control berries were able to accumulate a significantly higher content of TSA as compared to the two UV-screening treatments (Table 3). No significant differences were observed between the two UV screening treatments but in the last two years of trial slightly higher mean values were observed in Cannonau Vis + UV-A. Also, Spayd et al. [12] obtained inconsistent results while Martínez-Lüscher et al. [36] have reported that, in Tempranillo berries, although higher concentrations of extractable anthocyanins had been observed, UV-B light did not alter total anthocyanin concentration. Azuma et al. [31] studies have demonstrated that high temperature (>35 °C) severely decrease TSA in berry skin and that low temperature (15 °C) and light induce anthocyanin accumulation in a synergetic manner. In our study both varieties were exposed to high temperatures for long time during 2009, with a small permanence of low temperatures (<17 °C). As compared to the previous year, in 2010, direct light exposure promoted higher anthocyanin accumulation in Control berries of Bovale Grande but not in Cannonau. In this variety, Vis + UV-A and Vis berries TSA concentration was probably enhanced due to the effect of lower permanence of high temperatures and higher permanence of low temperatures (Figure 1).

Cannonau showed significantly lower TSA contents as compared to Bovale Grande in all three seasons. Yet, for Cannonau Vis + UVA treatment, we observed an increase of about 80% in TSA both in 2010 and 2011 as compared to the hot 2009. The same pattern was observed for the Vis treatment: an increase in TSA of 230% in 2010 and 95% in 2011 for Vis. In Bovale Grande, the variation in TSA content between 2009 and the other two years of trial was more relevant in absolute value, but not in percent variations, since for both Vis + UVA and Vis treatments the variation ranged from +50% to +92%.

**Table 3.** Effects of light regime on total skin anthocyanin content (mg malvidin $kg^{-1}$ berry) in Bovale Grande and Cannonau berries at harvest. Mean values ($n = 9$) and one-way ANOVA. Small letters indicate significant difference of mean values between treatments and ns refers to non-significant differences between treatment.

|  |  | Bovale Grande | | | Cannonau | | |
|---|---|---|---|---|---|---|---|
|  |  | 2009 | 2010 | 2011 | 2009 | 2010 | 2011 |
| Treatment | Control | 306.9 | 585.8 [a] | 654.1 | 119.5 [a] | 127.0 | 102.1 |
|  | Vis + UV-A | 298.8 | 450.3 [b] | 638.4 | 80.9 [a,b] | 182.1 | 143.2 |
|  | Vis | 258.3 | 496.5 [b] | 650.2 | 55.5 [b] | 146.4 | 108.6 |
|  | Sig. | ns | < 0.05 | ns | <0.05 | ns | ns |

Notes: Lower case letters in the same column indicate significant difference and ns refers to non-significant difference of mean values between treatment ($p < 0.05$) for each variety.

Though having important agronomic and oenological aptitudes [40,44], many accessions of Cannonau have shown low phenolic potential, namely regarding TSA. However, in sunny and warm climate conditions, using deficit irrigation strategies, this behavior can be partially compensated by an accumulation of higher proportion of more color stable forms of anthocyanins [45,46].

In our work, light regimes have influenced berry skin anthocyanin composition differently in the varieties and among seasons. In Figure 2 the proportion anthocyanin derivatives berry skin in Bovale Grande at harvest 2009, 2010 and 2011 are presented. In the hot season 2009, a higher proportion of cyanidin and peonidin glucosides was observed in Control and Vis + UV-A berries during ripening and at harvest, which is in accordance with previous studies suggesting a positive effect of light on dihydroxylated anthocyanins [33]. However, in the following years the treatments did not differ significantly in Bovale Grande and in Cannonau, the proportion of these derivatives was only significantly different between light treatments in 2010.

Besides, the exposure to natural UV light intensities did not induce differences in trisubtituted anthocyanins in Bovale Grande and a decrease in these forms was observed on Cannonau Control berries during 2009. In 2009, Bovale Grande Control and Vis + UV-A berries presented higher proportion of acetylglucoside forms (Figure 3), probably due to both the combined effect of UV light and higher permanence of elevated temperatures [12,13,17,34,36]. In 2010 and 2011, the differences between treatments were not so evident. Yet, Cannonau Control berries presented higher acylation degree with coumaric acid, and higher contents of all anthocyanin derivatives at harvest, except for malvidin glucosides (Figure 3), which can be ascribed to a combined effect of the light treatment and a higher permanence of low temperatures in those years [31].

Major differences were observed among seasons (Table 4). In the two cultivar, the elevated temperatures of 2009 lead to higher accumulation of delphinidin and petunidin, and less peonidin and malvidin derivatives (Figure 2) in accordance to the results obtained by other authors [12,13,17]. In addition, in 2009 a very significant increase in the proportion of acylated forms was evident right from the beginning of ripening, with much lower monoglucoside contents in both varieties and in every light treatment (Figure 3, Table 4). At harvest 2011, the variation in the proportion of anthocyanin derivatives showed a trend similar to that observed in 2009, with significantly higher proportion of delphinidin and petunidin derivatives than in 2010 in Bovale Grande, especially in Control berries, and a much lower proportion of peonidin and malvidin in all treatments as compared to 2010.

These results are in accordance to those reported by Azuma *et al.* [31] who observed an increasing peonidin and malvidin derivatives under light and low temperatures. As far as the acylation degree is concerned, again, in 2011 berry skin anthocyanin profile showed an intermediate content as compared to the other two seasons, with significantly higher anthocyanin acylation than in 2010, probably due two higher permanence of elevated temperatures (Table 4). Light regimes affected anthocyanin partitioning in the two varieties, but the influence of natural UV light intensities on anthocyanin metabolism can be largely surpassed by that of high temperatures, both via anthocyanin degradation and increased acetylation. A detailed analysis on this issue can be found in [32].

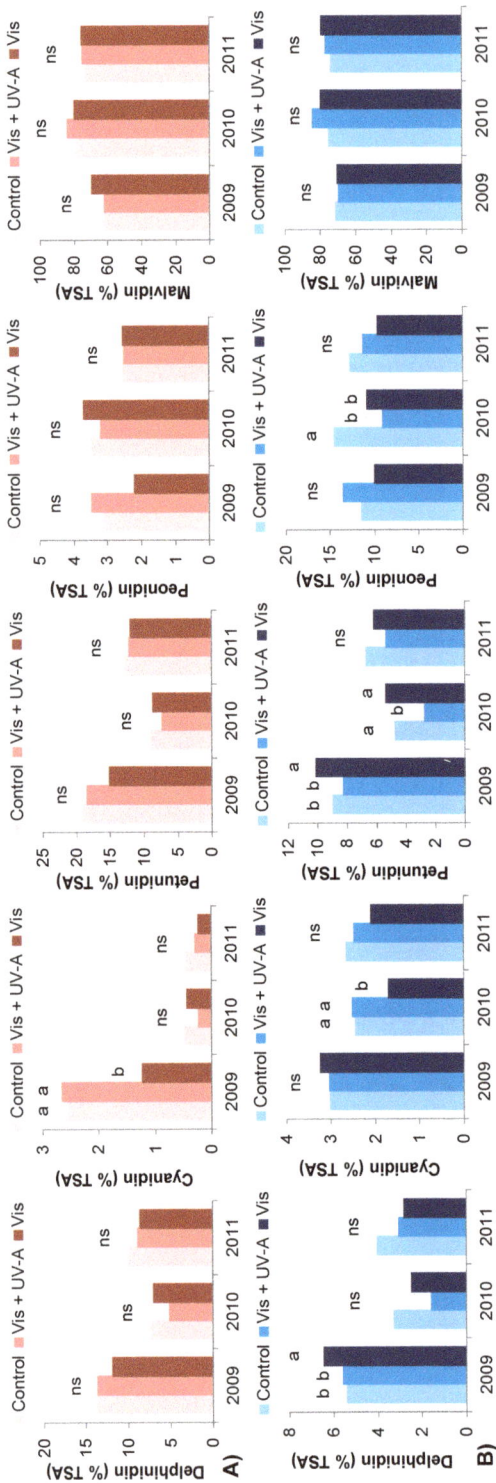

**Figure 2.** Proportion of TSA in delphinidin, cyanidin, petunidin, peonidin and malvidin based anthocyanins in Bovale Grande (**A**) and Cannonau (**B**) berries at harvest 2009, 2010 and 2011. Mean values ($n = 9$) and one-way ANOVA. Lower case letters (a and b) above the bars indicate significant difference and ns refers to non-significant difference of mean values between treatment ($p < 0.05$).

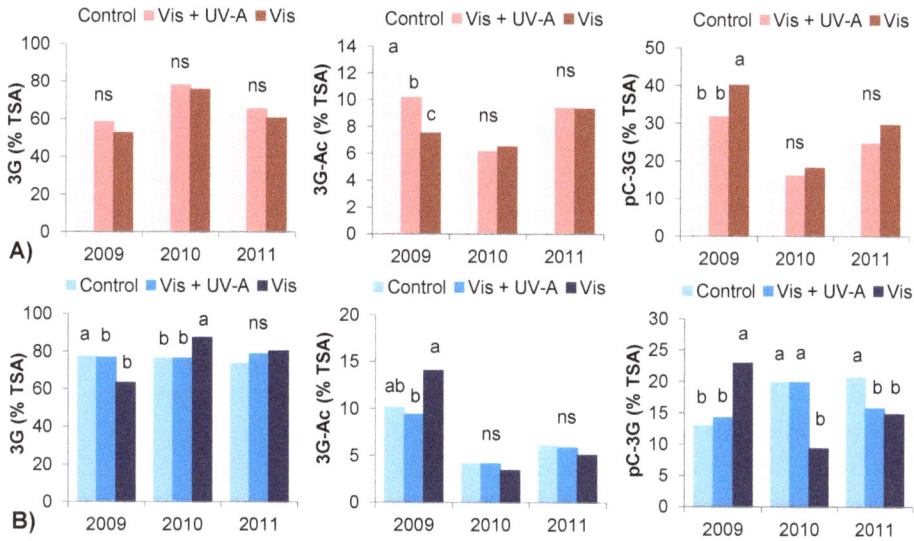

**Figure 3.** Proportion of TSA in 3-monoglucoside (3G), 3-acetyl-glucoside (3G-Ac) and 3-p-coumaroyl-glucoside (pC-3G) forms in Bovale Grande (**A**) and Cannonau (**B**) berries at harvest 2009, 2010 and 2011. Mean values ($n = 9$) and one-way ANOVA. Lower case letters (a and b) above the bars indicate significant difference and ns refers to non-significant difference of mean values between treatment ($p < 0.05$).

**Table 4.** Effect of light treatment on berry skin percent composition of monoglucoside and acylated anthocyanins in Bovale Grande and Cannonau at harvest 2009, 2010 and 2011. Mean values ($n = 9$) and one-way ANOVA.

|  |  | Bovale Grande | | | Cannonau | | |
|---|---|---|---|---|---|---|---|
|  |  | Control | Vis + UV-A | Vis | Control | Vis + UV-A | Vis |
| Monoglucosides | 2009 | 56.0 [b] | 58.2 [b] | 52.5 [b] | 77.0 | 76.4 | 63.1 [b] |
|  | 2010 | 78.0 [a] | 77.8 [a] | 75.5 [a] | 76.0 | 76.0 | 87.3 [a] |
|  | 2011 | 63.4 [b] | 65.2 [b] | 60.2 [ab] | 73.3 | 78.4 | 80.2 [a] |
|  | Sig. | <0.05 | <0.05 | <0.05 | ns | ns | <0.05 |
| Acetylglucosides | 2009 | 12.1 [a] | 10.2 [a] | 7.5 [b] | 10.1 [a] | 9.4 | 14.0 [a] |
|  | 2010 | 6.8 [b] | 6.1 [b] | 6.5 [b] | 4.2 [b] | 4.2 | 3.4 [b] |
|  | 2011 | 10.2 [a] | 9.4 [a] | 9.3 [a] | 6.1 [c] | 5.9 | 5.1 [b] |
|  | Sig. | <0.05 | <0.05 | <0.05 | <0.05 | <0.05 | <0.05 |
| Coumaroylglucosides | 2009 | 32.0 | 31.6 [a] | 40.0 [a] | 12.9 [b] | 14.2 | 22.9 [a] |
|  | 2010 | 15.3 | 16.0 [b] | 18.1 [b] | 19.8 [a] | 19.8 | 9.3 [c] |
|  | 2011 | 25.6 | 24.5 [ab] | 29.4 [ab] | 20.6 [a] | 15.7 | 14.8 [b] |
|  | Sig. | <0.05 | <0.05 | <0.05 | ns | ns | 0.05 |

Notes: Small letters in the same line indicate significant difference and ns refers to non-significant difference of mean values between treatment ($p < 0.05$) for each variety.

## 2.5. Thermal Efficiency for Berry Skin Anthocyanin Accumulation

Our results suggest that high and low temperatures were more effective than light treatment on influencing anthocyanin accumulation in Cannonau berry skin (Table 3, Figures 2 and 3). Greater sensitivity to anthocyanin decrease driven by high temperature was observed in Cannonau [32]. Contrary to Bovale Grande and many other varieties [30], it is extremely difficult to observe the typical two-phase relationship between dynamic of anthocyanin accumulation and that of sugars (°Brix) in Cannonau. In our study, we plotted TSA with total soluble solid (TSS) data from the three years of experiment, and we obtained two completely different scatterplots for the two varieties (Figure 4). Bovale Grande showed a classic linear relationship between data [30], with a first lag phase where TSS increases and very small changes occur in TSA, followed by a nearly linear phase where both compounds increase in parallel. Conversely, Cannonau data does not fit any geometrical curve, but present a quite random dispersion of points, much evident under high temperature conditions. Besides for its genetically feeble phenolic potential, such behavior could partially be explained by a high sensitivity to the permanence of critical ranges of temperature [30,31]. In fact, as compared to warm and low altitude sites, in Mediterranean mountain terroirs, where weather conditions are characterized by higher daily temperature ranges, the pattern of berry skin anthocyanin accumulation is linearly correlated with TSS increments and considerably greater TSA contents have long been reported [46].

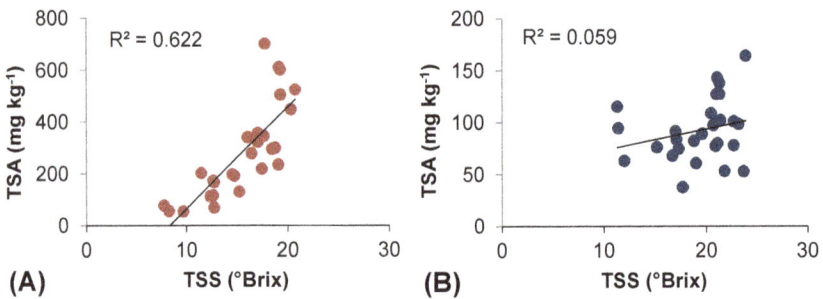

**Figure 4.** Relationships between total soluble solids (TSS) and total berry skin anthocyanin (TSA) content in (**A**) Bovale Grande and (**B**) Cannonau datasets.

In order to better understand the effects of temperature on berry skin anthocyanin accumulation, we calculated the accumulated thermal time for anthocyanin synthesis, using the normal heat hours (NHH) model [47,48] and we determined the permanence (in hours) of low (<15 °C and <17 °C) and high (>35 °C) temperatures, based on berry skin temperature recorded during ripening. We then tested the relationships between

TSA and the four predictor variables (NHH, $H_{T > 35\,°C}$, $H_{T < 15\,°C}$ and $H_{T < 17\,°C}$) for each variety.

Tables 5 and 6 show the main regression analysis estimates and the model performances of three models tested for Bovale Grande and Cannonau, respectively. The first model estimates TSA contents based on a single variable, the NHH, which was statistically highly significant and showed a very good fitting for Bovale Grande, with correlation (R) and determination ($R^2$) coefficients of 0.861 and 0.741, respectively. Conversely, despite being a highly significant variable in the simple linear regression for Cannonau model 1 (variable $p$-value = 0.004), the NHH was poorly correlated with TSA, with an R of 0.511 and a $R^2$ of only 0.235. In both varieties, the introduction of the second variable $H_{T > 35\,°C}$ resulted in an improvement of the regression model, especially for Cannonau, with no collinearity problems between predictor and dependent variables, as indicated by the values of tolerance and variance inflation factor. Nevertheless, as far as Bovale Grande is concerned, the second model only resulted in a very slight increase of R and $R^2$ as compared to model 1, and $H_{T > 35\,°C}$ $p$-value was indicative of non-statistical significant contribution of this variable for the overall regression. For Cannonau, the inclusion of NHH and $H_{T > 35\,°C}$ as predictor variables (model 2) improved considerably the modeling performances as compared to the simple linear model. Both variables were highly significant correlated to TSA and increased regression significance, R, $R^2$ and Adjusted $R^2$ (Adj.$R^2$), respectively to: 0.0001, 0.649, 0.421 and 0.378. The third model is divided into two alternatives, using: (a) $H_{T < 17\,°C}$ and (b) $H_{T < 15\,°C}$. The model including the three independent variables, NHH, $H_{T > 35\,°C}$ and $H_{T < 17\,°C}$ (model 3), was the one that explained the most of the variations in Bovale Grande TSA, about 75.9%, without showing collinearity problems. Also in Cannonau, this model accounted for a much higher proportion of total anthocyanin variation than the previous two, although total berry skin anthocyanin contents still remained quite weakly associated with the temperature driving variables (Adj.$R^2$ = 0.419).

In the last model, after adding NHH, the β coefficient of the variable representing prevalence of low temperatures assumed negative sign in the correlation with TSA in Bovale Grande. On the contrary, in Cannonau the permanence of low temperatures ($H_{T < 17\,°C}$ or $H_{T < 15\,°C}$) have demonstrated a positive role in TSA accumulation model 3 while $H_{T > 35\,°C}$ assumed negative influence, meaning that holding constant the other predictors, a variation of +1 in $H_{T > 35\,°C}$ results in a reduction of −0.209 in TSA.

**Table 5.** General model estimates of Bovale Grande berry TSA (mg malvidin kg$^{-1}$), linear regression analysis and model performance.

| Model | Predictors | Descriptive Statistics | | | | | | | Model Performance | | | | | |
| | | | | | | | | | Unstandardized Coefficients | | Variables Significance | | Collinearity Statistics | |
| | | N | df$_1$ | df$_2$ | Regression Sig. | R | R$^2$ | Adj. R$^2$ | β | Std. Error | T | Sig. | Tolerance | VIF |
|---|---|---|---|---|---|---|---|---|---|---|---|---|---|---|
| 1 | Intercept | 30 | 1 | 28 | 0.0001 | 0.861 | 0.741 | 0.732 | 50.348 | 29.272 | 1.72 | 0.096 | 1 | 1 |
| | NHH | | | | | | | | 0.594 | 0.066 | 8.845 | 0.000 | 1 | 1 |
| 2 | Intercept | 30 | 1 | 27 | 0.0001 | 0.865 | 0.748 | 0.730 | 44.92 | 29.988 | 1.498 | 0.146 | | |
| | NHH | | | | | | | | 0.552 | 0.081 | 6.79 | 0.000 | 0.671 | 1.489 |
| | H$_{T>35\,°C}$ | | | | | | | | 0.41 | 0.456 | 0.899 | 0.377 | 0.671 | 1.489 |
| 3 | Intercept | 30 | 1 | 26 | 0.0001 | | | | 43.206 | 29.741 | 1.524 | 0.140 | | |
| | NHH | | | | | | | | 0.808 | 0.153 | 5.538 | 0.000 | 0.187 | 5.359 |
| | H$_{T>35\,°C}$ (a) | | | | | 0.887 | 0.784 | 0.759 | 0.108 | 0.478 | 0.238 | 0.814 | 0.602 | 1.661 |
| | H$_{T<17\,°C}$ (b) | | | | | 0.873 | 0.762 | 0.735 | -1.084 | 0.527 | -2.059 | 0.050 | 0.244 | 4.106 |
| | H$_{T<15\,°C}$ | | | | | | | | -1.661 | 1.334 | -1.245 | 0.224 | 0.288 | 3.473 |

Notes: (a) Model 3 performance using the predictors NHH, HT > 35 °C and HT < 17 °C; (b) Model 3 performance using the predictors NHH, HT > 35 °C and HT < 15 °C.

296

**Table 6.** General model estimates of Cannonau berry TSA (mg malvidin kg$^{-1}$), linear regression analysis and model performance.

| Model | Predictors | Descriptive Statistics | | | | | | | Model Performance | | | | | | |
| | | N | df$_1$ | df$_2$ | Regression Sig. | R | R$^2$ | Adj. R$^2$ | Unstandardized Coefficients | | Variables Significance | | Collinearity Statistics | |
| | | | | | | | | | β | Std. Error | T | Sig. | Tolerance | VIF |
| 1 | **Intercept** | 30 | 1 | 28 | 0.004 | 0.511 | 0.261 | 0.235 | 71.013 | 7.776 | 9.133 | 0.000 | | |
| | **NHH** | | | | | | | | 0.055 | 0.017 | 3.145 | 0.004 | 1 | 1 |
| 2 | **Intercept** | 30 | 1 | 27 | 0.001 | 0.649 | 0.421 | 0.378 | 75.066 | 7.168 | 10.473 | 0.000 | | |
| | **NHH** | | | | | | | | 0.083 | 0.019 | 4.412 | 0.000 | 0.698 | 1.432 |
| | **H$_{T>35\,°C}$** | | | | | | | | −0.271 | 0.099 | −2.727 | 0.011 | 0.698 | 1.432 |
| 3 | **Intercept** | 30 | 1 | 26 | 0.001 | | | | 76.575 | 6.983 | 10.966 | 0.000 | | |
| | **NHH** | | | | | | | | 0.035 | 0.033 | 1.052 | 0.302 | 0.207 | 4.840 |
| | **H$_{T>35\,°C}$** [a] | | | | | 0.692 | 0.479 | 0.419 | −0.209 | 0.103 | −2.036 | 0.052 | 0.611 | 1.637 |
| | **H$_{T<17\,°C}$** [b] | | | | | 0.681 | 0.463 | 0.401 | 0.235 | 0.138 | 1.706 | 0.100 | 0.271 | 3.695 |
| | **H$_{T<15\,°C}$** | | | | | | | | 0.42 | 0.292 | 1.44 | 0.162 | 0.434 | 2.306 |

Notes: [a] Model 3 performance using the predictors NHH, HT > 35 °C and HT < 17 °C; [b] Model 3 performance using the predictors NHH, HT > 35 °C and HT < 15 °C.

297

For both varieties, the introduction of $H_{T < 15 °C}$ instead of $H_{T < 17 °C}$ produced a smaller improvement on the regression models 1 and 2, reducing correlation and determination coefficients as compared to model 3(a) and adding no statistical significance to the prediction. In Figure 5, the relationships between the total berry skin anthocyanin and the NHH and the linear regression model 3(a), for Bovale Grande (A) and Cannonau (B) datasets, are represented. In both varieties, the addition of prevalence of high (T > 35 °C) and low (T < 17 °C) temperatures during ripening as predictors of TSA increased the model efficiency goodness in the linear regression, almost doubling the $R^2$ for the Cannonau dataset.

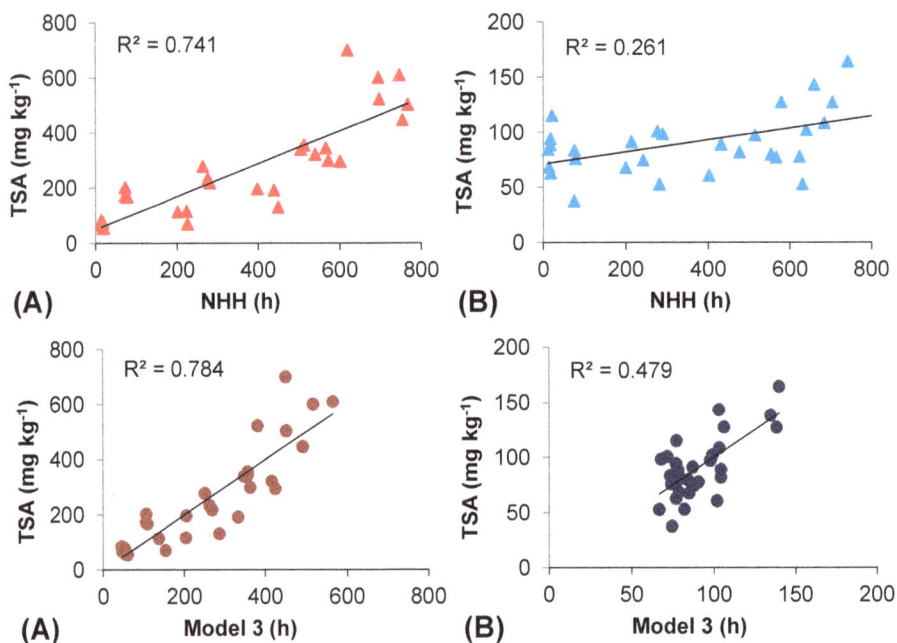

**Figure 5.** Relationships between total berry skin anthocyanin (TSA) content and the Normal Heat Hours (NHH) and between TSA and the regression model 3 in (**A**) Bovale Grande and (**B**) Cannonau datasets.

A regression considering only NHH and $H_{T < 17 °C}$, excluding $H_{T > 35 °C}$, would not improve much the model performance in Bovale Grande (R = 0.885; $R^2$ = 0.783, Adj.$R^2$ = 0.767) and in Cannonau (R = 0.629; $R^2$ = 0.396, Adj.$R^2$ = 0.351) as compared to the NHH model. Despite the influence of other factors on anthocyanin accumulation in berry skin, the permanence of high and low temperatures can help explain the TSA accumulation pattern in Cannonau berries under warm climate conditions. In fact, when modeling TSA based on NHH and the permanence of temperatures higher than 35 °C and lower than 17 °C (Table 6) we observed much better fitting and model performance for Cannonau data than using only the NHH. Additionally, our results showed that the NHH model alone already represents a good estimator of TSA for Bovale Grande, also under warm climate conditions (Table 5).

## 3. Experimental Section

### 3.1. Plant Material and Experimental Site

During 2009, 2010 and 2011 growing seasons, an experimental trial was conducted in the red grapevine clonal collection of the University of Sassari, in Oristano, Italy (39°54'N, 8°37'E). In this field, the three-year-old Bovale Grande and Cannonau vines, grafted on 779 rootstock and spaced 2.5 m × 1.0 m, had North-South row orientation and were trained to a vertical trellis, Guyot pruned and drip-irrigated. From fruit set until harvest, three randomized blocks, with the two grapevine varieties arranged in two adjacent rows, were set covering 16 contiguous plants of each variety per block with two plastic films (UVA diffused and UV-A AV diffused, 1.5 mm thickness, Ginegar, Plastic Products, Ginegar, Israel) of different sunlight transmittance: visible + UV-A (Vis + UV-A) and visible (Vis). In every block, the UV-screening treatments were compared to 8 vines directly exposed to natural sunlight (Control). Each plastic tunnel entirely covered vines canopy and were completely opened at north and south sides and left opened laterally at the trunk base level to allow for better air circulation inside the tunnel. Inside the tunnels average air temperature differed less than 1.2 °C from the external ambient. Light spectral properties of the plastic films were tested using both portable spectroradiometer (FieldSpec®3, range 350–1800 nm) and UV-A (315–400 nm) and UV-B (280–315 nm) single sensors coupled to a portable datalogger (Skye Instruments Ltd., Llandrindod Wells, UK) and a ceptometer (Sunscan SS1 and BF3, Delta T Devices Ltd., Cambridge, UK) for measuring PAR interception throughout the canopy and reference ambient PAR above the canopy, respectively.

### 3.2. Light Microclimate Conditions

PAR interception transversal profiles at the fruit zone were measured at solar midday using a portable ceptometer connected to a total and diffuse PAR and sunshine sensor (Sunscan SS1 and BF3, Delta T Devices Ltd.). The mean values of PAR attenuation into the canopy layers were express as percentage of ambient photon flux density outside the canopy (% reference PAR). For monitoring UV-A and UV-B intensity interception at the fruit zone the single sensors were placed close to the clusters, transversely to its vertical axis, facing the east and west sides of row. Both light microclimate measurements were taken at the same time, in 12 replicates per treatment, along the canopy wall, and under clear sky conditions.

### 3.3. Season Thermal Conditions

Table 1 shows the temperature conditions during the three experimental seasons, gathered from the closest weather station (Capo Frasca, 39°45'41"N, 8°28'01"E) [43,44]. Maximum, mean and minimum monthly average temperatures of the period from June to September 2009, 2010 and 2011, as well as the variation between the values recorded during these months and those of the 30-year long term average, are reported.

### 3.4. Berry Temperature Monitoring and Determinations

Berry skin temperature was monitored during ripening, in four replicates of east and west side clusters, using fine-wire copper-constantan thermocouples (GMR Strumenti, Florence, Italy) collected to four-channel dataloggers (Zeta-tec Co., Cambridge, UK) for data registration at a 10 min intervals. In order to quantify the thermal effect of solar radiation on anthocyanin accumulation, the prevalence of given ranges of temperature (15 °C, 17 °C, and 35 °C) along each ripening season was then calculated and the thermal time was computed using the normal heat hours (NHH) model [46,47], which have been proven to give a explanation of vine thermal requirements for anthocyanin synthesis and accumulation [47,49].

### 3.5. Berry Composition Analysis

For each treatment, cluster fractions were randomly collected in two-week sampling intervals, from veraison until harvest, and berry weight and composition (total soluble solids, pH, titratable acidity, total skin anthocyanin and phenols content) was analyzed following OIV methodologies [50]. Berry anthocyanin composition was determined in a sample of 50 berry skin extracts per replicate previously frozen using HPLC, and following Di Stefano and Cravero method [51] as described in Fernandes de Oliveira and Nieddu [39].

### 3.6. Statistical Analysis

Berry composition data analysis was carried out using one-way ANOVA and the least significant difference (LSD) test to compare means at $p$-value of 0.05, using SPSS 16.0 (SPSS Inc., Chicago, IL, USA). Multiple linear regression was performed in order to examine the influence of accumulated NHH and of the permanence (in hours) of given temperature ranges (<15 °C, 17 °C and > 35 °C) on anthocyanin accumulation pattern in the two studied varieties over the three-year data set. For NHH calculation, 35 °C maximum, 10 °C minimum and 26 °C optimal temperatures for anthocyanin synthesis were considered [47], while the definition of the abovementioned temperature ranges took into consideration the observed 10th and 90th percentiles of berry skin temperatures along the duration of ripening in the three experimental seasons. The four independent variables tested for building the models (NHH, $H_{T < 15\,°C}$, $H_{T < 17\,°C}$ and $H_{T > 35\,°C}$) generally have high correlation degrees and therefore were sequentially added using hierarchical regression analysis and multicollinearity diagnostic (using tolerance and variance inflation factor, VIF statistics). The variables were tested for normality, and t-test coefficients were used to determine single variables explanatory power as well as the combination of variables that accounted for higher proportion of observed variance in berry total skin anthocyanin (TSA).

## 4. Conclusions

This study has demonstrated that natural UV light intensities can have a positive influence on total anthocyanin contents and may favor berry metabolism toward the accumulation of higher proportion of dihydroxylated derivatives and the more color stable acylated forms. Natural UV light intensities did not induce differences in trisubtituted anthocyanins as reported by other studies [36,37] but a positive effect of UV was observed in cyanidin and peonidin derivatives in Bovale Grande Control and Vis + UV-A in 2009 and in Cannonau in 2010. Both UV light and high permanence of elevated temperatures induced increased acylation levels though the major effect of solar radiation was driven by high temperatures, which were able to alter the proportion of anthocyanin derivatives and the degree of acylation to a greater extent, promoting higher accumulation of acetyl- and coumaroylglucosides, namely of the delphinidin and petunidin derivatives.

The relationship between TSS and TSA was quite different in the two varieties, since for Cannonau the increases in TSS were not followed by increasing TSA as observed in Bovale Grande. Such behavior could partially be explained by a high sensitivity to the permanence of critical ranges of temperature. In fact, besides the higher between-years variation in TSA contents observed in Cannonau, also, the modeling exercise provided evidence for a greater sensitivity to high and low temperature of berry anthocyanin contents in Cannonau.

In this work, we proved that NHH is a very good estimator of TSA for Bovale Grande but not for Cannonau. A positive contribution of the permanence of elevated temperatures (T > 35 °C) to anthocyanin accumulation was highlighted by the regression analysis for both varieties, especially for Cannonau dataset. The permanence of low (T < 17 °C) improved slightly TSA estimation, mostly for Cannonau.

Overall, the Normal Heat Hours model can explain a great part of variation in anthocyanin patterns, but in warm climate conditions and for varieties highly sensitive to temperature like Cannonau, the prediction of total skin anthocyanin contents can benefit from the addiction of variables that take into account the permanence of high and low temperatures until berry maturation.

Increasing the knowledge on how berry total skin anthocyanin and composition respond to natural UV radiation and temperature in Mediterranean areas can help improve cultivation practices, namely those affecting canopy microclimate, in order to favor the production of higher quality grapes. Likewise, assessing varietal sensitivity to these abiotic factors and enhancing the accuracy of anthocyanin accumulation estimation models represent an important contribution to a better vineyard management, starting from the varietal choices up to the harvest decisions.

**Acknowledgments:** This research was supported by the Regione Autonoma della Sardegna (project Legge Regionale 7 agosto 2007 n.7. Studio dell'influenza dei fattori microclimatici (luce e temperatura) sull'accumulo e composizione dei polifenoli dell'uva. Sviluppo di un biosensore per la determinazione dei polifenoli e valutazione della loro capacità antiossidante *in vitro* e *in vivo*). The authors would also like to thank the financial contribution of the Convisar—Vino e Sardegna Consortium (project SQFVS APQ P6).

**Author Contributions:** A.F.O., A.D.C. and G.N. designed and performed research; A.F.O., L.M., A.D.C. and L.P. analyzed the data; A.F.O., and G.N. wrote the paper. All authors read and approved the final manuscript.

**Conflicts of Interest:** The authors declare no conflict of interest.

## References

1.  Zepp, R.G.; Erickson, D.J., III; Paul, N.D.; Sulzberger, B. Interactive effects of solar UV radiation and climate change on biogeochemical cycling. In *The Environmental Effects Assessment Panel Report*; United Nations Environment Programme: Nairobi, Kenya, 2006; pp. 135–162.
2.  United Nations Environment Programme (UNEP). *Environmental Effects of Ozone Deplection: 2010 Assessment*; UNEP: Nairobi, Kenya, 2010; p. 328.
3.  Zepp, R.G.; Erickson, D.J., III; Paul, N.D.; Sulzberger, B. Interactive effects of solar UV radiation and climate change on biogeochemical cycling. *Photochem. Photobiol. Sci.* **2007**, *6*, 286–300.
4.  Caldwell, M.M.; Bornman, J.F.; Ballaré, C.L.; Flint, S.D.; Kulandaivelu, G. Terrestrial ecosystems, increased solar ultraviolet radiation and interactions with other climatic change factors. *Photochem. Photobiol. Sci.* **2007**, *3*, 252–266.
5.  Sullivan, J.H.; Teramura, A.H. Field study of the interaction between solar ultraviolet-B radiation and drought on photosynthesis and growth in soybean. *Plat Physiol.* **1990**, *2*, 141–146.
6.  Steel, C.C.; Keller, M. Influence of UV-B irradiation on the carotenoid content of Vitis vinifera tissues. *Biochem. Soc. Trans.* **2000**, *28*, 883–885.
7.  Kolb, C.A.; Käser, M.A.; Kopecký, J.; Zotz, G.; Riederer, M.; Pfündel, E.E. Effects of natural intensities of visible and ultraviolet radiation on epidermal ultraviolet screening and photosynthesis in grape leaves. *Plant Physiol.* **2001**, *127*, 863–875.
8.  Julkunen-Tiitto, R.; Häggman, H.; Aphalo, P.J.; Lavola, A.; Tegelberg, R.; Veteli, T. Growth and defense in deciduous trees and shrubs under UV-B. *Environ. Pollut.* **2005**, *137*, 404–414.

9. Ruhland, C.T.; Fogal, M.J.; Buyarski, C.R.; Krna, M.A. Solae ultraviolet-B radiation increases phenolic contents and ferric reducing antioxidant power in Avena sativa. *Molecules* **2007**, *12*, 1220–1232.

10. Doupis, G.; Chartzoulakis, K.; Beis, A.; Patakas, A. Allometric and biochemical responses of grapevines subjected to drought and enhanced ultraviolet-B radiation. *Aust. J. Grape Wine Res.* **2011**, *17*, 36–42.

11. Gregan, S.M.; Wargent, J.J.; Liu, L.; Shinkle, J.; Hofmann, R.; Winefield, C.; Trought, M.; Jordan, B. Effects of solar ultraviolet radiation and canopy manipulation on the biochemical composition of Sauvignon Blanc grapes. *Aust. J. Grape Wine Res.* **2012**, *18*, 227–238.

12. Spayd, S.E.; Tarara, J.M.; Mee, D.L.; Ferguson, J.C. Separation of sunlight and temperature effects on the composition of Vitis vinifera cv. Merlot berries. *Am. J. Enol. Vitic.* **2002**, *53*, 171–182.

13. Downey, M.O.; Dokoozlian, N.K.; Krstic, M.P. Cultural practice and environmental impacts on the flavonoid composition of grapes and wine: A review of recent research. *Am. J. Enol. Vitic.* **2006**, *57*, 257–268.

14. Mori, K.; Goto-Yamamoto, N.; Kitayama, M.; Hashizume, K. Loss of anthocyanins in red-wine grape under high temperature. *J. Exp. Bot.* **2007**, *58*, 1935–1945.

15. Berli, F.J.; Fanzone, M.; Piccoli, P.; Bottini, R. Solar UV-B and ABA are involved in phenol metabolism of Vitis vinifera L. increasing biosynthesis of berry skin polyphenols. *J. Agric. Food Chem.* **2011**, *59*, 4874–4884.

16. Ristic, R.; Downey, M.O.; Iland, P.G.; Bindon, K.; Francis, I.L.; Herderich, M.; Robinson, S.P. Exclusion of sunlight from Shiraz grapes alters wine colour, tannin and sensory properties. *Aust. J. Grape Wine Res.* **2007**, *13*, 53–65.

17. Tarara, J.M.; Lee, J.; Spayd, S.E.; Scagel, C.F. Berry temperature and solar radiation alter acylation proportion, and concentration of anthocyanin in Merlot grapes. *Am. J. Enol. Vitic.* **2008**, *59*, 235–247.

18. Koyama, K.; Ikeda, H.; Poudel, P.R.; Goto-Yamamoto, N. Light quality affects flavonoid biosynthesis in young berries of Cabernet Sauvignon grape. *Phytochemistry* **2012**, *78*, 54–64.

19. He, F.; Liang, N-N.; Mu, L.; Pan, Q-H.; Wang, J.; Reeves, M.J.; Duan, C-Q. Anthocyanins and their variation in red wines I. Monomeric anthocyanins and their color expression. *Molecules* **2012**, *17*, 1571–1601.

20. Smart, R.E. Sunlight interception by vineyards. *Am. J. Enol. Vitic.* **1973**, *24*, 141–147.

21. Smart, R.E.; Robinson, M. Sunlight into wine. In *A Handbook for Winegrape Canopy Management*; Winetitles: Adelaide, Australia, 1991; p. 88.

22. Poni, S.; Lakso, A.N.; Intrieri, C.; Rebucci, B.; Filippetti, I. Laser scanning estimation of relative light interception by canopy components in different grapevine training systems. *Vitis* **1996**, *35*, 177–182.

23. Kliewer, W.M.; Dokoozlian, N.K. Leaf area/crop weight ratios of grapevines: Influence on fruit composition and wine quality. *Am. J. Enol. Vitic.* **2005**, *56*, 170–181.

24. Mabrouk, H.; Sinoquet, H. Indices of light microclimate and canopy structure determined by 3-D digitising and image analysis and their relationship to grape quality. *Aust. J. Grape Wine Res.* **1998**, *4*, 2–13.

25. Cohen, S.D.; Tarara, J.M.; Gambetta, G.A.; Matthews, M.A.; Kennedy, J.A. Impact of diurnal temperature variation on grape berry development, proanthocyanidin accumulation, and the expression of flavonoid pathway genes. *J. Exp. Bot.* **2012**, *63*, 2655–2665.

26. Downey, M.O.; Harvey, J.S.; Robinson, S.P. Analysis of tannins in seeds and skins of Shiraz grapes throughout berry development. *Aust. J. Grape Wine Res.* **2003**, *9*, 15–27.

27. He., F.; Mu, L.; Yan, G-L.; Liang, N-N.; Pan, Q-H.; Wang, J.; Reeves, M.J.; Duan, G-Q. Biosynthesis of anthocyanins and their regulation in colored grapes. *Molecules* **2010**, *15*, 9057–9091.

28. Bergqvist, J.; Dokoozlian, N.; Ebisuda, N. Sunlight exposure and temperature effects on berry growth and composition of Cabernet Sauvignon and Grenache in the central San Joaquin Valley of California. *Am. J. Enol. Vitic.* **2001**, *52*, 1–7.

29. Yamane, T.; Jeong, S.T.; Goto-Yamamoto, N.; Koshita, Y.; Kobayashi, S. Effects of temperature on anthocyanin biosynthesis in grape berry skins. *Am. J. Enol. Vitic.* **2006**, *57*, 54–59.

30. Sadras, V.O.; Moran, M.A. Elevated temperature decouples anthocyanins and sugars in berries of Shiraz and Cabernet Sauvignon. *Aust. J. Grape Wine Res.* **2012**, *18*, 115–122.

31. Azuma, A.; Yakushiji, H.; Koshita, Y.; Kobayashi, S. Flavonoid biosynthesis-related genes in grape skin are differentially regulated by temperature and light conditions. *Planta* **2012**, *236*, 1067–1080.

32. Fernandes de Oliveira, A.; Nieddu, G. Effects of natural ultraviolet radiation and high temperature on berry microclimate and anthocyanins in two red grape varieties Aust. *J. Grape Wine Res.* **2014**. submitted.

33. Downey, M.O.; Harvey, J.S.; Robinson, S.P. The effect of bunch shading on berry development and flavonoid accumulation in Shiraz grapes. *Aust. J. Grape Wine Res.* **2004**, *10*, 55–73.

34. Berli, F.J.; Moreno, D.; Piccoli, P.; Hespanhol-Viana, L.; Silva, M.F.; Bressan-Smith, R.; Cavagnaro, J.B.; Bottini, R. Abscisic acid is involved in the response of grape (*Vitis vinifera* L.) cv. Malbec leaf tissues to ultraviolet-B radiation by enhancing ultraviolet-absorbing compounds, antioxidant enzymes and membrane sterols. *Plant Cell Environ.* **2010**, *33*, 1–10.

35. Martinez- Lüscher, J.; Sánchez-Díaz, M.; Serge Delrot, S.; Aguirreolea, J.; Pascual, I.; Gomès, E. Ultraviolet-B radiation and water deficit interact to alter flavonol and anthocyanin profiles in grapevine berries. *Plant Cell Physiol.* **2014**, *55*, 1925–1936.

36. Martinez-Lüscher, J.; Torres, N.; Hilbert, G.; Richard, T.; Sánchez-Díaz, M.; Delrot, S.; Aguirreolea, J.; Pascual, I.; Gomès, E. Ultraviolet-B radiation modifies the quantitative and qualitative profile of flavonoids and amino acids in grape berries through Transcriptomic Regulation. *Phytochemistry* **2014**, *102*, 106–114.

37.	Carbonell-Bejerano, P.; Diago, M.-P.; Martínez-Abaigar, J.; Martínez-Zapater, J.; Tardáguila, J.; Núnez-Olivera, E. 2014 Solar ultraviolet radiation is necessary to enhance grapevine fruit ripening transcriptional and phenolic responses. *Plant Biol.* **2014**, *14*, 183.

38.	Matus, J.T.; Loyola, R.; Vega, A.; Peña-Neira, A.; Bordeu, E.; Arce-Johnson, P.; Alcalde, J.A. Post-veraison sunlight exposure induces MYB-mediated transcriptional regulation of anthocyanin and flavonol synthesis in berry skins of Vitis vinifera. *J. Exp. Bot.* **2009**, *60*, 853–867.

39.	Nieddu, G.; Chessa, I.; Cocco, G.F.; Nieddu, M.; Deidda, P. Caratterizzazione mediante marcatori RAPD dei vitigni tradizionali della Sardegna. *Acta Italus Hortus* **2006**, *13*, 275–280.

40.	Vacca, V.; del Caro, A.; Milella, G.G.; Nieddu, G. Preliminary characterisation of Sardinian red grape cultivars (*Vitis vinifera* L.) according to their phenolic potential. *S. Afr. J. Enol. Vitic.* **2009**, *30*, 93–100.

41.	TEMIS—Tropospheric Emission Monitoring Internet Service. UV station data based on satellite data (SCIAMACHY, GOME-2). Available online: http://www.temis.nl/uvradiation/SCIA/stations_uv.html (accessed on 4 August 2014).

42.	Aeronautica Militare. *Atlante Climatico D'Italia 1971–2000*; Centro Nazionale di Meteorologia e Climatologia Aeronautica: Rome, Italy, 2009; Volume III. Available online: http://clima.meteoam.it/AtlanteClim2/pdf/Atlas71-00%20-%20Vol%203.pdf (accessed on 16 August 2014).

43.	Tutiempo Network, S.L. Tutiempo. Net. Available online: http://www.tutiempo.net/clima/Capo_Frasca/165390.htm (accessed on 16 August 2014).

44.	Fernandes de Oliveira, A.; Mameli, M.G.; de Pau, L.; Satta, D.; Nieddu, G. Deficit irrigation strategies in *Vitis vinifera* L. cv. Cannonau under Mediterranean climate. Part I—Physiological responses, growth-yield balance and berry composition. *S. Afr. J. Enol. Vitic.* **2013**, *34*, 170–183.

45.	Fernandes de Oliveira, A.; Nieddu, G. Deficit irrigation strategies in *Vitis vinifera* L. cv. Cannonau under Mediterranean climate. Part II—Cluster microclimate and anthocyanin accumulation patterns. *S. Afr. J. Enol. Vitic.* **2013**, *34*, 184–195.

46.	Deidda, P.; Nieddu, G.; Pellizzaro, G.; Pia, G. Dynamics of anthocyanins in "Cannonau" grapevine as related to the environmental conditions. *Polyphenols* **1995**, *94*, 207–208.

47.	Rustioni, L.; Rossoni, M.; Cola, G.; Mariani, L.; Failla, O. Microclima termico e luminoso e accumulo di antociani in "Nebbiolo". *Quad. Vitic. Enol.* **2006**, *28*, 137–147.

48.	Cola, G.; Mariani, L.; Parisi, S.; Failla, O. Thermal time and grapevine phenology. *Acta Italus Hortus* **2012**, *3*, 31–34.

49.	Murada, G.; Zecca, O.; Cola, G.; Mariani, L.; Failla, O. Maturità fenolica del "Nebbiolo" in Valtellina: effetto dell'annata e del sito. *Quad. Vitic. Enol.* **2006**, *28*, 125–136.

50.	*OIV recueil des méthodes internationales d'analyse des vins et des moûts*; Office International de la vigne et du vin: Paris, France, 1990; p. 368.

51.	Di Stefano, R.; Cravero, M.C. The grape phenolic determination. *Riv. Vit. Ital.* **1991**, *49*, 37–45.

**Sample Availability:** *Sample Availability*: Samples are not available from the authors.

# Section 4:
# Anthocyanin Composition and their Biological Properties

# Mistaken Identity: Clarification of
# *Rubus coreanus* Miquel (Bokbunja)

Jungmin Lee, Michael Dossett and Chad E. Finn

**Abstract:** In the U.S., there has been a recent surge in Korean black raspberry products available and in the number of reports about this species appearing in the scientific literature. Despite this, the majority of products sold and the work carried out has been on *Rubus occidentalis* L., not *R. coreanus* Miquel. The importance of accurate recognition of all starting material is multiplied for research downstream, including genetics/genomics, plant breeding, phenolic identification, food processing improvements and pharmacokinetic investigations. An overview of distinguishing characteristics separating *R. coreanus* from *R. occidentalis* will be presented. Research conducted on correctly identified fruit will also be summarized to aid future studies that might showcase the unique qualities that bokbunja can offer.

Reprinted from *Molecules*. Cite as: Lee, J.; Dossett, M.; Finn, C.E. Mistaken Identity: Clarification of *Rubus coreanus* Miquel (Bokbunja). *Molecules* **2014**, *19*, 10524–10533.

## 1. Introduction

According to the 1867 records of Friedrich Miquel [1], wild *Rubus coreanus* Miq. (bokbunja native to eastern Asia) Chinese, Japanese and Korean [2] plants and fruit were collected in Korea by Richard Oldham and verified by Naohiro Naruhashi, as early as 1863. Within the *Rubus* genus, *R. coreanus* is in the subgenus, *Idaeobatus*, along with at least 99 other *Rubus* species, including other commercially harvested species, such as red raspberry (*R. idaeus* L.), the Japanese wineberry (*R. phoenicolasius* L.), the Andean blackberry (*R. glaucus* Benth.), Mysore raspberry (*R. niveus* Thunb.) and the black raspberry (*R. occidentalis* L.) [2]. In the late 1960s, commercial cultivation of what was thought to be *R. coreanus* (anonymous Korean commercial grower) started in South Korea. While *R. coreanus* (bokbunja) beverage products were marketed as traditional foods, they were unlike a true Korean traditional food (e.g., kimchi) in that they were not readily available in the marketplace until around the year 2004 (personal observation; [3]). A recent literature search showed an increase in *R. coreanus* research articles being published around the year 2007.

Identity concerns over *R. coreanus* plants [3–6] were initially brought to our attention from the fruit images utilized on bokbunja commercial products in the U.S. marketplace; *R. coreanus* (Korean black raspberry) fruit was misrepresented by images of *R. occidentalis* (native to eastern North America, [2]) fruit. Only a small fraction of commercially cultivated black raspberries in Korea are *R. coreanus*, while the majority (reported at >2,800 hectares in 2013 cultivated by >10,000 farmers; [3,7,8]) are actually *R. occidentalis* (personal observation; anonymous Korean commercial grower; [3–5,8]). Based on randomly amplified polymorphic DNA fragments and chloroplast markers, Eu *et al.* [3–5] demonstrated that commercially grown black raspberry plants in Korea are more closely related to North American *R. occidentalis* cultivars than to native *R. coreanus* and, in fact, are *R. occidentalis* not *R. coreanus*. Currently, production of *R. coreanus* in Korea is unable to meet the demand for bokbunja products. Identifying the best *R. coreanus* selections or breeding cultivars for commercial plantings is underway by Kim *et al.* [8,9], where Kim *et al.* [9] already has identified promising cultivars (Jungkeum 1, Jungkeum 2, Jungkeum 3, Jungkeum 4 and Jungkeum 5).

Phenolic profiles have become a valuable laboratory tool in small fruit research: our own studies of species, cultivar and genotype in blueberries (*Vaccinium corymbosum* L., *V. deliciosum* Piper, *V. membranaceum* Douglas ex Torr., *V. ovalifolium* Sm. and *V. ovatum* Pursh.), strawberries (*Fragaria* spp. L.), elderberries (*Sambucus canadensis* L. and *S. nigra* L.), black raspberries (*R. occidentalis* and *R. coreanus*) and lingonberries (*V. vitis-idaea* L.) were greatly aided by the ability to contrast phenolic profiles [10–21]. This collective phenolic literature directly assists ingredient assurance and product quality control and can be used in authenticity and adulteration monitoring, phenolic degradation, pharmacokinetics, *etc.*, but when misidentified fruit (thought to be that of *R. coreanus*) is harvested, all work downstream becomes misinformation that only causes further disorder. For example, our *Rubus* phenolic review article [22] was written before access to authenticated *R. coreanus* fruit samples existed [6], and it summarized some scientific papers that were conducted on incorrectly identified *R. coreanus* fruit. The health benefits of *R. coreanus* fruit might be uniquely different from *R. occidentalis*, but this is difficult to gauge based on the current confusion among growers, producers and scientific communities.

A one-page fact-sheet with photos depicting leaves, flowers, fruit and anthocyanin profiles is available for download to help growers, ingredient suppliers, food processors, and researchers distinguish between these two black raspberries [23]. The objective of this review is to reduce future mistakes by highlighting this issue, to provide a guide to clearly differentiate these species and to provide a summary of phenolic research conducted on the actual *R. coreanus* fruit.

## 2. History of Commercialization of *Rubus coreanus* and *R. occidentalis* Plants

*Rubus occidentalis* has been widely grown commercially in eastern North America, where it is native, since the mid-late 1800s [24] and has been used in a variety of food products because of its dark color and unique flavor [22]. While *R. coreanus* is not cultivated commercially in North America, as early as 1937, Darrow [25] recognized its value as a source of resistance to a variety of disease pathogens for breeding. Unfortunately, this potential has not been fully realize; while *R. coreanus* has been valuable in breeding red raspberry [26], hybrids with *R. occidentalis* are highly sterile [27]. It is not clear when *R. occidentalis* was first introduced to Korea. We are unaware of any work comparing the agronomic qualities of these two species as grown in Korea; however, in North America, *R. coreanus* is vigorous and resistant to many of the diseases that cause problems for black and red raspberry growers. Despite this, its fruit tends to be smaller and softer and lack the distinctive flavor of *R. occidentalis*. These reasons, combined with its vigor and thornier canes that may make *R. coreanus* more difficult to manage, could be part of the reason why it is not as commonly grown on a commercial scale.

## 3. Morphological and Phenological Differences

*Rubus coreanus* flowers are a light to dark purple-pink color [3,6,8,28] compared to the white colored flowers of *R. occidentalis*. *Rubus coreanus* plants typically have two or more additional leaflets compared to *R. occidentalis*; *R. coreanus* typically has five to nine leaflets that are always pinnately-arranged, while *R. occidentalis* usually has three (ternate) or occasionally five palmately-arranged leaflets (Lee *et al.* [6]). *Rubus coreanus* fruit are superficially similar to those of *R. occidentalis*; genotypes of both species produce fruit that ranges from albino (orange), purple to black in color, and the fruit is hollow, as the torus remains on the plant when the fruit is picked [6]. However, well-formed fruit of *R. occidentalis* have smaller drupelets, leading to a smoother surface contour, and usually have some degree of fine white pubescence. This pubescence may occur across the epidermis of the *R. occidentalis* fruit, but is usually concentrated around the edges of the drupelets and is less evident in *R. coreanus*, leading to a somewhat glossier appearance. Fruit of *R. coreanus* can have an unusual bicolored appearance, where anthocyanins concentrate into dark spots on the tip of each drupelet, at the base of the style, against an orange background (see Figure 1c; orange with dark spots on the top of each drupelet of aggregate fruit). Clear images of the leaves, flowers and fruit can be found in Lee *et al.* [6], Eu *et al.* [3,4], Kim *et al.* [9] and in Figure 1. Plant size, vigor, leaf morphology, cane morphology and fruit ripening dates can be found in Lee *et al.* [6], Keep *et al.* [28] and Miquel [1]. *Rubus coreanus* fruit ripen in late July and early August, whereas *R. occidentalis* fruit ripen a few weeks earlier (in June/July) [6,8].

**Figure 1.** There are clear distinguishing morphological differences between *Rubus coreanus* and *R. occidentalis*. A photo of leaves can be found in Lee *et al.* [6]. Again, *R. coreanus* has pink flowers (**a**) and appears glossy, as there is less white hair (pubescence) on the fruit (**c**). *Rubus occidentalis* has white flowers (**b**) and white hair on the fruit (**d**), which make the fruit appear dull.

## 4. Anthocyanin Profiles

Besides their unique vegetative traits, the two species have distinctive anthocyanin profiles (Figure 2 and Table 1). *Rubus coreanus* fruit contains fewer anthocyanins (up to three) compared to *R. occidentalis* (up to seven) [6,10–12,18–21]. A list of the individual anthocyanins can be found in Table 1. A clear anthocyanin profile of 'Munger' fruit overlaid with *R. coreanus* fruit is shown in Figure 2. In the U.S., the cultivar, Munger (*R. occidentalis*), is the most widely grown, and 'Munger' fruit has a reliable anthocyanin profile over varying growing seasons (comparing Figure 2 to Dossett *et al.* [18,19]). While both species contain glycosides of cyanidin and pelargonidin [6,10], trace levels of peonidin-3-rutinoside are only reported in some *R. occidentalis* fruit [18–21].

Our findings [6,10] confirm the identification correctly reported by Kim *et al.* [29], Heo *et al.* [30] and Lee *et al.* [31]. The two anthocyanins Kim *et al.* [29] found in *R. coreanus* fruit were glucoside and rutinoside of cyanidin, and cyanidin-3-rutinoside was the main pigment, followed by cyanidin-3-glucoside. In samples from CRUB 1634 16-1 fruit (*R. coreanus* genotype at USDA-ARS), cyanidin-3-rutinoside (lightest colored fruit) was also the chief anthocyanin, though fruit from two other *R. coreanus* genotypes (CRUB 1634 19-28 and CRUB 1634 19-23) from the same population had more cyanidin-3-glucoside and less cyanidin-3-rutinoside [6]. Heo *et al.* [30] and Lee *et al.* [31] also reported only two measurable anthocyanins in *R. coreanus*. Heo *et al.* [30] described cyanidin-3-rutinoside content being greater than cyanidin-3-glucoside in mature fruit, but found the order reversed in immature fruit. Since *R. coreanus* does not contain xylose-containing pigments (*i.e.*, cyanidin-3-xylosylrutinoside and/or cyanidin-3-sambubioside; see Figure 2. and Table 1), their detection indicates the presence of *R. occidentalis* fruit or another unknown contaminant and that the sample is not pure *R. coreanus*.

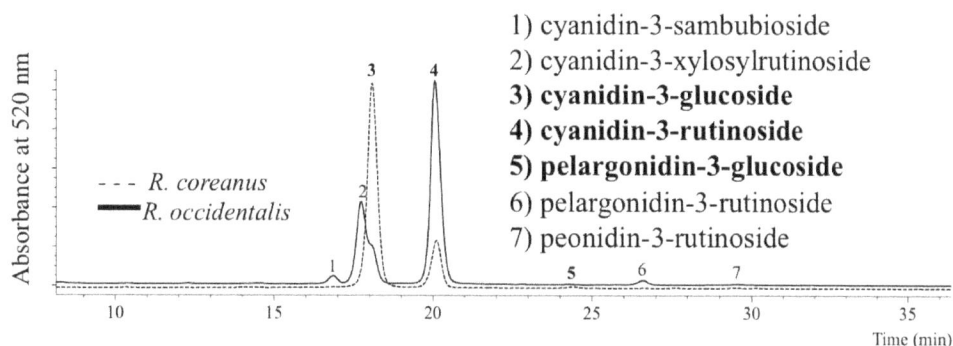

1) cyanidin-3-sambubioside
2) cyanidin-3-xylosylrutinoside
**3) cyanidin-3-glucoside**
**4) cyanidin-3-rutinoside**
**5) pelargonidin-3-glucoside**
6) pelargonidin-3-rutinoside
7) peonidin-3-rutinoside

**Figure 2.** Anthocyanin profile of *Rubus occidentalis* cv. Munger (solid line) and *R. coreanus* (dotted line) fruits. Anthocyanin peak identifications in bold are the ones found in *R. coreanus* fruit [6,10].

**Table 1.** Anthocyanins found in *Rubus coreanus versus R. occidentalis* fruit. Anthocyanins listed in the order of HPLC elution. '+' indicates present. '−' indicates not present. '+/−' indicates both cases have occurred. A clear recent example of additional anthocyanin profiles of the two species can be found Lee *et al.* [6], Dossett *et al.* [18] and Lee [10–12].

| Peak Numbering in Figure 2. | Anthocyanin | *R. coreanus* | *R. occidentalis* |
|---|---|---|---|
| 1 | cyanidin-3-sambubioside | − | + |
| 2 | cyanidin-3-xylosylrutinoside | − | + * |
| 3 | cyanidin-3-glucoside | + | + |
| 4 | cyanidin-3-rutinoside | + | +/− |
| 5 | pelargonidin-3-glucoside | +/− | +/− |
| 6 | pelargonidin-3-rutinoside | − | +/− |
| 7 | peonidin-3-rutinoside | − | +/− |

* Cyanidin-3-xylosylrutinoside was found lacking in the fruit of one wild collected *R. occidentalis* plant out of >1,000 genotypes analyzed in our laboratory [6,10–12,18–21]. Lacking cyanidin-3-xylosylrutinoside in *R. occidentalis* fruit occurs rarely [20].

Due to this difference in the anthocyanin profile (chemotaxonomical distinction), products from these species can be identified in the absence of the vegetative attributes described above. For example, a Korean commercial bokbunja juice sample was obtained, and analysis showed that it had the anthocyanin profile of *R. occidentalis* fruit, not *R. coreanus* fruit [10]. This commercial juice contained cyanidin-3-sambubioside and cyanidin-3-xylosylrutinoside, not found in *R. coreanus*. Researchers should be aware that after processing (*i.e.*, freeze drying, juicing, concentrating, heating), the proportion of the individual anthocyanin peaks might be altered, and unknown polymeric anthocyanins may be formed and appear in the chromatograms, as pointed out by Lee *et al.* [32], Lee and Wrolstad [33], Lee [11], Sadilova *et al.* [34] and Novotny *et al.* [35]. Techniques for improved retention of color using food processing methods, ideal storage condition, *etc.*, will result in differing response between *R. coreanus* and *R. occidentalis*, since the predominant cyanidin-based anthocyanins in their fruits have different colors, tinctorial strengths (visual detection threshold), spectral characteristics, thermal degradation kinetics, *etc.*, due to independent structures [34–36]. Different cyanidin-based anthocyanins exhibit altered bioavailability in human subjects [37–39], so the potential health benefits of *R. coreanus* fruit might be unique and different from *R. occidentalis*, as the dominant anthocyanin and the ratio of the individual anthocyanins are characteristic for each species.

## 5. Phenolics Other Than Anthocyanins

From published bokbunja data, two studies that worked with correctly identified *R. coreanus* fruit [30,31] have reported the non-anthocyanin phenolic profile in *R. coreanus* fruit. Phenolic acids (ellagic acid and coumaric acid hexose), flavonol-glycosides (quercetin-glucoside, quercetin-rutinoside, quercetin-glucuronide and kaempferol-glucoside), flavanol polymers (numerous procyanidins; tentatively identified) and hydrolyzable tannins (numerous ellagic acid derivatives; tentatively identified) are in *R. coreanus* fruit [30,31]. It is certain that *R. coreanus* fruit contains ellagic acid derivatives [30,31], since they are widely distributed in *Rubus* fruit [22], but that group of phenolics remains challenging to identify and quantify [22].

*Rubus occidentalis* fruit has been reported to contain the same non-anthocyanin phenolic classes as *R. coreanus*, but with some differences in the individual phenolics within: phenolic acids (ellagic acid, ferulic acid, caffeic acid, *p*-coumaric acid, dihydroxybenzoic acid, *etc.*), flavonol-glycosides (quercetin-glucoside, quercetin-rutinoside, myricetin-glucoside, dihydrokaempferol-glucoside), flavanol monomers (epicatechin) and hydrolyzable tannins (numerous ellagic acid derivatives) [22,39,40].

Phenolics other than anthocyanins in *R. coreanus* and *R. occidentalis* fruits remain a much-needed area of research [22]. Due to the lack of available non-anthocyanin phenolic standards (especially for the larger compounds, like ellagitannins), and the challenges to extract, isolate and analyze these compounds [22], utilizing anthocyanin profiles for authenticity and adulteration is easier and clearer [11,16,41,42]. Examples of using anthocyanin for the authenticity of fruit products, cranberry (*V. macrocarpon* Ait.) juice and *R. occidentalis* fruit sold as dietary supplements are provided in Lee [11,16]. Again, randomly amplified polymorphic DNA fragments and other genetic markers can also be used to distinguish these two species, as illustrated by Eu *et al.* [3,4].

## 6. Studies Reporting on the Incorrect Species

The unique phytochemical composition (specifically anthocyanin), as explained above, is the principal reason why it is crucial that bokbunja processing, storage and pharmacokinetic work be done on the correct species, especially if companies or researchers hope to find that *R. coreanus* fruit and products offer exclusive benefits compared to the more widely available *R. occidentalis*; otherwise, our knowledge of *R. coreanus* fruit benefits will only add to the findings of consuming *R. occidentalis* fruit or potentially create confusing and/or conflicting results. A list of incorrectly identified *R. coreanus* fruit used in further research was summarized before [10], though the three examples below emphasize the misunderstandings created from incorrectly identifying the subject species. An interesting note is that Examples 2 and 3 obtained samples from commercial fields and food processors.

(1) Hyun *et al.* [7] actually reports on the anthocyanin biosynthetic genes involved in *R. occidentalis*, not *R. coreanus* fruit, despite what is reported in the paper. In the fruit image provided by Hyun *et al.* [7], the pubescence on the aggregate fruit is clearly present, and they report the presence of cyanidin-3-xylosylrutinoside, which is an indicator that these fruits are that of *R. occidentalis*, not *R. coreanus*. This study [7] examined cultivated black raspberry from Gochang, Korea.

(2) Ku and Mun [43] used black raspberry liquor (cordial) press cake (from commercial liquor processor, Gochang, Korea) for additional phenolic extractions in value-added product development, but the extraction optimizations were conducted on *R. occidentalis* press cakes, not *R. coreanus*, as indicated by the presence of cyanidin-3-sambubioside and cyanidin-3-xylosylrutinoside, which are not found in *R. coreanus* fruit.

(3) Kim *et al.* [44] used misidentified *R. coreanus* fruit to conduct a phytochemical identification study (reported cyanidin-3-sambubioside presence, which is not found in *R. coreanus* fruit), then used those fruit to conduct a study on whether these (misidentified *R. coreanus*) fruit could aid in reducing DNA damage to cigarette smokers [45]. This study [44] obtained samples from a commercial field from Gokseong, Korea.

## 7. Conclusions

Most cultivated *R. coreanus* fruit in Korea are that of *R. occidentalis* based on vegetative traits, fruit anthocyanin profiles and DNA profiling, as discussed above. Commercial bokbunja product ingredient listings need to be corrected to *R. occidentalis* to prevent further confusion. Since there is nothing wrong with growing *R. occidentalis* in Korea for the functional food market, we only propose that the correct species name is utilized on labeling and documentation to prevent confusion in the marketplace and research community. We are hopeful that future work on *Rubus* fruit will be clear, whether *R. coreanus*, *R. occidentalis* or a mix of the two is used. It is helpful to have the fruit authenticated by a well-trained plant taxonomist prior to further examining its processing stability, health benefits, *etc.* Genetic fingerprinting has become a relatively inexpensive service provided by commercial laboratories, and the information produced by Eu *et al.* [3,4] would allow any of these laboratories to confirm which species they are using in their study. If a well-trained taxonomist is not available, then this review article and several papers referenced in this work will provide guidance for clear identification.

**Acknowledgments:** We thank an anonymous South Korean researcher for their insight into this issue. This project was funded by USDA-ARS CRIS Numbers 5358-21000-047-00D and 5358-21220-002-00D and the Specialty Crop Research Initiative (SCRI) Grant Number 2011-51181-30676 from the USDA-National Institute of Food and Agriculture (NIFA). Mention of trade names or commercial products in this publication is solely for the purpose of providing specific information and does not imply recommendation or endorsement by the U.S. Department of Agriculture.

**Author Contributions:** Jungmin Lee was lead in compiling research literature and writing of this article with the aid of Michael Dossett and Chad E. Finn.

**Conflicts of Interest:** The authors declare no conflict of interest.

## References

1. Miquel, F.A.W. *Rubus coreanus* Miquel. *Ann. Mus. Bot. Lugduno-Batavi.* **1867**, *3*, 34.
2. USDA, ARS, National Genetic Resources Program. *Germplasm Resources Information Network-(GRIN)* [Online Database]. National Germplasm Resources Laboratory, Beltsville, Maryland, USA. Available online: http://www.ars-grin.gov/cgi-bin/npgs/html/splist.pl?18606 (accessed on 4 February 2014).
3. Eu, G.S.; Chung, B.Y.; Bandopadhyay, R.; Yoo, N.H.; Choi, D.G.; Yun, S.J. Phylogenic relationships of *Rubus* species revealed by randomly amplified polymorphic DNA markers. *J. Crop Sci. Biotech.* **2008**, *11*, 39–44.
4. Eu, G.S.; Park, M.R.; Yun, S.J. Internal transcribed spacer (ITS) regions reveals phylogenic relationships of *Rubus* species cultivated in Korea. *Korean J. Med. Crop Sci.* **2009**, *17*, 165–172.
5. Eu, G.S.; Park, M.R.; Baek, S.H.; Yun, S.J. Phylogenic relationships of *Rubus* cultivated in Korea revealed by chloroplast DNA spacers. *Korean J. Med. Crop Sci.* **2010**, *18*, 266–272.

6.   Lee, J.; Dossett, M.; Finn, C.E. Anthocyanin fingerprinting of true bokbunja (*Rubus coreanus* Miq.) fruit. *J. Funct. Foods* **2013**, *5*, 1985–1990.

7.   Hyun, T.K.; Lee, S.; Rim, Y.; Kumar, R.; Han, X.; Lee, S.Y.; Lee, C.H.; Kim, J. *De-novo* RNA sequencing and metabolite profiling to identify genes involved in anthocyanin biosynthesis in Korean black raspberry (*Rubus coreanus* Miquel). *PLoS One* **2014**, *9*, e88292.

8.   Kim, S.; Kim, M.; Jang, Y.; Kim, H.; Lee, D. Morphological characteristics and classification of selected population of *Rubus coreanus* Miq. *Life Sci. J.* **2013**, *10*, 144–151.

9.   Kim, S.H.; Chung, H.G.; Han, J. Breeding of Korean black raspberry (*Rubus coreanus* Miq.) for high productivity in Korea. *Acta Hort.* **2008**, *777*, 141–146.

10.  Lee, J. Establishing a case for improved food phenolic analysis. *Food Sci. Nutr.* **2014**, *2*, 1–8.

11.  Lee, J. Marketplace analysis demonstrates quality control standards needed for black raspberry dietary supplements. *Plant Foods Hum. Nutr.* **2014**, *69*, 161–167.

12.  Lee, J.; Dossett, M.; Finn, C.E. Anthocyanin rich black raspberries can be made even better. *Acta Hort.* **2014**, *1017*, 127–133.

13.  Lee, J.; Finn, C.E. Anthocyanins and other polyphenolics in American elderberry (*Sambucus canadensis*) and European elderberry (*S. nigra*) cultivars. *J. Sci. Food Agric.* **2007**, *87*, 2665–2675.

14.  Lee, J.; Finn, C.E.; Wrolstad, R.E. Comparison of anthocyanin pigment and other phenolic compounds of *Vaccinium membranaceum* and *Vaccinium ovatum* native to the Pacific Northwest of North America. *J. Agric. Food Chem.* **2004**, *52*, 7039–7044.

15.  Lee, J.; Finn, C.E.; Wrolstad, R.E. Anthocyanin pigment and total phenolics content of three *Vaccinium* species native to the Pacific Northwest of North America. *HortScience* **2004**, *39*, 959–964.

16.  Lee, J. Proanthocyanidin A2 purification and quantification of American cranberry (*Vaccinium macrocarpon* Ait.) products. *J. Funct. Foods* **2013**, *5*, 144–153.

17.  Finn, C.E.; Moore, P.P.; Yorgey, B.M.; Lee, J.; Strik, B.C.; Kempler, C.; Martin, R.R. 'Charm' strawberry. *HortScience* **2013**, *48*, 1184–1188.

18.  Dossett, M.; Lee, J.; Finn, C.E. Inheritance of phenological, vegetative, and fruit chemistry traits in black raspberry. *J. Am. Soc. Hort. Sci.* **2008**, *133*, 408–417.

19.  Dossett, M.; Lee, J.; Finn, C.E. Variation of anthocyanins and total phenolics in black raspberry populations. *J. Funct. Foods* **2010**, *2*, 292–297.

20.  Dossett, M.; Lee, J.; Finn, C.E. Characterization of a novel anthocyanin profile in wild black raspberry mutants: An opportunity for studying the genetic control of pigment and color. *J. Funct. Foods* **2011**, *3*, 207–214.

21.  Dossett, M.; Lee, J.; Finn, C.E. Anthocyanin content of wild black raspberry germplasm. *Acta Hort.* **2012**, *946*, 43–47.

22.  Lee, J.; Dossett, M.; Finn, C.E. *Rubus* fruit phenolic research: The good, the bad, and the confusing. *Food Chem.* **2012**, *130*, 785–796.

23.  Developing the Genomic Infrastructure for Black Raspberries. Available online: http://black-raspberries.com (accessed on 17 July 2014).

24. Jennings, D.L. *Raspberries and Blackberries: Their Breeding, Diseases and Growth*; Academic Press: San Diego, CA, USA, 1988.

25. Darrow, G.M. Blackberry and raspberry improvemen. In *Yearbook of The United States Department of Agricultur*; U.S. Government Printing Office: Washington, DC, USA, 1937; pp. 496–553.

26. Keep, E.; Knight, W.H.; Parker, J.H. *Rubus coreanus* as donor of resistance to cane disease and mildew in red raspberry breeding. *Euphytica* **1977**, *26*, 505–510.

27. Williams, C.F. Influence of parentage in species hybridization of raspberries. *J. Amer. Soc. Hort. Sci.* **1950**, *56*, 149–156.

28. Keep, E.; Knight, W.H.; Parker, J.H. The inheritance of flower colour and vegetative characters in *Rubus coreanus*. *Euphytica* **1977**, *26*, 185–192.

29. Kim, S.J.; Lee, H.J.; Kim, B.S.; Lee, D.; Lee, S.J.; Yoo, S.H.; Chang, H.I. Antiulcer activity of anthocyanins from *Rubus coreanus* via association with regulation of the activity of matrix metalloproteinase-2. *J. Agric. Food Chem.* **2011**, *59*, 11786–11793.

30. Heo, S.; Lee, D.Y.; Choi, H.K.; Lee, J.; Kim, J.H.; Cho, S.M.; Lee, H.J.; Auh, J.H. Metabolite fingerprinting of bokbunja (*Rubus coreanus* Miquel) by UPLC-qTOF-MS. *Food Sci. Biotech.* **2011**, *20*, 567–570.

31. Lee, D.Y.; Heo, S.; Kim, S.G.; Choi, H.K.; Lee, H.J.; Cho, S.M.; Auh, J.H. Metabolomic characterization of the region- and maturity-specificity of *Rubus coreanus* Miquel (bokbunja). *Food Res. Int.* **2013**, *54*, 508–515.

32. Lee, J.; Durst, R.W.; Wrolstad, R.E. Impact of juice processing on blueberry anthocyanins and polyphenolics: Comparison of two pretreatments. *J. Food Sci.* **2002**, *67*, 1660–1667.

33. Lee, J.; Wrolstad, R.E. Extraction of anthocyanins and polyphenolics from blueberry processing waste. *J. Food Sci.* **2004**, *69*, C564–C573.

34. Sadilova, E.; Stintzing, F.C.; Carle, R. Thermal degradation of acylated and nonacylated anthocyanins. *J. Food Sci.* **2006**, *71*, C504–C512.

35. Novotny, J.A.; Clevidence, B.A.; Kurilich, A.C. Anthocyanin kinetics are dependent on anthocyanin structure. *Br. J. Nutr.* **2012**, *107*, 504–509.

36. Stintzing, F.C.; Stintzing, A.S.; Carle, R.; Frei, B.; Wrolstad, R.E. Color and antioxidant properties of cyanidin-based anthocyanin pigments. *J. Agric. Food Chem.* **2002**, *50*, 6172–6181.

37. Aura, A.; Martin-Lopez, P.; O'Leary, K.A.; Williamson, G.; Oksman-Caldentey, K.M.; Poutanen, K.; Santos-Buelga, C. *In vitro* metabolism of anthocyanins by human gut microflora. *Eur. J. Nutr.* **2005**, *44*, 133–142.

38. Charron, C.S.; Kurilich, A.C.; Clevidence, B.A.; Simon, P.W.; Harrison, D.J.; Britz, S.J.; Baer, D.J.; Novotny, J.A. Bioavailability of anthocyanins from purple carrot juice: Effects of acylation and plant matrix. *J. Agric. Food Chem.* **2009**, *57*, 1226–1230.

39. Stoner, G.D. Food stuffs for prevention cancer: The preclinical and clinical development of berries. *Cancer Prev. Res.* **2009**, *2*, 87–194.

40. Paudel, L.; Wyzgoski, F.J.; Scheerens, J.C.; Chanon, A.M.; Reese, R.N.; Smiljanic, D.; Wesdemiotis, C.; Blakeslee, J.J.; Riedl, K.M.; Rinaldi, P.L. Nonanthocyanin secondary metabolites of black raspberry (*Rubus occidentalis* L.) fruits: Identification by HPLC-DAD, NMR, HPLC-ESI-MS, and ESI-MS/MS analyses. *J. Agric. Food Chem.* **2013**, *61*, 12032–12043.

41. Hong, V.; Wrolstad, R.E. Detection of adulteration in commercial cranberry juice drinks and concentrates. *J. Aoac. Int.* **1986**, *69*, 208–213.

42. Penman, K.G.; Halstead, C.W.; Matthias, A.; de Voss, J.J.; Stuthe, J.M.; Bone, K.M.; Lehmann, R.P. Bilberry adulteration using the food dye amaranth. *J. Agric. Food Chem.* **2006**, *54*, 7378–7382.

43. Ku, C.S.; Mun, S.P. Optimization of the extraction of anthocyanin from bokbunja (*Rubus coreanus* Miq.) marc produced during traditional wine processing and characterization of extracts. *Bioresour. Technol.* **2008**, *99*, 8325–8330.

44. Kim, H.S.; Park, S.J.; Hyun, S.H.; Yang, S.O.; Lee, J.; Auh, J.H.; Kim, J.H.; Cho, S.M.; Marriott, P.J.; Choi, H.K. Biochemical monitoring of black raspberry (*Rubus coreanus* Miquel) fruits according to maturation stage by [1]H-NMR using multiple solvent systems. *Food Res. Int.* **2011**, *44*, 197–1987.

45. Suh, H.W.; Kim, S.H.; Park, S.J.; Hyun, S.H.; Lee, S.K.; Auh, J.H.; Lee, H.J.; Cho, S.M.; Kim, J.H.; Choi, H.K. Effect of Korean black raspberry (*Rubus coreanus* Miquel) fruit administration on DNA damage levels in smokers and screening biomarker investigation using [1]H-NMR-based metabolic profiling. *Food Res. Int.* **2013**, *54*, 1255–1262.

# Anti-Inflammatory Effect of the Blueberry Anthocyanins Malvidin-3-Glucoside and Malvidin-3-Galactoside in Endothelial Cells

Wu-Yang Huang, Ya-Mei Liu, Jian Wang, Xing-Na Wang and Chun-Yang Li

**Abstract:** Blueberry fruits have a wide range of health benefits because of their abundant anthocyanins, which are natural antioxidants. The purpose of this study was to investigate the inhibitory effect of blueberry's two main anthocyanins (malvidin-3-glucoside and malvidin-3-galactoside) on inflammatory response in endothelial cells. These two malvidin glycosides could inhibit tumor necrosis factor-alpha (TNF-$\alpha$) induced increases of monocyte chemotactic protein-1 (MCP-1), intercellular adhesion molecule-1 (ICAM-1), and vascular cell adhesion molecule-1 (VCAM-1) production both in the protein and mRNA levels in a concentration-dependent manner. Mv-3-glc at the concentration of 1 $\mu$M could inhibit 35.9% increased MCP-1, 54.4% ICAM-1, and 44.7% VCAM-1 protein in supernatant, as well as 9.88% MCP-1 and 48.6% ICAM-1 mRNA expression ($p < 0.05$). In addition, they could decrease I$\kappa$B$\alpha$ degradation (Mv-3-glc, Mv-3-gal, and their mixture at the concentration of 50 $\mu$M had the inhibition rate of 84.8%, 75.3%, and 43.2%, respectively, $p < 0.01$) and block the nuclear translocation of p65, which suggested their anti-inflammation mechanism was mediated by the nuclear factor-kappa B (NF-$\kappa$B) pathway. In general malvidin-3-glucoside had better anti-inflammatory effect than malvidin-3-galactoside. These results indicated that blueberry is good resource of anti-inflammatory anthocyanins, which can be promising molecules for the development of nutraceuticals to prevent chronic inflammation in many diseases.

Reprinted from *Molecules*. Cite as: Huang, W.-Y.; Liu, Y.-M.; Wang, J.; Wang, X.-N.; Li, C.-Y. Anti-Inflammatory Effect of the Blueberry Anthocyanins Malvidin-3-Glucoside and Malvidin-3-Galactoside in Endothelial Cells. *Molecules* **2014**, *19*, 12827–12841.

## 1. Introduction

Due to their antioxidant capacity and nutritional quality fruits are an economical potential resource of functional foods and nutraceuticals, so they play an important role in human nutrition and health [1,2]. Fruits possess diverse phenolic compounds (e.g., phenolic acids and polyphenols which include flavonoids (anthocyanins, flavanols, and catechins) and tannins), carotenoids, and vitamin C and E, which are well-known dietary antioxidants beneficial to the endogenous antioxidant defense strategies [3]. These bioactive phytochemicals contribute the major health-promoting function of fruits [4,5]. Rabbiteye blueberry (*Vaccinium ashei*) has a wide range of bioactivities, such as super antioxidant ability, anti-diabetic capacity, anti-proliferative quality, anti-inflammatory effects, and protective effects protect against cancer and stroke [6,7]. Besides, blueberries may alleviate conditions occurring in Alzheimer's disease and aging [8]. Blueberries also help to maintain healthy blood flow via LDL oxidation, normal platelet aggregation, and endothelial function improvement [9,10]. Anthocyanins in blueberries are mainly responsible for those health benefits [11]. Accumulation of reactive oxygen species generated by inflammatory cells is thought to be one of the major factors contributing to chronic inflammation in many diseases, such as atherosclerosis [12], so antioxidants could be considered to have potential capacity to prevent and treat chronic inflammation. Compared with the synthetic antioxidants, natural antioxidants are higher efficient and economical. Besides, synthetic antioxidants may exhibit toxicity and side-effects. Thus, it's urgent to search for more natural antioxidant resources. Anthocyanins are considered as most potent natural hydrophilic antioxidants and their properties extend well beyond suppressing free radicals [13]. According to an investigation at Tufts University in the USA, the content of anthocyanins in blueberry was reported as the highest among 40 vegetables and fruits [14]. We previously demonstrated the presence of nine anthocyanins in blueberry and found that malvidin-3-glucoside (Mv-3-glc) and malvidin-3-galactoside (Mv-3-gal) were the most abundant [15]. Their structures are shown in Figure 1. In addition, malvidin has been found to possess inhibition to TNF-$\alpha$-induced inflammatory responses [16], and malvidin-3-glucoside in wild blueberries significantly reduces the expression of pro-inflammatory genes *in vitro* [17]. Potentiating interactions can be additive, concommitant, inhibitory, supraadditive, or synergistic. In light of the observed synergism of antioxidants, such as vitamin E and vitamin C, catechin and malvidin 3-glucoside, and tea polyphenols and vitamin E [18,19], it has been suggested that combinations of antioxidants with different radical scavenging efficiency may show a different antioxidant activity from the expected one based on the sum of their individual effects [20]. In the present study, we further investigated the anti-inflammatory properties of Mv-3-glc and Mv-3-gal, as well as their synergistic effect in human vascular umbilical endothelial cells.

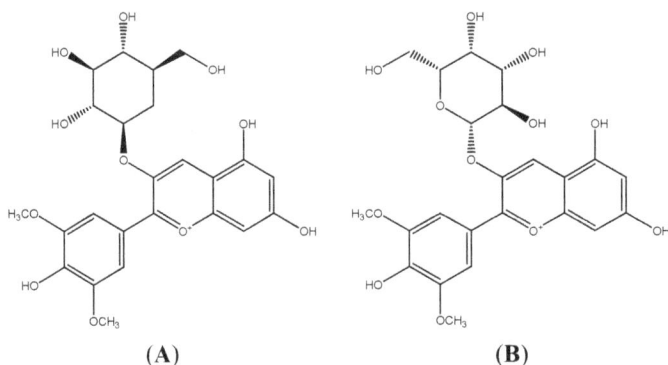

**(A)**                    **(B)**

**Figure 1.** The structures of malvidin-3-glucoside (**A**) and malvidin-3-galactoside (**B**).

## 2. Results and Discussion

*2.1. Effect of Mv-3-glc and Mv-3-gal on TNF-α-Induced MCP-1, ICAM-1, and VCAM-1 Production in Supernatant*

As shown in Table 1, the surface protein expression of MCP-1, ICAM-1, and VCAM-1 was low in unstimulated cells. Exposure of cells to TNF-α (10 µg/L) for 6 h significantly induced up-regulation of surface expression of MCP-1, ICAM-1, and VCAM-1, although expression levels were not same for Mv-3-glc, MCP-1 protein expression increased from 0.023 µg/L to 0.519 µg/L, ICAM-1 increased from 0.111 µg/L to 0.605 µg/L, and VCAM-1 increased from 0.021 µg/L to 0.314 µg/L. For Mv-3-gal, MCP-1 increased from 0.023 µg/L to 0.264 µg/L, ICAM-1 increased from 0.102 µg/L to 0.337 µg/L, and VCAM-1 increased from 0.013 µg/L to 0.124 µg/L. For Mv-3-glc and Mv-3-gal mixture, MCP-1 increased from 0.015 µg/L to 0.507 µg/L, ICAM-1 increased from 0.161 µg/L to 0.625 µg/L, and VCAM-1 increased from 0.014 µg/L to 0.445 µg/L. Pretreatment with Mv-3-glc, Mv-3-gal and their mixture partially inhibited this up-regulation in a concentration-dependent manner. In most time, the inhibitory effect of Mv-3-glc was better than Mv-3-gal. Mv-3-glc at the concentration of 1 µM could inhibit 35.9% increased MCP-1 production (0.341 µg/L), 54.4% increased ICAM-1 (0.336 µg/L) and 44.7% increased VCAM-1 (0.183 µg/L) (all $p < 0.01$), while 1 µM Mv-3-gal could only inhibit 18.2% MCP-1 (0.216 µg/L) ($p < 0.05$), 54.9% ICAM-1 (0.208 µg/L) and 19.9% VCAM-1 (0.106 µg/L) (both $p < 0.01$). Mv-3-glc at the concentration of 10 µM could inhibit 66.3% increased MCP-1 production (0.175 µg/L) and 63.8% increased VCAM-1 (0.127 µg/L) (both $p < 0.001$), and 69.6% increased ICAM-1 (0.261 µg/L) ($p < 0.01$), while 10 µM Mv-3-gal could inhibit 64.3% MCP-1 (0.109 µg/L), 52.3% VCAM-1 (0.066 µg/L), and 64.7% ICAM-1 (0.185 µg/L) (all $p < 0.01$), respectively. In addition, 50 µM and 100 µM Mv-3-glc both inhibited more than 90% increased three protein expressions ($p < 0.001$). The Mv-3-glc and Mv-3-gal mixture at the concentration of 10 µM inhibited 86.8% increased MCP-1 (0.080 µg/L), 98% increased ICAM-1 (0.170 µg/L), and 70.1% increased VCAM-1 (0.143 µg/L) (all $p < 0.001$), which was greatly stronger than Mv-3-glc and Mv-3-gal respectively. This indicated that Mv-3-glc might have additive effect with Mv-3-gal. For ICAM-1, all the three treatments at different concentrations got the inhibition rate more than 50%. Even, 1 µM Mv-3-glc and Mv-3-gal mixture could inhibit more than 90% increased ICAM-1. ICAM-1 expression was less than the control when cells were treated with high concentration of these two malvidin glycosides, which indicated that they might directly decrease ICAM-1 production without stimulation of TNF-α.

**Table 1.** Effects of Mv-3-glc and Mv-3-gal on TNF-α-induced MCP-1, ICAM-1, and VCAM-1 in supernatant.

| Treatment | Protein in Supernatant (µg/L) | | |
|---|---|---|---|
| | MCP-1 | ICAM-1 | VCAM-1 |
| Control | 0.023 ± 0.001 *** | 0.111 ± 0.005 *** | 0.021 ± 0.016 *** |
| 10 µg/L TNF-α | 0.519 ± 0.014 | 0.605 ± 0.020 | 0.314 ± 0.009 |
| 1 µM Mv-3-glc + TNF-α | 0.341 ± 0.009 ** | 0.336 ± 0.021 ** | 0.183 ± 0.005 ** |
| 10 µM Mv-3-glc + TNF-α | 0.175 ± 0.011 *** | 0.261 ± 0.017 ** | 0.127 ± 0.001 *** |
| 50 µM Mv-3-glc + TNF-α | 0.039 ± 0.001 *** | 0.098 ± 0.026 *** | 0.049 ± 0.001 *** |
| 100 µM Mv-3-glc + TNF-α | 0.033 ± 0.002 *** | 0.049 ± 0.001 *** | 0.039 ± 0.001 *** |
| Control | 0.023 ± 0.003 *** | 0.102 ± 0.003 *** | 0.013 ± 0.002 ** |
| 10 µg/L TNF-α | 0.264 ± 0.009 | 0.337 ± 0.010 | 0.124 ± 0.006 |
| 1 µM Mv-3-gal + TNF-α | 0.216 ± 0.010 * | 0.208 ± 0.009 ** | 0.106 ± 0.001 * |
| 10 µM Mv-3-gal + TNF-α | 0.109 ± 0.001 ** | 0.185 ± 0.009 ** | 0.066 ± 0.001 ** |
| 50 µM Mv-3-gal + TNF-α | 0.079 ± 0.001 ** | 0.133 ± 0.020 ** | 0.028 ± 0.003 ** |
| 100 µM Mv-3-gal + TNF-α | 0.048 ± 0.001 *** | 0.044 ± 0.022 *** | 0.017 ± 0.001 ** |
| Control | 0.015 ± 0.003 *** | 0.161 ± 0.005 *** | 0.014 ± 0.001 *** |
| 10 µg/L TNF-α | 0.507 ± 0.011 | 0.625 ± 0.018 | 0.445 ± 0.005 |
| 1 µM (Mv-3-glc + Mv-3-gal) + TNF-α | 0.421 ± 0.016 ** | 0.199 ± 0.008 *** | 0.241 ± 0.004 *** |
| 10 µM (Mv-3-glc + Mv-3-gal) + TNF-α | 0.080 ± 0.002 *** | 0.170 ± 0.001 *** | 0.143 ± 0.004 *** |
| 50 µM (Mv-3-glc + Mv-3-gal) + TNF-α | 0.023 ± 0.001 *** | 0.058 ± 0.005 *** | 0.048 ± 0.003 *** |
| 100 µM (Mv-3-glc + Mv-3-gal) + TNF-α | 0.018 ± 0.001 *** | 0.013 ± 0.002 *** | 0.037 ± 0.002 *** |

Note: *, **, and *** indicate $p < 0.05$, $p < 0.01$, and $p < 0.001$ respectively compared to TNF-α alone.

## 2.2. Effects of Mv-3-glc and Mv-3-gal on TNF-α-Induced ICAM-1 and VCAM-1 Proteins in Cell

Like supernatant, Mv-3-glc, Mv-3-gal and their mixture had different inhibitory effects on TNF-α-induced protein expression levels of endothelial ICAM-1 and VCAM-1 in a concentration-dependent manner (Figure 2). None of the samples had any significant inhibitory effects at a concentration of 1 µM, except for the mixture that inhibited increased VCAM-1 protein expression by 16.5% ($p < 0.05$).

Since proteins in supernatant are secreted from cells, it indicated that malvidin glycosides affect more protein secretion inhibition than protein production. For most conditions, Mv-3-glc showed stronger effect than Mv-3-gal. Mv-3-glc at low concentration (1 μM and 10 μM) could not inhibit the increased ICAM-1 protein level, but at high concentration it showed significant inhibitory effects (for 50 μM, $p < 0.05$; and for 100 μM, $p < 0.01$, respectively). In addition, Mv-3-glc and Mv-3-gal at high concentration all inhibited increased ICAM-1 and VCAM-1 protein expression by more than 50%. Mv-3-glc at a concentration of 50 μM could inhibit 81.8% ICAM-1 and 54.6% VAM-1 (both $p < 0.05$), and Mv-3-glc at concentration of 100 μM could inhibited 115.4% ICAM-1 and 93.8% VCAM-1 (both $p < 0.01$); while 50 μM and 100 μM Mv-3-gal could inhibited 60.6% and 76.2% ICAM-1 (both $p < 0.05$), and 52.5% and 58.0% VCAM-1(both $p < 0.01$), respectively.

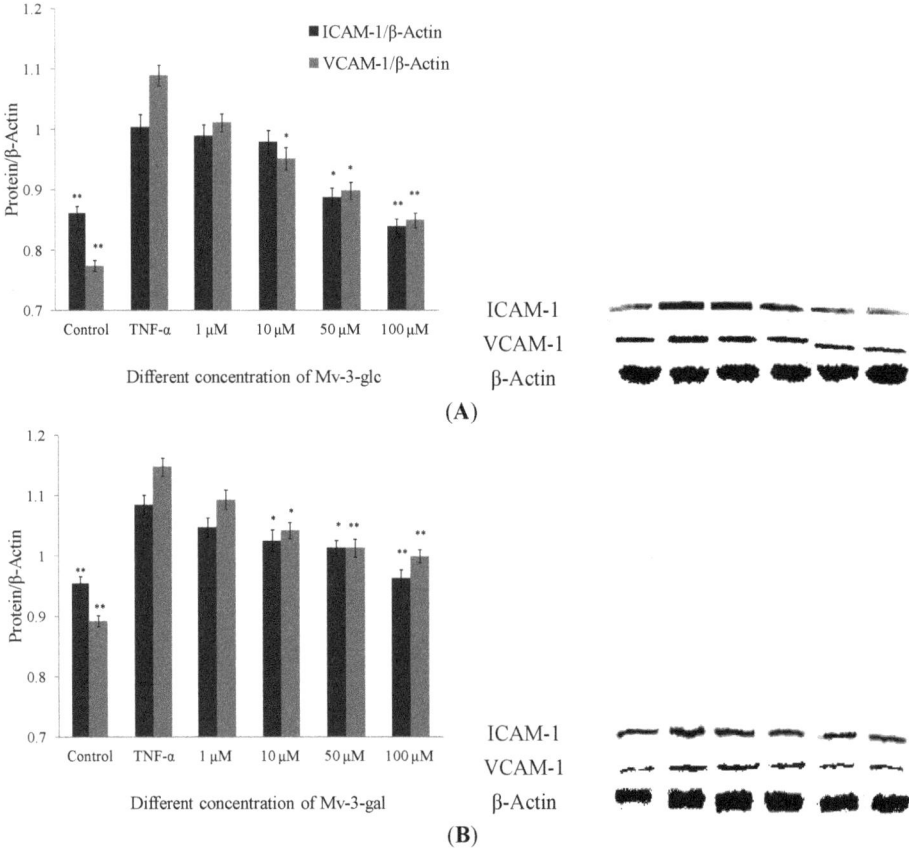

(A)

(B)

Figure 2. *Cont.*

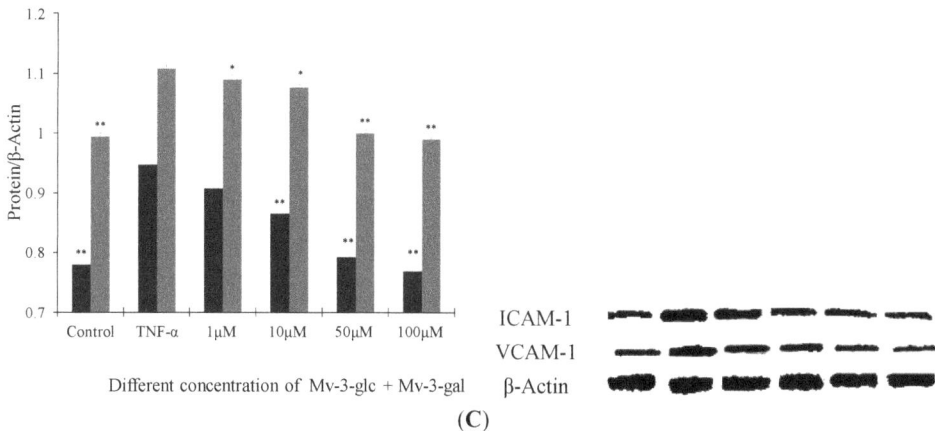

**Figure 2.** Effects of Mv-3-glc (**A**), Mv-3-gal (**B**), and their mixture (Mv-3-glc + Mv-3-ga) (**C**) on TNF-α-induced ICAM-1 and VCAM-1 protein expression. * and ** indicate $p < 0.05$ and $p < 0.01$ respectively compared to TNF-α alone.

As to ICAM-1 level in cells treated with Mv-3-glc at concentration of 100 μM, and ICAM-1 and VCAM-1 protein expression levels in cells treated with 100 μM Mv-3-glc and Mv-3-gal mixture were less than the control (all $p < 0.01$), which also indicated their strong anti-inflammatory effects. Mv-3-glc had synergistic effect with Mv-3-gal, with their mixture's inhibition rate was larger than that of each one, except for VCAM-1 expression in cells treated with the low concentrations of mixture (1 μM and 10 μM).

### 2.3. Effects of Mv-3-glc and Mv-3-gal on TNF-α-Induced MCP-1 and ICAM-1 mRNA in Cells

Low concentration (1 μM) of Mv-3-gal had no significant inhibitory effect on MCP-1 mRNA expression. All the other treatments produced significant inhibition ($p < 0.05$) on MCP-1 and ICAM-1 mRNA expression in a concentration-dependent manner (Figure 3).

Mv-3-glc exhibited better inhibitory effects than Mv-3-gal. Except for 1 μM (inhibition rate was 9.88%, $p < 0.05$), Mv-3-glc at all the other three concentrations had inhibition rates of more than 90% ($p < 0.001$) on MCP-1, and Mv-3-glc at all the four concentrations also had greatly significant inhibitory effects to ICAM-1 ($p < 0.001$). At the low concentration (1 μM), the inhibition rates of the mixture (53.8% for MCP-1 and 86.9% for ICAM-1) were much higher than Mv-3-glc (9.88% and 48.6%) and Mv-3-gal (3.04% and 39.2%), respectively. However, the mixture's inhibition rates were not always more than each one for ICAM-1. High concentration of Mv-3-glc, Mv-3-gal, and their mixture expressed lower lever MCP-1 or ICAM-1 mRNA expressions than the control. These indicated that Mv-3-glc and Mv-3-gal could directly decrease MCP-1 and ICAM-1 mRNA expression without TNF-α-stimulation at the high concentration.

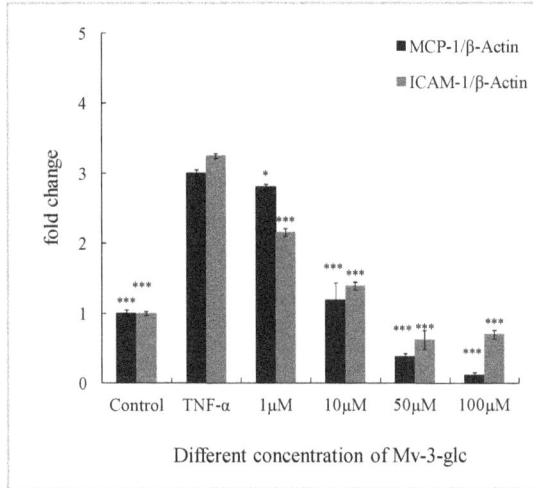

(A)

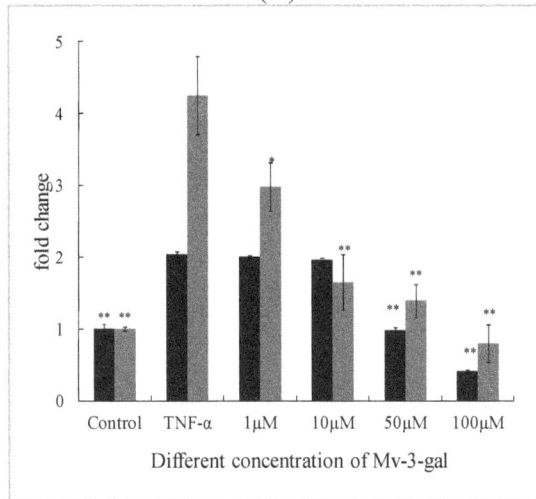

(B)

**Figure 3.** *Cont.*

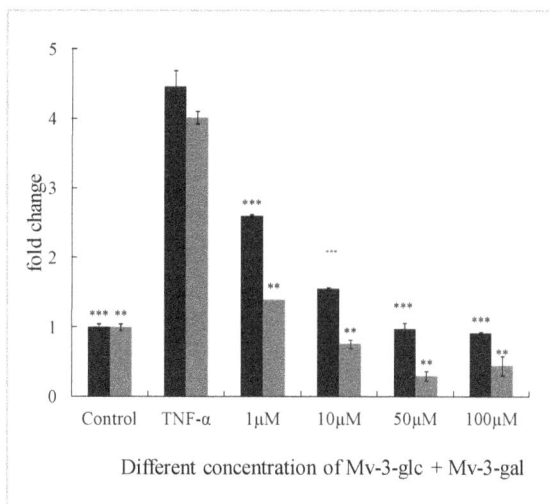

(**C**)

**Figure 3.** Effects of Mv-3-glc (**A**), Mv-3-gal (**B**), andtheir mixture (Mv-3-glc + Mv-3-ga) (**C**) on TNF-α-induced MCP-1 and ICAM-1 mRNA expression. *, **, and *** indicate $p < 0.05$, $p < 0.01$, and $p < 0.001$ respectively compared to TNF-α alone. β-Actin is used as internal standard, and mRNA levels are expressed as fold increase over the control.

## 2.4. Effects of Mv-3-glc and Mv-3-gal on TNF-α-Induced IκB Degradation

In this study, the amount of IκBα was greatly reduced after exposure to 10 μg/L TNF-α ($p < 0.01$). The inhibitory effect of Mv-3-glc, Mv-3-gal and their mixture at different concentrations on TNF-α-induced expression of IκBα degradation were all significant. They also were in the concentration-dependent manner. Mv-3-glc always exhibited better inhibitory capacity than Mv-3-gal in all the four concentrations. Mv-3-glc and Mv-3-gal mixture at the concentration of 10 μM and 50 μM had the inhibition rate of 78.0% and 84.8%, respectively, which were more than those of Mv-3-glc (63.7% and 75.3%) or Mv-3-gal (34.1% and 43.2%). Mv-3-glc and the mixture at the concentration of 100 μM could completely inhibit the IκBα degradation ($p < 0.01$), since the IκBα expression was even more than the control (Figure 4).

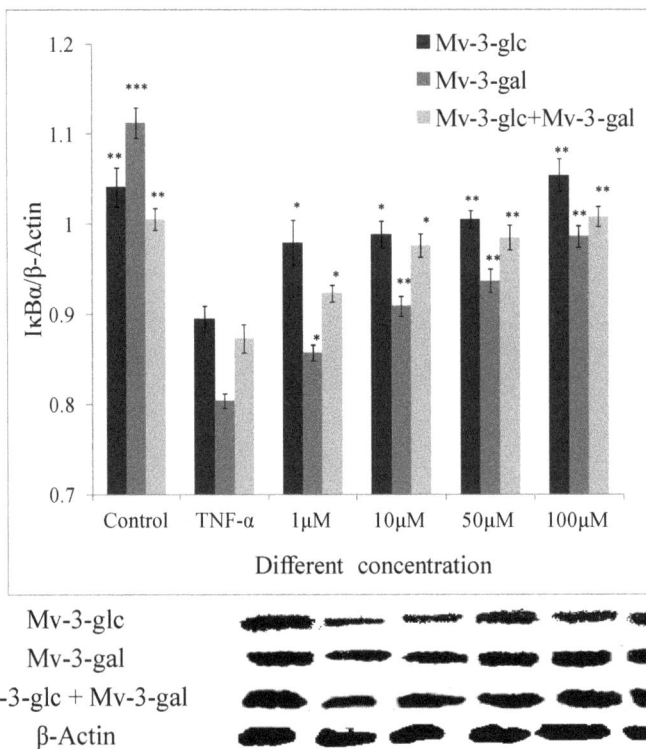

**Figure 4.** Effects of Mv-3-glc, Mv-3-gal, and their mixture (Mv-3-glc + Mv-3-ga) on IκBα protein levels in endothelial cells. *, **, and *** indicate $p < 0.05$, $p < 0.01$, and $p < 0.001$ respectively compared to TNF-α alone.

## 2.5. Effects of Mv-3-glc and Mv-3-gal on TNF-α-Induced NF-κB Translocation

On activation of the NF-κB pathway, the p65 protein is released from the cytosol and migrates into the cell nucleus where it interacts with the promoter regions of various proteins which up-regulated in inflammation [21]. Immunocytochemistry was performed by using NF-κB and fluorescein isothiocyanate (FITC)-conjugated antibody. In un-stimulated cells, the levels of p65 in the nucleus were very low. Upon simulation of TNF-α, the levels of p65 in the nucleus were increased. On the other hand, upon treatment of the cells with Mv-3-glc, Mv-3-gal, and their mixture, the fluorescence intensity levels of p65 were decreased in the nucleus (Figure 5). The high concentration of Mv-3-glc, Mv-3-gal, and their mixture inhibited the nuclear translocation of p65, with the p65 protein level in the cell nucleus similar to the control.

## 2.6. Discussion

Anthocyanins are one of the largest and most important groups of water-soluble pigments in most fruits. Berries, as colored fruits, are highly chemoprotective because of their bioactive anthocyanins [22,23]. Different anthocyanins offer different antioxidant capacity. A theoretical study evaluated the antioxidant character of three widespread anthocyanidins (cyanidin, delphinidin, and malvidin) [24] according to different parameters (bond dissociation enthalpy, ionization potential, proton affinity, and electron transfer enthalpy) and the atomic charges corresponding to the O atoms of the hydroxyl groups. It is found that antioxidant effect of anthocyanins is based on the free radical scavenging by means of the OH groups.

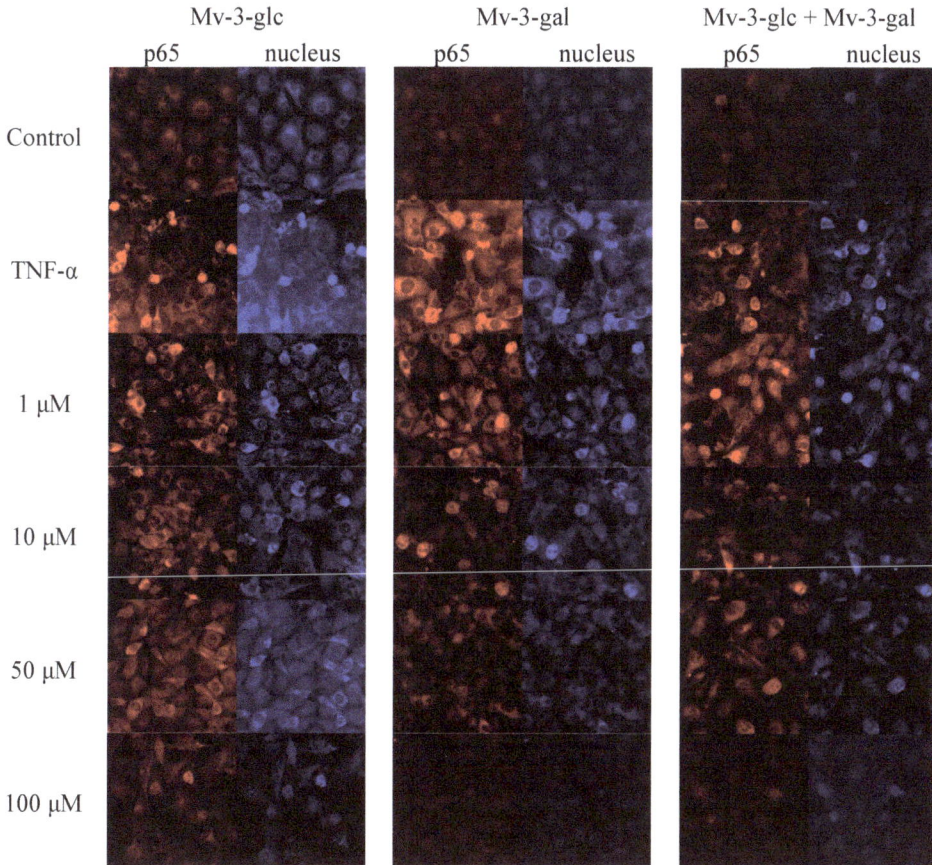

**Figure 5.** Effects of Mv-3-glc, Mv-3-gal, and their mixture (Mv-3-glc + Mv-3-ga) on NF-κB pathway (endothelial p65 translocation) in HUVECs.

Wild lowbush blueberries (*Vaccinium angustifolium* Ait) are a rich source of anthocyanins and other flavonoids with anti-inflammatory activities, in which malvidin-3-glucoside was significantly more effective than epicatechin or chlorogenic acid in reducing the expression of pro-inflammatory genes *in vitro* [17]. In our studied Rabbiteye blueberry (*V. ashei*), Mv-3-glc and Mv-3-gal which are the two main glycoside forms of malvidin were the most abundant anthocyanins [15,25]. Malvidin possesses great antioxidant activity, and cytotoxicity against human monocytic leukemia cells and HT-29 colon cancer cells [26,27], and anti-hypertensive activity by inhibiting angiotensin I-converting enzyme (ACE) [28]. Our previous study reported the inhibitory effect of malvidin on TNF-$\alpha$ induced inflammatory response [16]. This study showed that Mv-3-glc and Mv-3-gal also possessed potential anti-inflammatory capacity. Mv-3-glc had better inhibitory effect than Mv-3-gal, indicating the differences in the two glycosides' effects. In addition, Mv-3-glc and Mv-3-gal sometimes showed additive effects. An isobologram could be constructed in the future over a range of concentrations to verify whether they have synergy or not. Previous studies confirmed that Mv-3-glc could increase NO bioavailability, as well as inhibit peroxynitrite-induced NF-$\kappa$B activation, which supported its benefits in cardiovascular health. A treatment of cyanidin 3-rutinoside and cyanidin 3-glucoside from the berry *Morus alba* L. also resulted in an inhibition on the activation of c-Jun and NF-$\kappa$B, therefore it could decrease the *in vitro* invasiveness of cancer cells [29]. The anti-invasive activity on human colon cancer cells of the anthocyanins from fruits of *Vitis coignetiae Pulliat* was associated with modulation of constitutive NF-$\kappa$B activation through suppression I$\kappa$B$\alpha$ phosphorylation [30]. Lipopolysaccharide-induced NF-$\kappa$B p65 translocation to the nucleus was markedly attenuated by anthocyanins of blueberry, blackberry, and blackcurrant, which mostly included malvidin-3-glucoside, cyanidin-3-glucoside and delphinidin-3-rutinoside. Their anti-inflammatory effects in macrophages were relative to their antioxidant capacity [31], so the antioxidant capacity made anthocyanins promising to develop nutraceuticals to improve endothelial function [32]. These indicate the anthocyanins could be the potential alternatives to prevent and treat the inflammation in many diseases [7]. Recent reports indicated action mechanisms of protecting vascular endothelium including modulation of crucial signaling pathways and gene regulation [33,34]. The present study showed that Mv-3-glc and Mv-3-gal also possessed anti-inflammatory properties by NF-$\kappa$B pathway in endothelial cells.

Chronic inflammation is a common factor linking various pathologies in many diseases such as atherosclerosis and cancer. Vascular inflammation is a complex process, including the accumulation and activation of immune suppressor cells, pro-inflammatory cytokines, chemokines, growth and angiogenic factors and activation of several inflammatory signaling pathways mediated predominantly by NF-$\kappa$B transcription factors [7]. The transcriptional activation of NF-$\kappa$B plays a key role in the development of the inflammatory response [35]. It is well established that NF-κB is normally in an inactive form bound to inhibitory proteins, the IκBs. Exposed to external stimuli such as TNF-α, IKB kinase (IKK) phosphorylates IκBα, which lead to ubiquitination-dependent degradation of IκBα [36]. It is well accepted that the activation of NF-κB in endothelial cells is associated with mononuclear cell infiltration and an increased transcription of adhesion molecules, chemokines, and cytokines [21,37]. TNF-α can activate NF-κB, and then MCP-1, ICAM-1, and VCAM-1 were over expressed in vascular endothelial cells [38,39]. In this study, Mv-3-glc and Mv-3-gal inhibited TNF-α-induced MCP-1, ICAM-1 and VCAM-1 production as well as IκBα degradation. In addition, Mv-3-glc and Mv-3-gal were able to inhibit the nuclear translocation of p65, one subunit of NF-κB, suggesting the mechanism by which Mv-3-glc and Mv-3-gal could block pro-inflammatory signaling downstream of IκB.

## 3. Experimental Section

### 3.1. Chemicals and Reagents

Dulbecco's PBS, M199 medium, TNF-α, trypsin, and Mv, Mv-3-glc, and Mv-3-gal were bought from Sigma Chemical Co., Ltd (Nanjing, China). Fetal bovine serum was purchased from Gibco/Invitrogen (Shanghai, China). Penicillin and streptomycin were obtained from Life Technologies (Shanghai, China). MCP-1, ICAM-1, and VCAM-1 ELISA Kit were purchased from Boster Biotechnology Inc. (Wuhan, China). Trizol reagent, PrimeScript RT master mix, and SYBR Green 2-step qRT-PCR kit were got from TaKaRa Bio Inc. (Dalian, China). NF-κB activation and nuclear translocation Assay Kit were bought from Beyotine Technology Inc. (Nanjing, China). All the chemicals and reagents were of the analytical grade.

### 3.2. Antibodies

Rabbit monoclonal primary antibody against ICAM-1, rabbit polyclonal primary antibodies against VCAM-1, mouse polyclonal primary antibody to the β-Actin antibody, and goat anti-rabbit/mouse HRP-conjugated secondary antibody, were purchased from Boster Biotechnology Inc. Rabbit monoclonal primary antibody to IκBα was bought from Beyotine Technology Inc. Primary antibodies were used at 1:200 dilutions, and secondary antibodies were used at 1:1000 dilutions.

### 3.3. Endothelial Cell Culture and Treatment

Human umbilical vein endothelial cells (HUVECs) are a representative model system for studying inflammation and oxidative stress in the vasculature [40]. HUVECs were saved in National Technical Research Centre of Veterinary Biological Products (Nanjing, China). The second to 6th passage cells were used for all experiments at 80%–90% confluence. HUVECs were quiesced in a reduced serum medium for 4 h prior to experiment. In a separate set of experiments, the cells were treated with 1, 10, 50, and 100 μM Mv-glc, Mv-gal, or their mixture for 18 h, followed by TNF-α (10 μg/L) stimulation for 6 h. DMSO was used as control. The supernatants were collected for ELISA analysis. The cells were prepared for western blotting.

### 3.4. ELISA Analysis and Western Blotting

The levels of MCP-1, ICAM-1, and VCAM-1 in the supernatants were quantified using ELISA kits. The assay procedure was employed according to the kit protocol booklet instructions. The absorbance of the resulting yellow color was measured at 450 nm on a StatFax-2100 Microplate Reader (Awareness Technology Inc., Palm City, FL, USA). The reader was controlled via Hyper Terminal Applet ELISA software. Western blotting was performed on the HUVEC lysates as described before [41]. Beside ICAM-1 and VCAM-1, IκBα was also analyzed by western blotting. Data were normalized by re-probing the membrane with an antibody against β-Actin which was used as a loading control.

### 3.5. Real-Time qRT-PCR

The total RNA was isolated from HUVECs using the Trizol reagent (TaKaRa Bio Inc.). It was reverse-transcribed into cDNA using PrimeScript RT master mix. For reverse-transcription, the SYBR Green 2-step qRT-PCR kit (TaKaRa Bio Inc.) was used. The real-time quantitative PCR analysis was carried out using the LightCycler 480 (Roche Diagnostics Inc., Rotkreuz, Switzerland). The primers for amplification were followed: forward primer 5-GTTGTCCCAAAGAAGCTGTGA-3 and reverse primer 5-AATCCGA ACCCACTTCTGC-3 for MCP-1 (83 bp); forward primer 5-CCACAGTCACCTATGGC AAC-3 and reverse primer 5-AGTGTCTCCTGGCTCTGGTT-3 for ICAM-1 (124 bp); forward primer 5-TGGACTTCGAGCAAGAGATG-3 and reverse primer 5-GAAG GAAGGCTGGAAGAGTG-3 for β-actin (137 bp). The reaction was conducted with an initial denaturing at 94 °C for 30 s, then involved 40 cycles of 60 °C for 20 s, and at 65 °C for 15 s in the end. Relative gene expression data was analyzed using the $2^{-\Delta\Delta Ct}$ method.

### 3.6. Immunofluorescence

Nuclear translocation of p65 is widely used as a measure for NF-κB activation. HUVECs were fixed in 3.75% paraformaldehyde, and permeabilized with 0.1% Triton-X-100 incubated overnight with primary antibody against p65. On the following day, the cells were incubated with secondary antibody (goat anti-rabbit conjugated with FTTC) for 1 h. The cell nuclei were stained with DAPI (4',6-diamidino-2-phenylindole), and visualized under an Axiovision 4 Fluorescent Microscope (Zeiss, Oberkochen, Germany). All images presented are in (×100) magnification.

## 3.7. Statistical Analysis

All data presented are mean value ± standard deviation (SD). The data were analyzed by a one-way ANOVA using the SPSS 19.0 Statistical Software. Differences were considered significant with $p$ value < 0.05.

## 4. Conclusions

In the present study, treatment with Mv-3-glc, Mv-3-gal, and their mixture significantly attenuated monocyte adhesion in TNF-α-stimulated HUVECs by inhibiting MCP-1, ICAM-1, and VCAM-1 protein and mRNA expressions both in endothelial cell and supernatants. In addition, they affected IκBα degradation and the nuclear translocation of p65, indicating that they possessed anti-inflammatory effects by blocking the NF-κB pathway mechanism. Mv-3-glc, Mv-3-gal, and their mixture all showed their inhibitory effects on TNF-α-induced inflammatory response in a concentration-dependent manner. Mv-3-glc had better potential anti-inflammatory effect than Mv-3-gal, and they showed synergistic effect sometimes. Blueberries are a good source of anthocyanins such as Mv-3-glc and Mv-3-gal, which can be a promising molecules for the development of nutraceuticals to improve endothelial function and thereby to prevent the progression of chronic inflammation.

**Acknowledgments:** This research was supported by grants from National Natural Science Foundation of China (NSFC31101264), Natural Science Foundation Program of Jiangsu Province (BK20141386), Jiangsu Province Agricultural Technology Independent Innovation Fund cx(12)5029. We thank the help of Zheng Q.S. from National Technical Research Centre of Veterinary Biological Products, Jiangsu Academy of Agricultural Science.

**Author Contributions:** Wu-Yang Huang wrote the whole paper; Ya-Mei Liu did most experiments, such as Western Blotting and Real-Time qRT-PCR; Jian Wang did some experiment, such as ELISA; Xing-Na Wang help to data analysis, and Chun-Yang Li arranged the whole experiments.

**Conflicts of Interest:** The authors declare no conflict of interest.

## References

1.  Li, T.S.C. *Vegetables and Fruits: Nutritional and Therapeutic Values*; CRC Press: Boca Raton, FL, USA, 2008.
2.  Tulipani, S.; Mezzetti, B.; Capocasa, F.; Bompadre, S.; Beekwilder, J.; de Vos, C.H.R.; Capanoglu, E.; Bovy, A.; Battino, M. Antioxidants, phenolic compounds, and nutritional quality of different strawberry genotypes. *J. Agric. Food Chem.* **2008**, *56*, 696–704.
3.  Shahidi, F., Ho, C.T., Eds.; *Phenolic Compounds in Foods and Natural Health Products*; American Chemical Society: Washington, DC, USA, 2005.
4.  Liu, R.H. Health benefits of fruits and vegetables are from additive and synergistic combination of phytochemicals. *Am. J. Clin. Nutr.* **2003**, *78*, S517S–S520.

5.  Huang, W.Y.; Cai, Y.Z.; Zhang, Y.B. Natural phenolic compounds from medicinal herbs and dietary plants: Potential use for cancer prevention. *Nutr. Cancer* **2010**, *62*, 1–20.

6.  Zhou, Z.; Liu, Y.; Miao, A.D.; Wang, S. Protocatechuic aldehyde suppresses TNF-a-induced ICAM-1 and VCAM-1 expression in human umbilical vein endothelial cells. *Eur. J. Pharmacol.* **2005**, *513*, 1–8.

7.  Kanterman1, J.; Sade-Feldman1, M.; Baniyash, M. New insights into chronic inflammation-induced immunosuppression. *Semin. Cancer Biol.* **2012**, *22*, 307–308.

8.  Krikorian, R.; Shidler, M.D.; Nash, T.A.; Kalt, W.; Vinqvist-Tymchuk, M.R.; Shukitt-Hale, B.; Joseph, J.A. Blueberry supplementation improves memory in older adults. *J. Agric. Food Chem.* **2010**, *58*, 3996–4000.

9.  Shaughnessy, K.S.; Boswall, I.A.; Scanlan, A.P.; Gottschall-Pass, K.T.; Sweeney, M.I. Diets containing blueberry extract lower blood pressure in spontaneously hypertensive stroke-prone rats. *Nutr. Res.* **2009**, *29*, 130–138.

10. Kalt, W.; Foote, K.; Fillmore, S.A.; Lyon, M.; van Lunen, T.A.; Mc Rae, K.B. Effect of blueberry feeding on plasma lipids in pigs. *Br. J. Nutr.* **2008**, *100*, 70–78.

11. Castrejon, A.D.R.; Eichholz, I.; Rohn, S.; Krohn, L.W.; Huykens-Keil, S. Phenolic profile and antioxidant activity of highbush blueberry (*Vaccinium corymbosum* L.) during fruit maturation and ripening. *Food Chem.* **2008**, *109*, 564–572.

12. Li, H.Y.; Deng, Z.Y.; Zhu, H.H.; Hu, C.L.; Liu, R.H.; Young, J.C.; Tsao, R. Highly pigmented vegetables: Anthocyanin compositions and their role in antioxidant activities. *Food Res. Int.* **2012**, *46*, 250–259.

13. Srivastava, A.; Akoh, C.C.; Fischer, J.; Krewer, G. Effect of anthocyanin fractions from selected cultivars of Georgia-grown blueberries on apoptosis and phase II enzymes. *J. Agric. Food Chem.* **2007**, *55*, 3180–3185.

14. Moyer, R.A.; Hummer, K.E.; Finn, C.E.; Frei, B.; Wrolstad, R. Anthocyanins, phenolics, and antioxidant capacity in diverse small fruits: Vaccinium, rubus, and ribes. *J. Agric. Food Chem.* **2002**, *50*, 519–525.

15. Huang, W.Y.; Zhang, H.C.; Liu, W.X.; Li, C.Y. Survey of antioxidant capacity and phenolic composition of blueberry, blackberry, and strawberry in Nanjing. *J. Zhejiang Univ. Sci. B* **2012**, *13*, 94–102.

16. Huang, W.Y.; Wang, J.; Liu, Y.M.; Zheng, Q.S.; Li, C.Y. Inhibitory effect of Malvidin on TNF-$\alpha$-induced inflammatory response in endothelial cells. *Eur. J. Pharmacol.* **2014**, *723*, 67–72.

17. Esposito, D.; Chen, A.; Grace, M.H.; Komarnytsky, S.; Lila, M.A. Inhibitory effects of wild blueberry anthocyanins and other flavonoids on biomarkers of acute and chronic inflammation *in vitro*. *J. Agric. Food Chem.* **2014**, *62*, 7022–7028.

18. Jia, Z.S.; Zhou, B.; Yang, L.; Wu, L.M.; Liu, Z.L. Synergistic antioxidant effect of green tea polyphenols with a-tocopherol on free radical initiated peroxidation of linoleic acid in micelles. *J. Chem. Soc. Perkin Trans.* 2 **2000**, 785–791.

19. Rossetto, M.; Vanzani, P.; Mattivi, F.; Lunelli, M.; Scarpa, M.; Rigo, A. Synergistic antioxidant effect of catechin and malvidin-3-glucoside on free radical-initiated peroxidation of linoleic acid in micelles. *Arch. Biochem. Biophys.* **2002**, *408*, 239–245.

20. Saucier, C.T.; Waterhouse, A.L. Antioxidant activity of green tea polyphenols against lipid peroxidation initiated by lipid-soluble radicals in micelles. *J. Agric. Food Chem.* **1999**, *47*, 4491–4494.

21. Lentsch, A.B.; Jordan, J.A.; Czermak, B.J.; Diehl, K.M.; Younkin, E.M.; Sarma, V.; Ward, P.A. Inhibition of NF-kappaB activation and augmentation of IkappaBbeta by secretory leukocyte protease inhibitor during lung inflammation. *Am. J. Pathol.* **1999**, *154*, 239–247.

22. Aqil, F.; Gupta, A.; Munagala, R.; Jeyabalan, J.; Kausar, H.; Sharma, R.J.; Singh, I.P.; Gupta, R.C. Antioxidant and antiproliferative activities of anthocyanin/ellagitannin-enriched extracts from *Syzygium cumini* L. ('Jamun', the Indian blackberry). *Nutr. Cancer Int. J.* **2012**, *64*, 428–438.

23. Nie; Zhao, Z.P.; Chen, G.P.; Zhang, B.; Ye, M.; Hu, Z.L. *Brassica napus* possesses enhanced antioxidant capacity via heterologous expression of anthocyanin pathway gene transcription factors. *J. Plant Physiol.* **2013**, *60*, 108–115.

24. Pop, R.; Stefanut, M.N.; Cata, A.; Tanasie, C.; Medeleanu, M. Ab initio study regarding the evaluation of the antioxidant character of cyanidin, delphinidin and malvidin. *Cent. Eur. J. Chem.* **2012**, *10*, 180–186.

25. Hosseinian, F.S.; Beta, T. Saskatoon and wild blueberries have higher anthocyanin contents than other Manitoban berries. *J. Agric. Food Chem.* **2007**, *55*, 10832–10838.

26. Hyun, J.W.; Chung, H.S. Cyanidin and malvidin from *Oryza sativa* cv. Heugjinjubyeo mediate cytotoxicity against human monocytic leukemia cells by arrest of G(2)/M phase and induction of apoptosis. *J. Agric. Food Chem.* **2004**, *52*, 2213–2217.

27. Patterson, S.J.; Fischer, J.G.; Dulebohn, R.V. DNA damage in HT-29 colon cancer cells is enhanced by high concentrations of the anthocyanin malvidin. *FASEB J.* **2008**, *22*, 890.

28. Lee, C.; Han, D.; Kim, B.; Baek, N.; Baik, B.K. Antioxidant and anti-hypertensive activity of anthocyanin-rich extracts from hulless pigmented barley cultivars. *Int. J. Food Sci. Technol.* **2013**, *48*, 984–991.

29. Chen, P.-N.; Chu, S.-C.; Chiou, H.-L.; Kuo, W.-H.; Chiang, C.-L.; Hsieh, Y.-S. Mulberry anthocyanins, cyanidin 3-rutinoside and cyanidin 3-glucoside, exhibited an inhibitory effect on the migration and invasion of a human lung cancer cell line. *Cancer Lett.* **2006**, *235*, 248–259.

30. Yun, J.W.; Lee, W.S.; Kim, M.J.; Lu, J.N.; Kang, M.H.; Kim, H.G.; Kim, D.C.; Choi, E.J.; Choi, J.Y.; Kim, H.G.; *et al.* Characterization of a profile of the anthocyanins isolated from Vitis coignetiae Pulliat and their anti-invasive activity on HT-29 human colon cancer cells. *Food Chem. Toxicol.* **2010**, *48*, 903–909.

31. Lee, S.G.; Kim, B.; Yang, Y.; Pham, T.X.; Park, Y.-K.; Manatou, J.; Koo, S.I.; Chun, O.K.; Lee, J.-Y. Berry anthocyanins nsuppress the expression and secretion of proinflammatory mediators in macrophages by inhibiting nuclear translocation of NF-κB independent of NRF2-mediated mechanism. *J. Nutr. Biochem.* **2014**, *25*, 404–411.

32. Joana, P.; Teresa, C.P.D.; Leonor, M.A. Malvidin-3-glucoside protects endothelial cells up-regulating endothelial NO synthase and inhibiting peroxynitrite-induced NF-κB activation. *Chem. Biol. Interact.* **2012**, *199*, 192–200.

33. Tsoyi, K.; Park, H.B.; Kim, Y.M.; Chung, J.I.; Shin, S.C.; Lee, W.S.; Seo, H.G.; Lee, J.H.; Chang, K.C.; Kim, H.J. Anthocyanins from black soybean coats inhibit UVB-induced inflammatory cyclooxygenase-2 gene expression and PGE2 production through regulation of the nuclear factor-κB and phosphatidylinositol 3-kinase/AKt pathway. *J. Agric. Food Chem.* **2008**, *56*, 8969–8974.

34. Paixao, J.; Dinis, T.C.P.; Almeida, L.M. Dietary anthocyanins protect endothelial cells against peroxynitrite-induced mitochondrial apoptosis pathway and Bax nuclear translocation: An *in vitro* approach. *Apoptosis* **2011**, *16*, 976–989.

35. Cao, L.H.; Lee, Y.J.; Kang, D.G.; Kim, J.S.; Lee, H.S. Effect of *Zanthoxylum schinifolium* on TNF-α-induced vascular inflammation in human umbilical vein endothelial cells. *Vasc. Pharmacol.* **2009**, *50*, 200–207.

36. Umetani, M.; Nakao, H.; Doi, T.; Lwaski, A.; Ohtaka, M.; Nagoya, T. A novel cell adhesion inhibitor, K-7174, reduces the endothelial VCAM-1 induction by inflammatory cytokines, acting through the regulation of GATA. *Biochem. Bioph. Res. Commun.* **2000**, *272*, 270–274.

37. Hatada, E.N.; Do, R.K.; Orlofsky, A. NF-kappa B1 p50 is required for BLyS attenuation of apoptosis but dispensable for processing of NF-kappa B2 p100 to p52 in quiescent mature B cells. *J. Immunol.* **2003**, *171*, 761–768.

38. Yang, L.; Froio, R.M.; Sciuto, T.E.; Dvorak, A.M.; Alon, R.; Luscinskas, F.W. ICAM-1 regulates neutrophil adhesion and transcellular migration of TNF-alpha-activated vascular endothelium under flow. *Blood* **2005**, *106*, 584–592.

39. Sarvesh, K.; Brajendra, K.S.; Anil, K.P.; Ajit, K.; Sunil, K. A chromone analog inhibits TNF-α induced expression of cell adhesion molecules on human endothelial cells via blocking NF-κB activation. *Bioorg. Med. Chem.* **2007**, *15*, 2952–2962.

40. Finkenzeller, G.; Graner, S.; Kirkpatrick, C.J.; Fuchs, S.; Stark, G.B. Impaired *in vivo* vasculogenic potential of endothelial progenitor cells in comparison to human umbilical vein endothelial cells in a spheroid-based implantation model. *Cell Prolif.* **2009**, *42*, 498–505.

41. Stewart, K.G.; Zhang, Y.; Davidge, S.T. Estrogen decreases prostaglandin H synthase products from endothelial cells. *J. Soc. Gynecol. Investig.* **1999**, *6*, 322–327.

**Sample Availability:** *Sample Availability*: Samples are available from the authors.

# Effect of Standardized Cranberry Extract on the Activity and Expression of Selected Biotransformation Enzymes in Rat Liver and Intestine

**Hana Bártíková, Iva Boušová, Pavla Jedličková, Kateřina Lněničková, Lenka Skálová and Barbora Szotáková**

**Abstract:** The use of dietary supplements containing cranberry extract is a common way to prevent urinary tract infections. As consumption of these supplements containing a mixture of concentrated anthocyanins and proanthocyanidins has increased, interest in their possible interactions with drug-metabolizing enzymes has grown. In this *in vivo* study, rats were treated with a standardized cranberry extract (CystiCran®) obtained from *Vaccinium macrocarpon* in two dosage schemes (14 days, 0.5 mg of proanthocyanidins/kg/day; 1 day, 1.5 mg of proanthocyanidins/kg/day). The aim of this study was to evaluate the effect of anthocyanins and proanthocyanidins contained in this extract on the activity and expression of intestinal and hepatic biotransformation enzymes: cytochrome P450 (CYP1A1, CYP1A2, CYP2B and CYP3A), carbonyl reductase 1 (CBR1), glutathione-S-transferase (GST) and UDP-glucuronosyl transferase (UGT). Administration of cranberry extract led to moderate increases in the activities of hepatic CYP3A (by 34%), CYP1A1 (by 38%), UGT (by 40%), CBR1 (by 17%) and GST (by 13%), while activities of these enzymes in the small intestine were unchanged. No changes in the relative amounts of these proteins were found. Taken together, the interactions of cranberry extract with simultaneously administered drugs seem not to be serious.

Reprinted from *Molecules*. Cite as: Bártíková, H.; Boušová, I.; Jedličková, P.; Lněničková, K.; Skálová, L.; Szotáková, B. Effect of Standardized Cranberry Extract on the Activity and Expression of Selected Biotransformation Enzymes in Rat Liver and Intestine. *Molecules* **2014**, *19*, 14948–14960.

## 1. Introduction

Cranberry (*Vaccinium macrocarpon*, Ericaceae), a native plant of North America, is among the top selling dietary supplements around the world. The juice as well as dietary supplements derived from this berry exert various beneficial effects on human health, including prevention and treatment of urinary tract infections, anti-cancer and antioxidant activities [1,2]. However, the positive effect of cranberry juice in the prevention of urinary tract infections has recently been disputed [3,4]. The biological activities of cranberry can be attributed to a diverse group of phytochemicals, including phenylpropanoids (such as flavonoids and resveratrol), phenolic acids and isoprenoids such as lutein and ursolic acid [5,6]. Cranberries also represent a very good source of vitamin C [7]. Flavonoid constituents found in cranberry fruit belong primarily to three classes: proanthocyanidins, anthocyanins and flavonols [8]. Proanthocyanidins (also known as polyflavan-3-ols) found in cranberry fruits are primarily dimers, trimers and larger oligomers of (-)-epicatechin, containing two types of linkages between epicatechin units: the more common $C4\beta \rightarrow C8$ (B-type) linkage and a less common A-type linkage featuring both $C4\beta \rightarrow C8$ and $C2\beta \rightarrow O7$ interflavanoid bonds [9]. Among anthocyanins, glycosides of cyanidin, peonidin and petunidin prevail. The most abundant flavonol aglycone is quercetin, followed by myricetin and kaempferol. Cranberry fruits contain up to 91.5 mg of anthocyanins, 40 mg of flavonols and 180 mg of proanthocyanidins (with degrees of polymerization $\leqslant 10$) per 100 g of ripe fruit [5].

Consumption of cranberry juice and/or various dietary supplements containing cranberries have always been considered safe, however, several studies have suggested that cranberry juice is capable of interacting with drug-metabolizing enzymes and thus may elicit clinically relevant interactions with certain drugs. For instance, cranberry juice inhibited the activity of cytochrome P450 3A (CYP3A) *in vitro* in human liver and rat intestinal microsomes [10]. Moreover, in rats, cranberry juice was as effective as grapefruit juice in enhancing the systemic exposure of nifedipine, the calcium channel antagonist and CYP3A substrate. In both cases, the area under the curve (AUC) of nifedipine was increased by 60% compared to saline [10]. Indeed, enteric, but not hepatic, CYP3A-mediated first-pass metabolism of midazolam was inhibited by cranberry juice in healthy volunteers [11]. In contrast, no interaction has been found in the human study involving CYP3A substrate cyclosporine and cranberry juice [12]. Long-term treatment with three cranberry extracts had no effect on glutathione S-transferase (GST) activity in rats [13]. Besides, weak inhibitory activity of cranberry extract towards UDP-glucuronosyl transferase 1A9 (UGT1A9) in human liver microsomes was reported [14]. Individual anthocyanidins were able to significantly inhibit human and rat carbonyl reductase (CBR) and UGT *in vitro* [15].

The abovementioned data show that information about the effect of cranberry extract/juice on drug-metabolizing enzymes has been inconsistent and insufficient. The inconsistence of the results from various studies is mainly due to the use of different experimental conditions. There are differences not only in the model systems, but also in the administered substances (undefined cranberry juice, defined cranberry extract, individual anthocyanidins, *etc.*), and dosage schemes (various doses, various treatment durations, *etc.*). Moreover, most of the studies have focused only on selected enzymes in one organ (mostly liver) and a comprehensive view has been lacking.

Therefore, the present *in vivo* study was designed to test the effect of cranberry extract (in two different dosage schemes) on the activity and expression of a panel of drug-metabolizing enzymes (a total of seven) in rats. In addition to the hepatic enzymes most studied in the literature, intestinal enzymes were also included in this project to obtain more complete information. With the aim to increase reproducibility and repeatability of our experiments, CystiCran®, a patented and standardized cranberry extract, was used in this study.

## 2. Results and Discussion

Cranberry standardized extract CystiCran® (CC, Decas Botanical Synergies, Carver, MA, USA) contains 1.6 mg of anthocyanidins and 36 mg of proanthocyanidins in one tablet, mainly those comprising A-type linkages, which have been associated with preventing adhesion of uropathogenic bacteria to uroepithelial cells [16]. In the present study, male rats were orally treated with CC to simulate consumption of cranberry juice and/or dietary supplements containing cranberry extract. Cranberry juice as well as dietary supplements are usually taken orally, therefore *p.o.* administration of CC (by gastric gavage) was chosen. The two dosage schemes employed represented two possible situations: regular long-term consumption of dietary supplement as a prevention of urinary tract infection (14 days, 0.5 mg of proanthocyanidins/kg/day, *i.e.*, 1 tablet of CC/day in human therapy) and short-term overdose by dietary supplement (1 day, 1.5 mg of proanthocyanidins/kg/day, *i.e.*, 3 tablets of CC/day in human therapy). In order to obtain more comprehensive information about the impact of cranberry extract on drug-metabolizing enzymes, their activities and expressions were studied not only in liver, but also in small intestine. Moreover, anthocyanins and proanthocyanidins are polyphenols with low absorption [17], and thus their systemic bioavailability is low, but intestinal mucosa exposure is high.

Proanthocyanidins are stable during gastric transit. While degradation of polymeric proanthocyanidins to the corresponding monomers is negligible in the gastrointestinal tract *in vivo*, proanthocyanidin oligomers with a degree of polymerization lower than 5 are absorbable [18]. Upon absorption, all dietary polyphenols in the human body are subject to substantial transformation catalyzed by drug-metabolizing enzymes [19]. Unlike extensive phase II metabolism of absorbed monomers, which are glucuronidated, sulfated and methylated [19], phase II metabolism of dimers appeared to be limited because glucuronidated or sulfated metabolites of dimers were not detected in biological fluids after intestinal perfusion in rats [18]. Nevertheless, the majority of proanthocyanidins reach the colon intact and are degraded into phenylvalerolactones and phenolic acids by colon microbiota. These microbial metabolites may contribute to the health-promoting properties of proanthocyanidins *in vivo* [18].

## 2.1. Effect of CystiCran® on the Activities of Phase I Biotransformation Enzymes

The phase I of drug metabolism includes oxidation, reduction or hydrolytic reactions, which lead to introduction or uncovering of functional groups (e.g., –OH, –COOH, –SH, –O– or –NH$_2$ group) resulting in new chemical entities with different physico-chemical and biological properties. Reactions carried out by phase I enzymes usually transform active drugd into their inactive metabolite(s), but in certain cases, phase I biotransformation causes bioactivation of xenobiotics. Enzymes responsible for phase I biotransformation of drugs and other xenobiotics are abundantly present in the liver, gastrointestinal tract, lungs and kidneys. Many of them are inducible by various xenobiotics. Induction of drug metabolism may arise as a consequence of increased synthesis, decreased degradation, activation of enzymes or a combination of these three processes, although it should be emphasized that the majority of enzymes are induced at the transcriptional activation level. Thus, enzyme induction takes place only after its prolonged exposure to the xenobiotic [20]. Cytochromes P450 (CYP), a superfamily of the most important drug-metabolizing enzymes, are involved in the metabolism of about 75% of all drugs. These enzymes utilize one molecule of oxygen and produce oxidized substrates and a molecule of water [21].

In this study, specific activities of four CYP isoforms (*i.e.*, CYP1A1, 1A2, CYP2B, and CYP3A) were assessed in microsomal fractions of rat liver (Figure 1) and small intestine using alkoxyresorufins as substrates. Long-term treatment with low dose of cranberry extract (CC-14L) caused increase in the catalytic activity of hepatic CYP1A1 and CYP3A by 38% and 34%, respectively. Non-significant elevation in hepatic CYP2B activity was observed in CC-14L as well as in CC-1H groups. In rat small intestine, no specific activities of selected CYP isoforms were detected either in control or in CC-treated rats.

While in a study using healthy volunteers cranberry juice/extract caused inhibition of CYP3A [11], our results showed mild increases in CYP1A1 and CYP3A activities. This discrepancy may be caused by the different experimental models. Moreover, cranberry extract also contains other constituents such as the flavonoid quercetin and the natural stilbene resveratrol, whose *in vivo* inductive effects on CYP3A activity were described [22,23]. As the CYP3A subfamily is involved in the oxidative biotransformation of numerous drugs, modulation of CYP3A activity can have profound clinical consequences, but only with drugs that have narrow therapeutic windows [24]. Induction of CYP1A1/2 is also considered to be undesirable as these isoforms carry out bioactivation of polycyclic aromatic hydrocarbons (e.g., benzo[a]pyrene), recognized carcinogens in humans and rodents. Fortunately, induction of CYP1A1/2 and CYP3A by cranberry extract was only mild and thus serious consequences could be excluded.

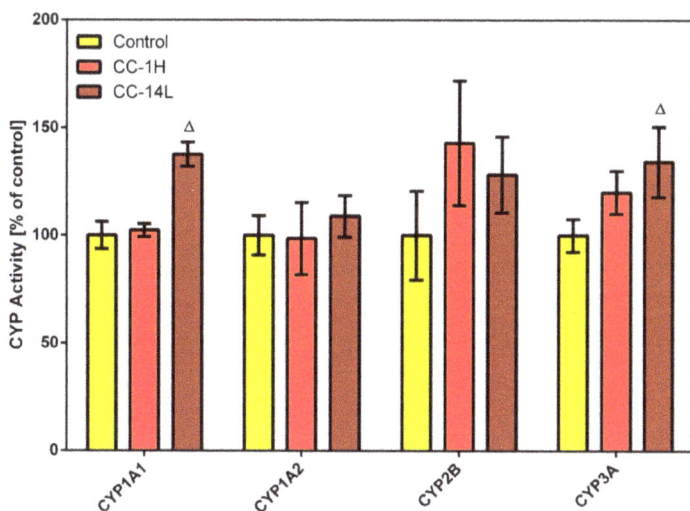

**Figure 1.** Effect of cranberry extract administered in two dosage schemes on the specific activities of cytochrome P450 isoforms 1A1, 1A2, 2B and 3A in rat liver microsomes. Specific activities were determined fluorimetrically using alkoxyresorufines as substrates. Data represent the mean $\pm$ SD of three independent experiments. The triangle indicates a significant difference from the control ($p < 0.05$).

Phase I reducing reactions of xenobiotics are often catalyzed by carbonyl reductase 1, a member of the short-chain dehydrogenases/reductases family, that reduces a wide variety of carbonyl compounds including quinones, prostaglandins, menadione, and various xenobiotics [24]. In our study, specific activity of CBR1 was measured in cytosolic fractions of liver and small intestinal mucosa homogenates using menadione as a specific substrate. Treatment of rats with CystiCran® with both dosage schemes caused an elevation in CBR1 activities in the liver and small intestine, although only the changes found in the liver were statistically significant. Activity of hepatic CBR1 was increased by 17% and 30% in the CC-1H and CC-14L groups compared to untreated control, respectively (Figure 2).

CBRs, particularly isoform CBR1, are leading biotransformation enzymes catalyzing deactivation of carbonyl-bearing drugs, including cytostatics. Therefore, regulation of their activity has been intensively studied and by this way, several flavonoids (e.g., quercetin, quercitrin and rutin) and anthocyanidins (e.g., delphinidin, cyanidin and malvidin) have been recognized as *in vitro* inhibitors of CBR1 [15,25]. To our knowledge, no *in vivo* studies covering this topic have been performed yet. However, these compounds may act as inducers of CBR1 or may not to influence its activity *in vivo*. Their inductive/inhibitory effect is dependent also on the concentration used [26,27].

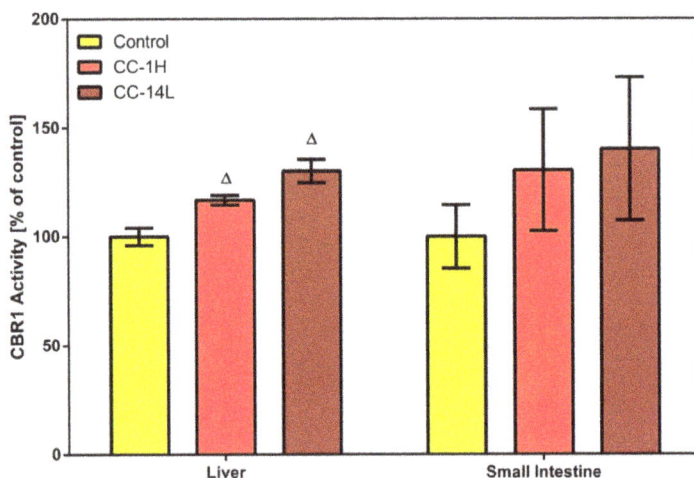

**Figure 2.** Effect of cranberry extract administered in two dosage schemes on the specific activity of carbonyl reductase 1 in the cytosolic fractions of rat liver and small intestine. Specific activities were determined fluorimetrically using menadione as a substrate. Data represent the mean $\pm$ SD of three independent experiments. Triangles indicate a significant difference from the control ($p < 0.05$).

## 2.2. Effect of CystiCran® on the Activities of Phase II Conjugating Enzymes

In phase II, xenobiotics or their phase I metabolites undergo conjugation reactions with endogenous compounds such as glutathione or UDP-glucuronic acid. Conjugation usually introduces hydrophilic ionizable functional groups onto the molecule of a xenobiotic, thus making it more polar and facilitating its renal excretion. Almost all phase II biotransformation reactions lead to detoxification/deactivation of the parent xenobiotic or metabolites formed in phase I. Glucuronidation, catalyzed by UGT, is the most common conjugation pathway in humans. Typical UGT substrates are xenobiotic or eobiotic alcohols, phenols or carboxylic acids. On the other hand, electrophilic compounds are mostly conjugated with glutathione through the action of GSTs [28].

In this study, homogenates of rat liver and small intestine were analyzed for UGT and GST activities. Specific activity of UGT was tested in microsomal fractions using $p$-nitrophenol as a substrate. In the liver, UGT activity was significantly increased in CC-treated rats compared to the control group (Figure 3A).

On the other hand, specific UGT activity in the small intestine was at the detection limit, therefore no changes in this activity were detected. Catalytic activity of GST was assessed in cytosolic fractions by measuring its conjugation activity with the universal substrate 1-chloro-2,4-dinitrobenzene (CDNB). GST activity was elevated by 13% in the liver of both CC-treated groups compared to the control group, while an insignificant decrease in GST activity was observed in the small intestine of rats treated with CC (Figure 3B).

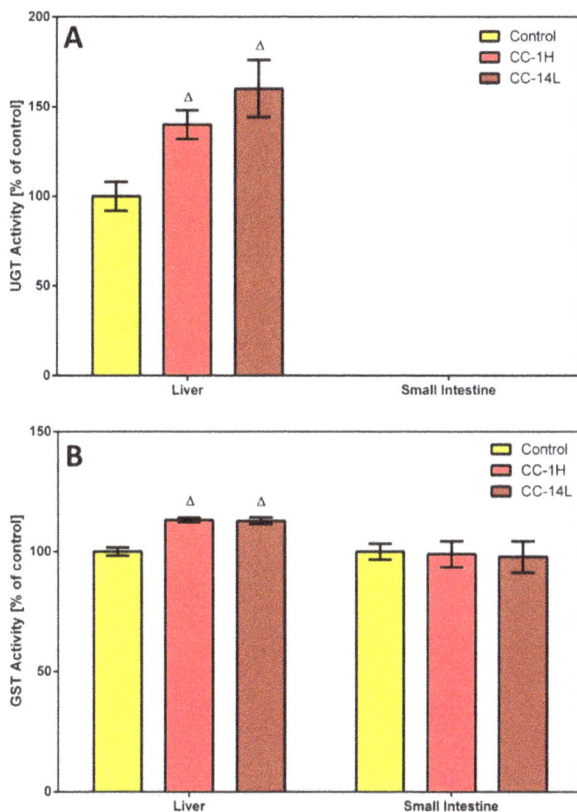

**Figure 3.** Effect of cranberry extract administered in two dosage schemes on the specific activities of UDP-glucuronosyl transferase in the microsomal fractions of rat liver and small intestine (**A**) and glutathione S-transferase in the cytosolic fractions of rat liver and small intestine (**B**). Specific activities were determined spectrophotometrically using $p$-nitrophenol and CDNB as UGT and GST substrates, respectively. Data represent the mean $\pm$ SD of three independent experiments. Triangles indicate a significant difference from the control ($p < 0.05$).

Several constituents found in cranberry may be responsible for increased GST and UGT activities in rat liver. Consumption of various fruits and plants, which are rich sources of anthocyanidins as well as proanthocyanidins, was shown to elevate UGT and GST activities in rat liver [29,30]. Observed induction of UGT activity may be caused also by quercetin, whose content in cranberry is estimated to be in the range of 83–121 mg/kg (*i.e.*, about 50 µg per 0.5 mg cranberry supplement capsule) [31]. This assumption was confirmed by the study of van der Logt *et al.* [32], in which significant induction of $p$-nitrophenol glucuronidation was found in rat liver and small intestine after 14-day intake of quercetin. Another constituent capable of UGT and GST activities induction is the stilbene resveratrol. This compound caused 1.8-fold and 1.3-fold increases of UGT and GST activities in the liver of male rats upon 28-day intake of resveratrol, respectively [33]. Moreover, 28-day administration of resveratrol to healthy volunteers led to induction of GSTP protein levels and UGT1A1 activity in individuals with low baseline enzyme level/activity, while overall GST and UGT1A1 activities were minimally affected by this intervention [22]. Induction of GST activity by cranberry extract can be considered protective, because increased levels/activity of GST, especially in the gastrointestinal tract, can prevent organisms against harmful electrophiles [26].

Certainly, it is necessary to take into consideration that the observed effects of cranberry extract may not be ascribed to only one chemical constituent as it is a complex mixture of various phytonutrients. Individual components of this mixture may interact with each other and the resulting biological activity of the cranberry extract is therefore dependent on their synergistic/antagonistic behavior. Moreover, biological effects of cranberry juice/extract are concentration-dependent. Thus, its inhibitory activity observed *in vitro* need not to be found also *in vivo* because the concentrations of active compounds in organism is influenced by their bioavailability. While high concentrations of bioactive compounds directly influence the activities of the studied enzymes in the *in vitro* experiments, the activity as well as expression of these enzymes are affected by low concentrations of these compounds in the *in vivo* study.

## 2.3. Expression of Biotransformation Enzymes

In our study, the activities of drug-metabolizing enzymes were assessed primarily as the activities which are more important than protein and mRNA levels from a pharmacological/toxicological point of view. When elevated activity of a certain enzyme was found, the corresponding protein was quantified using an immunoblotting technique. The amounts of liver CYP1A1, CYP3A4, and UGT1A proteins were detected in the microsomal fractions of control as well as CC-treated rats, while levels of hepatic CBR1 and GSTP proteins were quantified in the cytosolic fractions. Calnexin and β-actin were used as loading controls in microsomal and cytosolic fraction, respectively. Protein quantification did not revealed any significant changes in the proteins amount after CC treatment (Table 1).

Although the protein quantification was not in full agreement with the activities of biotransformation enzymes, similar trends were found in several studied enzymes (e.g., CYP3A4, CBR1 and GST alpha). Some discrepancies may be explained by differences in enzyme activity assessment and performance of immunoblotting. In the case of UGT and GST, enzyme activity was assessed using universal substrates covering most of the isoforms, while immunoblotting was performed using antibody specific for one isoform (GST alpha) or one enzyme family (UGT1A). Therefore, observed changes in enzyme activity may be caused by different isoform(s) (e.g., GST pi). Moreover, modulation of various biotransformation enzymes by cranberry extract may occur at several levels (transcriptional, posttranscriptional *etc.*) [34].

**Table 1.** Effect of cranberry extract administered in two dosage schemes on the protein level of CYP1A1, CYP3A, UGT (microsomal fractions), CBR1 and GST (cytosolic fractions) in rat liver. Protein expressions were detected by immunoblotting using specific antibodies and normalized to the amount of loading controls calnexin (microsomal fractions) and β-actin (cytosolic fractions). Data represent the mean ± SD of three independent experiments.

| Enzyme | Mean (%) ± SD | | |
|---|---|---|---|
| | Control | CC-1H | CC-14L |
| CYP1A1 | 100.0 ± 12.3 | 89.8 ± 17.1 | 100.2 ± 9.3 |
| CYP3A4 | 100.0 ± 9.6 | 108.2 ± 12.7 | 116.0 ± 10.4 |
| CBR1 | 100.0 ± 14.2 | 106.3 ± 8.6 | 111.5 ± 15.9 |
| UGT | 100.0 ± 12.1 | 108.2 ± 5.9 | 100.7 ± 11.0 |
| GST alpha | 100.0 ± 7.9 | 104.5 ± 8.3 | 112.8 ± 7.6 |

## 3. Experimental Section

### 3.1. Chemicals and Reagents

Benzyloxyresorufin, 7-ethoxyresorufin, 7-methoxyresorufin, 7-pentoxyresorufin, menadione, 1-chloro-2,4-dinitrobenzene (CDNB), 4-nitrophenol (NP), reduced glutathione (GSH), UDP-glucuronic acid, NADPH, and chemicals used for realization of electrophoresis were products of Sigma-Aldrich (Prague, Czech Republic). Precision Plus molecular weight standard and non-fat dry milk were purchased from Bio-Rad (Bio-Rad Laboratories, Hercules, CA, USA). For immunoblotting, rabbit polyclonal anti-UGT antibody (Cell Signaling, Leiden, The Netherlands), rabbit polyclonal anti-GST alpha antibody, rabbit monoclonal anti-CBR1 antibody, rabbit polyclonal anti-calnexin antibody, rabbit polyclonal anti-beta actin antibody (Abcam, Cambridge, UK), rabbit polyclonal anti-CYP1A1 antibody (Novus Biologicals, Cambridge, UK), rabbit polyclonal anti-CYP 3A4 antibody (Sigma-Aldrich), bovine anti-rabbit IgG antibody conjugated with horseradish peroxidase and chemiluminescence kit (Santa Cruz Biotechnology, Santa Cruz, CA, USA) were used. All other chemicals used were of HPLC or analytical grade.

### 3.2. Laboratory Animals

Male Wistar rats were obtained from Meditox (Konarovice, Czech Republic). They were housed in air-conditioned animal quarters with a 12 h light/dark cycle.

Food (a standard rat chow diet) and water were provided *ad libitum*. The animal protocols used in this work were evaluated and approved by the Ethic Committee of the Ministry of Education, Youth and Sports (Protocol 20363/2011-30). They are in accordance with the Guide for the Care and Use of Laboratory Animals (Protection of Animals from Cruelty Act No. 246/92, Czech Republic). At 12 weeks of age, rats were randomly divided into three groups of four. Rats of the first group (CC-14L) were orally administered with therapeutic dose of CystiCran® once daily for a period of 2 weeks; three times higher dose of CystiCran® was administered at once by gastric gavage to the second group of rats (CC-1H; 24 h before the end of experiment), and the third group represents untreated controls. At the end of experiment, animals were sacrificed by decapitation. Livers and small intestines were removed immediately, the intestinal contents were washed out with cold 0.9% saline solution and mucosa was scraped. Both tissues were stored at −80 °C until preparation of subcellular fractions.

*3.3. Preparation of Microsomal and Cytosolic Fractions*

Frozen liver or mucosa from small intestine were thawed at room temperature up to 15 min and processed to microsomal and cytosolic fractions. The subcellular fractions were isolated by differential centrifugation of the tissue homogenate [35] and stored at −80 °C. Protein concentrations were assayed using the bicinchoninic acid (BCA) assay according to manufacturer's instructions (Sigma-Aldrich).

*3.4. Enzyme Assays*

Enzyme activities were assayed in the cytosolic and microsomal fractions from homogenates of liver and small intestinal mucosa of control and treated rats. The enzyme assays (each performed in 4–8 replicates) were repeated three times. The amount of organic solvents in the final reaction mixtures did not exceed 1% (v/v).

The activities of several CYP isoforms were assessed. The activities of 7-ethoxyresorufin *O*-dealkylase (EROD; specific for CYP1A1), 7-methoxyresorufin *O*-dealkylase (MROD; CYP1A2), 7-pentoxyresorufin *O*-dealkylase (PROD; CYP2B) and 7-benzyloxyresorufin *O*-dearylase (BROD; CYP3A) were measured at 37 °C using fluorimetric determination of arising resorufin [36]. Each substrate dissolved in dimethylsulphoxide (DMSO) was added at a final concentration of 5 µM. Assays were conducted at the excitation/emission wavelengths of 530/585 nm using luminescence spectrophotometer LS50B (Perkin-Elmer, Cambridge, UK).

Carbonyl reductase 1 activity was measured in the cytosolic fractions using menadione as a substrate [37]. Consumption of NADPH was determined at excitation/emission wavelength of 380/460 nm using a Perkin Elmer LS50B luminescence spectrophotometer at 37 °C.

The cytosolic glutathione S-transferase activities were assessed by standard colorimetric assay using CDNB as an electrophilic substrate [38], which was adapted for measurement in 96-well plates. The absorbance of rising product S-(2,4-dinitrophenyl)glutathione was detected at 340 nm by Tecan Infinite M200 multimode microplate reader (Tecan Group, Männedorf, Switzerland).

The microsomal UDP-glucuronosyltransferase activities towards $p$-nitrophenol were assayed as described by Mizuma *et al.* [39]. Absorbance of unconjugated $p$-nitrophenol was measured at 405 nm by the Tecan Infinite M200.

### 3.5. Western Blotting

Microsomal proteins of rat liver were separated by SDS-PAGE (10% stacking gel) [40] and subsequently transferred onto nitrocellulose membranes (0.45 μm) using Trans-Blot® TurboTM Transfer System (Bio-Rad, Hercules, CA, USA). Protein concentrations were determined using the BCA protein assay (Sigma-Aldrich). The membranes were blocked in 5% non-fat dry milk/TBS-Tween-20 for 2 h. Immunodetection of biotransformation enzymes was performed using corresponding primary antibodies (described in the Chemicals and Reagents section). The bands were visualized with respective horseradish peroxidase-conjugated secondary antibodies using the chemiluminescence kit according to manufacturer's instructions. Calnexin and β-actin served as the loading controls for microsomal and cytosolic fraction, respectively. Intensity of bands was evaluated using a C-DiGitTM Blot Scanner (Li-Cor, Lincoln, NE, USA).

### 3.6. Statistical Analysis

All calculations were done using Microsoft Excel and GraphPad Prism 6. All values were expressed as mean $\pm$ SD. One-way Anova was used for the statistical evaluation of differences between control and treated groups, and differences were considered as significant when $p < 0.05$.

## 4. Conclusions

In conclusion, *in vivo* administration of standardized cranberry extract in both studied dosage schemes caused only mild changes of some activities of drug-metabolizing enzymes in rat liver, while those in small intestine were not affected. Interestingly, long-term consumption of regular dose has more pronounced effects on drug-metabolizing enzymes' activities than short-term overdose by cranberry extract. Nevertheless, consumption of cranberry juice/extract in reasonable amounts seems to be safe and serious supplement–drug interactions do not seem probable.

**Acknowledgments:** This work was supported by Czech Science Foundation (grant No. P303/12/G163) and by Charles University in Prague (Research Project SVV 260 065).

**Author Contributions:** Conceived and designed the experiments: B.S. and L.S. Performed the experiments: H.B., P.J., K.L. and B.S. Analyzed the data: P.J., H.B. and B.S. Contributed reagents/materials/analysis tools: B.S. and L.S. Wrote the paper: I.B., L.S. and B.S. Revised paper: I.B. and B.S.

**Conflicts of Interest:** The authors declare no conflict of interest.

## References

1.  Klein, M.A. Cranberry (*Vaccinium macrocarpon*) aiton. In *Encyclopedia of Dietary Supplements*, 1st ed.; Coates, P.M., Blackman, M.R., Cragg, G.M., Levine, M., Moss, J., White, J.D., Eds.; Marcel Dekker: New York, NY, USA, 2005; pp. 143–149.
2.  Kresty, L.A.; Howell, A.B.; Baird, M. Cranberry proanthocyanidins mediate growth arrest of lung cancer cells through modulation of gene expression and rapid induction of apoptosis. *Molecules* **2011**, *16*, 2375–2390.
3.  Barbosa-Cesnik, C.; Brown, M.B.; Buxton, M.; Zhang, L.; DeBusscher, J.; Foxman, B. Cranberry juice fails to prevent recurrent urinary tract infection: Results from a randomized placebo-controlled trial. *Clin. Infect. Dis.* **2011**, *52*, 23–30.
4.  Freire Gde, C. Cranberries for preventing urinary tract infections. *Sao Paulo Med. J.* **2013**, *131*, 363.
5.  Dao, C.A.; Patel, K.D.; Neto, C.C. Phytochemicals from the fruit and foliage of cranberry (*Vaccinium macrocarpon*)-potential benefits for human health. *ACS Symp. Ser.* **2012**, *1093*, 79–94.
6.  Wang, Y.; Catana, F.; Yang, Y.; Roderick, R.; van Breemen, R.B. An LC-MS method for analyzing total resveratrol in grape juice, cranberry juice, and in wine. *J. Agric. Food Chem.* **2002**, *50*, 431–435.
7.  Borges, G.; Degeneve, A.; Mullen, W.; Crozier, A. Identification of flavonoid and phenolic antioxidants in black currants, blueberries, raspberries, red currants, and cranberries. *J. Agric. Food Chem.* **2010**, *58*, 3901–3909.
8.  Deziel, B.A.; Patel, K.; Neto, C.; Gottschall-Pass, K.; Hurta, R.A.R. Proanthocyanidins from the american cranberry (*Vaccinium macrocarpon*) inhibit matrix metalloproteinase-2 and matrix metalloproteinase-9 activity in human prostate cancer cells via alterations in multiple cellular signalling pathways. *J. Cell. Biochem.* **2010**, *111*, 742–754.
9.  Foo, L.Y.; Lu, Y.R.; Howell, A.B.; Vorsa, N. A-type proanthocyanidin trimers from cranberry that inhibit adherence of uropathogenic P-fimbriated Escherichia coli. *J. Nat. Prod.* **2000**, *63*, 1225–1228.
10. Uesawa, Y.; Mohri, K. Effects of cranberry juice on nifedipine pharmacokinetics in rats. *J. Pharm. Pharmacol.* **2006**, *58*, 1067–1072.
11. Ngo, N.; Yan, Z.X.; Graf, T.N.; Carrizosa, D.R.; Kashuba, A.D.M.; Dees, E.C.; Oberlies, N.H.; Paine, M.F. Identification of a cranberry juice product that inhibits enteric CYP3A-mediated first-pass metabolism in humans. *Drug Metab. Dispos.* **2009**, *37*, 514–522.
12. Grenier, J.; Fradette, C.; Morelli, G.; Merritt, G.J.; Vranderick, M.; Ducharme, M.P. Pomelo juice, but not cranberry juice, affects the pharmacokinetics of cyclosporine in humans. *Clin. Pharmacol. Ther.* **2006**, *79*, 255–262.

13.  Palikova, I.; Vostalova, J.; Zdarilova, A.; Svobodova, A.; Kosina, P.; Vecera, R.; Stejskal, D.; Proskova, J.; Hrbac, J.; Bednar, P.; *et al.* Long-term effects of three commercial cranberry products on the antioxidative status in rats: A pilot study. *J. Agric. Food Chem.* **2010**, *58*, 1672–1678.

14.  Mohamed, M.E.F.; Frye, R.F. Inhibitory effects of commonly used herbal extracts on UDP-glucuronosyltransferase 1A4, 1A6, and 1A9 enzyme activities. *Drug Metab. Dispos.* **2011**, *39*, 1522–1528.

15.  Szotáková, B.; Bártíková, H.; Hlaváčová, J.; Boušová, I.; Skálová, L. Inhibitory effect of anthocyanidins on hepatic glutathione S-transferase, UDP-glucuronosyltransferase and carbonyl reductase activities in rat and human. *Xenobiotica* **2013**, *43*, 679–685.

16.  Howell, A.B.; Reed, J.D.; Krueger, C.G.; Winterbottom, R.; Cunningham, D.G.; Leahy, M. A-type cranberry proanthocyanidins and uropathogenic bacterial anti-adhesion activity. *Phytochemistry* **2005**, *66*, 2281–2291.

17.  Manach, C.; Williamson, G.; Morand, C.; Scalbert, A.; Remesy, C. Bioavailability and bioefficacy of polyphenols in humans. I. Review of 97 bioavailability studies. *Am. J. Clin. Nutr.* **2005**, *81*, 230s–242s.

18.  Ou, K.; Gu, L. Absorption and metabolism of proanthocyanidins. *J. Funct. Foods* **2014**, *7*, 43–53.

19.  Spencer, J.P.E.; el Mohsen, M.M.A.; Rice-Evans, C. Cellular uptake and metabolism of flavonoids and their metabolites: implications for their bioactivity. *Arch. Biochem. Biophys.* **2004**, *423*, 148–161.

20.  Kramer, S.D.; Testa, B. The biochemistry of drug metabolism–an introduction: Part 6. Inter-individual factors affecting drug metabolism. *Chem. Biodivers.* 2008; 5, 2465–2578.

21.  Guengerich, F.P. Cytochrome p450 and chemical toxicology. *Chem. Res. Toxicol.* **2008**, *21*, 70–83.

22.  Chow, H.H.; Garland, L.L.; Hsu, C.H.; Vining, D.R.; Chew, W.M.; Miller, J.A.; Perloff, M.; Crowell, J.A.; Alberts, D.S. Resveratrol modulates drug-and carcinogen-metabolizing enzymes in a healthy volunteer study. *Cancer Prev. Res.* **2010**, *3*, 1168–1175.

23.  Duan, K.M.; Wang, S.Y.; Ouyang, W.; Mao, Y.M.; Yang, L.J. Effect of quercetin on CYP3A activity in Chinese healthy participants. *J. Clin. Pharmacol.* **2012**, *52*, 940–946.

24.  Testa, B.; Kramer, S.D. The biochemistry of drug metabolism-An introduction-Part 2. Redox reactions and their enzymes. *Chem. Biodivers.* **2007**, *4*, 257–405.

25.  Carlquist, M.; Frejd, T.; Gorwa-Grauslund, M.F. Flavonoids as inhibitors of human carbonyl reductase 1. *Chem. Biol. Interact.* **2008**, *174*, 98–108.

26.  Boušová, I.; Skálová, L. Inhibition and induction of glutathione S-transferases by flavonoids: Possible pharmacological and toxicological consequences. *Drug Metab. Rev.* **2012**, *44*, 267–286.

27.  Bártíková, H.; Skálová, L.; Dršata, J.; Boušová, I. Interaction of anthocyanins with drug-metabolizing and antioxidant enzymes. *Curr. Med. Chem.* **2013**, *20*, 4665–4679.

28.  Testa, B.; Kramer, S.D. The biochemistry of drug metabolism-an introduction Part 4. Reactions of conjugation and their enzymes. *Chem. Biodivers.* **2008**, *5*, 2171–2336.

29. Ajiboye, T.O.; Salawu, N.A.; Yakubu, M.T.; Oladiji, A.T.; Akanji, M.A.; Okogun, J.I. Antioxidant and drug detoxification potentials of Hibiscus sabdariffa anthocyanin extract. *Drug Chem. Toxicol.* **2011**, *34*, 109–115.

30. Boateng, J.; Verghese, M.; Shackelford, L.; Walker, L.T.; Khatiwada, J.; Ogutu, S.; Williams, D.S.; Jones, J.; Guyton, M.; Asiamah, D.; *et al.* Selected fruits reduce azoxymethane (AOM)-induced aberrant crypt foci (ACF) in Fisher 344 male rats. *Food Chem. Toxicol.* **2007**, *45*, 725–732.

31. Hakkinen, S.H.; Karenlampi, S.O.; Heinonen, I.M.; Mykkanen, H.M.; Torronen, A.R. Content of the flavonols quercetin, myricetin, and kaempferol in 25 edible berries. *J. Agric. Food Chem.* **1999**, *47*, 2274–2279.

32. Van der Logt, E.M.J.; Roelofs, H.M.J.; Nagengast, F.M.; Peters, W.H.M. Induction of rat hepatic and intestinal UDP-glucuronosyltransferases by naturally occurring dietary anticarcinogens. *Carcinogenesis* **2003**, *24*, 1651–1656.

33. Hebbar, V.; Shen, G.; Hu, R.; Kim, B.R.; Chen, C.; Korytko, P.J.; Crowell, J.A.; Levine, B.S.; Kong, A.N. Toxicogenomics of resveratrol in rat liver. *Life Sci.* **2005**, *76*, 2299–2314.

34. Maier, T.; Guell, M.; Serrano, L. Correlation of mRNA and protein in complex biological samples. *FEBS Lett.* **2009**, *583*, 3966–3973.

35. Gillette, J. Techniques for studying drug metabolism *in vitro*. In *Fundamentals of Drug Metabolism and Drug Disposition*; La Du, B.N., Mandel, H.G., Way, E., Eds.; The Williams and Wilkins Company: Baltimore, MA, USA, 1971; pp. 400–418.

36. Weaver, R.J.; Thompson, S.; Smith, G.; Dickins, M.; Elcombe, C.R.; Mayer, R.T.; Burke, M.D. A comparative-study of constitutive and induced alkoxyresorufin *O*-dealkylation and individual cytochrome-P450 forms in cynomolgus monkey (macaca-fascicularis), human, mouse, rat and hamster liver-microsomes. *Biochem. Pharmacol.* **1994**, *47*, 763–773.

37. Maté, L.; Virkel, G.; Lifschitz, A.; Ballent, M.; Lanusse, C. Hepatic and extra-hepatic metabolic pathways involved in flubendazole biotransformation in sheep. *Biochem. Pharmacol.* **2008**, *76*, 773–783.

38. Habig, W.H.; Jakoby, W.B. Glutathione S-transferases (rat and human). *Methods Enzymol.* **1981**, *77*, 218–231.

39. Mizuma, T.; Machida, M.; Hayashi, M.; Awazu, S. Correlation of drug conjugative metabolism rates between *in vivo* and *in vitro*: Glucuronidation and sulfation of *p*-nitrophenol as a model compound in rat. *J. Pharmacobiodyn.* **1982**, *5*, 811–817.

40. Laemmli, U.K. Cleavage of structural proteins during assembly of head of bacteriophage-T4. *Nature* **1970**, *227*, 680–685.

**Sample Availability:** *Sample Availability*: Not available.

# Chemically Synthesized Glycosides of Hydroxylated Flavylium Ions as Suitable Models of Anthocyanins: Binding to Iron Ions and Human Serum Albumin, Antioxidant Activity in Model Gastric Conditions

Sheiraz Al Bittar, Nathalie Mora, Michèle Loonis and Olivier Dangles

**Abstract:** Polyhydroxylated flavylium ions, such as 3',4',7-trihydroxyflavylium chloride (P1) and its more water-soluble 7-$O$-β-D-glucopyranoside (P2), are readily accessible by chemical synthesis and suitable models of natural anthocyanins in terms of color and species distribution in aqueous solution. Owing to their catechol B-ring, they rapidly bind $Fe^{III}$, weakly interact with $Fe^{II}$ and promote its autoxidation to $Fe^{III}$. Both pigments inhibit heme-induced lipid peroxidation in mildly acidic conditions (a model of postprandial oxidative stress in the stomach), the colorless (chalcone) forms being more potent than the colored forms. Finally, P1 and P2 are moderate ligands of human serum albumin (HSA), their likely carrier in the blood circulation, with chalcones having a higher affinity for HSA than the corresponding colored forms.

Reprinted from *Molecules*. Cite as: Al Bittar, S.; Mora, N.; Loonis, M.; Dangles, O. Chemically Synthesized Glycosides of Hydroxylated Flavylium Ions as Suitable Models of Anthocyanins: Binding to Iron Ions and Human Serum Albumin, Antioxidant Activity in Model Gastric Conditions. *Molecules* **2014**, *19*, 20709–20730.

## 1. Introduction

Anthocyanins are responsible for the colors of numerous flowers, fruits, vegetables and even cereals. Colors expressed by anthocyanins vary from red to blue depending on pH, self-association (especially, in the case of acylated anthocyanins) and interactions with metal ions ($Al^{3+}$, $Fe^{3+}$, $Mg^{2+}$) and phenolic copigments, such as flavones, flavonols and hydroxycinnamic acids [1–5]. Through their coloring properties, anthocyanins strongly contribute to food quality and appeal to consumers. They may also contribute to the health benefits of diets rich in plant products [6]. For instance, anthocyanins with an electron-rich B-ring, in particular an $o$-dihydroxylated B-ring (catechol), are intrinsically good antioxidants, either by acting as electron donors to reactive oxygen species or by chelating transition metal ions (potential inducers of oxidative stress) as inert complexes [7,8].

351

Dietary anthocyanins can be partly absorbed along the gastrointestinal (GI) tract (from stomach to colon) [9] but have an overall poor bioavailability in humans, at least based on the very low circulating concentrations of the native forms and their conjugates [10]. In fact, anthocyanins may be relatively unstable in the intestine [11–14] and, as polyphenols in general [15], undergo an extensive catabolism by intestinal glucosidases and by the enzymes of the colonic microbiota. In particular, hydrolysis of the anthocyanins' glycosidic bond at C3-OH, which releases highly unstable anthocyanidins, must be a critical step toward cleavage of the C-ring. Consequently, a large part of the health benefits of anthocyanins is expected to be mediated by their degradation products and their conjugates [16].

On the other hand, anthocyanins, as ubiquitous dietary polyphenols, can accumulate under their native form in the GI tract and possibly protect dietary lipids and proteins against oxidation. Indeed, in gastric conditions (high $O_2$ content, acidic pH), lipid peroxidation induced by dietary heme iron could be very significant but efficiently inhibited by polyphenols [17–21]. Through reduction of high-valence heme iron, polyphenols could preserve the nutritional value of the dietary bolus and prevent the formation of toxic lipid peroxidation products. This hypothesis of an early antioxidant protection by dietary polyphenols, including anthocyanins, is gaining evidence from *in vivo* studies [22].

Once in the general blood circulation, polyphenols and their metabolites, typically bound to human serum albumin (HSA) [23,24], are delivered to tissues for specific biological effects [15].

3-Deoxyanthocyanidins and their glucosides have been identified in cereals such as red sorghum [25]. Lacking the C3-OH group of anthocyanidins, which is critically involved in their degradation, 3-deoxyanthocyanidins express more stable colors [26]. They are also promising pigments in terms of potential health benefits, expressed by antioxidant and cell-specific effects [27–29]. So far, little is known about their bioavailability but it may be speculated that it is higher than that of anthocyanins, as 3-deoxyanthocyanidins are probably less prone to catabolism in the GI tract. For future development as food ingredients, it is also noteworthy that mutagenesis-assisted breeding can dramatically increase 3-deoxyanthocyanidin accumulation in sorghum leaves [30].

Interestingly, 3-deoxyanthocyanidins and their glucosides, in particular simplified analogs lacking the C5-OH group, are far more accessible by chemical synthesis than even the simplest anthocyanins. In a previous paper [31], we reported the chemical synthesis, structural transformations, aluminium binding and radical-scavenging (DPPH test) of 3',4',7-trihydroxyflavylium chloride (P1) and its 7-*O*-β-D-glucoside (P2) (Figure 1). In this work, their capacity to bind iron ions and inhibit heme-induced lipid peroxidation in mildly acidic conditions (a model of postprandial oxidative stress in the stomach) will be quantitatively studied as well as their affinity for HSA, their likely carrier in the blood circulation. In each model, the activity of the colored and colorless forms will be discriminated.

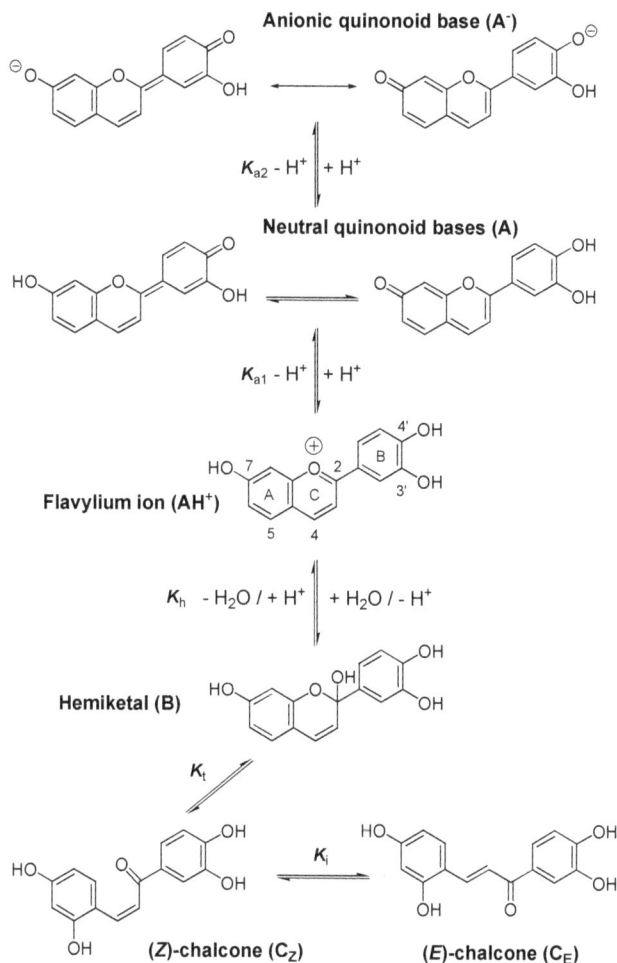

**Figure 1.** Structural transformations of the 3',4',7-trihydroxyflavylium ion (P1).

The aim of this work is to emphasize, through detailed quantitative physico-chemical analyses, that readily available 3-deoxyanthocyanidins—a relatively overlooked class of natural pigments—are interesting colorants and antioxidants deserving further examination for future applications.

## 2. Results and Discussion

As a general comment, interpretation of our data rests on the well-established scheme of structural transformation for the flavylium ion of anthocyanins (Figure 1) [32,33]. However, flavylium ions lacking the glycosyloxy substituent of natural anthocyanins at C3 display some peculiarities: dehydration of hemiketal B into the highly planar flavylium ion is faster as well as its sequential conversion into $C_Z$ and $C_E$. $C_E$ is also much more stable than $C_Z$ ($K_i \approx 75$ for P2 [32]) whereas the two isomers display close stability with natural anthocyanins. Consequently, B and $C_Z$ can be regarded as transient (non-accumulating) intermediates in the overall conversion of the flavylium ion into the corresponding (E)-chalcone.

### 2.1. Iron-Pigment Binding

Together with copigmentation and self-association, metal-anthocyanin binding is one of the most important mechanisms for varying and stabilizing natural colors [1]. In our previous work [31], both P1 and P2 were shown to bind $Al^{III}$ in mildly acidic solutions, thereby forming chelates having a quinonoid chromophore as the result of the simultaneous loss of the two protons at C3'-OH and C4'-OH. Interestingly, the $Al^{III}$-P2 complex is more resistant than the $Al^{III}$-P1 complex toward water addition leading to the free (unbound) (E)-chalcone.

In this work, P1 and P2 are compared for their ability to bind $Fe^{III}$ and $Fe^{II}$. As iron ions take part in the production of reactive oxygen species (e.g., via the Fenton reaction [34]), their binding as redox-inert chelates can be considered a potential antioxidant mechanism. Moreover, transition metal ions such as iron and copper ions being present in our diet [35], metal-anthocyanin binding could also take place in the upper GI tract (in mildly acidic conditions) and modulate the properties and stability of anthocyanins in this biological site.

### 2.1.1. Pigment P1

The successive addition of P1 and $Fe^{III}$ (0.5–5 equiv.) to a pH 4 acetate buffer results in the fast decay of A(470 nm) and the development of a broad visible band in the range 450–750 nm with an absorption maximum at *ca.* 510 nm (Figures 2 and 3). Those spectral changes can be interpreted by the formation of a P1-$Fe^{III}$ complex having a quinonoid chromophore that acts as a donor in a charge transfer interaction with the $Fe^{III}$ empty orbitals.

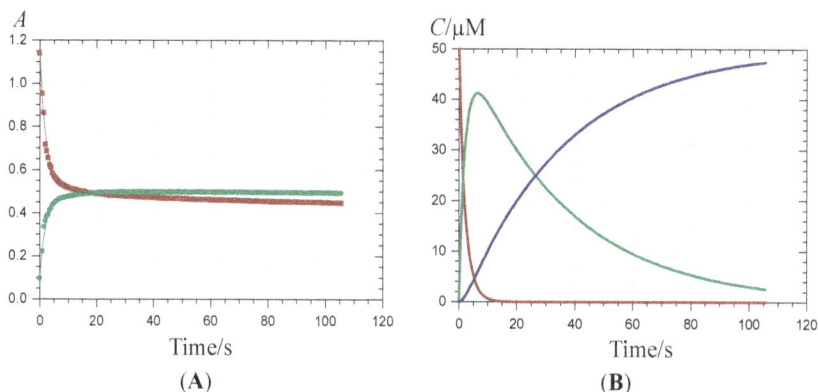

**Figure 2.** Kinetics of $Fe^{III}$-P1 binding (pH 4 acetate buffer, 25 °C, 4 equiv. $Fe^{III}$). **(A)** Time-dependence of the visible absorbance at 470 (■) and 620 nm (●); **(B)** time-dependence of the free pigment (──) and the kinetic (──) and thermodynamic (──) complexes.

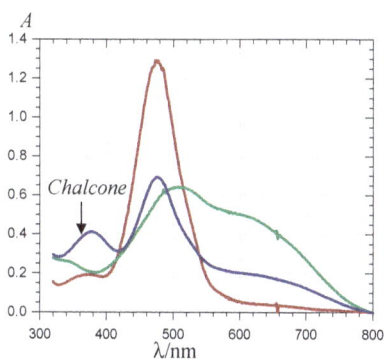

**Figure 3.** UV-visible spectra of P1 (──), the P1-$Fe^{III}$ complex (──, *ca.* 30 s after addition of 2 equiv. $Fe^{III}$) and the complex formed *ca.* 400 s after addition of 5 equiv. $Fe^{II}$ (──) (pH 4 acetate buffer, 25 °C, pigment concentration = 50 µM).

Over 2 min, free chalcone formation (typical absorption at $\lambda_{max}$ = 375 nm) is negligible (confirmed by HPLC-MS analysis), even when P1 is in excess (0.5 equiv. $Fe^{III}$). When compared with $Al^{III}$-P1 binding [31], $Fe^{III}$-P1 binding is much faster and quasi-irreversible as the final maximal absorbance at 620 nm is reached with 1 equiv. $Fe^{III}$. The time dependence of A(470 nm) and A(620 nm) can be interpreted by the fast formation of a first complex (rate constant of binding $k_b$) followed by its slower first-order conversion into a second complex (rate constant of rearrangement $k_r$) (Figure 2). A simultaneous curve-fitting of both curves (Equations (1)–(3)) gives access to the corresponding rate constants and molar absorption coefficients (Table 1).

$$-\frac{d}{dt}[Fe^{III}] = -\frac{d}{dt}[L] = k_b[Fe^{III}][L] \tag{1}$$

$$\frac{d}{dt}[Fe^{III}L_1] = k_b[Fe^{III}][L] - k_r[Fe^{III}L_1] \tag{2}$$

$$\frac{d}{dt}[Fe^{III}L_2] = k_r[Fe^{III}L_1] \tag{3}$$

The $k_r$ values, which suggest a quasi-total consumption of the first complex over 2 min, are much higher than those obtained for water addition to free P1 (chalcone formation) and its $Al^{III}$ complex [31]. Moreover, at the end of the kinetics, the broad absorption band, almost covering the visible spectrum and still well visible after several hours, is not compatible with a $Fe^{III}$-chalcone complex. Addition of $Fe^{III}$ (5 equiv.) to an equilibrated solution of P1 in which $C_E$ is the dominant species shows the binding of the minor colored forms with little impact on the chalcone band over 2 min (data not shown), thus indicating that $C_E$ does not bind $Fe^{III}$ in mildly acidic solution. The hypothesis of $Fe^{III}$ reduction and concomitant oxidation of P1 is also not consistent with the spectrum obtained after acidification to pH < 2 (total recovery of free P1) and the HPLC-MS analysis (no oxidation product detected). Finally, one can propose the formation of a kinetic product (complex 1) evolving into a thermodynamic product (complex 2), possibly by additional coordination of acetate ions. Similar kinetic patterns were previously observed with other phenols in their binding to $Fe^{III}$ [36]. Thus, starting from the flavylium ion, addition of $Fe^{III}$ results in the fast binding of the colored forms (in fast acid-base equilibrium, collectively noted L in Equations (1)–(3)). Concomitantly, the fraction of free flavylium in solution is greatly lowered so that water addition (and subsequent chalcone formation) is quenched.

When $Fe^{II}$ is added in an equimolar concentration, a slow decay of A(470 nm) paralleled by a slow increase of A(375 nm) is observed. As the corresponding absorption bands are not shifted in comparison to free P1, it can be concluded that $Fe^{II}$-P1 binding is negligible and the spectral changes are fully ascribed to water addition to free P1 with concomitant chalcone formation. A double first-order curve-fitting at 470 and 375 nm yields: $k_h{}^{obs} = 140 \; (\pm 1) \times 10^{-5} \; s^{-1}$, in reasonable agreement with the value in the absence of $Fe^{II}$ ($k_h{}^{obs} \approx 120 \times 10^{-5} \; s^{-1}$, half-life of free P1 at pH 4 ≈ 10 min). However, addition of an excess $Fe^{II}$ (5 equiv.) causes the slow development of a broad visible band in the range 500–750 nm, again with no shift in the band at 470 nm (in contrast to $Fe^{III}$, see Figures 3 and 4). Moreover, a relatively fast accumulation of free chalcone reaching saturation after 300–400 s is also observed. In a pH 4 acetate buffer, $Fe^{II}$ titration (ferrozine test, data not shown) shows that $Fe^{II}$ autoxidation is negligible. However, the broad visible band appearing in the range 500–750 nm is evidence for the formation of a $Fe^{III}$-P1 complex [36,37].

Thus, it is proposed that a weak $Fe^{II}$–P1 binding occurs that promotes a slow $Fe^{II}$ autoxidation (apparent first-order rate constant $k_{autox}$) without totally quenching water addition to P1. Then, the $Fe^{III}$–P1 slowly accumulates. Using this kinetic model (detailed below with P2), the corresponding rate constants can be estimated (Table 1) and the different concentrations plotted as a function of time (Figure 4).

**Table 1.** Kinetic analysis of P1-$Fe^{III}$ binding. Simultaneous curve-fitting of the A(470 nm) and A(620 nm) *vs.* time curves according to a simple model assuming irreversible 1:1 binding (rate constant $k_b$) followed by first-order rearrangement of complex 1 into complex 2 (rate constant $k_r$) (pH 4 acetate buffer, 25 °C, pigment concentration = 50 µM).

| $M_t/L_t$, $\lambda$/nm | $k_b$/M$^{-1}\cdot$s$^{-1}$ | $10^3 k_r$/s$^{-1}$ | $\varepsilon_1$/M$^{-1}\cdot$cm$^{-1}$ | $\varepsilon_2$/M$^{-1}\cdot$cm$^{-1}$ |
|---|---|---|---|---|
| 1, 470 | 17,890 ($\pm$180) | 4 ($\pm$2) | 10,340 | 7910 |
| 620 | | | 12,270 | 10,810 |
| 2, 470 | 7190 ($\pm$250) | 44 ($\pm$4) | 13,190 | 10,320 |
| 620 | | | 12,180 | 12,780 |
| 3, 470 | 4450 ($\pm$60) | 16 ($\pm$3) | 12,010 | 10,300 |
| 620 | | | 13,160 | 13,060 |
| 4, 470 | 2670 ($\pm$30) | 29 ($\pm$2) | 10,360 | 8900 |
| 620 | | | 9710 | 9930 |
| 5, 470 | 3370 ($\pm$50) | 59 ($\pm$2) | 12,830 | 10,300 |
| 620 | | | 10,680 | 11,580 |
| 5, 470 [a] | | | 10,700 | |
| 630 | 250 ($\pm$30) | - | 11,800 | - |
| 375 | | | 4200 | |

Notes: [a] $Fe^{II}$, apparent first-order autoxidation of $Fe^{II}$: $k_{autox}$ = 58 ($\pm$6) $\times$ 10$^{-5}$ s$^{-1}$; chalcone formation: $k_h{}^{obs}$ = 95 ($\pm$4) $\times$ 10$^{-5}$ s$^{-1}$, $\varepsilon_{CE}$ = 33,800 M$^{-1}\cdot$cm$^{-1}$ at 375 nm.

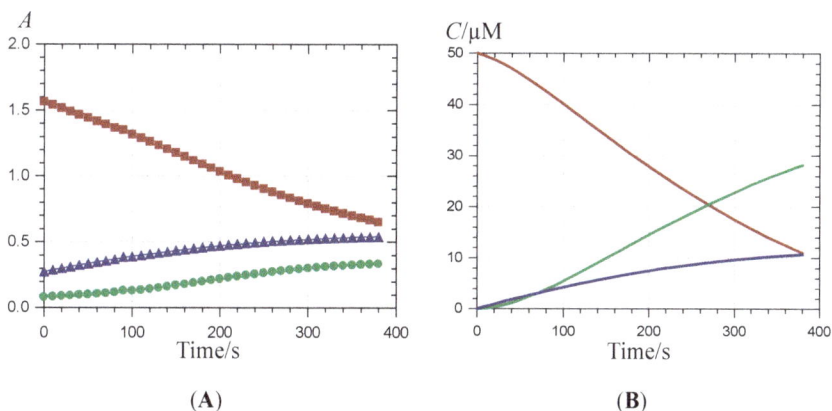

**Figure 4.** Kinetics of $Fe^{II}$–P1 binding (pH 4 acetate buffer, 25 °C, 5 equiv. $Fe^{II}$). (A) time-dependence of the visible absorbance at 470 (■), 620 (●) and 375 nm (▲); (B) time-dependence of the free pigment (━), the metal complex (━) and the free chalcone (━).

357

## 2.1.2. Pigment P2

The successive addition of P2 and $Fe^{III}$ (0.5–5 equiv.) to a pH 4 acetate buffer results in spectral changes (Figure 5) that are close to the ones observed with P1. They are consistent with the formation of a P2-$Fe^{III}$ complex having a quinonoid chromophore that acts as a donor in a charge transfer interaction with $Fe^{III}$.

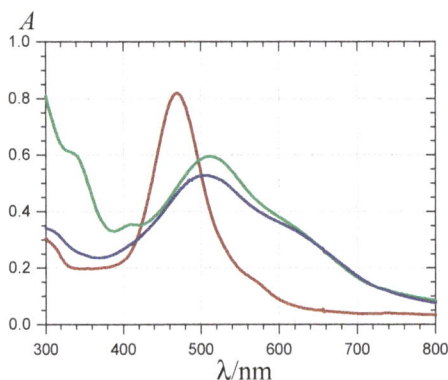

**Figure 5.** UV-visible spectra of P2 (──), the P2-$Fe^{III}$ complex (━━, *ca.* 10 s after $Fe^{III}$ addition) and the complex formed *ca.* 10 min after addition of $Fe^{II}$ (──) (pH 4 acetate buffer, 25 °C, pigment concentration = 50 μM, iron-P2 molar ratio = 5).

Unlike $Al^{III}$ [31], $Fe^{III}$ binds P2 even more rapidly than P1, so that an accurate kinetic analysis is not possible by conventional UV-visible spectroscopy. However, assuming irreversible 1:1 binding, a lower limit can be proposed for the second-order rate constant of P2-$Fe^{III}$ binding: $k_b > 5 \times 10^3$ $M^{-1} \cdot s^{-1}$. Free chalcone formation is negligible (confirmed by HPLC-MS analysis), even when P2 is in excess (0.5 equiv. $Fe^{III}$). However, a slight decay of A(650 nm) is observed with 0.5–1 equiv. $Fe^{III}$. Although fast, $Fe^{III}$-P2 binding is reversible and the final maximal absorbance at 650 nm is only reached with an excess $Fe^{III}$ (*ca.* 5 equiv.).

The plot of $\Delta A = A_{max} - A_0$ (at 650 nm) as a function of the total metal concentration $M_t$ can be successfully analyzed according to a 1:1 reversible binding model (Equations (4) and (5)), thereby allowing the determination of the $Fe^{III}$-P2 binding constant: $K_b = 21 (\pm 6) \times 10^3$ $M^{-1}$, $\Delta\varepsilon = \varepsilon_{complex} - \varepsilon_{P2} = 6500 (\pm 500)$ $M^{-1} \cdot cm^{-1}$ ($r = 0.995$). This $K_b$ value is identical to the one estimated for the $Al^{III}$-P2 complex [31]. Thus, the two trivalent hard metal cations $Fe^{III}$ and $Al^{III}$ have the same affinity for the P2 catechol nucleus. However, the $Fe^{III}$-P2 binding is much faster, the equilibrium being reached in a few seconds *vs.* several minutes with $Al^{III}$.

$$\Delta A = \Delta\varepsilon(Fe^{III}_{total} - [Fe^{III}]) \qquad (4)$$

$$Fe^{III}_{total} = [Fe^{III}](1 + \frac{K_b L_{total}}{1 + K_b [Fe^{III}]})$$ (5)

$L_{total}$: total ligand concentration, $Fe^{III}_{total}$: total metal concentration, $K_b$: metal-pigment binding constant, $\Delta\varepsilon = \varepsilon_{FeL}^{650} - \varepsilon_L^{650}$.

The observation that $Fe^{III}$-P2 binding is faster than $Fe^{III}$-P1 binding may be ascribed to different binding species in solution at pH 4. Indeed, the higher acidity of the P1 flavylium ion [31] probably indicates that P1 deprotonation at C7-OH is more favorable than at C4'-OH while P2 deprotonation can only occur at C4'-OH (Figure 6). Thus, $Fe^{III}$-P1 binding probably requires a thermodynamically unfavorable change in quinonoid tautomer that is not needed with P2.

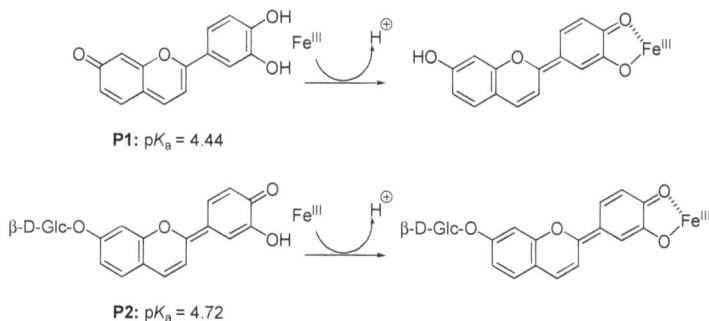

**P1**: $pK_a = 4.44$

**P2**: $pK_a = 4.72$

**Figure 6.** Iron-pigment binding.

Like P1, P2 apparently binds $Fe^{II}$ much more slowly than $Fe^{III}$. For instance, while $Fe^{III}$-P2 binding reaches equilibrium in a few seconds, $Fe^{II}$-P2 binding requires *ca.* 4 min with 5 equiv. $Fe^{II}$ (Figure S1). With 1 equiv. $Fe^{II}$, the equilibrium is not even achieved after 10 min. Interestingly, with 5 equiv. iron, the final spectra characteristic of the complexes are very close, except for a strong absorption band developing below 360 nm for the $Fe^{III}$-P2 complex (shoulder at 340 nm) that is characteristic of free $Fe^{III}$ (Figure 4). It can thus be proposed that the same $Fe^{III}$-P2 complex is ultimately formed after addition of $Fe^{III}$ or $Fe^{II}$. In other words, $Fe^{II}$ slowly binds P2 with simultaneous conversion into $Fe^{III}$, while free $Fe^{II}$ in excess remains stable in solution. In particular, the broad absorption band beyond 600 nm (not observed with the $Al^{III}$-P2 complex) is characteristic of a catechol-to-$Fe^{III}$ charge transfer interaction.

The curves showing the time dependence of A(470 nm) and A(650 nm) display short lag phases (Figure S1) suggesting that a preliminary slow autoxidation of $Fe^{II}$ (apparent first-order rate constant $k_{autox}$) must take place to trigger the binding (second-order rate constant $k_b$). Hence, both curves could be fitted against the following model (Equations (6)–(10), Table 2).

$$-\frac{d}{dt}[Fe^{II}] = k_{autox}[Fe^{II}] \tag{6}$$

$$\frac{d}{dt}[Fe^{III}] = k_{autox}[Fe^{II}] - k_b[Fe^{III}][L] \tag{7}$$

$$-\frac{d}{dt}[L] = k_b[Fe^{III}][L] + k_h^{obs}[L] \tag{8}$$

$$\frac{d}{dt}[LFe^{III}] = k_b[Fe^{III}][L] \tag{9}$$

$$\frac{d}{dt}[C_E] = k_h^{obs}[L] \tag{10}$$

**Table 2.** Kinetic analysis of the spectral changes following addition of $Fe^{II}$ to a P2 solution (pH 4 acetate buffer, 25 °C, pigment concentration = 50 μM).

| $M_t/L_t$, $\lambda/nm$ [a] | $10^5 k_{autox}/s^{-1}$ | $k_b/M^{-1} \cdot s^{-1}$ | $\varepsilon_{ML}/M^{-1} \cdot cm^{-1}$ |
|---|---|---|---|
| 0.5, 470 ($r = 0.9978$) | 215 ($\pm 2$) | n.d. [b] | 8800 [c] |
| 650 ($r = 0.9975$) | 13.7 ($\pm 0.2$) [d] | | 7200 [c] |
| 1, 470 ($r = 0.9985$) | 169 ($\pm 1$) | n.d. [b] | 8800 [c] |
| 650 ($r = 0.9985$) | 13.3 ($\pm 0.4$) [d] | | 7200 [c] |
| 2, 470 ($r = 0.9992$) | 181 ($\pm 5$) | 473 ($\pm 34$) | 8870 |
| 650 ($r = 0.9993$) | | | 7320 |
| 3, 470 ($r = 0.9998$) | 154 ($\pm 2$) | 663 ($\pm 29$) | 8850 |
| 650 ($r = 0.9996$) | | | 7110 |
| 4, 470 ($r = 0.9998$) | 141 ($\pm 2$) | 593 ($\pm 23$) | 8670 |
| 650 ($r = 0.9999$) | | | 7070 |
| 5, 470 ($r = 0.9988$) | 163 ($\pm 7$) | 785 ($\pm 75$) | 8810 |
| 650 ($r = 0.9994$) | | | 7340 |

Notes: [a] Each $A$ vs. time curve is a mean of 2 experimental curves; [b] Steady-state assumed for $Fe^{III}$; [c] Set constant; [d] Apparent rate constant of water addition ($k_h^{obs}$).

With an excess $Fe^{II}$, chalcone formation can be neglected with P2 ($k_h^{obs} = 0$), while it is detectable with P1 (Figures 3 and 4).

In summary, $Fe^{III}$ rapidly binds both P1 and P2 in mildly acidic solutions, thereby quenching their conversion into the corresponding chalcones. With P2, the binding is faster but reversible. By contrast, P1 and P2 only weakly interact with $Fe^{II}$, thereby promoting its autoxidation with subsequent fast binding of $Fe^{III}$.

## 2.2. Pigment-Serum Albumin Binding

HSA, the major plasma protein (*ca.* 0.6 mM), is responsible for the transport of a large variety of ligands [38], including drugs and dietary components such as fatty acids and polyphenols [23,24,39]. The heart-shaped structure of HSA consists of three helical domains I (1–195), II (196–383) and III (384–585), each being divided into sub-domains A and B [38]. The main binding sites of drugs and polyphenols are site 1 and site 2 (respectively located in sub-domains IIA and IIIA), which consist in hydrophobic pockets lined by positively charged aminoacid residues (Arg, Lys).

Whereas glycoside hydrolysis prior to intestinal absorption seems the rule with polyphenols in general, native anthocyanins (glycosides) have been detected in the blood circulation, although in very low (sub-micromolar) concentration [10]. Moreover, under physiological conditions, delphinidin, cyanidin and pelargonidin 3-$O$-β-D-glucosides have been reported to bind to HSA site 1 with thermodynamic binding constants in the range 69–144 $\times$ $10^3$ $M^{-1}$ [40]. So far, no work has discriminated the colored and colorless forms by their affinity for HSA, despite the fact that the colorless forms are expected to largely prevail at equilibrium in neutral conditions.

In this study, pigment–HSA binding was first evidenced by UV-visible spectroscopy. For instance, the visible band of P1 at pH 7.4 shifts from 540–570 nm when an excess HSA (2 equiv.) is added (Figure 7). However, this is not so with P2 (unchanged $\lambda_{max}$ = 530 nm). The bathochromic shift specifically observed for P1 suggests a role for the free C7-OH group. At pH 7.4, the anionic quinonoid form makes a substantial contribution. In the case of P1, the binding to HSA could even favor the formation of the anionic quinonoid base, in agreement with the high density of positive charges (protonated Lys and Arg residues) present in sub-domain IIA, the typical binding site of flavonoids [39]. To check this hypothesis, the pH dependence of the visible spectrum of P1 around neutrality was evaluated in the presence and absence of HSA. Very similar titration curves were obtained in agreement with a $pK_{a2}$ value of $ca.$ 7.1 (Table 3, Figure S2). Thus, binding to HSA does not significantly shift the equilibrium between the neutral and anionic quinonoid bases. Hence, the HSA-induced bathochromic shift may be rather ascribed to perturbation in the molecular orbitals specifically involved in the visible band, e.g., the HOMO of the anionic quinonoid base (due to possible charge transfer interactions) with no impact on the global stability.

After equilibration for $ca.$ 24 h, the titration curves were modified by the gradual appearance of the chalcone (Figure S2). The residual color is approximately the same in the absence or presence of HSA. Thus, HSA has a minor impact on the quinonoid bases-chalcone equilibrium, which is indicative that the different forms have close affinities for the protein. The residual color at pH 7.4 is consistent with a $K_i = (C_E)/(A)$ value of $ca.$ 10, in agreement with the $pK_{a1}$ and $pK'_h$ values previously determined for P1 ([31], 4.44 and 3.45, respectively). From the $K_i$ and $K_{a2}$ values, a distribution diagram of the different P1 species can be plotted around neutrality in the presence or absence of HSA (Figure S3).

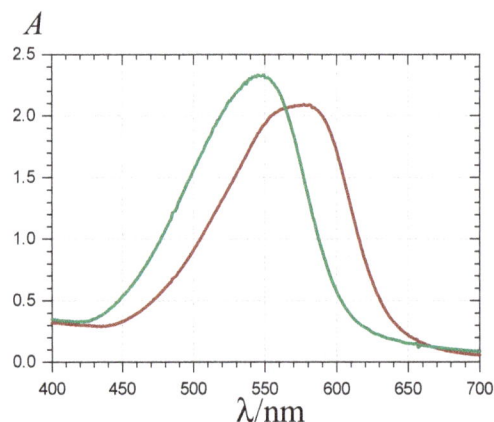

**Figure 7.** UV-visible spectra of P1 (━) and the P1-HSA complex (━) (pH 7.4 phosphate buffer, 25 °C, pigment concentration = 50 μM, HSA-P1 molar ratio = 2).

**Table 3.** Kinetic and thermodynamic parameters for the structural transformations of P1 and P2 in neutral conditions with and without HSA.

|  | P1 | P2 |
|---|---|---|
| $pK_{a2}$, $r_a$ (550 nm), no HSA | 7.12 ($\pm$0.05), 6.3 ($\pm$0.6) [a] | n.a. [b] |
| $pK_{a2}$, $r_A$ (580 nm), 5 equiv. HSA | 7.11 ($\pm$0.04), 3.1 ($\pm$0.1) [a] | n.a. [b] |
| $k_h^{obs}$ (s$^{-1}$), 530 nm, no HSA | n.a., too slow<br>*ca.* $-10\%$ color loss after 45 min | 88 ($\pm$1) $\times$ 10$^{-5}$ [c] |
| $k_h^{obs}$ (s$^{-1}$), 530 nm, 2 equiv. HSA | n.a., too slow<br>*ca.* $-10\%$ color loss after 45 min | 81 ($\pm$1) $\times$ 10$^{-5}$ [c] |

Notes: [a] From the curve-fitting of the *A vs.* pH curves at equilibrium ($r_A$ = ratio of the molar absorption coefficients of the anionic to neutral quinonoid bases); [b] No proton loss in the pH range 6–8, total conversion of colored forms into chalcone; [c] From a first-order curve-fitting of the color loss at pH 7.4.

In the case of P2, the situation is simpler as no anionic quinonoid base can form. Monitoring the decay of the color over time shows that the apparent first-order rate constant of water addition ($k_h^{obs}$) is only weakly affected by HSA (Table 3). Moreover, whether HSA is present or not, the color loss can be considered complete. Thus, the quinonoid base concentration at equilibrium is negligible ($K_i > 10$).

For an accurate estimation of the corresponding binding constants, the pigment-HSA binding was investigated by fluorescence spectroscopy. The intrinsic HSA fluorescence at 340 nm (excitation at 295 nm) is due to its single Trp residue (Trp-214) located in sub-domain IIA. Its strong quenching by P1, P2 and their chalcones (Figure 8) is evidence that the binding actually takes place to or near this site.

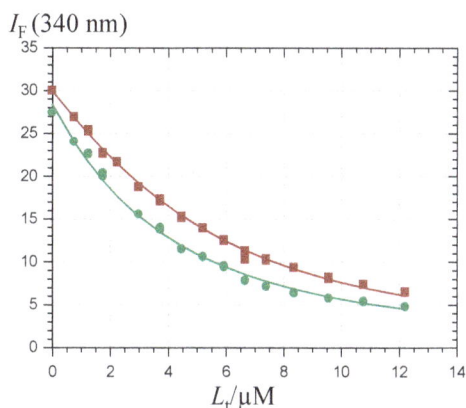

**Figure 8.** Quenching of the HSA fluorescence by the P1 quinonoid bases (■) and chalcone (●). HSA concentration = 2 µM, pH 7.4 phosphate buffer, 37 °C, excitation at 295 nm.

The excitation wavelength was selected so as to maximize the fluorescence of the single Trp residue of HSA. However, the pigments, especially in their chalcone form, substantially absorb light at the excitation (295 nm) or/and emission (340 nm) wavelengths so that an inner filter correction is necessary. Hence, the protein fluorescence intensity is expressed in Equation (11).

$$I_F = f_P [P] \exp(-\varepsilon_L l\, L_t) \tag{11}$$

$$L_t = [L] (1 + K_b [P]) \tag{12}$$

$$P_t = [P] (1 + K_b [L]) \tag{13}$$

In Equation (11), $f_P$ is the molar fluorescence intensity of HSA and $\varepsilon_L$ stands for the sum of the molar absorption coefficients of the ligand at the excitation and emission wavelengths (Table 4). Its value is determined independently by UV-visible spectroscopy from a Beer's plot. Finally, $l$ is the mean distance travelled by the excitation light at the site of emission light detection. For the spectrometer used in this work, $l$ is estimated to be 0.65 cm. Beside the expression of $I_F$, the relationships used in the curve-fitting procedures were combinations of the mass law for the complex and mass conservation for the ligand L (pigments) and protein P (Equations (12) and (13), $L_t$: total ligand concentration, $P_t$: total protein concentration).

The $K_b$ values (Table 4) illustrate two major points:

(1)   The Glc moiety strongly destabilizes the complexes, especially for the colored forms ($K_b$ value reduced by a factor 15–16).

(2)   The chalcones, with their open more linear structure, display a higher affinity for HSA ($K_b$ raised by a factor *ca.* 3 for P2) than the corresponding colored forms, although this increase is marginal with P1 in agreement with the investigation by UV-visible spectroscopy. This suggests that the very low circulating concentration of anthocyanins (in comparison to other flavonoids) [10,15] could be partly due to their conversion in colorless forms that may have escaped detection.

**Table 4.** Binding constant ($K_b$) of pigments and their chalcones to HSA (2 µM) in a pH 7.4 phosphate buffer at 25 °C ($n = 2$).

| | $10^3 K_b/M^{-1}$ | $10^6 f_P/M^{-1}$ | $\varepsilon_L/M^{-1}\ cm^{-1}$ [a] | $r$ |
|---|---|---|---|---|
| P1 colored forms | 273 (±7) | 15.5 (±0.1) | 8900 + 5800 | 0.998 |
| P1 chalcone | 344 (±12) [b] | 14.2 (±0.1) | 15,800 + 16,400 [b] | 0.997 |
| P2 colored forms | 17.5 (±0.5) | 14.1 (±0.1) | 3800 + 2800 | 0.999 |
| P2 chalcone | 58.4 (±1.9) | 13.5 (±0.1) | 7200 + 7000 | 0.998 |

Notes: [a] First value at 295 nm (excitation wavelength), second value at 340 nm (emission wavelength); [b] Apparent values including a minor contribution of the residual colored forms present at equilibrium. Assuming a 3:1 chalcone-to-colored forms molar ratio (see Figure S3), the true value for the sole chalcone can be estimated: $K_b = 368 \times 10^3\ M^{-1}$.

Interestingly, the $K_b$ values for anthocyanidin 3-*O*-β-D-glucosides [40] are intermediates between the values for the P2 and P1 colored forms. Thus, P1 is a better HSA ligand than common anthocyanins, while the reverse is true for P2.

### 2.3. Inhibition of the Heme-Induced Peroxidation of Linoleic Acid

Given their poor bioavailability and extensive catabolism [10], anthocyanins are expected to exert their antioxidant activity in humans (in the restricted sense of electron donation to reactive oxygen species involved in oxidative stress) prior to intestinal absorption, *i.e.*, in the gastro-intestinal tract, where they can accumulate in substantial concentrations and in their native forms following the consumption of plant products. On the other hand, in the gastric compartment, acidity, dioxygen and pro-oxidant species present in foods (iron, lipid hydroperoxides, $H_2O_2$) can provide suitable conditions for the oxidation of dietary polyunsaturated acids (PUFAs) in postprandial conditions [17–22]. This oxidation results in a loss of essential lipids and in the accumulation of potentially toxic lipid oxidation products. These lipid hydroperoxides and aldehydes can also alter dietary proteins and may even contribute to increasing the concentration of circulating minimally modified lipoproteins that are more prone to further oxidation and take part in atherogenesis. Based on simple *in vitro* models, our works suggest that heme-induced lipid oxidation is particularly fast in the first period of gastric digestion (pH 4–6) but efficiently inhibited by plant antioxidants (polyphenols, α-tocopherol, carotenoids) [21,41,42]. Recently, the pertinence of our model was confirmed by gastric fluid analysis in minipigs [22].

With linoleic acid (LH) as a model of dietary PUFA, conjugated dienes (CDs) are acceptable markers of the early phase of lipid oxidation and can be approximately identified with lipid hydroperoxides (LOOH), the corresponding alcohols (LOH) making only a minor contribution. CD accumulation is easily followed by UV-visible spectroscopy in the presence or absence of antioxidant.

A simple visual comparison of the curves featuring CD accumulation in the presence of a fixed pigment concentration (Figure 9) shows that P2, whether in its colored or chalcone form, is a poorer antioxidant than P1, in agreement with our preliminary investigation of the DPPH radical-scavenging activity [31]. Interestingly, the chalcone forms come up as more potent inhibitors than the corresponding colored forms.

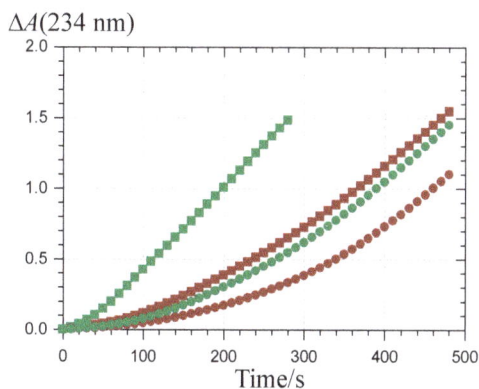

**Figure 9.** Inhibition of the metmyoglobin-induced peroxidation of linoleic acid. Pigment concentration = 2.5 µM, ■: P1 colored forms, ■: P2 colored forms, ●: P1 chalcone, ●: P2 chalcone (pH 5.8 phosphate buffer + Brij®35, 37 °C).

The metmyoglobin-induced peroxidation of linoleic acid is initiated via a $Fe^{III}$-$Fe^{IV}$ redox cycle involving small concentrations of PUFA hydroperoxides inevitably contaminating any PUFA sample [43–45]. As hydrophilic antioxidants, polyphenols typically inhibit lipid peroxidation at the initiation stage by reducing hypervalent heme iron ($Fe^{IV}$), instead of significantly scavenging lipid peroxyl radicals, as lipophilic antioxidants (α-tocopherol, carotenoids) do [21,41,42,46].

The reactions involved in the heme-induced peroxidation of linoleic acid in the presence of an antioxidant are summed up in Figure 10 with the corresponding rate constants.

In the absence of antioxidant, the short lag phase is better reproduced without assuming a steady-state for $Fe^{IV}$. On the other hand, the two initiation rate constants can be taken equal ($k_{i1} = k_{i2}$) so as to restrict the total number of adjustable parameters. Thus, in a first step, the curves of uninhibited lipid peroxidation are analyzed so as to estimate a value for $k_{i1}$ (rate constant of LOOH cleavage by low-valence heme) that will be used in all curve-fitting experiments related to inhibited peroxidation with the following adjustable parameters (see Appendix A for details): $r_2 = \frac{k_p}{\sqrt{2k_t}}$, a measure of PUFA oxidizability, AE $= \frac{k_a}{k_{i2}}$, the antioxidant efficiency and the antioxidant stoichiometry $n$, defined as the number of hypervalent iron species reduced per antioxidant molecule. For all four antioxidants (the two pigments and their chalcones), excellent curve-fittings ($r > 0.999$) were obtained.

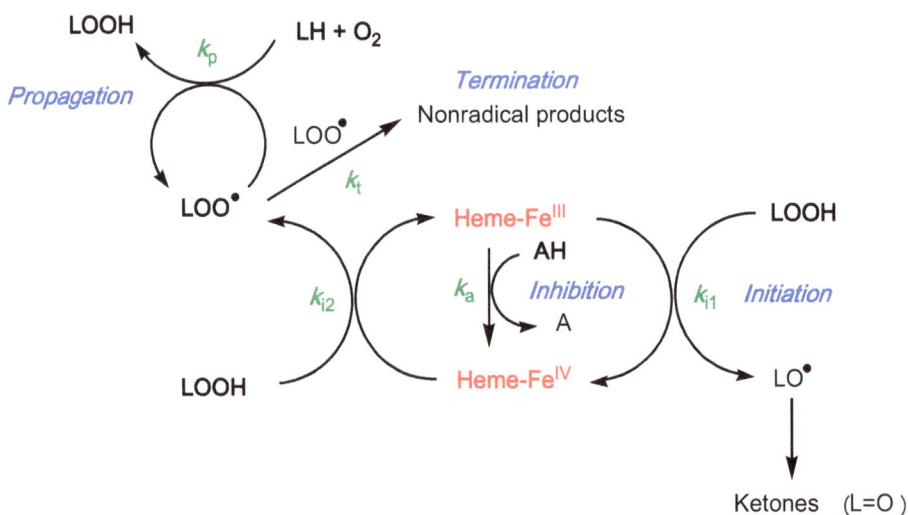

**Figure 10.** Metmyoglobin-induced peroxidation of linoleic acid and its inhibition by polyphenols (LH: PUFA, LOOH: PUFA hydroperoxide, AH: antioxidant, $H^+$ and $HO^-$ ions omitted).

From the parameter values (Table 5), the following comments can be made:

(1)  the antioxidant efficiency, which lies in the range 10–100, does not allow a clear discrimination between antioxidants. Its drift toward lower values when the antioxidant concentration increases suggests that modelling an antioxidant (stoichiometry $n$) as $n$ independent sub-units, each capable of transferring one electron to $Fe^{IV}$ with the same rate constant ($k_a$), may be too crude and/or that antioxidant–metmyoglobin binding can take place (resulting in two populations of free and bound antioxidant molecules with distinct reactivities).

(2)  the antioxidant stoichiometry suggests that a catechol B-ring favors repeated electron transfer to $Fe^{IV}$ (probably through $o$-quinone intermediates) and thus prolonged inhibition. By contrast, the P2 quinonoid base displays a B-ring that is deactivated by the keto group at C4'.

(3)  at high antioxidant concentration, the lipid oxidizability tends to decrease. This drift is ascribed to partial heme degradation and to the accumulation of phenolic oxidation products retaining a weak antioxidant character. The latter point is consistent with the structure of P1 oxidation products already determined by us [47].

**Table 5.** Kinetic analysis of the metmyoglobin-induced peroxidation of linoleic acid. Curve-fitting of the $A$(234 nm) $vs.$ time curves (CD accumulation). Rate constant of lipid hydroperoxide cleavage by metmyoglobin: $k_{i1} = 3 \times 10^3\ M^{-1} \cdot s^{-1}$ (see Figure 10 & Appendix A).

| Pigment/µM | $r_2/M^{-1/2}\,s^{-1/2}$ | AE | $n$ |
|---|---|---|---|
| P1, 0.5 | 2.8 ($\pm$0.1) | 137 ($\pm$16) | 3.0 ($\pm$0.1) |
| 1 | 2.6 ($\pm$0.1) | 40 ($\pm$2) | 2.5 ($\pm$0.1) |
| 1.5 | 2.3 ($\pm$0.1) | 38 ($\pm$5) | 2.5 ($\pm$0.1) |
| 2 | 2.2 ($\pm$0.1) | 29 ($\pm$3) | 3.2 ($\pm$0.2) |
| 2.5 | 2.1 ($\pm$0.1) | 11 ($\pm$1) | 4.0 ($\pm$0.3) |
| P1-$C_E$, 0.5 | 2.7 ($\pm$0.1) | 108 ($\pm$6) | 4.4 ($\pm$0.1) |
| 1 | 2.4 ($\pm$0.1) | 59 ($\pm$3) | 5.6 ($\pm$0.1) |
| 1.5 | 2.3 ($\pm$0.1) | 40 ($\pm$2) | 4.3 ($\pm$0.1) |
| 2 | 2.1 ($\pm$0.1) | 29 ($\pm$1) | 5.2 ($\pm$0.1) |
| 2.5 | 1.9 ($\pm$0.1) | 28 ($\pm$1) | 3.9 ($\pm$0.1) |
| P2, 1.5 | 2.4 ($\pm$0.1) | 95 ($\pm$6) | 0.9 ($\pm$0.1) |
| 2.5 | 2.3 ($\pm$0.1) | 76 ($\pm$14) | 0.5 ($\pm$0.1) |
| 5 | 1.9 ($\pm$0.1) | 15 ($\pm$2) | 1.4 ($\pm$0.1) |
| 6.25 | 1.6 ($\pm$0.1) | 17 ($\pm$2) | 1.2 ($\pm$0.1) |
| 7.5 | 1.0 ($\pm$0.1) | 24 ($\pm$1) | 0.9 ($\pm$0.1) |
| P2-$C_E$, 1.25 | 2.5 ($\pm$0.1) | 31 ($\pm$3) | 3.1 ($\pm$0.1) |
| 2.5 | 2.1 ($\pm$0.1) | 19 ($\pm$1) | 3.5 ($\pm$0.2) |
| 3.75 | 1.6 ($\pm$0.1) | 21 ($\pm$2) | 1.9 ($\pm$0.1) |
| 5 | 1.2 ($\pm$0.1) | 29 ($\pm$2) | 1.3 ($\pm$0.1) |

## 3. Experimental Section

### 3.1. Chemicals

FeSO$_4$, 7H$_2$O (98%) and CH$_3$CO$_2$Na, 3H$_2$O (99%) were purchased from Alfa-Aesar. Fe (NO$_3$)$_3$ (99%) was from Acros. HSA (fraction V, 96%–99%, MW = 66,500 g·mol$^{-1}$), Na$_2$HPO$_4$, 7H$_2$O, NaH$_2$PO$_4$, 2H$_2$O, polyoxyethyleneglycol 23 lauryl ether (Brij$^{\circledR}$35), (9Z, 12Z)-octadecadienoic acid (linoleic acid >99%), myoglobin from equine heart (type II, MW $ca.$ 17,600 g·mol$^{-1}$) were from Sigma-Aldrich. Phosphate and acetate buffers were prepared with non-mineralized water C-23597 405 purchased from VWR to limit metal contamination. 3',4',7-Trihydroxyflavylium (P1) and its 7-O-β-D-glucoside (P2) were chemically synthesized as described in our previous work [31].

## 3.2. UV-Spectroscopy

An Agilent 8453 UV-visible spectrometer equipped with a 1024-element diode-array detector was used to record the absorption spectra over the wavelength range 190–1100 nm. A water thermostated bath was used to control the cell temperature with an accuracy of $\pm 0.1$ °C. The spectroscopic measurements were carried out with a quartz cell of 1 cm optical path length.

## 3.3. Fluorescence Spectroscopy

Steady-state fluorescence spectra were recorded on a thermostated *Safas Xenius* fluorimeter. The excitation and emission slit widths were set at 10 nm. All studies were performed at 37 ($\pm 1$) °C, excitation at 295 nm (HSA Trp residue), emission light collected between 270 and 410 nm.

## 3.4. Iron-Pigment Binding

To 2 mL of 0.1 M acetate buffer at pH 4.0 placed into the spectrometer cell at 25 °C were successively added 50 μL of a freshly prepared 2 mM pigment solution in acidified MeOH (0.1 M HCl) and 50 μL of freshly prepared iron solution in 0.05 M HCl (concentration range: 1–10 mM). The final iron/pigment molar ratios were in the range 0.5–5. Spectra were typically recorded every 0.5 s over 2 min (binding kinetics) or every 15 s over 15 min (complex stability).

## 3.5. Inhibition of the Heme-Induced Peroxidation of Linoleic Acid

The experimental conditions used were adapted from an already published procedure [21]. Metmyoglobin (17.6 mg) was dissolved in 20 mL of phosphate buffer (20 mM, pH 6.8). After filtration through 0.45 μm filter, its concentration was standardized at 50 μM using $\varepsilon = 7700$ $M^{-1} \cdot cm^{-1}$ at 525 nm. Given volumes (20 μL) of daily prepared solutions of linoleic acid (70 mM) in MeOH and pigment (0.05–0.25 mM) were added to 2 mL of Brij®35 (4 mM) solution in phosphate buffer (20 mM, pH 5.8). The concentrated solutions of pigments were (a) prepared in 0.1 M HCl in MeOH for investigating inhibition by the colored forms or (b) incubated in the buffer for 24 h at 37 °C to ensure maximal conversion into the corresponding chalcones. The non-ionic surfactant Brij®35 was chosen for its good stability and very low content of hydroperoxides, which could react with iron. The final concentrations in the cell were 0.7 mM linoleic acid and 0.5–2.5 μM pigment. Oxidation was initiated by adding 20 μL of the 50 μM metmyoglobin solution (final concentration in the cell: 0.5 μM) to the sample under constant magnetic stirring in open air at 37 °C. Each experiment was run in duplicate. Lipid peroxidation was followed by monitoring the concentration of conjugated dienes (CDs) at 234 nm using $\varepsilon = 24 \times 10^3 \ \mathrm{M^{-1} \cdot cm^{-1}}$.

### 3.6. Influence of HSA on the Structural Transformations of Pigments

Aliquots (50 µL) of 2 mM solution of pigments prepared in acidified MeOH (0.1 M HCl) were added to 2 mL of pH 7.4 phosphate buffer (50 mM $Na_2HPO_4$ + 100 mM NaCl) in the presence or absence of HSA (0–2 equiv.) at 37 °C. Spectra were recorded every 30 s over 7000 s. All experiments were carried out twice.

Similar experiments were also carried out after varying the pH of the phosphate buffer in the range 6–8. The spectra were recorded immediately after pigment addition and after equilibration over *ca.* 24 h.

### 3.7. Pigment-HSA Binding

Solutions were prepared daily by dissolving HSA in a pH 7.4 buffer (50 mM phosphate + 100 mM NaCl). Aliquots of a 0.5 mM (P1) or 2 mM (P2) solutions were added via syringe to 2 mL of a 2 µM HSA solution placed in a quartz cell (path length: 1 cm) at 37 °C. The concentrated solutions of pigments were (a) prepared in 0.1 M HCl in MeOH for investigating flavylium–HSA binding (MeOH concentration ⩽2.5%) or (b) incubated in the buffer for 24 h at 37 °C to ensure maximal conversion into the corresponding chalcones.

For investigating flavylium-HSA binding, a single addition was carried out with subsequent recording of the fluorescence spectrum and renewal of the sample for the next pigment concentration. In such conditions, the flavylium-to-chalcone conversion is negligible.

### 3.8. Data Analysis

All curve-fittings were carried out with the Scientist software (MicroMath, Salt Lake City, UT, USA) through least square regression. They yielded optimized values for the parameters implemented in the models (see Text & Appendix A). Standard deviations are reported.

## 4. Conclusions

In this work, 3',4',7-trihydroxyflavylium chloride (P1) and its more water-soluble 7-*O*-β-D-glucopyranoside (P2), come up as suitable models for investigating important properties of anthocyanins: binding of iron ions and serum albumin, inhibition of lipid peroxidation induced by dietary iron in model gastric conditions.

Binding of $Fe^{III}$ is typically fast, especially with the glucoside, and promotes both color variation (due to B-ring deprotonation and additional ligand-to-iron charge transfer) and stabilization (due to the quenching of chalcone formation). Binding of $Fe^{II}$ by itself is not detectable at pH 4 but both pigments promote $Fe^{II}$

autoxidation (followed by the binding of $Fe^{III}$ thus formed), a phenomenon that can be considered protective as $Fe^{II}$ is a potential pro-oxidant through the Fenton reaction. Here again, the glucoside appears superior in accelerating $Fe^{II}$ autoxidation, so that the competing chalcone formation is barely detectable. Binding of serum albumin is weaker with the glucoside, probably because of steric repulsion. It is noteworthy that the chalcone forms a good HSA ligand. In particular, the chalcone glucoside binds HSA three times more tightly than the corresponding colored forms. Consequently, our study suggests that 3-deoxyanthocyanins could partly circulate under their chalcone form in the blood.

Finally, the chalcone forms appear as better inhibitors of heme-induced lipid peroxidation, especially in the case of the glucoside (poorly reactive in its colored form). This prevailing role of the colorless forms in the antioxidant protection afforded by anthocyanins is original and probably important as the physical conditions occurring in the GI tract (temperature, pH, interactions with dietary proteins) could well favor the conversion of the colored forms into the colorless forms.

Overall, 3-deoxyanthocyanins and their chalcones are potentially attractive colorants and antioxidants. Their stability and accessibility by chemical synthesis could foster industrial developments. For instance, iron–3-deoxyanthocyanin chelates could be used in the preparation of colored gels for applications in the food and cosmetic industries [48]. 3-Deoxyanthocyanins could also be developed as natural pH indicators, e.g., for food packaging [49]. They deserve additional investigation of their health-related properties (e.g., their bioavailability).

**Supplementary Materials:** Supplementary materials can be accessed at: http://www.mdpi.com/1420-3049/19/12/20709/s1.

**Author Contributions:** S.A.B., 40% (experimental work & first version of manuscript); N.M, 10% (aid in experimental work); M.L., 10% (aid in experimental work); O.D., 40% (physico-chemical analyses, final version of manuscript and revision).

# Appendix A.

*Mathematical Treatment for the Inhibition of Heme-Induced Lipid Peroxidation*

The reactions and the corresponding rate constants are displayed in Figure 10. The peroxidation rate can be written as:

$$R_p = d(LOOH)/dt = k_p(LOO^\bullet)(LH) - k_{i1}(LOOH)(Fe^{III}) - k_{i2}(LOOH)(Fe^{IV}) = R_p - R_{i1} - R_{i2}$$

The rate of lipid consumption is: $-d(LH)/dt = R_p$

The rate of antioxidant consumption is: $R_a = -d(AH)/dt$

Assuming a steady-state for the lipid peroxyl radicals, we may write: $R_{i2} = 2k_t(LOO^\bullet)^2$

We thus deduce: $R_p = r_2(LH)R_{i2}^{1/2} - R_{i1} - R_{i2}$ with $r_2 = k_p/(2k_t)^{1/2}$

Finally, one has: $-d(Fe^{III})/dt = d(Fe^{IV})/dt = R_{i1} - R_{i2} - R_a$

In the absence of antioxidant, the short lag phase is better reproduced without assuming a steady-state for $Fe^{IV}$. On the other hand, the two initiation rate constants can be taken equal ($k_{i1} = k_{i2}$) so as to restrict the total number of adjustable parameters. We thus estimate $k_{i1}$ (rate constant of LOOH cleavage by low-valence heme): $k_{i1} = 3 \times 10^3 \, M^{-1} \cdot s^{-1}$.

In the presence of an antioxidant, a steady-state for $Fe^{IV}$ can be assumed: $R_{i1} = R_{i2} + R_a$

This relationship can be written as: $k_{i1}(LOOH)(Fe^{III}) = [k_{i2}(LOOH) + k_a(AH)](Fe^{IV})$

We thus deduce: $R_{i2} = \dfrac{R_{i1}}{1 + \frac{AE(AH)}{(LOOH)}}$

with $AE = k_a/k_{i2}$ (antioxidant efficiency at inhibiting initiation).

Using the $k_{i1}$ value previously determined, the curves of inhibited lipid peroxidation are analyzed to estimate the oxidizability $r_2$, antioxidant efficiency $AE$ and stoichiometry $n$. Parameter $n$ is defined as the number of hypervalent iron species reduced per antioxidant molecule. It is implemented in the program by the following initial condition: AH concentration = $n \times$ total antioxidant concentration.

**Conflicts of Interest:** The authors declare no conflict of interest.

## References

1. Yoshida, K.; Mori, M.; Kondo, T. Blue flower color development by anthocyanins: From chemical structure to cell physiology. *Nat. Prod. Rep.* **2009**, *26*, 857–964.
2. Cavalcanti, R.N.; Santos, D.T.; Meireles, M.A.A. Non-thermal stabilization mechanisms of anthocyanins in model and food systems—An overview. *Food Res. Int.* **2011**, *44*, 499–509.
3. Gonzalez-Manzano, S.; Duenas, M.; Rivas-Gonzalo, J.C.; Escribano-Bailon, M.T.; Santos-Buelga, C. Studies on the copigmentation between anthocyanins and flavan-3-ols and their influence in the colour expression of red wine. *Food Chem.* **2009**, *114*, 649–656.
4. Malien-Aubert, C.; Dangles, O.; Amiot, M.J. Color stability of commercial anthocyanin-based extracts in relation to the phenolic composition. Protective effects by intra- and intermolecular copigmentation. *J. Agric. Food Chem.* **2001**, *49*, 170–176.
5. Galland, S.; Mora, N.; Abert-Vian, M.; Rakotomanomana, N.; Dangles, O. Chemical synthesis of hydroxycinnamic acid glucosides and evaluation of their ability to stabilize natural colors via anthocyanin copigmentation. *J. Agric. Food Chem.* **2007**, *55*, 7573–7579.
6. Tsuda, T. Dietary anthocyanin-rich plants: Biochemical basis and recent progress in health benefits studies. *Mol. Nutr. Food Res.* **2012**, *56*, 159–170.
7. Goupy, P.; Bautista-Ortin, A.-B.; Fulcrand, H.; Dangles, O. Antioxidant activity of wine pigments derived from anthocyanins: Hydrogen transfer reactions to the DPPH radical and inhibition of the heme-induced peroxidation of linoleic acid. *J. Agric. Food Chem.* **2009**, *57*, 5762–5770.

8.  Deng, J.; Cheng, J.; Liao, X.; Zhang, T.; Leng, X.; Zhao, G. Comparative study on iron release from soybean (glycine max) seed ferritin induced by anthocyanins and ascorbate. *J. Agric. Food Chem.* **2010**, *58*, 635–641.

9.  Fernandes, I.; de Freitas, V.; Reis, C.; Mateus, N. A new approach on the gastric absorption of anthocyanins. *Food Funct.* **2012**, *3*, 508–516.

10. Kay, C.D. Aspects of anthocyanin absorption, metabolism and pharmacokinetics in humans. *Nutr. Res. Rev.* **2006**, *19*, 137–146.

11. Bouayed, J.; Hoffmann, L.; Bohn, T. Total phenolics, flavonoids, anthocyanins and antioxidant activity following simulated gastro-intestinal digestion and dialysis of apple varieties: Bioaccessibility and potential uptake. *Food Chem.* **2011**, *128*, 14–21.

12. Fleschhut, J.; Kratzer, F.; Rechkemmer, G.; Kulling, S.E. Stability and biotransformation of various dietary anthocyanins *in vitro. Eur. J. Nutr.* **2006**, *45*, 7–18.

13. Vitaglione, P.; Donnarumma, G.; Napolitano, A.; Galvano, F.; Gallo, A.; Scalfi, L.; Fogliano, V. Protocatechuic acid is the major human metabolite of cyanidin-glucosides. *J. Nutr.* **2007**, *137*, 2043–2048.

14. Kay, C.; Kroon, P.; Cassidy, A. The major intestinal metabolites of anthocyanins are unlikely to be conjugates of their parent compounds but metabolites of their degradation products. *Proc. Nutr. Soc.* **2008**, *67*, E309.

15. Del Rio, D.; Rodriguez-Mateos, A.; Spencer, J.P.E.; Tognolini, M.; Borges, G.; Crozier, A. Dietary (poly)phenolics in human health: Structures, bioavailability, and evidence of protective effects against chronic diseases. *Antioxid. Redox Signal.* **2013**, *18*, 1818–1892.

16. Edwards, M.; Czank, C.; Cassidy, A.; Kay, C.D. Vascular bioactivity of anthocyanin degradants: Inhibition of endothelial superoxide production. *Proc. Nutr. Soc.* **2013**, *72*, E228.

17. Kanner, J.; Lapidot, T. The stomach as a bioreactor: Dietary lipid peroxidation in the gastric fluid and the effects of plant-derived antioxidants. *Free Radic. Biol. Med.* **2001**, *31*, 1388–1395.

18. Lapidot, T.; Granit, R.; Kanner, J. Lipid peroxidation by "free" iron ions and myoglobin as affected by dietary antioxidants in simulated gastric fluids. *J. Agric. Food Chem.* **2005**, *53*, 3293–3390.

19. Dangles, O. Antioxidant activity of plant phenols: Chemical mechanisms and biological significance. *Curr. Org. Chem.* **2012**, *16*, 1–23.

20. Lorrain, B.; Dangles, O.; Loonis, M.; Armand, M.; Dufour, C. Dietary iron-initiated lipid oxidation and its inhibition by polyphenols in gastric conditions. *J. Agric. Food Chem.* **2012**, *60*, 9074–9081.

21. Goupy, P.; Vulcain, E.; Caris-Veyrat, C.; Dangles, O. Dietary antioxidants as inhibitors of the heme-induced peroxidation of linoleic acid: Mechanism of action and synergism. *Free Radic. Biol. Med.* **2007**, *43*, 933–946.

22. Gobert, M.; Remond, D.; Loonis, M.; Buffiere, C.; Sante-Lhoutellier, V.; Dufour, C. Fruits, vegetables and their polyphenols protect dietary lipids from oxidation during gastric digestion. *Food Funct.* **2014**, *5*, 2166–2174.

23. Khan, M.K.; Rakotomanomana, N.; Dufour, C.; Dangles, O. Binding of flavanones and their glucuronides and chalcones to human serum albumin. *Food Funct.* **2011**, *2*, 617–626.

24. Galland, S.; Rakotomanomana, N.; Dufour, C.; Mora, N.; Dangles, O. Synthesis of hydroxycinnamic acid glucuronides and investigation of their affinity for human serum albumin. *Org. Biomol. Chem.* **2008**, *6*, 4253–4260.

25. Awika, J.M.; Rooney, L.W.; Waniska, R.D. Anthocyanins from black sorghum and their antioxidant properties. *Food Chem.* **2004**, *90*, 293–301.

26. Yang, L.; Dykes, L.; Awika, J.M. Thermal stability of 3-deoxyanthocyanidin pigments. *Food Chem.* **2014**, *160*, 246–254.

27. Awika, J.M.; Rooney, L.W. Sorghum phytochemicals and their potential impact on human health. *Phytochemistry* **2004**, *65*, 1199–1221.

28. Carbonneau, M.-A.; Cisse, M.; Mora-Soumille, N.; Dairi, S.; Rosa, M.; Michel, F.; Lauret, C.; Cristol, J.-P.; Dangles, O. Antioxidant properties of 3-deoxyanthocyanidins and polyphenolic extracts from Cote d'Ivoire's red and white sorghums assessed by ORAC and *in vitro* LDL oxidizability tests. *Food Chem.* **2014**, *145*, 701–709.

29. Taylor, J.R.N.; Belton, P.S.; Beta, T.; Duodu, K.G. Increasing the utilisation of sorghum, millets and pseudocereals: Developments in the science of their phenolic phytochemicals, biofortification and protein functionality. *J. Cereal Sci.* **2014**, *59*, 257–275.

30. Petti, C.; Kushwaha, R.; Tateno, M.; Harman-Ware, A.E.; Crocker, M.; Awika, J.; DeBolt, S. Mutagenesis breeding for increased 3-deoxyanthocyanidin accumulation in leaves of sorghum bicolor (L.) moench: A source of natural food pigment. *J. Agric. Food Chem.* **2014**, *62*, 1227–1232.

31. Mora-Soumille, N.; al Bittar, S.; Rosa, M.; Dangles, O. Analogs of anthocyanins with a 3',4'-dihydroxy substitution: Synthesis and investigation of their acid-base, hydration, metal binding and hydrogen-donating properties in aqueous solution. *Dyes Pigments* **2013**, *96*, 7–15.

32. Petrov, V.; Gavara, R.; Dangles, O.; al Bittar, S.; Mora-Soumille, N.; Pina, F. Flash photolysis and stopped-flow UV-visible spectroscopy study of 3',4'-dihydroxy-7-O-β-D-glucopyranosyloxyflavylium chloride, an anthocyanin analogue exhibiting efficient photochromic properties. *Photochem. Photobiol. Sci.* **2013**, *12*, 576–581.

33. Pina, F. Chemical applications of anthocyanins and related compounds. A source of bioinspiration. *J. Agric. Food Chem.* **2014**, *62*, 6885–6897.

34. Moran, J.F.; Klucas, R.V.; Grayer, R.J.; Abian, J.; Becana, M. Complexes of iron with phenolic compounds from soybean nodules and other legume tissues: Prooxidant and antioxidant properties . *Free Radic. Biol. Med.* **1997**, *22*, 861–870.

35. Tokalioglu, S.; Gurbuz, F. Selective determination of copper and iron in various food samples by the solid phase extraction. *Food Chem.* **2010**, *123*, 183–187.

36. Nkhili, E.; Loonis, M.; Mihai, S.; el Hajji, H.; Dangles, O. Reactivity of food phenols with iron and copper ions: Binding, dioxygen activation and oxidation mechanisms. *Food Funct.* **2014**, *5*, 1186–1202.

37. Perron, N.R.; Wang, H.C.; DeGuire, S.N.; Jenkins, M.; Lawson, M.; Brumaghim, J.L. Kinetics of iron oxidation upon polyphenol binding. *Dalton Trans.* **2010**, *39*, 9982–9987.

38. Varshney, A.; Sen, P.; Ahmad, E.; Rehan, M.; Subbarao, N.; Khan, R.H. Ligand binding strategies of human serum albumin: How can the cargo be utilized? *Chirality* **2010**, *22*, 77–87.

39. Dufour, C.; Dangles, O. Flavonoid-serum albumin complexation: Determination of binding constants and binding sites by fluorescence spectroscopy. *Biochim. Biophys. Acta* **2005**, *1721*, 164–173.

40. Tang, L.; Zuo, H.; Shu, L. Comparison of the interaction between three anthocyanins and human serum albumin by spectroscopy. *J. Lumin.* **2014**, *153*, 54–63.

41. Vulcain, E.; Goupy, P.; Caris-Veyrat, C.; Dangles, O. Inhibition of the metmyoglobin-induced peroxidation of linoleic acid by dietary antioxidants: Action in the aqueous *vs.* lipid phase. *Free Radic. Res.* **2005**, *39*, 547–563.

42. Sy, C.; Caris-Veyrat, C.; Dufour, C.; Boutaleb, M.; Borel, P.; Dangles, O. Inhibition of iron-induced lipid peroxidation by newly identified bacterial carotenoids in model gastric conditions. Comparison with common carotenoids. *Food Funct.* **2013**, *4*, 698–712.

43. Roginsky, V.; Zheltukhina, G.A.; Nebolsin, V.E. Efficacy of metmyoglobin and hemin as a catalyst of lipid peroxidation determined by using a new testing system. *J. Agric. Food Chem.* **2007**, *55*, 6798–6806.

44. Reeder, B.J.; Wilson, M.T. The effects of pH on the mechanism of hydrogen peroxide and lipid hydroperoxide consumption by myoglobin: A role for the protonated ferryl species. *Free Radic. Biol. Med.* **2001**, *30*, 1311–1318.

45. Baron, C.P.; Skibsted, L.H. Prooxidative activity of myoglobin species in linoleic acid emulsions. *J. Agric. Food Chem.* **1997**, *45*, 1704–1710.

46. Hu, M.; Skibsted, L.H. Kinetics of reduction of ferrylmyoglobin by (−)-epigallocatechin gallate and green tea extract. *J. Agric. Food Chem.* **2002**, *50*, 2998–3003.

47. Dangles, O.; Fargeix, G.; Dufour, C. Antioxidant properties of anthocyanins and tannins: A mechanistic investigation with catechin and the 3′,4′,7-trihydroxyflavylium ion. *J. Chem. Soc. Perkin Trans.* **2000**, *2*, 1653–1663.

48. Buchweitz, M.; Brauch, J.; Carle, R.; Kammerer, D.R. Application of ferric anthocyanin chelates as natural blue food colorants in polysaccharide and gelatin based gels. *Food Res. Int.* **2013**, *51*, 274–282.

49. Pereira, V.A.; de Queiroz Arruda, I.N.; Stefani, R. Active chitosan/PVA films with anthocyanins from Brassica oleraceae (red cabbage) as time-temperature indicators for application in intelligent food packaging. *Food Hydrocoll.* **2015**, *43*, 180–188.

**Sample Availability:** *Sample Availability*: Samples of compounds P1 and P2 are available from the authors.

MDPI AG

St. Alban-Anlage 66

4052 Basel, Switzerland

Tel. +41 61 683 77 34

Fax +41 61 302 89 18

http://www.mdpi.com

*Molecules* Editorial Office

E-mail: molecules@mdpi.com

http://www.mdpi.com/journal/molecules

www.ingramcontent.com/pod-product-compliance
Lightning Source LLC
Chambersburg PA
CBHW051925190326
41458CB00026B/6406